Ceac

D1497864

THE HISTORY OF HUMAN SPACE FLIGHT

UNIVERSITY PRESS OF FLORIDA

Florida A&M University, Tallahassee
Florida Atlantic University, Boca Raton
Florida Gulf Coast University, Ft. Myers
Florida International University, Miami
Florida State University, Tallahassee
New College of Florida, Sarasota
University of Central Florida, Orlando
University of Florida, Gainesville
University of North Florida, Jacksonville
University of South Florida, Tampa
University of West Florida, Pensacola

THE HISTORY
OF HUMAN
SPACE FLIGHT

TED SPITZMILLER

Foreword by Sid Gutierrez

University Press of Florida

Gainesville · Tallahassee · Tampa · Boca Raton

Pensacola · Orlando · Miami · Jacksonville · Ft. Myers · Sarasota

Copyright 2017 by Ted Spitzmiller
All rights reserved
Printed in the United States of America on acid-free paper

This book may be available in an electronic edition.

22 21 20 19 18 17   6 5 4 3 2 1

Library of Congress Control Number: 2016956656
ISBN 978-0-8130-5427-8

The University Press of Florida is the scholarly publishing agency for the State University
System of Florida, comprising Florida A&M University, Florida Atlantic University,
Florida Gulf Coast University, Florida International University, Florida State University,
New College of Florida, University of Central Florida, University of Florida, University of
North Florida, University of South Florida, and University of West Florida.

University Press of Florida
15 Northwest 15th Street
Gainesville, FL 32611-2079
http://upress.ufl.edu

To the visionaries of the past; to the engineers, flight controllers, managers, and technicians who made it happen; and to the intrepid flyers who risked it all to be a part of the greatest of all human endeavors

CONTENTS

FOREWORD

I am still both inspired and amazed when I gaze at the Moon on a clear evening and contemplate the fact that twelve Americans walked on its surface. Entering the twentieth century, man had never gone aloft in a heavier-than-air craft. Less than seven decades later, we landed on the Moon. In a blend of history, science, politics, and personalities, Ted Spitzmiller tells the story of this incredible era and the events that led up to it as well as the challenges that followed. If you missed the "space race," you will have a chance to live it. If you lived it, you will learn things you never knew before.

The History of Human Space Flight is written so everyone can understand complex topics like rendezvous. Yet Ted also brings to life many of the unique personalities, such as Dr. John Stapp, "the fastest man on earth." He takes on a number of controversial issues, offering the most thorough and unbiased explanation I have read of Wernher von Braun's involvement with the Nazi Party. He provides a new perspective on other events, including astronaut Gus Grissom's infamous "blown hatch." You will learn why "JPL" was called the Jet Propulsion Laboratory when they actually built rockets. This book is complete and balanced—the product of extensive research. But what I like best is the style. It is reminiscent of Bill Bryson's *A Short History of Everything*. This approach puts both science and history in context.

I am a product of the Sputnik generation and was fortunate to live the dream and become a space shuttle pilot and commander. But as I read Ted's work, I found I had much to learn about this great adventure in human exploration. Never before have humans advanced so quickly into the unknown, to then seemingly almost stop. My greatest fear as a young person hoping to become an astronaut was that all of the habitable planets would have been explored by the time I got to fly. Instead, nearly half a century after walking on the Moon, we have not returned, much less gone on to Mars. Ted provides answers to this mystery as he explains the cost of human spaceflight in national treasure and human sacrifice. He ends with a glimpse into the future of human spacecraft—not a Buck Rogers future with ray guns but a real future driven by commercial space companies. *The History of Human Space Flight* is an informative and thoroughly enjoyable read about our journey beyond our world.

SID GUTIERREZ

Sidney M. Gutierrez is a former Air Force F-15 fighter pilot, F-16 test pilot, NASA astronaut and space shuttle pilot and commander. Today he is the chairman and CEO of Rocket Crafters, Inc.—a new commercial space flight company.

ACKNOWLEDGMENTS

The original research by the many authors listed in the bibliography has allowed me to capture the essence of this great adventure. I am indebted to former shuttle commander Sid Gutierrez and former NASA flight controller Marianne Dyson for reviewing the manuscript and providing guidance to ensure technical accuracy. However, any errors the reader may find are solely mine alone.

I would also like to recognize the considerable assistance of UPF editor Patti Bower for her extraordinary effort in helping to make the manuscript more readable.

PROLOGUE

The desire of humans to venture into the space above our planet has been recorded for more than a millennium. Even the definition of "space" has undergone a transformation over the past few hundred years as humanity has learned more about what lies beyond the atmosphere surrounding our planet.

The term "outer space," first coined by the noted English author H. G. Wells (1866–1946) in his novel *First Men in the Moon* in 1901, is often shortened simply to "space." The word typically identifies the region beyond Earth's atmosphere established (in recent times) as 62 miles—328,000 feet (100 km) above Earth's surface. This somewhat arbitrary value, called the "Kármán line" (in tribute to aerospace engineer Theodore von Kármán [1881–1963]), was chosen for purposes of space treaties and the validation of space records by the Fédération aéronautique internationale.

The U.S. government recognizes a somewhat lower altitude of 50 miles as the start of "space" and qualifying for the status of "astronaut." Yet another

---

Fédération aéronautique internationale

The Fédération aéronautique internationale (FAI) was created in 1905, when members of the Aero Club of France, the German Airship League, and the Aero Club of Belgium proposed an organization to recognize "the special importance of aeronautics" and to create "an Association for regulating the sport of flying," and to form "a Universal Aeronautical Federation to regulate the various aviation meetings and advance the science and sport of Aeronautics." The group's primary purposes were "to methodically catalogue the best performances achieved, so that they be known to everybody; to identify their distinguishing features so as to permit comparisons to be made; and to verify evidence and thus ensure that record-holders have undisputed claims to their titles." (FAI 2016) definition of space might be considered—the "Armstrong limit" of the atmosphere. Above this altitude, the lack of atmospheric pressure requires the use of either a sealed pressure vessel or spacesuit to maintain life. This places the space threshold at about 12 miles (63,000 feet or 19.3 km). This region is often referred to as "the edge of space" and the "space boundary."

Yet those who were the first to break the bounds of Earth's gravity—the eighteenth-century balloonists—might also be considered as space travelers because, to these early explorers, anything above Earth was "space."

By expanding the definition then, there are those who went into orbit or beyond and those who flew balloons, ballistic trajectories, or rocket-powered aircraft on relatively short up-and-down forays into space. For our purposes, the term "spacecraft" defines any vehicle that carries its human cargo into one or more of these definitions of space.

Anyone traveling into space is referred to generically as an astronaut in the English-speaking world, or cosmonaut in the Russian sphere of influence. The Chinese, the only other nation that has developed the ability to travel above the Kármán line, uses "*yǔ háng yuán*" (space navigating personnel) in official Chinese-language texts, but the popular press refers to their voyagers as *taikonauts*.

The journey through these pages begins with the earliest success in traveling off our world. Progressing from the first aerostats that took humans aloft, the adventure ascends with the technologies of the twentieth century to include rocket-powered aircraft and orbiting spacecraft. Experiments of space medicine pioneers that defined the physical limits of human tolerance to acceleration, microgravity, lack of oxygen, and atmospheric pressure are included. Also identified is the balance of autonomy between the human occupants of the spacecraft, their automated systems, and the ground-based flight controllers.

This book examines the pioneering iterations of manned spacecraft, the individuals who flew in them, and those who created, directed, and supported them. While the development of the launch vehicles is briefly reviewed, the emphasis is on the spacecraft and their human occupants.

INTRODUCTION

Attempting to document a topic as vast as the epic of human space flight presents many obvious problems for an author, such as what events to include. Certainly all of the prominent "firsts" have to be covered, but then the question becomes to what detail? Another quandary is the delicate balance between the often dry technologies and engineering aspects and the more personal human side—virtually all who have traveled in space have experiences that demand telling.

As the reader will discover, there were many more people involved than those who went aloft, and their stories need to be included. Likewise, because many of the events interacted with the politics of the era, that aspect has to be considered. Yet there is a limit to how much can be told within a restricted number of pages; determining the exclusion of many events and people was painful.

However, I also brought my own biases to the project. Recalling the news of the last V-2 launch from White Sands, New Mexico, in 1952 was an episode that led me to become a dedicated enthusiast of space exploration at an early age. I was fortunate to be able to follow most all of the early events, and many became the focus of what is contained within these pages.

Selected technical descriptions are highlighted to provide a more in-depth focus on certain concepts. For obvious reasons these have been simplified and may not present complete detail. The Internet can provide more exposure for those topics that the reader may find intriguing.

Another element is that many of the programs overlapped and some interacted. Trying to keep the chronology of events while holding the programs together is a challenge. Thus, there are segments that the reader may perceive as somewhat disjointed with respect to the timeline.

The vast amount of information now available from many sources makes research on this topic a daunting task. This author is indebted to those whose work is referenced in the bibliography. However, therein lies another problem. Much of the literature about the Soviet program, which was published before the fall of the Soviet Union, was often incomplete and subject to misinformation. While the reader is encouraged to choose from this list for more in-depth exposure to areas of specific interest, information in some of the older publications may be suspect. Transcripts of communication between mission control and the astronauts cited in this book can be found on the NASA website, http://nasa.gov/topics/history.

While the reader should not feel obligated to begin reading at the beginning, most of the chapters are woven into the story of space flight, and there is some dependency in the chronology of events.

1

AEROSTATS OPEN THE VISTA

Human space exploration began in a subtle manner, with the use of aero-stats—balloons, as we more commonly call them today. These quiet, fragile creations, fully at the mercy of the winds, played an important role in the conquest of space. The balloon was an invention of an earlier time, and within the period of a few months in 1783, two forms of buoyancy were used to lift the first humans into space.

A New Invention: The Balloon

In 1766 Englishman Henry Cavendish discovered the properties of a gas that he called "phlogiston," which was one-fourteenth as light as air. When enclosed in a lightweight container, it satisfied the principles that Archimedes had defined regarding buoyancy; it floated. When it was recognized that this colorless, odorless, and nontoxic element produced water when it was burned, it received the Greek name hydrogen—for "water-former." Because of its most dangerous property, most people referred to it as "inflammable air."

In France, starting in June 1783, Joseph-Michel Montgolfier and his brother, Jacques-Étienne Montgolfier, experimented with the ability to fly small silk sacks filled with hydrogen. But, as the silk allowed the hydrogen to quickly escape because of its porous nature, they moved on to using "smoke." The particulates in the smoke tended to seal the envelope and retain the hot gas. These sacks, too, defied gravity—supported by the buoyant properties of rising hot air. Floating bags of hot air had been noted in both the Chinese and Indian cultures in past centuries. However, when the smoke cooled, it also descended.

The Montgolfier brothers filled inverted paper bags with a light cloth fabric with the smoke from a large hay fire. They did not actually understand the physics involved and believed that some properties of the smoke,

The Archimedes Principle

Archimedes (c. 287 BC–c. 212 BC), the Greek mathematician and philosopher, is credited with being the first to define the principles of buoyancy in his treatise *On Floating Bodies*. He writes: "An object immersed in a fluid is buoyed up by a force equal to the weight it displaces." Thus, a container (balloon) of a substance lighter than air (air being a fluid) will rise until its "weight" is equal to the volume of air it displaces.

not the hot air, provided the lifting action. Over a period of several months, they experimented with ever-larger bags (a product of their family's business). Their success generated not only notoriety but also an enthusiasm for ballooning that was caught by many—and the science of aerostats was born. It is from this word that the term "aeronaut" was also coined.

In an effort to validate scientifically the work of the Montgolfiers, the Academy of Sciences in Paris authorized French inventor, scientist, and mathematician Jacques Alexandre Cesar Charles to duplicate their work. However, Charles believed they had continued to use hydrogen and thus the validation effort actually expanded on this original path to lighter-than-air flight—and a friendly rivalry.

Adding to the rapid advance in the art of ballooning was the invention of a rubber coating for silk (by the brothers Jean and Noël Robert), that allowed Charles' envelope to retain the hydrogen. The "inflammable air" was generated by pouring sulfuric acid over iron filings. This produced hydrogen gas, which filled the envelope of his balloon that he aptly named the Globe. America's Ben Franklin (then U.S. ambassador to France) was one of those notables who watched the ascent of these as-yet unmanned aerostats with great interest.

In Versailles, the Montgolfiers' next spectacular balloon was *Aérostat Réveillon*, named for Jean-Baptiste Réveillon—the manufacturer who produced the beautifully adorned paper from which Jacques-Étienne fashioned the envelopes. This colorful 40 ft. tall balloon was sent aloft in September 1783. The passengers aboard this pioneering flight were a duck, a sheep, and a rooster. Each animal had a scientific purpose, with the sheep representing a close approximation of the human physiology. There was concern about possible negative effects of a flight into the "upper atmosphere"—thus the caution.

Competition developed between the brothers and Jacques Charles to send aloft a human. Several tethered tests were made with Jacques-Étienne on board to gain experience and confidence in the balloon. Techniques for feeding the central fire basket to supply the smoke (hot air) were also developed.

King Louis XVI had become an ardent supporter of these experiments and suggested that two prisoners be the first aeronauts because of the risk. However, the brothers prevailed, citing the high probability of success and the fame the first fliers would acquire. Thus, Jean-François Pilâtre de Rozier, a French chemistry and physics teacher, and François Laurent le Vieux d'Arlandes made the first human free balloon flight in November 1783 in a Montgolfier balloon.

Launched near Paris in the presence of the king and Marie Antoinette, their twenty-five-minute flight traveled more than 5 mi. while attaining an altitude of about 3,000 ft. The 75 ft. tall balloon had a volume of 56,000 cu. ft., which compares with a typical modern-day four-passenger hot air balloon.

While the event is not celebrated to the extent of the first powered heavier-than-air flight by the Wright brothers, Orville and Wilber Wright, in 1903, it was the first major milestone for humans to break away from the gravity well that constrains us to the planet. However, with the ability to seal the envelope, hydrogen became the buoyant gas of choice. The use of "hot air" faded for two hundred years until the use of propane burners brought a resurgence to the sport of hot air ballooning in the 1970s.

From this point forward, the space above Earth could now be explored and its attributes measured. However, most of the flights for the next hundred years were made to set records and exploit the moneymaking potential of balloon exhibition flights rather than to advance science.

A month following the first human ascension, in December 1783, Jacques Charles and Nicolas-Louis Robert launched their balloon in Paris. The 13,000 cu. ft. hydrogen-filled balloon used sand ballast to control altitude. They ascended to a height of about 1,800 ft. and landed after a two-hour flight covering 16 mi. Sand was the ballast of choice, being easily dropped without hurting anyone below, as the new aeronauts learned their craft by trial and error.

De Rozier died in 1785 when another balloon he was piloting crashed during an attempt to fly across the English Channel. He and his companion, Pierre Romain, became the first known fatalities in the conquest of space.

The first balloon flight in America took place a decade later, in January

Ballast and Buoyancy

Gas balloons use ballast to control buoyancy during flight. Sand or water is carried aloft at launch and released in controlled measures by the pilot to adjust the balloon's altitude by reducing the gross weight of the balloon, which will then rise to a new pressure altitude.

To descend, lifting gas is released by a valve atop the envelope. The balloon will remain at its equilibrium altitude until there is another dynamic change in the lift equation. The descent to the ground is a critical operation as the trade-off between the valving of lifting gas and the descent rate is controlled by releasing the remaining ballast. Hot air balloons simply allow the envelope to cool to descend, and heat the air to stop the decent or to re-ascend.

Jean-François Pilâtre de Rozier and François Laurent le Vieux d'Arlandes made the first free balloon flight using hot air as a lifting medium in November 1783. Courtesy of Bibliothèque nationale, Paris.

1793, when the noted French balloonist Jean-Pierre Blanchard ascended from Philadelphia, Pennsylvania, and landed in Deptford Township in New Jersey. Witnessing the ascent that day were President George Washington and future presidents John Adams, Thomas Jefferson, James Madison, and James Monroe.

Ascending to New Heights

Although the emphasis of these early flights was to entertain by setting records rather than to advance true science, the almost tragic by-product of many stunts revealed the lack of knowledge of this new high-altitude environment. By 1804 several balloonists had topped 20,000 ft. The 1862 ascent of Henry Coxwell and James Glaisher of England claimed 29,000 ft., and the pair narrowly avoided death from hypoxia and freezing.

The year 1875 saw the deaths of Théodore Sivel and Joseph Croce-Spinelli in their record attempt to 28,000 ft. A third member of the crew of the balloon named *Zenith*, Gaston Tissandier, barely survived. There was a lot to learn about traveling in this region high above Earth.

In 1896 French meteorologist Léon P. Tisserenc de Bort undertook a significant investigation of the upper atmosphere by sending several unmanned balloons to 50,000 ft. In the days before electronic telemetry, Tisserenc de Bort's effort relied on mechanical recording devices, which revealed significant changes in the character of the regions of the atmosphere through which the balloon passed. He termed the areas closest to Earth (within 6–8 mi.) the troposphere and the area above 40,000 ft. the stratosphere, where he perceived there was virtually no weather phenomena.

Arthur Berson, a Polish meteorologist and pioneer of aerology, and Reinhard Süring, a German meteorologist, set a record of 35,424 ft. in 1901 (although the actual height was doubted by many authorities of the time). This represents the altitude at which current jetliners routinely operate and is about one-tenth the distance to the Kármán line. The world had to wait more than twenty-five years before that record would be surpassed.

By the late 1920s supercharged engines allowed airplanes to achieve brief flights of up to 30,000 ft., and heavy clothing and oxygen were now recognized as minimal equipment. Lacking pressure-breathing apparatus, U.S. Army captain Hawthorne C. Gray passed out at 27,000 ft. during his 1927 flight, and the balloon descended on its own—with the pilot regaining consciousness in time to avoid a catastrophic landing. The sand ballast alone weighed two tons.

Oxygen Systems

As aircraft achieved higher altitudes, these early pilots recognized that they were not getting enough oxygen into their blood, and they were suffering from hypoxia. Although oxygen levels at all altitudes are constant (about 20 percent of the gases in the atmosphere, with nitrogen making up about 78 percent), simply providing higher percentages of oxygen when flying above 12,000 ft. was not enough. As the altitudes went beyond 25,000 ft., the lack of atmospheric pressure, mandatory for the oxygen to enter the hemoglobin of the blood, required "pressure breathing." With discoveries such as this, science was now beginning to play a more significant role in enabling these higher flights.

In his second attempt two months later Gray reached 42,240 ft. but ran out of ballast and had to parachute from the rapidly descending gondola at 8,000 ft. Because he did not land with his aircraft, as Fédération aéronautique internationale (FAI) rules required, his record was not recognized. A similar situation confronted the first Soviet cosmonaut thirty-five years later.

By this time the goal of simply establishing a new record, as appealing as that may have been to the participants (and for national pride), could no longer be the sole justification for flights. Science had to be the fundamental rationale, especially when government sponsorship was involved—the U.S. Army, in Gray's situation. Mechanically driven instruments had to be perfected that would—just like the crew—function reliably in the low pressures and temperatures experienced.

On his third and fatal flight, despite extensive preparation, Gray scribbled a final hypoxic note at 40,000 ft. that described the deep blue sky and brilliant sun. He also indicated that he had depleted his ballast. With his radio antenna broken (while discarding an empty oxygen tank), there was no communication, and his fate was not known until the following day when his lifeless body was found in the remains of the gondola. The cause of death was apparently not from the hard landing but more likely due to hypoxia or heart failure. The recording instruments indicated 44,000 ft. Capt. H. C. Gray, the first human ever to exceed 40,000 ft., is interred in Arlington National Cemetery.

Despite the advanced planning and courage of men like Gray, the technology of the time had not kept pace. Effective life-support systems as well as more advanced envelopes and ballast were needed, along with more accurate and reliable scientific instruments.

Auguste Piccard: 1931 Flight

Swiss physics professor Auguste Piccard profoundly changed the character of exploration of the upper atmosphere—the fringe of space. In retrospect of his accomplishments, he wrote: "The sport of the scientist consist of utilizing all that he knows of foreseeing the dangers, of studying every detail with profound attention, in always using the admirable instrument of mathematical analysis where ever it could shed its magic light upon his work" (quoted in Ryan 1995, 40).

Piccard was an early ballooning enthusiast who, with his twin brother, Jean, had flown from Zurich into France in 1913. Both had served in the Swiss Army lighter-than-air service in 1915. He recognized the need to provide a safer life-support system for high-altitude flight and in 1931 had fashioned a hermetically sealed (airtight) crew capsule—the precursor of the true spacecraft. The spherical, 7 ft. diameter cabin had eight small double-pane glass portholes and could accommodate two observers and a few hundred pounds of scientific instrumentation.

Piccard's first record flight attempt took place in May 1931, from Augsburg, Germany, using a 500,000 cu. ft. hydrogen balloon. Ascending rapidly,

---

Pressurized Environments

By sealing the capsule, the occupants were maintained at the ambient pressure of the launch environment—sea-level pressure being 14.7 psi (760 mm) and the standard percentage of oxygen would suffice. This also allowed greater warmth as higher-pressure air can more effectively maintain heat. A small Dewar of liquid oxygen provided for the replacement of that vital life-sustaining gas. It was periodically poured out on the floor of the gondola and, due to its extremely low temperature (-297 degrees), it immediately evaporated into the cabin's atmosphere. To "scrub" the encapsulated air of the exhaled carbon dioxide, Piccard adapted a recent discovery of chemically removing the gas.

Auguste Piccard at age forty-seven, with the pressurized (hermetically sealed) gondola flown in 1931. Note the access hatch for entry. Courtesy of Jacques Piccard.

in less than an hour Piccard and companion Paul Kipfer reached 51,000 ft. Several problems occurred along the way, the most serious of which was a small leak caused by a broken wire that passed through the cabin to the equipment mounted outside the pressure hull. This vented the precious air out into the thin atmosphere. However, Piccard had anticipated the small leak problem and quickly patched it using cotton and petroleum jelly.

When it came time to descend, the rope attached to the hydrogen-venting valve became entangled with another and they could not descend. The balloon drifted throughout the day as the first humans to penetrate the stratosphere observed the brilliant display of colors that Gray had documented in his final notes a few years earlier. As they floated into the evening, the cooling of the envelope started their descent. It was not until eight o'clock, when they had descended to 15,000 feet, that they could open the porthole to allow the now fouled cabin air to be refreshed from the outside.

They descended over the Bavarian Alps and prepared for a hard landing on the rough, snow-covered peaks. They had brought with them crash helmets fashioned from wicker chicken baskets and a pillow that they placed over their heads. Luck was with them as the gondola settled easily in the deep snow and, using the rip panel rope, they opened the envelope to release the remaining hydrogen gas. They were fortunate that their landing site allowed them to hike down to civilization after spending a cold night in the capsule.

A second flight a year later was even more successful, reaching over 53,000 ft. on its twelve-hour journey from Zurich to Lake Ganda in Italy. Piccard and Charles Kippford were knighted for their pioneer journey and embarked on a lecture tour proclaiming, quite prophetically, that a closed-capsule system would one day be used to transport humans to the Moon.

Jean Piccard: 1933 World's Fair

Auguste Piccard was asked by organizers of Chicago's 1933 World's Fair to attempt another stratospheric flight, but he declined. He suggested that his brother Jean organize the effort. Jean, who had immigrated to the United States in 1926, enthusiastically accepted the challenge. However, he was not a licensed balloon pilot, so a second crewmember was needed to handle that demanding chore. Thomas "Tex" Settle, a U.S. Navy pilot who had extensive lighter-than-air experience, was chosen. Like others who previously declined the flight, Settle had serious reservations about flying with Jean Piccard, who had little aeronautical experience and was known as a difficult personality. (Writer Gene Roddenberry [1921–1971] paid homage to Jean Piccard, with Star Trek character Jean-Luc Picard, played by Patrick Stewart.)

Although the press emphasized the desire for a new altitude record, Piccard and Settle stressed science. Several new instruments were included. Among them was an ultraviolet ray experiment to determine if an ozone layer existed in the atmosphere, and a cosmic ray package (assembled by the famous physicists Arthur Compton and Robert Millikan) to determine the characteristics and origin of those mysterious particles.

The huge 660,000 cu. ft. envelope was made by the Goodyear-Zeppelin Company, and the spherical Piccard-designed gondola, by Dow Chemical Company. Dow used a magnesium alloy that was two-thirds the weight of Piccard's first aluminum, hermetically sealed cabin.

The balloon, in keeping with the theme of the 1933 World's Fair, was named *Century of Progress*. It was sponsored by the National Broadcasting Company (NBC) and the *Chicago Daily News* and was launched on August 5. However, when Settle attempted to release some gas passing through 5,000 ft., the valve stuck open and the balloon descended rapidly—landing just 2 mi. from its launch point before he was able to recycle the valve. There were no injuries or major damage.

Before Tex Settle could prepare the equipment for another launch, a balloon called *Stratostat* took to the skies in September 1933. This one, piloted by two Soviet army officers, Ernst Birnbaum and Konstantin Godunov, ascended to claim a new record of 62,304 ft. Unfortunately, the claim was not verified by the FAI (who sanctioned all record flights) as the USSR was not recognized by many international agencies at this time. However, the *New York Times* reported that the flight's effect on the Soviet people was the same as that of Charles Lindbergh with the American people following his epic flight across the Atlantic Ocean from New York to Paris in 1927.

Jean Piccard and Tex Settle had been struggling to complete the financial backing when the Soviet *Stratostat* made the headlines. The patriotism of NBC and the *Chicago Times* was prodded, and the support money flowed. However, Piccard was replaced by Air Corps officer Chester Fordney—who had never been in a balloon! Although an ardent academic who could operate the experimental apparatus, Fordney may have been selected because his father was commandant of the Marine Corps.

The textbook flight lifted off in November 1933 from Akron, Ohio, where the inflation took place inside the enormous Goodyear-Zeppelin hanger used to house the Navy dirigibles *Akron* and *Macon*.

As a precursor of the public interest that engulfed the first astronauts less than thirty years later, the communications between the crew and ground support were broadcast live on the radio. The flight went well until their return, when they miscalculated their descent rate. Running low on ballast, they threw out their radio batteries to help arrest the descent. Consequently, they were unable to report their landing location. They had to spend the night in the gondola—on a mud flat near the Delaware River in New Jersey, just a few miles from the home of Jean Piccard. The FAI declared the flight a record as it reached an official altitude of 61,237 ft.

As the first orbiting astronauts would lament in their pioneering flights, Settle commented, "We were seldom permitted the time to look out the windows of the Gondola . . . ballooning is chiefly work" (Ryan 1995, 50). His interests in the upper reaches of the unknown were further emphasized when he joined a local rocket society a few years later.

Perhaps annoyed that the Americans had received credit for a record that the Soviets had been denied, January 1934 saw another flight by the USSR. This time three civilians, Felosienko, Wasienko, and Vsyskin, reached an altitude of 72,178 ft. Tragically, all were killed during the descent, apparently because of poor ballast management. The *Osoaviakhim* (the balloon was named for a Soviet paramilitary training organization) used a different method of securing the cabin to the envelope, and these ropes failed under the loads imposed by the rapid descent. The remains of the heralded fliers were interred in the Kremlin, as would a trio of three cosmonauts thirty-four years later.

U.S. Army: National Geographic Explorer I

Several months after the epic flight of the *Century of Progress*, the U.S. Army teamed with the National Geographic Society (NGS) to push the frontier further by proposing a balloon with three times the capacity of the one flown by Settle and Fordney. The sixty-thousand-dollar, three-million-cubic-foot *Explorer* had a sealed gondola enlarged to an 8.5 ft. diameter to accommodate three men and weighing more than a ton.

While exhorting the prospects of another record, NGS assured its subscribers that science was the goal. Five specific objectives were set for *Explorer*: (1) collect air samples from various levels of the upper atmosphere; (2) determine if any life exists in the form of spores at these extreme altitudes; (3) locate and measure the ozone layer; (4) measure cosmic rays and try to pinpoint their source; and (5) determine if existing means of measuring altitude are accurate.

U.S. Army captain Albert Stevens was the primary force behind the effort and provided a small token of the sixty-thousand-dollar cost. Accompanying him was Capt. Orvil Anderson and Maj. William Kepner. After considerable searching, an ideal launch site was located in the Black Hills of South Dakota. Recognizing a good tourist draw, the town fathers of nearby

The U.S. Army / National Geographic *Explorer I*, showing some of the many sand bags used as ballast. Courtesy of U.S. Army.

Rapid City agreed to provide access roads and electricity to the 400 ft. deep canyon now known as the Stratobowl. The launch came at 6:45 a.m. on July 28, 1936, with more than thirty thousand spectators looking on. The balloon rose rapidly and the crew leveled temporarily at 30,000 ft. for some cosmic ray readings. Releasing more ballast, the balloon then continued to rise to 60,000 ft., when a tear was observed in the lower part of the envelope. With NBC providing a nationwide broadcast, Kepner dramatically described the problem: "the bottom of the balloon is pretty well torn out and it is just a big hole in the bottom. I don't know how long it would hold together" (Ryan 1995, 54).

The answer to that question was quick in coming as the envelope disintegrated and the balloon began a free fall. There could be no thought of escape by parachute until the pressurized gondola fell into the lower level of the atmosphere. The three crewmembers were well prepared and disciplined—strapping on their parachutes and waiting for the altimeter to let them know when it was safe to open the hatch. All of this was broadcast live over the commercial radio.

Passing through 20,000 ft., the access door was opened and the crew prepared to jump. The residual hydrogen in the now almost deflated envelope exploded. The men, shielded from the blast by the gondola, then began to bail out. Moving to the exit, Anderson prepared to jump at 3,000 ft. but the D-ring caught on a piece of equipment and the chute unfurled inside the gondola. Quickly gathering the fabric, Anderson squeezed through the opening and pushed off, throwing the arm full of silk into the slipstream.

Stevens was next, and what was left of the balloon was in a high-speed fall. He had to work hard to get out at about 2,000 ft. Kepner was right behind him pulling his ripcord at what was estimated at about 500 ft. The three survived the near tragedy, but the expensive gondola and most of its scientific equipment was a total loss. The barograph was recovered and showed that the expedition had missed setting a new record by just 624 ft.

Jean and Jeannette Piccard: 1934

The U.S. Army and the NGS went off to investigate what had occurred with the *Explorer* flight, but would return to compete. In the meantime, Jean Piccard reappeared with a proposal to fly the *Century of Progress* again. Although the envelope and gondola had survived its previous flight in good condition, Piccard could not get any of the previous sponsors (NGS, Goodyear, and Dow) to contribute. Perhaps the intent to have Jean's wife, Jeannette, pilot the balloon cast doubt on the possibility of success, as she had no experience ballooning and had to qualify for a lighter-than-air pilot's certificate. The Piccards were eventually able to get enough smaller donations, and Jeannette, known for her outspoken feminism, earned her pilot's credentials under the guidance of the renowned 1927 Gordon Bennett Cup (the oldest gas balloon race) winner, Ed Hill.

With physics instrumentation provided by William F. G. Swann and Robert Millikan, the intrepid team launched near Dearborn, Michigan, in October 1934. Although the flight failed to achieve any records, it reached 57,579 ft. The crew received considerable acclaim—primarily because of the pilot's gender—she was the first woman to venture into the stratosphere.

Hydrogen Verses Helium as a Lifting Gas

Air is 78 percent nitrogen (N^7). We'll set aside the 20 percent oxygen component to make the calculation a bit simpler. Nitrogen is a diatomic element, which means it requires two atoms to form a molecule with a resulting atomic mass of $14 \times 2 = 28$. Hydrogen (H^1), which is also diatomic, has an atomic mass of 2, so a hydrogen balloon has a net lifting force of about $(28 - 2) / 28 = .93$ the weight of the air it displaces.

Helium (H^2) is not diatomic and thus has a molecular atomic mass of 4. It produces a net lifting force of about $(28 - 4) / 24 = .86$ the weight of the air it displaces. The net lifting force on the hydrogen balloon is about 10 percent more; therefore, to lift the same weight, a helium balloon must be proportionally larger.

National Geographic *Explorer II*: 1935

While the Piccards took center stage during the summer of 1935, the NGS, with Dr. Gilbert Grosvenor as its legendary head, moved forward with *Explorer II*. There were several changes; the most obvious was the switch from hydrogen to helium as the lifting gas. Because helium is less buoyant, the envelope now grew in size to an astounding 300 ft. tall and 3.7 million cu. ft. However, it was felt that this was a small price to pay to avoid the extremely hazardous inflation operation, let alone the flight itself. (It would be another year before the *Hindenburg* disaster virtually grounded all hydrogen lighter-than-air craft).

The second major change was to reduce the crew back to two. The gondola was just too confining for three humans to operate efficiently—this also reduced the weight and allowed a larger load of scientific instrumentation—including one experiment to assess the biological effects of cosmic rays. A significant improvement in the radio communications reflected a higher level of electronics technology.

Although the sponsors had hoped for a launch during the summer of 1935, there were just too many hurdles to overcome. It was not until November, when the equipment and weather cooperated, that *Explorer II* rose from the *Stratobowl* in South Dakota. An estimated twenty thousand spectators along with hundreds of ground crew cheered as the shimmering balloon rose rapidly into space.

STRATOSPHERE BALLOON
IN STRATO BOWL
30 MINUTES BEFORE TAKE-OFF
ALTITUDE ATTAINED: 14 MILES

The National Geographic *Explorer II* launches from the Stratobowl in 1935. The balloon expanded to an almost spherical shape in the upper atmosphere. Courtesy of National Geographic.

It took four and a half hours for the balloon to ascend to a new record of 72,000 ft. From his vantage point, Stevens reported that highways, towns, and railroads were invisible and that "no sign of actual life on Earth could be detected. To us it was a foreign and lifeless world" (Ryan 1995, 59).

Another interesting aspect was the inability to rotate the balloon so they could observe in specific directions—the sunny side made viewing almost impossible because of the extreme brightness. A small electrically driven propeller had been installed on the outside of the gondola perpendicular to the vertical axis of the balloon. Unfortunately, there was not enough atmosphere for the airfoil of the propeller to produce sufficient thrust to rotate the balloon.

Unlike many of the previous ascents, the clear day and lack of wind allowed those on the ground to follow almost the entire flight. They watched as the tall, oblong envelope expanded to a nearly round ball, clearly visible from 15 mi. below.

As the balloon descended and neared the ground, several large experiments were thrown out with parachutes, to function as released ballast (slowing the descent), and were later recovered. Wearing football helmets, Stevens and Anderson braced for the landing, which came as the sun was setting at 4:13 p.m. However, the landing was described by Anderson as "soft as a feather" (Ryan 1995, 60).

Explorer II marked the end of an era of human balloon exploration. At this point there was little scientific return for the dollars spent, and, like the coming space age, unmanned probes could return most of the data without the risk of life. Nevertheless, a precedent had been set and essential groundwork laid such that, little more than twenty-five years in the future, the public would again be fascinated by watching men risk their lives to explore again the unknown of space. There were many similarities in the overall conduct and competition between the United States and the USSR in the aviation arena of the 1930s that played out in a more serious tone in the future. The sealed capsules and life-support systems of the aerostats were a precursor to the later rocket-launched spacecraft to come.

The First Space Suit

With the advent of supercharged engines, some notable aircraft flights ventured into the extreme altitudes previously the purview of the aerostat. Wiley Post was a famous flier during the early 1930s, known for his various record-breaking attempts in a single-engine Lockheed Vega named *Winnie Mae*. Post believed there were high-altitude winds (now called the "jet stream"), and he wanted to take advantage of them. However, the *Winnie Mae* could not be pressurized, and the lack adequate oxygen systems of the day, coupled with the intense cold at altitudes above 25,000 ft., thwarted Post.

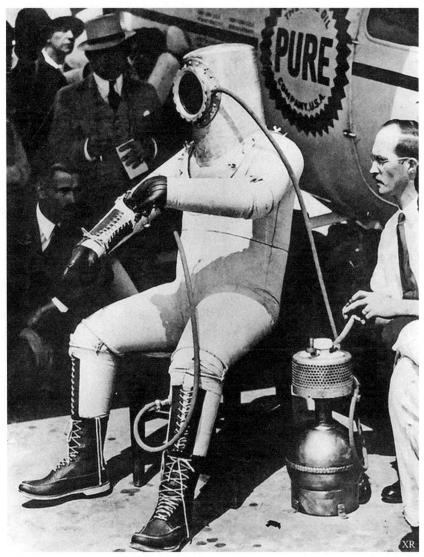

Wiley Post models the first pressure suit prior to a flight in 1934 of the *Winnie Mae*. Walter E. Burton, Akron, Ohio, Wiley Post Collection, Courtesy of the Oklahoma Historical Society, 19336.2.

In 1934, with support from Phillips Petroleum Company, Post worked with the B. F. Goodrich Company to develop the first practical pressure suit. Three suits were ultimately made for Post. The first ruptured during a pressure check. The redesigned second suit was too tight and they had to cut him out of it. The third worked.

The suit had three layers: first, a pair of long underwear, then an inner black-rubber air-pressure bladder, and finally an outer layer made of rubberized parachute fabric. The arm and leg joints of the outer layer were glued to a frame and had limited flexibility, but enough for him to operate the flight controls and get himself into and out of the aircraft. Attached to the frame were pigskin gloves and rubber boots. The helmet was aluminum and plastic with a removable faceplate that provided for earphones and a microphone. The suit itself, which closely resembled a deep-sea diver's outfit, was pressurized to 17,000 ft.—an unacceptable value today.

In the first flight with the suit in September 1934, Post reached 40,000 ft., and subsequent flight ultimately climbed as high as 50,000 ft. Post confirmed the presence of the jet stream. Four attempts, between February and June of 1935, were made to achieve the first high-altitude nonstop flight from Los Angeles to New York. However, mechanical problems brought him down early each time, and the farthest attempt reached Cleveland, Ohio—2,035 mi.

Post met an untimely death two months later in Alaska during a flight to Russia with his good friend—the famous humorist and social commentator Will Rogers.

2

CREATING ROCKET SCIENCE

The Role of Science Fiction: Exciting the Imagination

Space travel by the inhabitants of Earth was the subject of French author Jules Verne, who sent three men to the Moon in his 1865 science fiction novel *From the Earth to the Moon*. As with his previous novels, *Journey to the Center of the Earth* and *Twenty Thousand Leagues Under the Sea*, Verne's primary objective was a visionary story line with plausible technology as an interesting secondary appeal.

In *From the Earth to the Moon*, he avoided the problem of onboard propulsion by using a large 900 ft. long cannon (a well-understood technology) and 400,000 lb. of gun cotton (a relatively new form of explosive). The 9 ft. diameter, 20,000 lb. projectile (the spacecraft) was to be made of lightweight aluminum—an element so rare at the time that chemists had produced only a few small bars weighing just ounces.

Verne recognized the devastating effects of the crushing forces of acceleration (gravity forces, or g-forces) resulting from the cannon on the aluminum structure and provided its occupants with "water buffers" to enclose the passengers and enable them to survive the shock of firing. This sounded plausible to the average person; it would ease the acceleration on the body by distributing the g-force evenly. However, the actual load factor would have killed the three travelers.

Perhaps the greatest contribution that Verne made to the dream of space travel was to acquaint the general populace with the idea that a trip to the Moon (or to the planets, for that matter) was primarily a question of attaining a certain velocity. This velocity had to be great enough to allow an object to escape the strong hold of Earth's gravity. Verne made an effort to determine the velocity needed to accomplish the flight (using an energy balance equation). This speed is now termed "escape velocity," which is

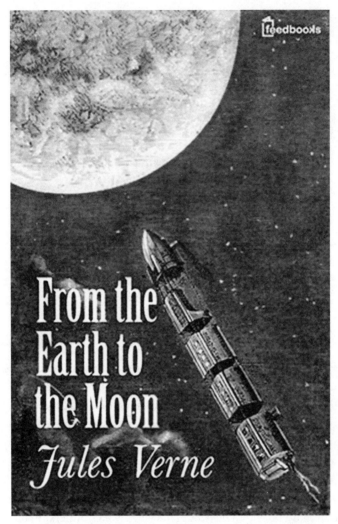

Jules Verne's science fiction novel opened the possibility of space travel to the masses. Courtesy of Thomas Y. Crowell, Publishers.

approximately 25,000 mi. per hour (7 mi. per second). He also recognized the apparent absence of gravity (weightlessness) during the flight. After the acceleration forces (caused by the propellant) are no longer present, the projectile is essentially in a free fall, and the human body cannot perceive the (still present) force of gravity. His book was a popular success, and it awakened the dreams of many in the years to come.

H. G. Wells produced several space-based sci-fi stories at the turn of the century that, like Verne's books, got the attention of the common person. In his *The War of the Worlds* (1897), Wells introduced the idea of extraterrestrials invading Earth from Mars—whose alleged canals were being hotly

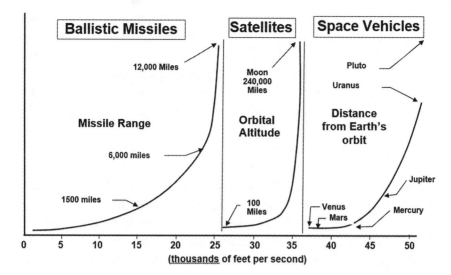

Jules Verne's premise that space travel was simply a matter of velocity is illustrated here with velocity on the horizontal axis and distance traveled on the vertical. Courtesy of Ted Spitzmiller.

debated among astronomers of the day. Earthlings became the invaders of an inhabited Moon in his *The First Men in the Moon* (1901).

While Verne and Wells had been among the first to popularize science fiction, it was Hugo Gernsback who brought it to the masses with his magazine *Amazing Stories* beginning in 1926. He credited Percival Lowell's writings as his inspiration. Gernsback was responsible for bringing many like-minded science fiction aficionados together with his publications (Lasser 2002, 8).

The late 1920s saw sci-fi enter the newspaper comic strips, and then radio and film in the 1930s, with the exploits of Flash Gordon and Buck Rogers. Although the special effects were not well done compared to today's films, the Saturday movie serials were popular with the younger kids and inspired a new generation to dream of the future.

The First Rockets

Reaction motors working on the principles of motion defined by physicist Sir Isaac Newton were actually demonstrated almost two thousand years earlier by Heron of Alexandria, a Greek mathematician and engineer. The actual date of the invention as well as the specific configuration of the device, now called the aeolipile, is not known, but it may be accurately traced to the time of Christ. The device, a small hollow copper sphere suspended over a fire, allowed the water within to be boiled. The resulting steam was expelled through curved tubes (action) to spin the device (reaction). It was simply a steam-driven rotor that, while providing a demonstration of reactive motion, was apparently carried no further in application. Use of action–reaction engines (rockets) had to wait for another millennium and a new age of visionaries.

A form of gunpowder, reportedly used in China late in the third century AD, propelled the earliest rockets. Bamboo tubes filled with a mixture of saltpeter (potassium nitrate), sulfur, and charcoal and sealed at both ends were tossed into ceremonial fires during religious festivals in the belief that the noise of the explosion would frighten evil spirits. It is conjectured that some of these bamboo tubes, instead of bursting, shot into the air by the rapidly burning gunpowder being exhausted (action) from a faulty seal on one end (reaction). The next obvious step, by some clever person, was to deliberately produce deliberately the same effect, and "fire arrows" were invented.

As is often the case with early tests, experimental apparatuses are not always built with much forethought and planning. According to one ancient legend, a Chinese official named Wan-Hoo, about AD 1000, attempted the first human rocket flight, perhaps with the Moon as his ultimate goal. Using a large wicker sedan chair propelled by forty-seven fire arrows (solid-fuel rockets) lit simultaneously by forty-seven assistants with torches, Wan-Hoo disappeared in a tremendous roar and billowing clouds of smoke, presumably on a successful flight from which he has yet to return—perhaps the first spacecraft?

The rocket found its way to Europe as a weapon of war by the fourteenth century but remained as an adjunct to cannon. British colonel Sir William Congreve began his work on improving the accuracy and power of the solid-fuel war rocket. These rockets did not use fins for stabilization in flight but rather long poles called guidesticks. The guidestick effectively ensured that the center of gravity was far enough ahead of the center of pressure to provide stability. The guidestick was at first mounted to the side of the rocket, but by 1815 the guidestick was screwed to the middle of a base plate around five equidistant exhaust holes. Rockets weighing up to 40 lb. could attain distances of three thousand yards. These were used against Napoleon's army in 1805 and again in 1806, when small boats slipped into the harbor at Boulogne and caused heavy damage to the fortress and ships there.

A Review of Newton's Laws of Motion

Sir Isaac Newton was born in Woolsthorpe, England, on Christmas Day 1642 (the year Galileo died). He published his laws in *Philosophiae Naturalis Principia Mathematica* (1687) and used them to prove the movement of objects. Isaac Newton is best remembered for his three laws of motion.

The first: "A body remains at rest, or moves in a straight line (at a constant velocity), unless acted upon by a net outside force." This is why a spacecraft continues to move in the direction resulting from its last encounter with a force acting on it.

The second: "The acceleration (a) of an object of constant mass (m) is proportional to the force (F) acting upon it: $F = ma$." This law determines the g-forces an astronaut experiences during both the powered and atmospheric reentry portions of the flight by dividing the mass into the force ($F/m = a$). When the rocket lifts off the launchpad, its weight (m) is being acted upon by the thrust of the engines (F). Assume the rocket weighs 1 million lb. at liftoff and its engines are generating 1.5 million lb. of thrust; the astronaut is experiencing 1.5 gs. If the rocket weighs 250,000 lb. when the fuel is consumed, the equation results in 6 gs of force just prior to engine shutdown.

The third law: "Whenever one body exerts force upon a second body, the second body exerts an equal and opposite force upon the first body." It is often expressed as "For every action there is an equal and opposite reaction." This reactive force is what generates the thrust of a rocket motor. The pressure within the combustion chamber is pushing equally on all sides but the exhaust portal. This can be simply demonstrated by allowing the air to escape from a balloon. It is important to understand that the exhaust gas is not pushing against the atmosphere.

During the War of 1812, a British rocket bombardment of Fort McHenry near Baltimore, Maryland, inspired Francis Scott Key to write America's national anthem, *The Star Spangled Banner*. He observed the American flag illuminated by "the rocket's red glare." The experience of being on the receiving end of the projectiles inspired the American military to adopt the rocket, and they were used in the Mexican War of 1846–50.

Englishman William Hale, looking to improve the accuracy of the Congreve rocket, devised the "stickless," or rotary, rocket, which he patented in 1844. This device obtained its stability in flight by imparting a spin to the rocket by slightly deflecting the exhaust gases with curved vanes in its three exhaust ports. Hale's 24 lb. rocket was 23 in. long and 2.5 in. in diameter and made from rolled sheet iron. The average range was about 1,200 yd.—although distances up to 4,000 yd. were possible. The Hale rocket was obsolete by 1890 but not completely withdrawn from most military inventories until 1919.

As the nineteenth century dawned, the dreamers and visionaries had a stable environment that was understood and could be proved mathematically. Only the imagination limited what was possible, and for some, that imagination would lead humanity into space.

Konstantin Tsiolkovsky

While there have been several notable theoretical prophets of space travel over the years, few have been able to transform their vision into the written word and receive a place of honor in the archives of history. One such early prophet who was a prolific writer and who achieved that status was the Russian Konstantin Tsiolkovsky. Born into poverty, Tsiolkovsky recalled being excited at an early age by a small balloon, a rare device in those days.

He lost most of his hearing by the age of thirteen due to scarlet fever. It was difficult for him to attend public school, so he was largely home-schooled and self-taught. Tsiolkovsky's interest in mathematics and physics, spurred by his reading of Jules Verne and a high degree of intelligence, allowed him to master the basics of these topics to the point that he satisfied the requirements for teaching in public schools—despite his hearing impairment. Tsiolkovsky was interested in astronomy and the associated physics.

In 1895 Tsiolkovsky mentioned space travel in an article for the first time. Believing that the space between the planets lacked any atmosphere, he foresaw that a space vessel must be completely sealed with oxygen reserves and air purification for the human travelers. By 1898 he completed a preliminary study of the basic problems of space flight, and in 1903 he published *Exploring Space by Means of Reaction Devices* in the Russian periodical *Science Survey Journal*. From 1911 to 1913 he published a series of articles in the Russian technical magazine *Aviation Reports*.

Tsiolkovsky was the first to record the need for a reaction engine to operate in the airless environment of space. The only known mode of propulsion that worked in a vacuum was the Newtonian principal of action

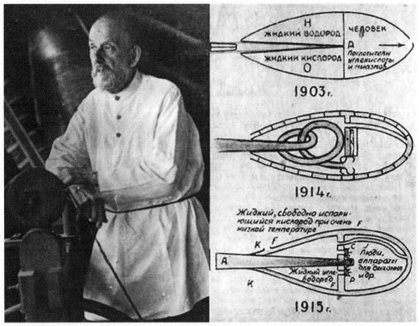

Н
ЖИДКИЙ ВОДОРОД | ЧЕЛОВЕК
А
ЖИДКИЙ КИСЛОРОД | Поглотители
углекислоты
О | и миазмов.

1903г.

1914г.

Жидкий, свободно испаря-
ющийся кислород при очень F
низкой температуре

К f

А

Жидкий или
водород
F

с
люди,
аппараты
для обихода
и др.

1915г.

Konstantin Tsiolkovsky and some of his sketches reveal the maturing of his vision of the rocket engine. Courtesy of Energiya.

and reaction—the rocket. He also introduced the concept of the multistage rocket. Recognition of his work earned him the title of "Father of Astronautics." Lack of funding restricted Tsiolkovsky's theoretical work from progressing to the experimental stages.

The World War (1914–1918) interrupted Tsiolkovsky's work, but fortunately the Russian Revolution (1917) did not appear to have any impact. His first publication under the Communist regime was a novel entitled *Outside of the Earth*, a fictional account of a journey into space. However, Tsiolkovsky's works were not widely published until after 1923. Tsiolkovsky was honored by the Soviet Union on his seventy-fifth birthday in 1932 for his contributions to astronautics. He died three years later, in 1935.

Robert Hutchings Goddard

The early twentieth century also saw another rising star in the field of astronautics in the person of Robert Hutchings Goddard, a professor at Clark University in Massachusetts. Goddard independently discovered virtually all the physics aspects involved in space flight formulated by Tsiolkovsky and was one of the first to publish his observations and mathematical proofs in English, and to move from theory to experimental applications.

Born in Worcester, Massachusetts, Goddard suffered from a variety of health problems most of his life (he was diagnosed with tuberculosis in 1913) and was fortunate to have a supportive and reasonably affluent family.

A 1908 graduate of Worcester Polytechnic Institute, Goddard became a physics instructor there and received a Ph.D. in physics from Clark University in 1911. In 1912, while a research instructor in physics at Princeton University, Goddard proved mathematically that a device using rocket power could achieve escape velocity and travel to the Moon and the planets. In 1914 he received a U.S. patent for the idea of multistage rockets.

While his contemporaries and predecessors were primarily theorists, Goddard was a practical experimenter. He pursued and acquired funding, typically in the form of grants, from a variety of sources that included his university as well as government and private capital.

He returned to Clark University in 1916 to teach physics and received his first grant of five thousand dollars from the Smithsonian Institution, with further grants awarded through 1934. His theoretical work to that point culminated in the 1919 Smithsonian Miscellaneous Publication No. 2540, *A Method of Reaching Extreme Altitudes* (Clary 2003, 88). In this paper Goddard detailed methods of raising weather-recording instruments higher than that possible with "sounding" balloons (sounding is the act of probing various aspects of the atmosphere using scientific instruments).

While a valuable piece of work, it was to be his most controversial, as Goddard was severely criticized by the press (no less than the prestigious *New York Times*) and by many in the scientific community for his views. The main point of discord was the statement he made during an interview that he believed rockets could be built to reach the Moon—a preposterous assumption by the scientific standards of the day. This experience led Goddard (and others such as Charles Lindbergh) to abhor the popular press for their ignorance and lack of understanding and sensitivity. He was so shaken by this experience that it influenced his interaction with the outside world from then on. (The *Times* issued a somewhat humorous apology some fifty years later during the launch of the first expedition to the Moon by *Apollo 11*).

Following World War I, Goddard turned his attention to developing rocket motors that used liquid propellants instead of solids—liquefied oxygen (LOX) and gasoline. On December 6, 1925, his rocket produced sufficient thrust (during a static test) to lift its own weight for the first time.

Measuring the Power of a Rocket

Thrust is the measure of the basic impulse provided by the rocket engine and is approximately comparable to the horsepower rating of internal combustion engine. Originally measured in pounds-force, it is now expressed in a metric equivalent called "newtons" (1 lb. of thrust equals 4.45 N). [This book will use the earlier expression of pounds as it is more clearly related to the common unit of measure in the United States.]

Robert Goddard and his first successful liquid-fuel rocket, 1926. The launch support is the darker tubular area, which Goddard is holding. The rocket engine is on the top of the vehicle in a tractor arrangement. Courtesy of National Air & Space Museum.

Regenerative Thrust Chamber Cooling

One of the many problems that had to be solved to enable high-thrust and long-duration burn times for liquid-fuel rocket engines was how to keep the combustion chamber from losing its strength or even melting from the extremely high temperatures. Although water was used in ground tests, it was not a practical solution for airborne rockets. Goddard resolved the problem by circulating the fuel through a set of tubes that surrounded the thrust chamber before injecting it for combustion. The process is called regenerative cooling.

The static test firing was followed on March 16, 1926, with the first flight of an 11 lb. liquid-fuel rocket in Auburn Massachusetts. As the flight time was a mere 2.5 seconds and the distance traveled was 184 feet (56 m), the achievement—although a milestone—was not spectacular (Clary 2003, 121). This success was followed by the first launch of a scientific payload (barometer and camera) in a rocket flight in 1929.

Goddard's rockets became larger, noisier, and hazardous. A flight on July 17, 1929, alarmed the quiet countryside of rural Massachusetts. The publicity caused Goddard's work to come to the attention of Charles Lindbergh, the noted aviator who had flown solo from New York to Paris two years earlier in 1927. Lindbergh visited Goddard to determine the focus and validity of his research and was impressed enough by his experiments to convince Harry Guggenheim (heir to a substantial fortune) to provide a series of grants from the Guggenheim Foundation beginning in 1930.

With the availability of generous funding, Goddard moved to a locale that provided more room for his experiments, choosing a small ranch on the outskirts of Roswell, New Mexico, to continue his work. He flew his rockets there from 1931 to 1941.

The advent of America's involvement in World War II in December 1941 caused Goddard's expertise to be redirected by the War Department. Moving from Roswell to Annapolis, Maryland, in 1942, he spent much of his time developing rocket-assisted take-off units (commonly called JATO) for the Navy's large flying boats. From 1943 until his death in 1945, he was an engineering consultant to the Curtiss-Wright Corporation in New Jersey. His team, some from the Roswell years, were employed by Curtiss-Wright to develop a throttleable rocket engine for the Bell X-2, a project that would

Goddard's work progressed at a slow but steady pace through the 1930s. Here is one of his A Series, flown between 1934 and 1936. Courtesy of National Air & Space Museum.

take a decade to accomplish. Goddard also served as director of the American Rocket Society during 1944 and 1945 (Clary 2003, 213).

After Goddard's death, some of his effort was carried on by those who worked with him. Lowell Randall, for example, was employed by Curtiss-Wright for seven years before returning to New Mexico to work at White Sands on the Rocketdyne engine of the Army's Redstone rocket.

Hermann Oberth

Hermann Julius Oberth credited two books by Jules Verne, *From the Earth to the Moon* and *Journey Around the Moon*, as arousing his interest in space travel at the early age of eleven. By the age of thirteen, Oberth, born in Austria-Hungary (today Romania) discovered that many of Verne's calculations were not plucked from thin air but had some validity. His youthful, unencumbered mind also grasped that interplanetary travel was not pure folly, as the scientific community assumed. He computed the gravitational forces that Jules Verne's space travelers would endure during acceleration in the gun barrel. His answer—47,000 times the Earth's gravity (Gs)—ensured that they would be squashed.

In 1913 Oberth went to Munich, Germany, to study medicine and later physics. In 1922 he submitted a doctoral thesis to the University of

Hermann Oberth's writings stirred the imagination of many in depression-torn Germany of the 1920s. Courtesy of U.S. Army.

Heidelberg about rocket-propelled space travel. It was rejected by the professional review committee because the idea of space travel was not realistic in academic circles. He never did earn a doctorate although he had the honorary title conferred on him.

Oberth assembled his findings for publication, but because no publisher would accept his work, he printed (at his own expense) a small, ninety-two-page booklet, *Die Rakete zu den Planetenräumen* (The rocket into interplanetary space) (Ley 1959, 108). This 1923 pamphlet contained four assertions: (1) Present technology allowed for the building of apparatus that can rise above the limits of Earth's atmosphere; (2) further developments would permit achieving velocities such that an apparatus could leave Earth's gravitational pull; (3) ultimately, these machines would carry humans; and (4) it might be possible to manufacture such apparatus for profit, and that all of these assertions would be achieved within the next few decades. Here was not only analysis and theory but also methods and applied purposes.

The success of his book encouraged Oberth to produce a less technical account of the possibilities of space flight. Another space flight enthusiast, Austrian Max Valier, a popular writer on scientific subjects, assisted in this endeavor by condensing Oberth's previous work and publishing it for him. This book inspired the formation of several rocket clubs in Germany.

Moviemaker Fritz Lang read Oberth's book and decided to produce an adventure story about space travel. The result was *Frau Im Mond* (Woman in the Moon), released in 1929. To add dramatic flair to the rocket flight, Lang invented the countdown to increase the tension for the audience—an artifact that is still an integral part of today's rocket launch sequence, and no less tension filled. The epic film enthralled moviegoers of the silent era and became a classic (Heppenheimer 1984, 9).

Society for Space Travel

During the summer of 1927 a group of Germans met in the back room of a restaurant to found a society that was to have a profound effect on rocketry—Verein fur Raumschiffahrt (Society for Space Travel), abbreviated VfR (Ley 1959, 117). Among the founding members were Johannes Winkler and Max Valier, who were subsequently joined in 1929 by Hermann Oberth and Dr. Walter Hohmann, author of *Die Erreichbarkeit der Himmelskorper* (*The Attainability of Heavenly Bodies*). Hohmann's mathematics on techniques for orbital changes became the key to achieving the rendezvous of two satellites thirty-five years into the future.

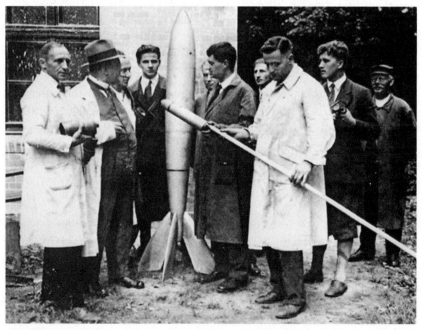

Members of the VfR pose with one of their creations. Von Braun is second from right, Oberth is fifth. Courtesy of National Air & Space Museum.

The VfR grew rapidly and within a year had over five hundred members to include Willy Ley and Wernher von Braun. In July 1930 the VfR demonstrated the Kegeldüse, a LOX- and gasoline-fueled rocket motor for the director of the Chemisch-Technische Reichsanstalt. The test ran for ninety seconds and generated 16 lb. of thrust but failed to impress the visiting dignitaries sufficiently for them to support the research.

The spring of 1932 brought the VfR to the attention of certain members of the Ordnance Department of the Reichswehr (the German army). The concept of the artillery rocket was being reexamined in light of the Versailles Treaty that prohibited artillery in the German army.

The VfR members were pleased to sign a contract with the army for the sum of one thousand marks. But von Braun recalls being fully aware that their toy-like Mirak II was a simplistic approach to a real liquid-fueled rocket capable of performing either a military or a scientific role. Its water-cooled motor produced 132 lb. of thrust. The problems of guidance and control, fuel pumps and valves, as well as the manufacturing costs made the efforts of the VfR completely inadequate to provide an experimental program needed by the military. Only the funds and facilities of the Reichswehr would make a large liquid-fuel rocket and space travel a reality.

The winter that followed the modest successes of 1932 blew cold with the rise to power of Adolf Hitler. The deepening world economic depression and the new demands of the Führer diminished the membership of the VfR and, consequently, the availability of funds. Several members of the VfR were offered positions with the German army to continue their work, although some balked at the notion of working for the army with the imminent rise to power of the Nazis. The Raketenflugplatz continued to operate briefly; however, the departure of von Braun to the army and a general decline in membership closed out active experimentation.

Wernher von Braun

While most of the pioneers of rocketry are not household names, one who was known to most Americans for a period of twenty years was Wernher von Braun. Von Braun's involvement was one that is easily partitioned into several phases, beginning with his association with the VfR and Hermann Oberth (1930s), the development of the German V-2 rocket (1940s), the launching of America's first satellite (1950s), and culminating in the building of the Saturn V Moon rocket (1960s) and the redesign of the space shuttle (1970s).

Wernher von Braun and Rudolf Nebel carry two early liquid-fuel rockets to the launch-pad. Courtesy of National Air & Space Museum.

Born in Wirsitz, Germany, von Braun's early life was comfortable. As a youngster, he was captivated by the exploits of Max Valier, who was setting speed records with autos, gliders, and iceboats. In 1925 the thirteen-year-old von Braun was entranced by an advertisement for Hermann Oberth's book *The Road to Interplanetary Space*. He was certain that humans would one day within his lifetime fly into outer space, and he wanted to be a part of the solution to the challenge of space travel. He was an early member of the VfR and worked tirelessly on the experiments. Von Braun's efforts, knowledge, and enthusiasm were noted by Capt. Walter Dornberger when the German army paid its first visit to the VfR.

In the spring of 1932 von Braun graduated from the Charlottenburg Institute of Technology with a bachelor's degree in mechanical engineering. That same year von Braun became employed by the German army and was charged with the development of liquid-fueled rockets (Dornberger 1954, 33). Two years later the twenty-two-year-old von Braun earned a doctorate in physics from the University of Berlin.

Dornberger respected von Braun's imagination but occasionally had to demand that he produce factual evidence to support his wide span of grand dreams. In this respect, Dornberger—von Braun's superior in the organization that created the V-2—helped create "von Braun the visionary." Dornberger was quoted saying of von Braun, "He was erratic at first and not completely persistent. He would go from one thing to another, but not until they had a clear idea of what he wanted to achieve. Then he would grow stubborn, and would not tolerate any impediments or deviations" (Bergaust 1976, 23).

While von Braun appeared to live a life free of vices that often corrupt those who achieve high positions of influence and power, his desire to avoid compromising his integrity may have been hardened by his brief association with the Nazi Party. While many in the VfR refused to follow von Braun as civilian employees in the pay of the Third Reich, von Braun notes that he was still a youngster in his early twenties and didn't realize the significance of the changes in political leadership as he was too wrapped up in rockets. In defense of his joining the Nazi Party in 1938, and the Schutzstaffel (SS) in 1943, his supporters note that it was simply a wartime expediency to help ensure that the A-4 (V-2) project held favor with those at the highest level of government.

Guggenheim Aeronautical Laboratory

In addition to funding Robert Goddard's work on rockets, The Daniel Guggenheim Fund for the Promotion of Aeronautics provided three hundred thousand dollars to the California Institute of Technology for the construction of a laboratory and the establishment of a graduate school of aeronautics. Dr. Robert A. Millikan, chairman of the California Institute's Executive Council, was the driving force in obtaining the grant. The eminent Hungarian scientist Dr. Theodore von Kármán was brought from Germany in 1930 to oversee the activities of this organization, which became known as the Guggenheim Aeronautical Laboratory California Institute of Technology (GALCIT).

A graduate student, Frank Malina, whose interest in space exploration was first aroused by his readings of Jules Verne, convinced Kármán to let him do his graduate research in the field of rocketry. From that beginning in 1934, research at GALCIT centered on both liquid and solid propellant types.

In March 1935 William Bollay, a graduate assistant to Kármán, became interested in rocket-powered aircraft because of a paper by the Austrian Eugen Sänger, who was working in Vienna at the time. Bollay became a proponent of rocket-powered aircraft as a promising method of reaching extreme altitudes and velocities. He thought it improbable that the rocket plane would be a contender with the airplane in ordinary air passenger transportation. For this, the stratosphere plane seemed much more suitable.

As rockets were still considered science fiction by the elite scientific establishment, Kármán decided to avoid the use of the word "rocket" and substituted "jet" in its place. Thus, in 1939 GALCIT's Rocket Research Project became the Air Corps' Jet Propulsion Research Project, and Kármán and Malina went on to found the Jet Propulsion Laboratory (JPL) a few years later, with Molina as its first director.

American Interplanetary Society

David Lasser, a science fiction writer for the *New York Herald Tribune*, and other enthusiasts formed the American Interplanetary Society (AIS) in 1930. Their stated objective was the "promotion of interest in and experimentation toward interplanetary expeditions and travel." Meeting biweekly in Manhattan's Museum of Natural History, they soon advanced to the next step—experimentation.

At Stockton, New Jersey, the AIS tested its first liquid-fueled rocket, based on the design of the German VfR Repulsor rocket. It produced 60 lb. of thrust for about thirty seconds but was damaged during the test and never flown. In May 1933 the AIS achieved its first successful flight to an altitude of 250 feet with the 6 ft. long Rocket No. 2 at Marine Park, Staten Island, New York. Rocket No. 4 reached 400 ft. in 1934.

Lasser's brief association with the AIS ended in 1933 when he resigned from the organization. However, his enduring contribution to the history of space exploration was his book *The Conquest of Space*. Within its pages, Lasser presented the practical concepts of space travel, far beyond the writings of Goddard. These thoughts of space flight, the essential equivalent of Oberth's book, were now available in English and proved to ignite the imagination of many, including a young Arthur C. Clarke.

Because some considered the word "interplanetary" in the title of their society as being too presumptuous, the name was changed the American Rocket Society (ARS) in 1934.

With war on the horizon in 1938, society members formed Reaction Motors Inc. to secure a government contract for providing a liquid-propellant rocket engine. With a meager start-up capital of five thousand dollars, the group set up shop in a garage but quickly outgrew it and moved to larger facility in Pompton Plains, New Jersey. With GALCIT (JPL) and the ARS (Reaction Motors), the foundation had been laid for the development in America of several important rocket engines as World War II began.

Sergei Korolev

Sergei Korolev was born in Ukraine and raised by his maternal grandparents. Initially trained in carpentry, his interest was in aviation following an air show he attended in 1913. In 1923 he joined the Society of Aviation and Aerial Navigation of Ukraine and the Crimea (also known as OAVUK) and had his first flying lesson. He attended the Kiev Polytechnic Institute where he studied aviation for two years before being accepted by the Bauman Moscow State Technical University. He flew gliders and powered aircraft and studied under Andrei Tupolev, the renowned aircraft designer.

Sergei Korolev (*left*) observes the launch preparation for an early liquid-fuel rocket of the former Soviet Union. Courtesy of Energiya.

After graduation, Korolev worked in aircraft design, completed the requirements for a pilot's license, and married Xenia Vincentini in 1931. At about that time Korolev's interest turned to liquid-fueled rocket engines for powering high-speed aircraft. Eventually he would become a key player in elevating the Soviet Union to a leading role in human space flight. Following World War II he was elevated to the position of chief designer in one of several Soviet rocket design bureaus.

Soviet Rocket Societies

Inspired by Tsiolkovsky, a group of space enthusiasts formed Gruppa Isutcheniya Reaktivnovo Dvisheniya (GIRD) in 1930. The organization's title translated to the Group for Investigation of Reaction Motion. Like the many other rocket societies fashioned during the period, GIRD built and tested rockets and promoted space exploration. Among GIRD's founding members was a pioneer in liquid-fuel rocket development, Fridrikh Arturovich Tsander and Sergei Korolev. By March of 1933 an 8 ft. long rocket was successfully flown and reached an altitude of 1,300 ft. after an eighteen-second flight (Brzezinski 2007, 107).

A parallel organization to GIRD was the Gas Dynamics Laboratory (GDL), which was likewise doing early research into the design and construction of rocket engines. Mikhail Tikhonravov and Valentin V. Glushko were early members who would play a significant role in Soviet space exploration. When the military determined that there was potential in the liquid-fuel rocket, GIRD and GDL were incorporated into the Reaction Propulsion Scientific Research Institute.

Several members of the organization were suspected of promoting a military coup against Soviet premier Joseph Stalin in 1937 and were executed. Many of the staff, including Korolev and Glushko, were imprisoned for "crimes against the state." As war with Germany loomed larger, Stalin recognized there were many talented people in the Gulags who were far more valuable to the war effort than serving time in prison (Siddiqi 2003b, 16). Korolev and selected others were "graduated" to a *sharashkas* (prison design bureau that were also called *sharagas*) (Oberg 1981, 20). Korolev's activities were recognized as being innovative and of high quality. The world would eventually be greatly influenced by the future work of Sergei P. Korolev and indirectly reshaped by it.

British Interplanetary Society

The British Interplanetary Society (BIS) was founded by space travel enthusiasts in October 1933 in Liverpool, England. The BIS focused more heavily on theoretical research than actual development of rockets because a British law enacted in 1875 banned the private use of explosives and rockets. One of the society's first significant efforts was to design a large, solid-propellant Moon rocket based on existing technologies of 1938. In the 1940s they proposed a more advanced Moon rocket that included a small module for descending to the Moon's surface. The American Apollo lunar landing program would use a similar technique twenty years later. Perhaps the most enduring of the societies discussed here, the BIS made its mark with its periodical, the *Journal of the British Interplanetary Society*, which was introduced in January 1934. It remains a valued publication today.

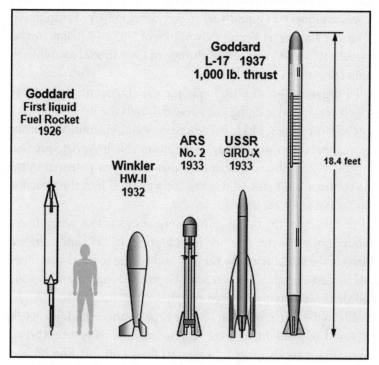

A size comparison of early liquid-fuel rockets. Courtesy of Ted Spitzmiller.

3

MATURING THE TECHNOLOGY

German Progress: The 1930s

Although solid-fuel rockets had been used as weapons for several centuries, they had fallen out of favor with the military as the development of rifled artillery (helical grooves in the gun barrel to impart a spin to the projectile) had proven extremely accurate and reliable. However, it was realized that new developments in rocketry might extend the reach of conventional artillery by a factor of ten—or more.

Capt. Walter Dornberger was tasked to review the progress being made by the various rocket societies in Germany. In the spring of 1932 Dornberger, dressed in civilian clothes, paid a visit to the VfR's Raketenflugplatz and witnessed their slow but potentially useful progress. He recognized that most of the VfR were primarily interested in space travel, and he made clear to them that the army was interested only in producing a usable weapon.

Dornberger was directed to develop a military capability using the liquid-fuel rocket with a range that should surpass any existing artillery, and its production was to be carried out in the greatest secrecy by industry. The conquest of space was about to receive a significant advance forward by a military application.

Under Dornberger's management, a program was established to create a rocket to send a one-ton warhead 200 mi. In the technology of the time, this was an order of magnitude greater than existing artillery, which at best could hurl a ton no more than 20 mi.

With Wernher von Braun's technical direction, the team laid out a plan for a progressive series of rockets to meet the goal established by Dornberger. These rockets were identified by the prefix "A" for the word "aggregate"—the sum total of their effort represented by the successive number that followed (Dornberger 1954, 37). Their first liquid-propellant rocket in the series was designated A-1 and took the new team six months to build. It underwent several static tests, but before the rocket could be flight-tested, it was destroyed by a "hard start." A hard start occurs when an incorrect propellant mixture ratio accumulates in the combustion chamber and explodes when the ignition source is introduced.

The team moved on to the A-2, incorporating what they learned. In December 1934 two A-2s, named Max and Moritz (the German names of the Katzenjammer Kids, two fictional misbehaving comic strip characters of the time), were successfully launched from the island of Borkum, in the North Sea. Unlike their namesakes, the rockets behaved perfectly and climbed to over 6,500 ft. They were crudely controlled by a spinning mass located between the propellant tanks to provide a gyroscopic effect to maintain a vertical ascent.

In December 1937 the A-3 was the first rocket launched from Greifswalder Oie, a tiny island in the Baltic near Peenemünde. The 3,300 lb. thrust motor (14.685 N) lifted the 21 ft. rocket briskly into the air. The original design called for supersonic velocities, but this was not to be achieved due to the excessive weight of the recording equipment aboard. With the A-3, the magnitude of Goddard's work had been eclipsed.

The A-3 used vanes in the exhaust plume to correct for any deviations sensed by the guidance system, much like the rudder of a boat. The vanes, made of molybdenum, had to withstand temperatures in excess of 5,000 °F.

While the A-3 was still under construction, the final specifications of the deliverable weapon, the A-4, were defined. The two most basic requirements were somewhat arbitrary in that it have twice the range of the Paris

---

Rocket Control

The control of a rocket by external aerodynamic fins can only be achieved after it has developed sufficient velocity for the air to exert adequate aerodynamic force. Some form of control needs to be used at low speeds and in the later stages of powered flight when the rocket is above 100,000 ft., where fins are of little value. Tsiolkovsky had suggested flat plates pressing against the exhaust gases, and Goddard had implemented such a scheme. Most contemporary large liquid-fueled rockets have no fins at all and typically use a technique of gimbaling (swiveling) the motor to change the thrust vector.

gun (80 mi.) used in World War I and ten times the explosive power of its warhead (200 lb.). Thus, the A-4 required a range of at least 160 mi. and a warhead weighing 2,000 lb.

However, the technology needed was still ahead of that provided by the A-3. So another rocket, designated A-5, was needed to prove various materials and techniques before moving to the A-4. The A-5 was twenty feet in length, two feet seven inches in diameter and weighed 2,000 lb. It used an improved version of the A-3 engine of 3,300 lb. of thrust (14,700 N). The first was launched in the summer of 1938, and ultimately twenty-five were a part of the flight test program.

Recruiting from the university campuses, von Braun, professor Walter Thiel (a propulsion specialist), Klaus Riedel (head of ground support equipment), and other members of the team made presentations on the types of problems facing the development of the A-4. They stressed that they were looking for short-term solutions that could be implemented within two years. This was to discourage those in academia whose sights were set too far into the future. Von Braun was concerned with solving problems in those technologies that could directly move the A-4 project ahead—space travel would have to wait until after the war.

The A-4: Pushing Technology

When Dr. Walter Thiel joined the Peenemünde staff in 1936, the largest liquid-propellant rocket motors of the time produced 3,000 lb. of thrust. To achieve greater thrust required higher propellant delivery rates. The rockets to that point had their tanks pressurized to deliver the fuel and oxidizer. The required thrust of the A-4 engine, almost twenty times that at of the A-5, dictated propellant delivery rates that could only be achieved with a high-performance pump.

Thiel's innovative approach quickly advanced the state of the art. The A-4's steam turbine (operating at 5,000 rpm) drove two pumps that injected the propellants (liquid oxygen and ethyl alcohol) into the combustion chamber at rates in excess of 50 gal./s. This produced the then phenomenal 56,000 lb. of thrust of the new engine.

By the time work had progressed to the A-4 in 1941, Hitler's war machine was doing well. His blitzkrieg (lightning war) had conquered virtually all of Europe. The German army was on the verge of completing its defeat of the Soviet Union, and England was an isolated island since the United States had yet to enter the war. Many in the Nazi high command saw no need for

Delivering the Propellants

The problem of how to power a pump of the size and capacity of larger rockets such as the A-4 was formidable. The pump must come up to a high operating speed within a few seconds to provide the required delivery rates and consistent pressure stability. The answer came with an understanding of hydrogen peroxide (H_2O_2). In levels of high purity (>80 percent), hydrogen peroxide decomposes rapidly when exposed to a catalyst such as sodium permanganate—producing water as a by-product. The decomposition releases energy in the form of heat, which turns the water into steam—the power to drive a turbine. Modern rockets use properties of the propellants themselves to drive the turbine pumps.

the A-4, and keeping the funding for the project was a constant challenge for Dornberger.

The first launch of an A-4 occurred in June 1942 and was a failure. The second attempt in August was more successful, with the rocket climbing to an altitude of about 7 mi. and passing through the sound barrier without incident—the first such airframe to successfully make the penetration. Many a heart hesitated as the flight proceeded through the lower level of the stratosphere, where the water vapor in the exhaust plume condensed to form a white trail. Although these vapor trails had been a common occurrence with high-flying aircraft, the apparent erratic shape the A-4 created by its rapid vertical advance through the horizontal high-altitude winds caused many on the ground to believe that it had gone out of control. However, the telemetry indicated all was well. The phenomenon was termed "frozen lightening" (Dornberger 1954, 20). However, the rocket soon disintegrated.

On October 3, 1942, the third rocket rose into a clear autumn sky, and observers noted good stability about all three axes. The end-of-burn occurred at sixty-three seconds, and it continued its now unpowered path arcing over the North Sea, reaching a height of 53 mi. into space, and traveling 120 mi. down range. By all standards, it was a successful test. Celebrating that evening at a party, Dornberger remarked to von Braun, "Do you realize what we accomplished today . . . the spaceship was born!" (Neufeld 2007, 137).

Numerous technical problems continued to plague the program, but one by one they were resolved. The ability to monitor what was going on in the rocket during flight was limited, and the capability to send back to Earth (by radio transmission) various readings from sensors of temperature and pressure was just being developed as a process called telemetry. Within the A-4 there were only four channels of telemetry being returned. As a comparison, the Saturn V Moon rocket in 1969 transmitted more than 3,500 measurements.

In March of 1944 the tide of war had turned against the Nazis, and the Waffen-SS (Schutzstaffel)—the armed component of the Nazi party—arrested von Braun and several other members of his team. They were charged with making statements that the A-4 was not intended as a weapon of war, but its primary intent was space travel. Dornberger defended his team members and insisted that without these scientists, the development of the A-4 would grind to a halt. Between the efforts of Dornberger and Albert Speer (the minister of armaments), the pride of the German rocket team were finally released to resume their work.

The A-4 began its assault on England and the Belgian port of Antwerp on September 8, 1944. The first application of a large liquid-fuel rocket, born from the dreams of countless visionaries and created with theory and work by persistent experimentalists, had arrived—but it probably was not what any of them had imagined. It was a destructive terror weapon of war, not a scientific leap into the cosmos.

Twenty-six A-4s hit London over the ensuing ten days, and Joseph Goebbels, Hitler's minister of propaganda, announced in a radio broadcast that "vengeance weapon number two," as promised by the Führer, was operational against England. It was through this announcement that von Braun learned his rocket was known to the world as the V-2. Ultimately,

Guidance Systems

Early guidance systems typically used a gyroscope for each axis of flight (pitch, yaw, and roll) and integrating accelerometers—devices that integrate time with acceleration to calculate velocity. By sensing the attitude and velocity of the rocket and comparing these values against a desired track, signals are sent to the control actuators to correct for any deviations.

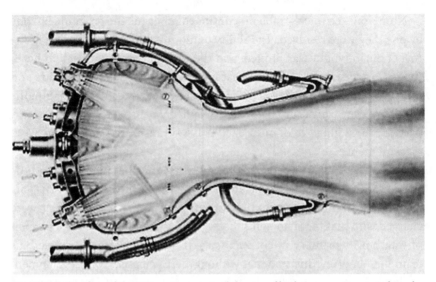

The thrust chamber of the V-2 engine required the use of high-temperature metals and engineering innovation. The tubular structure at top and bottom route fuel to cool the combustion chamber. Courtesy of National Air & Space Museum.

more than three thousand were launched against London. (The pulse-jet-powered Fieseler Fi-103 was named the V-1).

Toward the end of the war, as the Allies pushed the German army farther inland from the English Channel, the need to extend the 200 mi. range of the A-4 became apparent. As early as 1940 wind tunnel data showed that the range of the A-4 could be extended significantly with wings—to permit it to become a hypersonic glider in the upper atmosphere and doubling its range, with minimum cost and effort. However, it was not until October 1944 that serious design and development began on the project, when coastal launch sites for the A-4 in Holland were lost. Although several test flights of this rocket (designated A-4b) were flown, the problems of high temperatures and aerodynamic forces involved were beyond the crude adaptation being attempted. Years later the United States revived the hypersonic glider concept in the form of the X-20 Dyna-Soar program and later still, the space shuttle.

While the major emphasis of the German army's research at Peenemünde was developing the A-4, a small group of designers considered advanced rockets that rivaled even the Saturn V and the space shuttle.

In an effort to demonstrate futuristic projects, the von Braun team made studies toward missiles with intercontinental capabilities. A scaled-up, winged A-4, designated as the A-9, was envisioned as a second stage to an all-new rocket with a half-million pounds of thrust. The combination, known as the A9/A-10, had a range of 3,000 mi. to allow attacks on the East Coast of United States. Some drawings for the A-9 show a pressurized cockpit in place of a warhead, and tricycle landing gear.

The A-11, with a thrust of 3.5 million lb., never progressed beyond preliminary design studies. It would have been a lower stage for the A-9/A-10. It was estimated that this three-stage rocket could reach orbital velocity of 25,000 ft./s—the first engineering assessments of an orbital vehicle.

Not all the work being done at Peenemünde was by the von Braun team. An interesting study conducted by Dr. Eugen Sänger and German-born Dr. Irene Bredt analyzed the intriguing possibility of using the upper atmosphere to skip-glide a rocket-propelled bomber across great distances (similar in concept to the A-9/A-10). The work was based on an earlier study Sänger had conducted in 1933 while at the Technische Hochschule in Vienna, which proposed a liquid-fueled, rocket-powered, high-altitude airplane. The vehicle (which Sänger referred to as the *Silverbird*) was to be launched from a rocket-powered sled that accelerated the craft to 1,200 mph before using on-board rocket power. It would reach a velocity of

The exhaust nozzle of the V-2 shows the graphite vanes that provided stabilization and control, especially during the initial launch phase. Courtesy of National Air & Space Museum.

15,000 mph before exhausting its fuel and continuing to coast upward to an altitude of 160 mi. It was then to skip off the upper layers of the atmosphere much like a flat stone skipping across a pond.

The novel concept of launching discussed in the highly classified document, known as the *Sänger-Bredt Report*, was appropriated by the Germans during the war for use in a bomber. Called the *Raketenbomber*, it was to be driven by two A-4 engines. It was envisioned that it would bomb New York and land in Japan, where it would be refueled and relaunched on the reverse course for another mission. The work was put on hold in 1942, and Sänger spent the rest of the war working on ramjets for fighter aircraft.

While schemes such as these were too far-out for consideration during the closing year of the war, the report found its way to the United States and Russia following the war, where it was studied closely in light of the dawn of the atomic age.

Post World War II: Exploiting the V-2

Following the end of the war, both the United States and the Soviet Union made extensive efforts to acquire German technology. This resulted in sixteen Liberty cargo ships returning almost one hundred V-2s and extensive documentation to America. In addition, under the code name Operation Paperclip, 118 of the German scientists (including von Braun) were brought to America to work under contract for the U.S. government to ensure that the maximum technology transfer was achieved. Most of the team eventually became U.S. citizens (Bower 1987, 129).

The V-2 research program in America did not proceed as rapidly as the von Braun team had anticipated. All of the V-2s were transported to White Sands Proving Grounds in New Mexico. They were numbered and launched in approximately the numbered sequence. The first launch in April 1946, of V-2 No. 2, failed after nineteen seconds of flight when the graphite vane on fin IV disintegrated, causing the fin to tear itself from the rocket body. The third V-2 flight was more successful, reaching some 71 mi. into space.

With the realization that the number of V-2s was limited, an upper atmosphere research panel was organized. The V-2 Panel (as it was sometimes called) reviewed requests for experiments for each rocket, identified regions of particular scientific interest in the upper atmosphere, and assigned payload space to specific academic institutes. The V-2 designated TF-1 (a Hermes test flight) achieved the highest altitude, 132.6 mi., in August 1951. While the V-2 was a phenomenon for its day, it was quickly recognized that

Mass Ratio/Fraction
A factor in determining the performance capability of a rocket is the relationship between the mass of the structure relative to the propellants. The mass ratio of a rocket is the weight of the propellant divided by the rocket's total weight at liftoff. A higher mass fraction (percentage) means that greater proportion of the weight is propellant. The objective is to have the lightest possible structure containing the greatest possible fuel load. This results in a higher change in velocity termed Δv (pronounced "delta v").

it was not an efficient rocket in terms of its mass ratio. Its design would have to be significantly upgraded to make it a more capable weapon—or space booster.

At White Sands, the telemetry capabilities were significantly improved, and over fifty different channels of information were sent back from the missile, including the first heat transfer data from supersonic flight at over 3,400 mph. By the time the V-2s were expended, sixty-seven had been launched from White Sands, two from the new Joint Long Range Proving Grounds at Cape Canaveral in Florida, and another from the aircraft carrier USS *Midway* in Operation Sandy. The last V-2 launch occurred on September 19, 1952.

Institute Rabe and Operation Ost

The initial Soviet effort to acquire missile technology and expertise in the closing days of the war could be categorized as disorganized (as it had been for the Allies). Many of the officers in charge had no knowledge of missile technology and relied heavily on the interpretations of aviation engineers.

Stalin diverted combat troops to capture Peenemünde. That facility was essentially deserted, and the expected cache of information and equipment was meager. It soon became apparent that the majority of engineers and scientists had not only evacuated the major research and development centers, but most had also moved into the American and British occupation sectors of Germany. When it was realized that the principals, including von Braun, had willingly given themselves up to the Americans, Stalin was outraged.

In an effort to ensure that German technology was fully discovered and exploited, a central command for rocketry operations was established in Berlin following the end of hostilities. A special technical commission (abbreviated OTK in Russian), referred to as Institute Rabe was established. Soviet specialists in rocketry were flown to Germany, and a joint Soviet–German commission provided for the collection of information about the various rocket programs being uncovered.

When it was realized that the Americans had removed most of the completed A-4s from Mittelwerk ("Central Works," an underground factory), the Soviets decided to reopen that production plant to produce several dozen A-4s for testing and evaluation. German engineers were indispensable in this effort, and within a year after taking over Mittelwerk, the Soviets were ready to produce A-4s.

In the early fall of 1945 the Soviets formally began Operation Ost (the equivalent to the American's Operation Paperclip), an intense search for Germans with the talents needed to exploit the A-4 technology for the Institute Rabe. In the desperate days that followed Germany's surrender, most Germans were willing to work at any job that offered the opportunity to feed and house their families. Leading the Soviet's recruiting effort was Boris Chertok, of Institute Rabe, who achieved some success with those who had been passed over for Operation Paperclip and those who preferred to remain in Germany with their families. Virtually all of the activities were held in strictest secrecy, and the NKVD (Soviet People's Commissariat for Internal Affairs) secret police embedded informants at all levels.

The hunt for the more technically capable Germans, although intense, was kept as subtle as possible to avoid the possibility of alarming them. One of those who responded to the benign but intense solicitation of the Soviets was Helmut Gröttrup, who made several secret trips into the Soviet sector to discuss his future (Lardier and Barensky 2010, 16). A former assistant to the director for guidance and control at Peenemünde, Gröttrup was selected to participate in Institute Rabe in September 1945. Gröttrup's decision to join was motivated more by his desire to keep his family together than by science or politics. While many voluntarily joined with the Soviets, there were others whose participation was coerced through intimidation or outright threat. When it became apparent that America had the core of the von Braun team, a proclamation was issued on October 11, 1945, directing all scientists to register with the Soviet occupation forces. At that point a hard line was taken toward recruiting.

Eventually, two hundred German engineers were a part of the Soviet effort to provide expertise in aerodynamics, guidance and control, propellant chemistry, propulsion, and ballistics. While few of these people had played a primary role in developing the A-4, they would supply important capabilities to the Soviets in mastering production and testing of liquid-fuel rockets. The Russians also let it be known that they would pay large sums of money for von Braun or Dr. Ernst Steinhoff, who had been the principle engineer in formulating the V-2 guidance. It was reported that the Russians even initiated a kidnap plot for the latter. However, cooperation between the Germans and the Russians was generally amiable.

By October 1946 it was realized that Institute Rabe had achieved as much as it could, and it was decided to transfer all German and Soviet personnel to Soviet territory. This decision was concealed from the Germans, who were fearful of that possibility. The surprise notification of transfer was made following a technical meeting that the Russians turned into an all-night party—most likely to ensure that the Germans were well intoxicated before hearing the news. At about 4:00 a.m. the Germans were handed a document that essentially said their employment was to be continued in the USSR under a five-year contract. They were directed to gather their families and take food and clothing for a trip expected to last up to four weeks. The 152 rocket engineers were only a small part of an estimated 6,000 Germans from a wide variety of technologies that were eventually relocated to the Soviet Union.

The Germans arrived at the Science Research Institute No. 88 (NII-88) at Kaliningrad, about 10 mi. north of Moscow, where they found deplorable living conditions. The Soviets themselves were not living any better. The postwar conditions in the Soviet Union were about as bad as any civilization had experienced. Shortages of food, clothing, and housing were obvious. Medical treatment was almost nonexistent, and transportation was in great demand. Most everything had to be moved by rail since vehicles and gasoline were in short supply.

The basic hierarchy of who got the better accommodation was directly related to the educational standing of the worker. Dr. Gröttrup was housed in a large, six-room villa while most others lived in single-room dormitories, and some had to make do with tents. The workday was often twelve or more hours, six days a week. Some workers actually preferred to be on the job, as there was no real home to return to at the end of the day. Even the working conditions were primitive, with a lack the basics such as tables and chairs. The buildings were in disrepair and often leaked and were cold. As for pay, most of the Germans received better compensation than an equivalent Soviet worker.

For most of the Germans and their families, Operation Ost was a frightening experience. The work being done was often obscured by a wall of separation between the real Soviet rocket program and the activities that the Germans were given to perform. As von Braun had prophesied, these Germans were being "squeezed like lemons" for their knowledge and most would return to Germany after the five-year period with no knowledge of the current state of Soviet rocket technology that was of any value to Western intelligence (Bower 1987, 115).

Additionally, the German scientists were removed from any possible assistance to the Americans or British during this period of being "on ice." Unlike the Americans, there was never any intent of incorporating the Germans into the mainstream industry or social life of the Soviet Union. Likewise, there was no differentiation between "good Germans" or "bad Germans"—former Nazis. Thus, there was no moral concern over using all Germans who were available. Helmut Gröttrup was among the last group of Germans to be repatriated in November 1953.

VR-190: A Vision of the Future

The Soviets used their access to V-2 technology in the same manner as the Americans. While missile development reflected the military application that dominated the Soviet government in the post war years, the dreams of Korolev and others continued to focus on the stars. Mikhail Tikhonravov (a former member of GIRD/GDL) had designed the first liquid-fuel rocket in the Soviet Union. He convened an ad hoc group of engineers at the institute in 1945 to investigate the use of existing technology (the V-2) to develop a "vertical rocket" for carrying two passengers to an altitude of 190 km.—thus the designation VR-190.

The proposal was the first in the Soviet Union (and the world) for launching humans into space. The plan envisioned the use of a modified V-2 with a recoverable nose cone containing a pressurized cockpit for carrying two stratonauts (as the passengers were called). The passengers were in semireclined seats in the capsule from launch until touchdown. There were several objectives of the plan, including research on the effects of weightlessness and the separation and stabilization of the cockpit.

Much of the design was remarkably advanced for its time. It included a parachute system for the return to Earth, which was coupled with a breaking rocket for softening the landing of the cabin on the ground. The pressurized cockpit provided life support systems and attitude-control jets, allowing the vehicle to orient itself during its trajectory beyond the atmosphere. A heat shield on the bottom of the cabin was similar to that used on the Soyuz spacecraft twenty years later. Several other design elements of the VR-190 were incorporated into the first Soviet manned spacecraft in the early 1960s.

Recognizing that the hierarchy immediately above him would never approve such a plan, Tikhonravov appealed directly to Stalin. Although the response to his letter was somewhat positive in tone, the project never received funding. Korolev did not support the VR-190 project as he felt that a piloted rocket plane, an outgrowth of his interests and experiments of the 1930s, was the preferred method of exploring the upper atmosphere.

Most scientists relegated human orbital flight to "many years into the future" because of the obvious technological hurdles and costs. However, some in the Soviet Union expressed the desire to achieve the orbital goal before the United States did. A part of this mindset was the need to combat the perceived inferiority of Soviet science and industry by most of the "civilized" world.

Project Bumper: First Multistage Liquid-Fuel Rocket

In February 1946 JPL initiated a study for a two-stage 28,300 lb., 57 ft. long research rocket that used a V-2 first stage and a small American-built sounding rocket (the WAC Corporal) as a second stage. The project was approved and given the code name Project Bumper. As the first two-stage, liquid-fueled rocket ever assembled, it explored problems related to stage separation and rocket ignition at high altitude.

Bumper No. 5 achieved full success in February 1949, reaching an altitude of 244 mi. It is interesting to note the physics involved here. Had the second stage been fired at the high point of the V-2's flight (at best, 115 mi.), the capability of the WAC Corporal would have simply added another 40 mi. to the altitude (perhaps resulting in a 155 mi. altitude). However, by igniting the second stage at the end of the first stage burn (where it had achieved its greatest velocity), the cumulative velocity provided the record—which would hold for almost eight years in the United States.

The first multistage rocket takes flight at White Sands, New Mexico, with a V-2 carrying a WAC Corporal second stage. Courtesy of U.S. Department of Defense.

The Multistage Rocket

Hermann Oberth (as well as Konstantin Tsiolkovsky and Robert Goddard) understood that the higher the ratio between propellant mass and structural mass (mass ratio), the faster a rocket could travel. As a rocket expends fuel, its structural mass remains the same—the rocket has to carry along the dead weight of the partially empty fuel tanks and the structure that contains them. Additionally, the large and fuel-thirsty engines required to get the large vehicle off the ground are not necessarily the best choice once the initial acceleration into the upper atmosphere has been achieved (Zaehringer and Whitfield 2004, 62).

Oberth therefore reasoned that building the rocket in stages that could be separated (discarded) would improve the mass ratio of the rocket. He wrote, "the requirements for stages developed out of these formulas. If there is a small rocket on top of a big one, and if the big one is jettisoned and the small one is ignited, then their speeds are added" (Zaehringer and Whitfield 2004, 62). This concept became the key to attaining high velocities. The term "booster" is typically applied to the first stage of the vehicle.

It was decided that the remaining two Bumpers would fly trajectories for range, rather than altitude, which would result in higher velocities and attain additional aerodynamic heating data. These flight profiles required a testing range with a longer distance capability than available at White Sands. The area of Cape Canaveral at the Joint Long Range Proving Ground in Florida was chosen because it provided for firing southeast across the Atlantic Ocean. These tests were the first rockets launched there in 1950.

4

SPACE MEDICINE ON THE THRESHOLD

The end of World War II found Jean Piccard ready to renew balloon-centered high-altitude research, and he had gathered a small group of enthusiasts that included James Ford Bell, founder of the General Mills Corporation—known widely for breakfast cereals. Bell had been funding a research effort called Helios that included balloon fabricator Otto Winzen. Winzen, who had emigrated from Germany in 1937 but had spent a part of World War II in an allied internment camp, was an entrepreneur who pioneered the use of plastics in a variety of innovative products following the war. His process allowed the balloon's envelope to be extremely thin and lightweight.

Little more than ten years (and a world war) separated the civilian exploration by aerostats of the late 1930s from the research required by the military in the late 1940s for application to defense needs—and a corresponding change in the cast of characters occurred. While Winzen became an important part of this new generation of high-altitude research ballooning, Jean and Jeanette Piccard found no role in the new programs.

With the advent of the jet engine and the presence of nuclear weapons, the military saw a pressing need to understand the characteristics of the upper atmosphere beyond pure science. One of these was Lt. Cmdr. George W. Hoover, who at the time was attached to the Special Devices Center of the Office of Naval Research.

In the period when Capt. Charles "Chuck" Yeager broke the sound barrier in the X-1 rocket plane in 1947, the ability to gather information about the effect of supersonic flight on aerodynamic shapes was limited to the use of rockets—which were themselves unruly and expensive. One of Hoover's projects envisioned dropping a variety of shapes and airfoils from high-altitude balloons and allowing gravity to accelerate the vehicle through Mach 1. Electronics enabled the capture and transmission of data via telemetry to ground stations, in addition to parachute recovery of recording devices. Another pressing need was the desire to gain more information about high-altitude, high-speed escape from aircraft when control is lost.

Ultimately, the military decided to follow three tracks. The first, Project Skyhook, which used unmanned balloons, began operational flights in 1949 by dropping shapes called free-fall test vehicles from high altitudes at the White Sands Proving Grounds in New Mexico. The second track was a manned aerostat program called Project Strato-Lab. The third was the exploration of the physical stress on the human body caused by g-forces—using the rocket sled.

Rocket Sleds: Surviving High G-Forces

The subject of bailing out at a high speed and high altitude is an integral part of the history of space exploration. As ultimately demonstrated, the development of self-contained personal life-support systems was necessary. In addition, an understanding of the physiological limitations of the human body and psychological impression on the human mind of isolated flight are directly related to the space environment.

An individual who was pivotal to understanding the stresses on the human body under these circumstances had an unconventional entry into the field. Dr. John Paul Stapp was raised in the jungles of Brazil by his Baptist missionary parents. Although an American citizen, he could only speak Portuguese and some of the native dialects until he was twelve. Essentially homeschooled by his parents, he returned to the United States at the age of fifteen to complete high school and attend college. He graduated from Baylor University with a master's degree in chemistry and zoology—despite having no financial support, as his parents themselves barely eked out a living in the jungle.

A tragedy while he was in college involved an infant cousin whose clothes caught fire. The child lived less than three days while being constantly attended to by the young Stapp. The experience of watching his cousin die influenced him to become a medical doctor.

He completed his medical schooling at the University of Minnesota and then joined the Army in October 1944, where he served during the remainder of World War II. Following the war, Stapp was assigned to Wright Field in Dayton, Ohio, where the Army Air Forces were doing advanced research on aircraft and propulsion systems. It was also here, at the Aero Medical Laboratory of the Air Material Command, that Stapp found his calling. He was impressed by the need to expand medical research in the area

Solid-fuel rockets propelled the Earth-bound sled at speeds up to 600 mph. Courtesy of U.S. Department of Defense.

of aircrew survival. However, unlike some of his contemporaries, Stapp wanted to experience the environment firsthand—as the test subject.

One of his first experiments had him testing a liquid-oxygen breathing system in a B-17 especially modified for flight to 47,000 ft. Working in the unpressurized and essentially unheated aircraft, Stapp established the need to purge the blood of nitrogen by breathing 100 percent oxygen for thirty minutes before attempting flight into the stratosphere.

For his next assignment Stapp examined the various forces at work during high-speed bailout. It had already been found that leaving an aircraft at speeds as low as 250 mph exposed the pilot to several dangers—not the least of which was the probability of hitting the tail of the aircraft. The concept of rapidly ejecting a pilot strapped securely to his seat, using a modest propulsive charge, had been under taken by Germany and Britain during the war. The Germans had gone so far as to build a short, railroad track–like segment where rocket-powered sleds, propelled at high speeds for short distances, tested the ejection process and associated restraining harnesses.

To undertake his research, Stapp was transferred to Muroc Air Force Base (AFB) (soon to become Edwards AFB) to work with Northrop Aircraft Corporation. Northrop operated a two-thousand-foot track that previously was used to launch experimental unmanned JB-1 guided missiles (the American copy of the German V-1).

With a small staff, Stapp procured and built specialized equipment and instrumentation to study the effect of high-g acceleration and deceleration. The number of small solid-fuel rockets used determined the acceleration of the test sled. A water trough at the far end decelerated the sled. The number and size of scoops determined the decelerating g-force.

Following a series of unmanned tests, Stapp made his first run in December 1947. For the first sixteen runs, the test subject (Stapp being the most frequent) faced backward. The first forward deceleration occurred in August 1949. The facility was then moved to Holloman AFB, in New Mexico, located adjacent to the White Sands Missile Range and named to honor Col. George V. Holloman, who had pioneered automated flight controls for pilotless aircraft.

A variety of standard seat harnesses were tested and modified to find the best arrangement to distribute the g-forces evenly and with the least injury to the body. Some of this research would ultimately find its way into the design of automotive seatbelts.

Because of the desire to determine the effect of not only g-force but also the aircraft slipstream, Stapp and other volunteers were subjected to the

A sequence of three frames from a film showing the effect of the high-speed g-forces and windblast on Stapp's face. Courtesy of U.S. Department of Defense.

windblast of up to 570 mph. The research eventually extended into virtually all aviation and space life sciences. Stapp became the chief of the Aeromedical Laboratory in 1961 and endured twenty-nine rocket-sled runs.

At Holloman, Stapp's new facilities included a 3,500 ft. test track that he personally traversed three times in 1954, exposing his body to the rigors of g-force and windblast. The fourth test in December propelled him to 632 mph. During this ride, he momentarily endured 20 positive gs and 46 negative-gs. Solid-fuel rockets that generated 40,000 lb. of thrust propelled the sled named *Sonic Wind No. 1*. Stapp's body suffered no long-term effects except for his eyes. These had already experienced detached retinas from previous track runs, and this time the vision problem, though not entirely blinding, would stay with him the rest of his life. John Paul Stapp had made his last rocket-sled ride. The limits of human tolerance, if not precisely established, were now better understood because of his efforts. The media of the day referred to him as "the fastest man on Earth."

Aerostats Revived: An Upper-Atmosphere Research Tool

Stapp's work in addressing the problems of ejection led to investigating the concept of extended free fall following ejection to avoid the prolonged and potentially hazardous period "under chute." If the crew ejects at a high altitude, air currents could keep a pilot under-chute aloft at the extreme altitude for durations exceeding an hour, where the numbing cold and limited oxygen supply could easily kill. Thus, the concept of high-altitude low-opening (HALO) crew escape (free fall) was born.

By the mid-1950s the first iteration of supersonic fighters would soon be putting Stapp's research and innovations into practice. Although aircraft and unmanned rockets could penetrate into the region above 50,000 ft., they could do so only for a few minutes, and the instrumentation needed for experiments was often too heavy and bulky to be carried. The high-altitude balloon was less expensive—it was the research tool to embrace.

Stapp's rocket-sled experiments had taken him from Muroc in California to Holloman AFB in New Mexico in 1953 to work with the Balloon Research and Development Test Branch. It was here that tests were conducted first with living organisms such as fruit flies and mice, then with monkeys, and finally men.

Here, too, Stapp's path would cross several other prominent space researchers such as Dr. James Van Allen. Another individual was a young medical doctor named David Simons, who was involved in cosmic ray research—the same mysterious rays emanating from outer space that the pioneer balloonists had studied twenty years earlier.

Stapp proposed to advance the research by again employing piloted balloons to altitudes of over 100,000 ft. for periods greater than twenty-four hours. He asked Simons if he would volunteer to be the lone occupant for such a flight. Although Simon's background was in laboratory research and had been restricted to sending mice and monkeys, Simons accepted the challenge. After a hiatus of twenty years, the scientist would return to the upper-atmosphere field of battle. The U.S. Air Force, which had been separated from the Army in 1947 to become an independent service entity, established Project Manhigh, which would compete with a similar Navy program called Project Strato-Lab.

Stapp expanded the program beyond the radiation issue to include crew escape and the associated infrastructure involved. This program bridged the gap between the atmosphere and outer space in a region that Dr. Hubertus Strughold termed the "aeropause."

Strughold was an early researcher in aviation medicine in Germany, immigrating to the United States in 1947 to perform pioneering work for the U.S. Air Force and then the National Aeronautics and Space Administration (NASA). It was at a 1948 conference that Strughold first used the expression that characterized the efforts of researchers like Stapp—space medicine. Strughold's involvement with questionable inhumane experiments for the Nazis during World War II caused his fame to be significantly discredited following his death. He is the only individual to be removed from the International Space Hall of Fame in Alamogordo, New Mexico.

The Air Force established the School of Aviation Medicine at Randolph AFB, Texas, guided by Col. Harry G. Armstrong. The Armstrong limit (named to recognize his work) is that altitude (63,000 ft.) at which water (and blood) boils at the temperature of the human body—98.6 degrees Fahrenheit. Although the official scientific definition of space is sixty-two miles above Earth (the Kármán line), aircrews traveling above the Armstrong limit require the same physiological protection as those in space.

Project Manhigh: Kittinger

Before Dr. Simons could make the trip to 100,000 ft., he and a second test subject, a cocky Air Force fighter pilot, Capt. Joe Kittinger, would have to satisfy Stapp's prerequisites. The first required them to remain (earthbound) in the small gondola while wearing the MC-3—an unbelievably

Altitude Chamber

The need to simulate the effects of high altitudes on humans as well as aircraft components was recognized early in the history of aviation. The National Advisory Committee for Aeronautics built the first pressure chamber at the Bureau of Standards in 1917. Essentially, an altitude chamber (sometimes called a decompression chamber) is a structure from which the air is pumped out to create a low-pressure environment. The size of the pressure tank can vary depending on the primary use of the facility, but typically, for the human element, it is 15 ft. long and 10 ft. wide and can accommodate ten or more individuals.

uncomfortable, skin-tight, mechanical partial-pressure suit—for twenty-four hours to verify that they had no claustrophobic tendencies. The second was to complete a simulated flight to an altitude of 100,000 ft. in an altitude chamber while wearing the MC-3. The final two requirements were easy by comparison. Each had to make a parachute jump (Kittinger would make eight) and obtain a lighter-than-air free-balloon pilot license. The last required sixteen hours of flight time and sixteen takeoffs and landings.

Project Manhigh used a small pressurized capsule. However, unlike the shirtsleeve environment of the 1930s, the pilot would wear the MC-3 as a backup, should the pressure system fail—thus the need to confirm the pilot's tolerance to claustrophobia. The MC-3 was termed a mechanical pressure suit because it relied on the tight coupling of the fabric against the skin to keep the human body from expanding. It essentially used the pilot's own body fluids in a sausage-like casing held taught by an inflatable tube-like capstan that followed the contour of the torso and extremities.

Stapp decided that prior to Simon's attempt to ascend to 100,000 ft., a test flight should be accomplished by Captain Kittinger to verify the equipment and all the various procedures and techniques to be used. Simons was not pleased with the directive but recognized that Kittinger, as an experienced fighter pilot, had a level of familiarity in handling critical life-threatening situations.

Winzen Research Inc. was chosen to build the envelope, much to the consternation of General Mills. A large amount of the credit for Winzen's success was due to the efforts of Otto Winzen's wife, Vera, who trained the fabrication workers, oversaw the construction, and followed up on quality control. However, her interest in ballooning went deeper—she received her gas balloon pilot certificate in 1957.

While the use of balloons for high-altitude operations was less expensive and often not as hazardous as other means, there were drawbacks. Large gas balloons can only be inflated and landed in relatively calm wind conditions, and several techniques were used to avoid even the slightest gusts. For example, at Holloman, a long, 40 ft. trailer-like device was constructed called the Covered Wagon. This allowed the early stages of inflation to occur in a more protected environment. Because of the fragile nature of the balloon material, once inflated, the envelope had to be flown or scrapped—it could not be re-inflated. The balloon for Project Manhigh measured 300 ft. in height and, at altitude, it expanded to 200 ft. in diameter. These 10 million cu. ft. volume balloons were capable of lifting up to 3,000 lb. to 100,000 ft.

Winzen also built two pressurized capsules that measured eight feet in height and three feet in diameter. The atmosphere within was maintained at sea-level pressure but with a gas mixture of 60 percent oxygen, 20 percent helium, and 20 percent nitrogen. The carbon dioxide in the cabin was removed during the flight by chemical scrubbers, and the resupply of oxygen was sufficient to last for forty-eight hours of flight.

The preparation of Kittinger for the flight was not much different than the first astronauts, little more than four years into the future. A check of Kittinger's vital signs by Dr. Simons was followed by the rather torturous suiting-up process—to include a small microphone taped to the pilot's chest to record pulse and respiration.

In the predawn hours of June 2, 1957, Kittinger completed the preparation process and entered the gondola, which was then loaded onto a pickup truck and taken a few miles to Fleming Field, a small airport in South St. Paul, Minnesota. Shortly after 5:00 a.m. the balloon launched to begin its 500 ft./min. ascent.

Problems with the communications prohibited Kittinger from transmitting voice to ground control, although he could still use Morse code. The most dangerous part of the flight occurred at about 45,000 ft. when the jet stream was encountered. Had the envelope failed at any point, Kittinger could have elected to depressurize the gondola and take to his chute. A backup capability was that the entire gondola hung to the envelope with a large cargo-type parachute. Either Kittinger or ground personal via radio signal (Stapp) could cut the connections to the balloon and allow the large chute to take the gondola down to a safe landing.

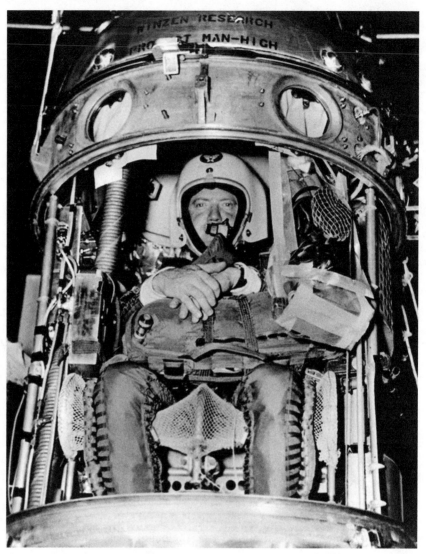

Captain Joe Kittinger prior to launch wearing the MC-3 mechanical pressure suit. Courtesy of Department of Defense.

Just before reaching his peak altitude of 96,000 ft., Kittinger discovered that the oxygen supply was half depleted—a technician had installed a component incorrectly, and the oxygen was being vented out into the atmosphere rather than into the gondola. Kittinger would have to begin his descent immediately. There could be no loitering for science or for personal reflections. However, the message Stapp received in response to the requirement for an immediate descent was the Morse code "C-O-M-E U-P A-N-D G-E-T M-E" (Ryan 1995, 83).

Simons was concerned that Kittinger had become the victim of the breakaway syndrome, a phenomenon occasionally encountered by individuals (particularly pilots) who are in an environment that causes them to become psychologically removed from reality. However, Kittinger was being a bit playful and shortly responded with "V-A-L-V-I-N-G G-A-S" (Ryan 1995, 84). Despite the problem, the descent went smoothly, as did the landing. The next flight was Simons,' and it would be for science.

Project Manhigh: Simons

Before Stapp would permit Simons to make his flight, he insisted on rehearsing the flight with the capsule and Simons in the altitude chamber. However, money was tight for this Air Force project, and the additional sum needed was about $14,000. Because Otto Winzen had so much invested in the project, he agreed to subsidize this test. One result of another successful test was more advertising for Winzen. In addition, Kittinger's flight called attention to the renewed effort in high-altitude research ballooning. Assistance, including physics instrumentation, was coming in from other sources who wanted to encourage the project. The National Bureau of Standards provided a calibrated and sealed barograph so the effort could be recognized as a record. Even *Life* magazine contributed $3,500 to pay the cost of having the FAI certify the results.

Simons would use the same gondola as Kittinger, but with many improvements. A new telemetry system would transmit heart and respiration rate as well as body temperature. Even a small telescope was installed to allow a look at some of the heavenly bodies without the blur of the atmosphere. As was the case with the astronauts a few years later, even Simon's food intake was carefully selected to provide low residue—there were no toilet facilities.

The launch was delayed because of weather—a problem the first astronauts would face repeatedly. Waiting for the perfect conditions was exasperating and some risk was inevitable. When a window of opportunity finally presented itself, there was the possibility that a fast-moving cold front would form. Meteorologist Bernard "Duke" Gildenberg knew that, if the flight went as planned, Simons would be well above the weather and the storm would simply slide underneath him. If he had to be brought back before the twenty-four-hour flight plan was completed, he could be in big trouble. However, the ability to get a full thirty-six-hour forecast for good weather was highly unlikely. Among the many pressures were the financial ones that could result in the project being canceled if it was not launched before the allocated money was depleted. The decision was made to accept the existing weather conditions and launch Simons.

Following his suit-up, Simons was sealed in the capsule and began the pre-breathing process to purge his blood of nitrogen. The capsule was trucked to the launch site—an open pit mine in Crosby, Minnesota—where it was attached to the envelope. The launch occurred at 9:22 a.m. on August 19, 1957. The balloon ascended at about 1,000 ft./min. and Simons kept a steady description of the various instruments and the unfolding view out the portholes—a verbal litany that the first astronauts would repeat in a few short years.

Two hours and eighteen minutes after launch, the balloon reached its maximum altitude for this stage of the flight—102,000 ft. The sky was a dark blue, and Simons could see the very slight curvature of the Earth. However, he was on a tight schedule, and viewing time was rationed as he worked to ensure that he could complete most of the twenty-five experiments on board *Manhigh II*.

All went well until just a few hours before sunset, when the high-frequency radio failed. This provided the data channel for the medical information from his body and a tracking signal for the ground stations. The very-high-frequency (VHF) radio was also showing some signs of losing its signal, which would cut off all voice communications. Stapp discussed whether they should end the flight immediately (to allow for a daytime landing) or continue and risk the bad weather that was moving in beneath Simons. After careful reflection, it was decided to fly through the night. Simons had already been awake for just over twenty-four hours, and his body ached from being so restricted in its movement, but the two years of planning and the possible cancellation of the project might mean he would never have this opportunity again.

Of all the sights that had presented themselves to Simons, the sunset left an indelible impression, with its startling array of colors. As night enclosed the balloon, the helium cooled and the balloon started to lose altitude.

Simons slept only for brief periods of perhaps thirty minutes without waking for some reason. By 1:00 a.m. he had descended to just under 70,000 ft. At this point he was concerned about the huge thunderheads that appeared to be reaching up to his altitude. However, he understood that even the massive cumulonimbus clouds could not reach that high—or could they? When he could no longer see the brilliant display of stars that had been his companion through the night, he realized that he must have descended into the upper reaches of the storm.

He had released some ballast earlier, but it was critical that he reserve a minimal amount for the landing phase. There was some discussion about releasing a set of batteries, but Wizen was reluctant to approve that action since the physical response would place a shock on the now frozen envelope (the plastic enclosure for the gas), which he feared might shatter. Simons prevailed, released the batteries, and the loss of weight of the batteries was just enough to allow the balloon to stop its slow descent and start an upward movement—and the balloon held together.

With the return of daylight, the balloon was now above 93,000 ft. again, and by 10:00 a.m. Simons released the last 60 lb. of ballast that was not reserved for the landing phase. The balloon responded by climbing to 121,000 ft.—a phenomenal record had been set. It was now time to start thinking about when to descend. Wizen favored waiting until later in the day to allow the storm to pass further east. Now yet another problem was identified—Simon's respiration rate was about forty-four breaths per minute, well above the average of twelve. He was asked to take a reading of the CO_2 and reported that it was at 4 percent. The preflight maximum had been set at 3 percent. An alarmed John Stapp immediately advised Simons and the ground support crew that Simons was no longer mentally capable of making decisions because of the impairment that high levels of CO_2 have on brain functions.

Stapp ordered Simons to begin valving gas and to close his helmet visor to allow his lungs to use the 100 percent oxygen of the suit to purge his blood of the high CO_2 level. Within minutes Simons regained a higher degree of mental awareness and realized how close to total incapacitation he had been. It was determined that the cold night had decreased the effectiveness of the CO_2 scrubbers. As the capsule continued to warm with the new day, the scrubbers went back to increased levels of efficiency.

However, the higher temperatures also made the pressure suit almost untenable. While Simons was encouraging a higher rate of valving, Gildenberg, who was calculating the valving effect on the rate of descent, had to ensure that the rate remained controllable. By 3:30 p.m. the balloon was still at 70,000 ft. with only twenty minutes of battery power estimated to remain; Simons turned off the radios to conserve power, turning them back on only for short periods to confirm critical information. He decided to increase the rate of descent. However, this action resulted in an extremely fast 1,300 ft./min. Now the release of the remaining ballast (the used 200 lb. battery packs) became critical. It was suggested that he release half at 4,000 ft. and the remainder at 300 ft.

The descent rate, although higher than desired, was acceptable and the gondola landed heavily on a plowed field in Elm Lake, South Dakota—405 miles from the launch site. The last actions by Simons released the capsule from the envelope, and then the bolts from the hatch were blown. Simons crawled out of his small confining space after enduring forty-four hours of incarceration.

For his efforts, Simons was awarded the Distinguished Flying Cross and the *New York Times* referred to him as "the first man in space" (Janson 1957, 17). As the initial results of the flight were being assessed, another event halfway around the world would have an immediate and long-term effect on the high-altitude balloon program—the Russians had launched the world's first artificial Earth satellite—*Sputnik*.

While in some respects the launch of *Sputnik* changed the mind-set of many in the government and the military toward research and development on the threshold of space, it also moved the emphasis from the high-altitude balloon to the rocket and the satellite. Stapp's efforts in developing much of the hardware to achieve a self-contained life-support environment as well as the psychological aspects of astronaut selection and the infrastructure needed to sustain long-duration, high-altitude flights was on the cutting edge. However, just as Stapp was now in a position to get recognition for his efforts and to expand them, he was moved back to Wright-Patterson AFB in Dayton, Ohio, and Col. Rufus Hessberg, became the new chief of the Aeromedical Field Laboratory at Holloman.

The planning and execution of a high-altitude balloon flight (or shortly to occur—space flight) was significantly different from that of the test flights of the experimental X-planes that flew from Edwards AFB. In the case of the aircraft, they were instrumented for a specific task and the pilot briefed on the flight profile. The flight was short and the test aspect often less than ten minutes. The amount of air–ground communication was relatively brief. The pilot was in complete command of the aircraft.

In the case of both the balloon and aircraft flight, detailed preparations were required for the instrumentation and the pilot. However, the balloon flight took place over many hours, and there was real-time communication, discussion, and changes in the flight plan. With the aircraft, the pilot had undisputed authority for the conduct of the flight. With Project Man-high, both Kittinger and Simons had encountered situations that caused the ground team director to question decisions made aloft and impose their own edicts. This dilemma would be encountered in the coming space age with more than a little acrimony occasionally being exchanged between the pilot and ground.

As planning got under way for *Manhigh III* in 1958, the selection for the pilot became of prime importance. Stapp's initial four requirements were greatly expanded, and six candidates were put through a grueling series of both physical and psychological tests that would, only months later, be applied to the first astronaut applicants of Project Mercury. These tests, in part, were performed by Dr. William R. "Randy" Lovelace II at his clinic in Albuquerque, New Mexico—as would those for Project Mercury.

The training program for 1 Lt. Clifton McClure, the "survivor" of the selection process for *Manhigh III*, was accelerated to allow the flight to be scheduled for the traditionally good weather of early fall in Minnesota. However, the last days of September did not provide the needed condi-tions. Recognizing the political situation that now existed with the nation's headlong "race" into space, with rockets as the dominant theme, the launch was moved to Holloman AFB, in New Mexico.

By 9:00 a.m. on the morning of October 6, 1958, McClure went through the same procedure that Simons knew well. These included attachment of the biomedical sensors, the suit-up, the insertion into the capsule, and the drive to the launch site. However, as the envelope was being filled with he-lium, a gust of wind caught it and the balloon burst. Fortunately, there was a spare on hand, but it would take the better part of the day to retrieve and prepare it for inflation—the flight was scrubbed for that day.

Therefore, it was on October 7 that McClure again went through the same ritual. Everything looked good as he was driven to the launch site, tightly packed in the 3 ft. by 8 ft. gondola. As the envelope was being in-flated, McClure inadvertently bumped against the small emergency para-chute that he could use if there was a problem that required him to leave the capsule. The chute opened and the parachute unfurled there inside the enclosed gondola. Because of the inability of the ground crew to look into that part of the capsule, only McClure knew what had occurred. He had not even wanted the parachute, as the large cargo chute attached between the gondola and the envelope would do the job, but Hessberg, Stapps replace-ment, had insisted on it. As there was a lot of time before the scheduled launch, McClure carefully worked the pile of silk back into the pack—no easy feat given the lack of space. It took him about two hours to complete the task.

The balloon took to the sky just before 7:00 a.m., and just two hours later McClure was at 99,000 ft., making observations and recording various instrument readings. At lunchtime, he consumed a specially prepared meal that was designed for astronauts eating in a weightless environment. Then a major problem occurred. The temperature inside the gondola and McClure's body temperature both reached unacceptable levels. To make matters worse, McClure could not get any water from the tube of the drinking container. There was discussion now on the ground as to the criticality of the situation, as it was still more than six hours before sundown and the natural cooling that would occur.

McClure assured them that, although it was hot, it was not unbearable, and he wanted to continue. It became obvious, as the inside temperature rose to 118 degrees, and McClure's body temperature to 105 degrees, that he had to start valving gas. By 5:00 p.m., he was still at more than seventy thousand feet when the communications failed. Now there was no way of knowing McClure's state of mind or consciousness. The telemetry continued to supply his vital signs, and although his temperature was now 107 degrees, his respiration and heart rate were good.

McClure thought about the options, including the parachute that he had so painfully repacked, but decided that the effort to attach the chute and blow the top off the gondola and exit was beyond his physical ability at this point. He could still have released the envelope and used the cargo chute but decided not to. Now the urge to urinate became an overriding factor and this required another set of contortions in the small capsule.

It was dark when the balloon made contact with the ground and McClure cut away the cables to avoid being dragged. A helicopter had followed his descent and arrived shortly to find McClure in good spirits despite a body temperature of 108.5 degrees.

Why did the capsule heat up? The cooling system that had been provided did not work well until it got up in altitude, so a large reservoir of dry ice had been positioned to keep the unit cool. However, someone had neglected to repack the dry ice before liftoff. There was also criticism of McClure for not advising the ground of the chute repacking episode—which some contend initiated the overheating by his activity in the confines of the pressure suit. The simple fact remained that McClure had survived a situation that few would have—either physically or emotionally. The medical and psychological tests that selected McClure had paid off. Project Mercury would build on that knowledge.

Project Manhigh was discontinued after McClure's journey, as the emphasis now was on orbital flight—"real space flight," as some would characterize it. However, those who were a part of Manhigh knew that they had paved the way. The technology, techniques, and concepts that it spawned continued to influence future programs. The creation of NASA in October 1958 and the goal to send a man into orbit would require addressing many of the same space medicine problems that had confronted Stapp.

Likewise, the system for life support that had been mastered—such as the pressurized capsule and its three-gas system—should have provided NASA with a less volatile alternative than the pure oxygen system that ultimately found its way into Mercury, Gemini, and, fatefully, Apollo. However, NASA's head, Robert Gilruth, and its chief spacecraft architect, Dr. Maxime A. Faget, citing critical time and weight considerations, refused to consider it as an option. While these constraints may have been valid for Mercury, they did not necessarily apply to the follow-on Gemini and Apollo. In fact, some members of the Apollo development team had never bothered to inquire about the technologies employed in Manhigh.

Manhigh had also brought into clearer focus the triad of the roles of the human occupant, the supporting ground team, and automated systems. The last aspect—automated systems—became a focal point with the coming high-tech rocket programs. However, for the aerostats, it was primarily simple timers and aneroid (pressure sensing) devices. Weather—with its winds and temperatures—was a wild card that would play a factor in many space flights.

Project Excelsior: Surviving High-Altitude Bailout

With respect to space medicine, the next challenge following Manhigh was the need to determine if the pilot of a high-performance aircraft (or spacecraft) could successfully eject at high speeds and altitudes following battle damage or a malfunction.

During World War II, crews flying at altitudes in excess of thirty thousand feet required new approaches to oxygen systems and heated flight suits. However, prolonged exposure to higher altitudes was only part of the issue. Bailing out of the aircraft presents a significant problem when the parachute opens at high altitude. This was explored by Col. Randy Lovelace in June 1943: "It was believed, prior to 1943, that if the parachute were opened at high altitude a severe opening shock would be avoided because of the decreased density of the atmosphere. Such a jump would require the prolonged use of oxygen during the descent in the parachute, as well as adequate clothing for protection against the extreme cold at altitude" (Lovelace 1943).

An experimental jump of this type from forty thousand feet was accomplished by Lovelace. However, he suffered an extremely hard opening shock, which rendered him unconscious and caused him to lose both gloves from the left hand. When he reached the ground, he had severe frostbite on his left hand and was suffering from shock. The nylon glove on the right hand served to protect it from frostbite.

It was realized that although the parachute opening occurs in the same amount of time, because of the significant difference between true airspeed when the opening is initiated and true airspeed after opening, the deceleration is greater, and the human body is subjected to extremely high opening shock. Lovelace would play a major role in space medicine for the next two decades.

As there was no aircraft that could lift the test subject and the required equipment to the altitudes (100,000 ft.) and speeds (Mach 1+) needed to make such tests, the aerostat was again called on.

Joe Kittinger, now a project engineer in the Escape Section of the Air Force Aero Medical Laboratory at Wright-Patterson AFB, was perfectly positioned to participate. However, the project presented a more difficult set of problems than Manhigh—where the aeronaut was simply encased in a spacesuit to survive the environment.

Leaving an aircraft at an altitude above the Armstrong Limit required not only an oxygen supply but also pressurization and warmth. In addition, experiments to that point indicated that crewmembers bailing out at extreme altitudes required a degree of free fall into the lower levels before deploying the parachute (HALO) to counter the cold and need for an extensive oxygen supply. Free fall usually created a situation in which the subject encountered significant tumbling and in many cases a flat spin that could incapacitate. It was obvious that a method of stabilization was needed that was simple, reliable, and effective.

Working with Francis Beaupre, inventor of a novel multistage parachute deployment system, the Excelsior team perfected a capability to provide for a small stabilizing chute (the first stage) to address the spinning motion and position the pilot in a feet-down attitude. Subsequent stages deployed the main chute in a manner to avoid the opening shock that Lovelace had

> ## Terminal Velocity
>
> An object in free fall will accelerate until it reaches a point where the atmospheric drag equals the gravitational acceleration value (32 ft./s/s). With those two values in equilibrium, the object will not fall any faster. That value, for the human body, is about 125 mph "spread-eagle" and up to 200 mph in a streamlined orientation—in the lower atmosphere.

encountered. Many test dummies and live subjects had been dropped from altitudes as high as 30,000 ft. However, to prove the Beaupre multistage parachute (BMSP), the ultimate test had to be human and from an exceptionally high altitude—100,000 ft. The lack of any significant atmosphere at that height causes the human body in free fall to accelerate to a terminal velocity of about 600 mph before entering the more dense layers of the atmosphere.

Kittinger had only made eleven jumps when he was assigned to the project in April 1958. He proceeded to become more proficient as the bailout altitudes increased to greater heights. However, to prove the system, two jumps were planned—one from 60,000 ft. and another from 100,000 ft. These would have to be accomplished from an aerostat.

An open gondola was used to avoid the cost and complexity of an enclosed and pressurized spacecraft. However, this required the test subject to be exposed to the space environment for much of the trip to the target altitude. The MC-3 partial-pressure suit was chosen over the full-pressure suit (as then worn in the X-15 rocket plane) in part to demonstrate that the standard Air Force suit was sufficient.

The 300 ft. tall polyurethane balloon was manufactured by General Mills and contained 3 million cu. ft. of helium. The lightweight, 800 lb. gondola provided for radio communications, associated electrical battery power, and the liquid-oxygen breathing supply for the several-hour flight. A small instrument panel included a pair of altimeters, a vertical speed indicator, a temperature sensor, an oxygen quantity, and a helmet pressure monitor as well as several switches for controlling the electrically heated flight suit.

The first flight test with Kittinger on board was launched in November 1958 from the small town of Truth or Consequences, New Mexico, about 80 mi. west of Holloman AFB. The aeronaut arrived in the predawn darkness amid the floodlit launch site to begin the suit-up in a small trailer. Medical sensors were attached to monitor his heartbeat and respiration. The BMSP and parachute harness were strapped over the bulky MC-3, and a reserve chute was then attached.

About forty-five minutes before launch, Kittinger was helped into the gondola—all 320 lb. of his body and equipment. The ground crew moved through their checklist as the countdown proceeded toward zero. When the restraining cables released, the aerostat rose rapidly, accompanied by several circling aircraft and tracked by radar. Capt. William Blanchard, the project medical officer, functioned as the capsule communicator and flight director in command of the mission. Because of Kittinger's limited ability to exercise any control over the activity, he was in no position to act as pilot-in-command.

As the gondola climbed into the bright morning sky, Kittinger read various numbers from the instruments to provide Blanchard with an indication of his physiological and psychological state of mind. The balloon had an inherent rotation that brought the sun into view periodically. As he rose into the Armstrong Limit, the sun's glare became distracting and periodically inhibited him from reading the instruments.

When his faceplate fogged up, Kittinger became concerned—he no longer had visual contact with the instruments or the outside world. At 58,000 ft., he noticed that the helmet was raising up on his head and threatened to break the seal to the pressure suit. If it did, explosive decompression would quickly kill him. The problem had been noted in ground tests in the pressure chamber, and additional lengths of nylon cord had been added to the clasps that held the helmet in place. He was confident that remedy would handle the problem, and it did.

When the faceplate began to clear, Kittinger noted that he was passing through the assigned jump altitude of 60,000 ft. He completed the steps necessary to disconnect himself from the gondola oxygen supply and arm the BMSP. As he tried to pull himself up into the doorway, it appeared that the instrument pack with which he was to jump, and on which he was sitting, was stuck to the gondola and kept him from standing. The pack had been attached to his harness and was used as a seat. With considerable effort, he was able to free himself and to activate the BMSP and the cameras that would record the fall. Finally, he was ready to step out of the doorway. He was now at 76,000 ft.

As he began his fall toward Earth, 15 mi. beneath, he recalled having no sense of motion. Without any atmosphere to create the rush of air, and being too high to have any noticeable increase in the size of objects on the Earth, Kittinger felt suspended in space. Following a sharp jolt, he sensed he was entering an increasingly accelerating flat spin. He grayed out and soon lost all peripheral vision. Kittinger recalls praying and then blacked out. Only the cameras now recorded the incidents that followed.

Because Kittinger was delayed in standing in the doorway (after initiating the BMSP, his exit was late), this upset the activation timers. The pilot chute deployed less than three seconds after he began his free fall—instead of fourteen seconds—well before any possible wind resistance could inflate it. It subsequently became entangled in his suit and was never able to be deployed—inhibiting the rest of the sequence to stabilize him and eventually extract the main chute.

It was only because of the aneroid barometric device on the reserve chute—that sensed him falling through 18,000 ft.—that the reserve chute deployed. However, the unconscious Kittinger was still in deep trouble as his spinning body caused the reserve chute to become entangled with the primary. Fortunately, Beaupre had designed a safety feature—weaker shroud lines for the main canopy such that they would break under the stress of a rapidly rotating body and allow the reserve to deploy finally. This occurred at only 6,000 ft. above the terrain. Kittinger was still unconscious when he hit the ground.

Only the Excelsior team knew of the problems—the press was not informed. Kittinger was not pleased with his performance and returned to Dayton to work through the series of glitches that had almost killed him. If the press had been kept in the dark, the Air Force had not been. It was only after significant pleading with its hierarchy that another jump was authorized to prove the BMSP.

With solutions implemented, the next free fall was launched thirteen months later in December 1959. The ascent encountered no moisture on the helmet faceplate and a shield kept the brilliant sun from obscuring the instrument panel. It took an hour for the balloon to reach the jump altitude—this time 75,000 ft.

Kittinger had no problems lifting himself into the doorway or starting the timers. The stabilizing chute came out as scheduled at fourteen seconds into the fall, keeping the speed to 100 mph and inhibiting any spinning tendency. At 18,000 ft., the main chute deployed and the mission was accomplished. The stage was set for the test from 100,000 ft.

The record flight was launched from near the small New Mexico town of Tularosa, 30 mi. north of Alamogordo, in August 1960. The takeoff checklist had now grown to over one thousand items, and each was completed in the predawn hours.

Duke Gildenberg, the meteorologist, had been tracking a storm system and finally decided to cancel the launch. However, the restraining lines were released just seconds before his decision reached the launch crew—Kittinger was on his way at 5:29 a.m. The *Excelsior III* gondola had a gross weight of 1,250 lb., and the balloon envelope was another 1,000 lb. The 360 ft. tall balloon accelerated rapidly at 1,200 ft./min.

The first sign of trouble occurred at 50,000 ft., when Kittinger realized that his right hand did not feel as it should in the tight pressurized glove. In fact, the tube that fed the pressure into the glove had cracked. He knew instinctively that without the pressure, the hand would expand as the balloon continued to rise above the Armstrong limit and would become excruciatingly painful. While he was concerned about his hand, he also realized that if he reported the problem the ascent would be scrubbed, and the approval for any future missions might not be forthcoming. Too much effort had been expended by many people, and he was not about to put an end to it.

He reviewed the tasks that remained and felt that he could accomplish them with only his left hand. Those monitoring his transmissions noted that there was a change in the tone of his voice at this point, although they had no idea of his problem. The weather that had threatened to cancel the launch caused a layer of cloud to form at about 20,000 ft. Kittinger could no longer see the ground.

As he reached the 100,000 ft. mark, he prepared for the jump by disconnecting himself from the gondola. Despite the extensive pain in his hand, he wanted those below to share his emotions:

> We are at 103,000 feet. Looking out over a very beautiful, beautiful world in a hostile sky . . . as you sit here, you realize that man will never conquer space. He will learn to live with it but he will never conquer it. Can see for over 400 miles. Beneath me I can see the clouds . . . they are beautiful. Looking through my mirror the sky is absolutely black [because of the restrictive movement imposed by the pressure suit and the helmet, several mirrors allowed Kittinger to view areas out of his line of sight] . . . void of anything. I can see the beautiful blue of the sky above it goes into a deep, deep dark, indescribable blue, which no artist can ever duplicate. It's fantastic. (Ryan 1995, 210)

At this point he revealed the problem with his unpressurized glove, which he said was "OK, no sweat" (Ryan 1995, 211). As he lifted himself to the doorway, his pulse increased from 106 to 136. With the disconnection of

Kittinger exits the gondola and begins his August 1960 free fall from 102,000 ft. Courtesy of Department of Defense.

the spacesuit from the gondola, connection was cut off to the outside world as capsule communicator Marvin Feldstein concluded the countdown.

Kittinger moved to the doorway, armed the BMSP, and stepped into space at 7:00 a.m. His heartbeat jumped to 156 as he spoke into the tape recorder, "Lord take care of me now" (Ryan 1995, p. 212), and he began his fall toward the Earth, 15 mi. beneath him.

Recalling his previous jump, where he had no sensation of falling, Kittinger rolled onto his back so he could look back at the balloon, which had now expanded in the rarified air to over 200 ft. in diameter. It appeared to be shooting away from him and soon appeared as a small white circle in the blackness of space.

The BMSP began its sequence sixteen seconds after Kittinger had armed it. The stabilizing chute deployed as he fell through 90,000 ft. traveling at 640 mph.

"Multi-stage is working perfectly," he reported into the recorder but then twice announced, "Can't get my breath." However, the sensation quickly passed as he was descending through 70,000 ft. "Multi-stage beautifully stabilizing . . . Multi-stage perfect" (Ryan 1995, 215). As he fell into the lower atmosphere, his speed (terminal velocity) dropped to 250 mph. Passing through 40,000 ft. he encountered the coldest region, and the faceplate fogged up a bit. Then the pressure suit relaxed, as it should.

Little cold on my legs. 34,000. Coming up on three minutes perfect stability. Under cast beneath me, override in my hand for the pack opening. Coming up on 20,000. Multi-stage is beautiful. Perfect stability. Four minutes. The undercast beneath me . . . the multi-stage going perfect. Beautiful. I can turn around perfect, can do anything and everything. Four minutes and 37 seconds of free fall . . . 18,000 feet. (Ryan 1995, 215)

Kittinger had never experienced falling into clouds; his heart rate reflected the occasion, and he raised his legs slightly as if to anticipate an impact. "Into the overcast the main chute just opened and right on the button. Thank you God, thank you for protecting me during that long descent. Thank you God" (Ryan 1995, 217).

The landing was uneventful, and it was several hours before the swelling of his right hand went down and its circulation returned to normal with no long-term damage. A second jump from one hundred thousand feet had been scheduled, but with this success it was canceled.

Project Strato-Lab: Long-Duration Experiments

During the same period as the Air Force Project Manhigh, the U.S. Navy began its Project Strato-Lab in 1954. An outgrowth of the rather modest Skyhook unmanned high-altitude research program, Strato-Lab was founded under the Office of Naval Research. Contracting with Charles D.

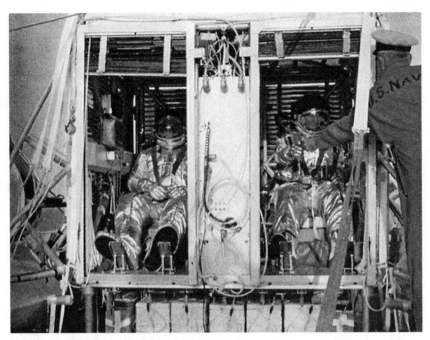

The Strato-Lab V was an open gondola requiring the two men to wear pressure suits. Courtesy of Department of Defense.

Moore of General Mills, the project had three objectives: (1) develop a high-altitude research platform; (2) conduct a study of human reaction to a high-altitude environment; and (3) conduct scientific experiments.

Strato-Lab I launched from the Stratobowl in the Black Hills of South Dakota in November 1956 and flew to 76,000 ft.—a record at the time for a piloted balloon. The pressurized gondola was crewed by Navy lieutenant commanders Malcolm Ross and Lee Lewis. However, a malfunction of the helium release valve caused the balloon to enter a rapid and disconcerting 4,000 ft./min. descent. As the craft descended below 15,000 ft., Lewis and Ross opened the ports and jettisoned everything that was not bolted down and were able to make a safe landing.

The next ascent in Strato-Lab II was in October 1957. It followed the record-breaking *Manhigh* with Simons to 100,000 ft., so an ascent to only 86,000 ft. received less attention in the press. The one hundred hours of space medicine flight experiments centered on mental and physical fatigue. Strato-Lab III flew in July 1958 and spent thirty-four hours at 82,000 ft.

Strato-Lab IV flew a 16 in. telescope mounted to the top of the gon-dola. The objective was to provide clear viewing of two planets above the atmospheric influence. Detecting water vapor on Venus and getting a closer view of the alleged canals of Mars were the objectives. The flight in November 1959 rose to 81,000 ft. The results were disappointing (no water vapor and no canals), but the concept of the telescope in space was achieved with orbiting satellites—most notably the Hubble space telescope—thirty years later.

Strato-Lab V, the final one, was carried by the largest balloon to that time—10 million cu. ft.—411 ft. tall by 300 ft. wide. The unpressurized gondola was a cube shape, 6 ft. wide by 5.5 ft. tall. Using what looked like a set of Venetian blinds on the exterior, the crew could manually affect the solar heating of the capsule by exposing the highly reflective aluminized Mylar for cooling or the black side for heat absorption. The open gondola was a test environment for the Navy's new Mark IV full-pressure suit, which served as the basis for the Mercury spacesuits.

As the mission launched on May 4, 1961, the day before Alan Shepard's suborbital rocket flight, it was totally overshadowed in the press reports of the day. However, it used a unique method of achieving a no-wind condition for launch. The balloon was set out on the deck of the aircraft carrier USS *Antietam* in the Gulf of Mexico. The ship's direction and speed was set to compensate for any wind to ensure a completely calm environment for the balloon's inflation.

Flying with Malcolm Ross was Lt. Cmdr. Victor Prather. The two were awestruck by the view from the new record of 113,000 ft. achieved after the 2.5-hour ascent. Tragically, during the water landing, Prather drowned when he slipped from the recovery helicopter back into the water. Just two months later, a similar recovery effort with astronaut Gus Grissom almost met a similar fate.

Human Centrifuge

A centrifuge is a machine that swings a capsule or "gondola" around in a circle on the end of a long arm to simulate gravitational forces on a test subject. The "g load" is a function of the speed at which the centrifuge spins and the length of the arm on which the gondola is mounted.

The first centrifuges were built by the English physician Erasmus Darwin (the father of Charles Darwin) for human research (ca. 1780). The first large-scale human centrifuge designed for aeronautical training was built in Germany in 1933.

A privately funded attempt in October 1964 to break Kittinger's free-fall record resulted in the death by depressurization of Nicholas Piantanida, when the faceplate of his helmet failed. The era of the piloted high-altitude aerostat flights had passed. More than half a century elapsed before Felix Baumgartner would eclipse Kittinger's record by leaping from 127,852 ft. in 2012.

Simulating High G-Forces

While Colonel Stapp was using the rocket sled for high-g and windblast experiments, another facility allowed researchers to gain longer exposure to the effects of acceleration and deceleration. The U.S. Navy operated the Aviation Medical Acceleration Laboratory (AMAL) at the Johnsville Naval Air Development Center in Pennsylvania. It was here, for over fifty years, that the most powerful human centrifuge was located.

When AMAL began operation in 1949, its centrifuge consisted of a gondola at the end of the 50 ft. arm. It could accelerate from a standstill to 178

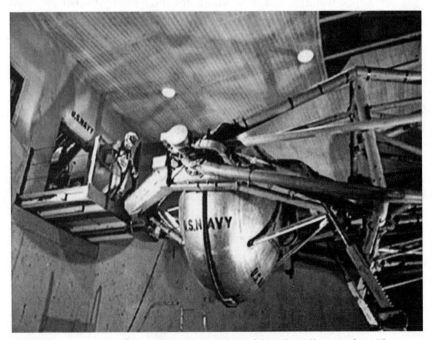

A test subject in a pressure suit enters the gondola of the Johnsville centrifuge. The gondola could be configured to replicate the cockpit, allowing the test subject to perform various functions while undergoing the high-g environment. Courtesy of Department of Defense.

mph in less than seven seconds and generate up to 40 gs. It was originally used to define and test the limits of human endurance to the high-g's generated by military fighter aircraft during combat maneuvers.

When in the sitting position, g-forces draw the blood out of the pilot's brain and into the lower extremities, causing gravity-induced loss of consciousness (referred to as GLOC). Researchers experimented with different equipment and techniques to mitigate the effects of g-forces. Among the garments developed was one worn around the pilot's lower abdomen and legs that was inflated with air or liquid when a sensor detected the onset of high gs. This applied pressure to these parts of the body to prevent the pooling of blood. The device is called a g-suit.

The average person will begin to "gray out" with 3 gs. Only with a g-suit can pilots of high-performance fighters remain conscious at 8 gs. In one experiment in 1958, researcher R. Flanagan Gray withstood 31 gs for five seconds while encased in a human-shaped capsule filled with water—Jules Verne's "water buffers."

In August 1959 the Mercury Seven astronauts came to Johnsville to begin centrifuge training, which is where, for the forty years that followed, most astronauts received their exposure to g-forces (Thompson 2004, 223). Centrifuge training is now primarily contracted out to private facilities, with the last run at Johnsville in 2005. Holloman AFB Aerospace Physiology Department currently operates a human centrifuge for training and evaluating prospective fighter pilots for high-g flight.

5

ROCKET-POWERED AIRCRAFT

Just as the aerostat ascended with the principle of buoyancy, and the rocket with Newton's laws, airplane enthusiasts sought to fly into space on wings using Bernoulli's principle to generate lift. During the late 1920s, aviation was experiencing exciting growth after being spurred on by Charles Lindbergh's dramatic flight to Paris in 1927. Over the years that followed, the concept of flying into space on winged aircraft was popular in science fiction—and in the plans of more than a few experimenters.

Germany Leads the Way

Germany was in the forefront of this activity. Fritz von Opel, heir to the Opel car company, was the first to successfully combine the airplane and the rocket engine while working with Max Valier (of the VfR) and Friedrich Sander, a pyrotechnics specialist. In 1928 von Opel purchased a tailless sailplane from Alexander Lippisch, a well-known builder of gliders in Germany. To the rear of its short, canoe-like fuselage, von Opel attached two 44 lb. thrust Sander solid-fuel rocket motors, creating the world's first rocket plane. Fritz Stamer, one of Lippisch's test pilots, was selected to fly it. Following two modest but successful flights, the third experienced an explosion of one of the rockets and Stamer quickly landed the airplane, which was then consumed by fire.

Nevertheless, von Opel was enthused. He had a custom rocket plane built with a wingspan of 36 ft. and length of 16 ft., which he named *Opel RAK.1*. With von Opel as the pilot, the rocket plane made a successful flight of almost 2 mi. in seventy-five seconds, reaching an estimated top speed of 90 mph.

While the von Opel experiments were essentially a dead end, they did influence Alexander Lippisch to work on a more advanced concept. Lippisch's interest in aviation was sparked at the age of fourteen, when he

Fritz von Opel (*right*) and Friedrich Sander (*left*) in front of *Opel RAK.1.* Flight attire in those days was quite formal, complete with a tie. Courtesy of National Air & Space Museum.

witnessed a flight by Orville Wright in Germany in September 1909. His motivation to create a tailless aircraft began when, in the early 1920s, a friend sent him a tropical plant seed that could glide long distances in the air currents. This seed had an arrow shaped wing, and Lippisch based the design of his tailless aircraft on this example from nature.

In 1937 Lippisch was informed that the German Reich Air Ministry (or Reichsluftfahrtministerium; Germany had recently annexed Austria) wanted an advanced flying wing design with a fuselage to allow the installation of a special power plant. He began design on the DFS 194, (DFS is the abbreviation for German Institute for Sailplane Flight) a single-seater with a large vertical fin and rudder. In 1939 Lippisch's team was placed under the Messerschmitt aircraft company, and work started on converting the DFS

194 to rocket power, with much of the effort being performed at the German rocket test facility at Peenemünde in October of that year.

The DFS 194 first flew as an unpowered glider. With an 882 lb. thrust Walter rocket motor at Peenemünde-West in August 1940, it achieved a maximum speed of 342 mph in level flight. The success of the flight-test program led directly to the creation of the Me-163, a rocket-powered interceptor.

The first prototype Me-163 commenced test flights in the summer of 1941. Powered by the 1,650 lb. thrust Walter HWK R11-203b rocket motor, extraordinary performance was achieved with speeds of up to 550 mph—200 mph faster than the best-performing fighters of the day. Because the Me-163 carried only a small quantity of fuel, it was difficult to explore the full performance envelope. To avoid using fuel for the basic takeoff and climb to altitude, one Me-163 was towed aloft as a glider, and then the engine was ignited. A speed of 623 mph was attained before the aircraft encountered stability problems, most likely the result of shock wave compressibility effects as the aircraft approached the speed of sound.

Introduced into limited combat late in World War II, the Me-163 did not achieve any success.

The Messerschmitt Me-163 rocket-propelled interceptor was a flying wing whose landing gear was released after takeoff to minimize weight. It then landed on a skid that composed the lower part of the fuselage. Courtesy of National Air & Space Museum.

First Flights in America

The first piloted flight of an aircraft propelled by rocket thrust alone was in California in August 1941. Using a small general aviation airplane (the Ercoupe) whose propeller had been removed, twelve low-thrust solid-fuel JATO rockets were installed under the wing. The plane was towed by a truck to a speed of about 25 mph when the pilot, Army captain Homer Boushey, released the towrope and fired the rockets. It was a modest beginning as the plane climbed to 20 ft. and then landed straight ahead on the runway.

In September 1942 Jack Northrop began a feasibility study for a rocket-powered interceptor. His radical flying wing design was interesting enough for the U.S. Army Air Forces to issue him a contract for three gliders, two of which were designated MX-334, and a powered version, the MX-324. In these gliders, the pilot lay on his stomach, allowing a true all-wing aircraft, with no significantly protruding cockpit, and permitting the pilot to withstand higher g-forces during maneuvers than the traditional sitting

The rocket-powered Northrop MX-324 (*top*) reveals the flying wing configuration with the pilot laying in a prone position. The Northrop XP-79 (*bottom*) shows its twin jet engine configuration, with the pilot again in a prone position. Courtesy of Department of Defense.

position. Although initially designed with no vertical surfaces, it was later determined that a dorsal fin was needed at higher speeds for directional stability.

The first flight tests of the MX-334 commenced in October 1943. After a series of gliding flights to determine that the handling characteristics were acceptable, it was time to move on to the rocket-powered MX-324. The Aerojet Corporation XCAL-200 liquid-fuel rocket engine used monoethyl-aniline and red fuming nitric acid to produce a mere 200 lb. of thrust. The proof-of-concept aircraft had provisions for only three minutes of fuel, at which time it would glide back for a landing.

Test pilot Harry Crosby flew the aircraft in July 1944. The MX-324 was towed to 8,000 ft. by a P-38, and the Aerojet motor was ignited. The flight of the small, 32 ft. wingspan craft lasted just four minutes and ended safely. While the project continued and ultimately produced the Northrop XP-79, the inadequacy of available rocket engines led to the installation of two Westinghouse J-30 turbojet engines. American rocket-engine technology had not been able to produce a truly flight-worthy product that could compare with the German Me-163, which had flown three years earlier and was in the process of being introduced into combat. The XP-79 had a short career because the only example became uncontrollable fifteen minutes into its maiden flight in September 1945. The pilot, who had bailed out, was struck by the rotating aircraft and killed.

Going Supersonic

During World War II, some existing aircraft, such as the propeller-driven Lockheed P-38, were capable of approaching the speed of sound (763 mph at sea level) in a dive, but unknown forces were at work that caused aircraft to become uncontrollable at those speeds.

While jet engines of that era were able to exceed the speed of piston-powered aircraft by a considerable margin, as aircraft approached Mach 1 (specifically, beyond Mach 0.8), the parasitic drag increased dramatically, as did a new supersonic component called "wave drag." In an effort to explore this unknown region, which had already killed several pilots, the U.S. Army and Navy, in collaboration with the National Advisory Committee for Aeronautics (NACA), began a program to explore high-speed flight. The rocket engine was selected as the power source for a series of experimental aircraft that would probe the mysterious region that became known as the sound or sonic barrier."

Mach Number

Because the speed of sound decreases with altitude, the term "Mach" is used to define that speed regardless of the flight condition. The term is derived from name of the Austrian physicist Ernst Mach, who theorized the effects of high speed on an object. Thus, Mach 0.9 is 90 percent the speed of sound at the altitude the aircraft is operating.

Several highly publicized accidents had given the sonic barrier a reputation of being a deadly, invincible wall. The director of British Scientific Research for Air had already declared that probing the sonic barrier was too risky for piloted operations and indicated a move to unmanned, instrumented, radio-controlled models. Flying into space on wings was not a simple proposition.

None of the European projects, however, approached the sound barrier with the same degree of engineering innovation and planning as did the coordinated effort of NACA and its director, Walter C. Williams, who was then head of the High-Speed Flight Research Station (HSFRS) at Muroc, California. Williams, who had come to NACA in 1939, would play an ever-expanding role in America's quest for supersonic flight development and its future space program.

Wind-tunnel technology of the era suffered from the inability to understand supersonic airflow around the airframe and to recreate it adequately in a laboratory environment. Therefore, it was left to building experimental aircraft to explore the transonic region (Mach .8 to 1.2). The first of these was designated as the XS-1. The "X" prefix designated that its purpose was

The Aeronautical Wind Tunnel

The wind tunnel is a tool used by researchers to understand the effect of airflow on an object. Varying in size from a few feet across to 80 ft., these enclosures use large fans to move air through them to simulate the environment of flight. Pressure measurements and flow patterns can be determined by sophisticated instruments that could not be packaged for flight. It also avoids exposing a new airframe to the possible unknown hazards of flight.

solely experimental and not intended as an operational aircraft, and the "S" represented its goal of supersonic flight (the "S" was soon dropped). The Bell Aircraft Corporation of Buffalo, New York, received a contract in 1944 to build three aircraft, each with slightly different wing airfoils, under the technical direction of NACA and funded by the Army Air Forces.

The engine chosen to power the X-1 was the only engine suitable—the Reaction Motors Inc. XLR-11. This small New Jersey company was founded by several members of the American Rocket Society in the late 1930s. The XLR-11 was composed of four combustion chambers that each produced 1,500 lb. of thrust. This was ideal for the X-1, allowing the flight-test profile to be ramped up by firing any combination of chambers to produce incremental thrust levels (throttleable rocket engines were still years away from being practical). The engine used ethyl alcohol and liquid oxygen as propellants, and the X-1 carried enough fuel for 150 seconds of powered flight. Best of all, the XLR-11 weighed only 200 lbs.

Because rocket motors devoured their propellant rapidly to generate high thrust levels, it was decided from the beginning to air launch the X-1 from the largest plane then available—the B-29. This procedure would allow more time at altitude for the test flight.

For the powered tests, the program used the Muroc Army Air Field on the west shore of Rogers Dry Lake in the Mojave Desert of southern California. Carrying almost 500 lbs. of test equipment to monitor a variety of test points for aerodynamic pressures and temperatures, the X-1's first powered flight was achieved in December 1946.

By the summer of 1947, the preliminary testing had been completed and negotiations between Bell and the Army were undertaken regarding the cost of the follow-on contract to attempt to go supersonic. Because of the hazards involved in testing the X-1, Chalmers "Slick" Goodlin, Bell's test pilot, had received a $10,000 bonus for the initial subsonic test flights. He was also to receive $150,000 for his participation in the supersonic portion of the program (what has been reported as a standard industry contract). However, the Army, in an effort to reduce costs, decided that they would take over the next stage of the X-1 tests.

The Air Force (which became a separate branch of the military that year) assumed responsibility for the X-1 flight-test program in July 1947, and Capt. Chuck Yeager became the prime test pilot. The drop sequence required the B-29 to climb to 25,000 ft. and pitch over to a shallow dive to achieve an indicated airspeed of 250 mph. The speed had to be that high because the fully loaded X-1 wing would stall at 240 mph.

The Bell X-1 was the first aircraft to exceed the speed of sound. The rocket engine exhaust often exhibits a diamond shock wave pattern, as is clearly seen here. Courtesy of Department of Defense.

Tuesday, October 14, 1947, was Yeager's thirteenth flight in the X-1—the last nine had been powered. During the preceding flights, he had carefully evaluated the handling characteristics while the engineers recorded the data from the sensors mounted on the aircraft. The program had steadily progressed in small increments of 10–15 mph toward the objective of Mach 1. That morning Yeager's ribs hurt due to a fall from a horse the previous Saturday. So much so that Jack Ridley, the engineer who was Yeager's alter ego, fashioned a small piece of broom handle so that Yeager could get the needed leverage to move the door latch into place in the cramped cockpit.

As the B-29 climbed through 12,000 ft., Yeager went aft to climb into the X-1. He wore only a cloth flight suit and the trademark of a military pilot—a brown leather jacket. He recalled: "Climbing down into the X-1 was never my favorite moment. The wind blast from the four bomber prop engines was deafening, and the wind chill was way below zero. . . . I had to bounce on the ladder to get it going, and be lowered into the slipstream . . . that bitch of a wind took your breath away and chilled you to the bone." Yeager admits to being nervous prior to each flight: "fear is churning around inside you whether you think of it consciously or not" (Yeager and Janos 1985, 111). Perhaps it was more the thought of embarrassing himself by doing something stupid in front of his peers than the prospect of killing himself that caused his fear.

The flight plan called for Yeager to move to the next plateau—Mach 0.97. He writes, "That moving tail really bolstered my morale and I wanted to get to that sound barrier" (Yeager and Janos 1985, 129). Bell equipped the X-1 with an adjustable-incidence horizontal stabilizer in an attempt to thwart the interference posed by the supersonic shock wave on the aircraft's pitch control surface—the elevator. The drop speed was a little lower than normal, and Yeager had to reduce the angle of attack slightly to maintain control as he lit the four chambers. "We climbed at .88 Mach and began to buffet, so I flipped the stabilizer switch and changed the setting two degrees . . . We smoothed right out" (Yeager and Janos 1985, 130).

Yeager's use of the word "we" to describe his oneness with the airplane is another trademark of the test pilot. Leveling out at 42,000 ft., with three chambers firing, he saw the Machmeter move through .96, and the aircraft was smooth and stable. Then the needle fluctuated and went off scale—"We were flying supersonic." Analysis of the data showed a maximum speed of Mach 1.07 was held for 20.5 s—about 700 mph at that altitude. The tracking van on the ground heard the distinct thunderclap—the characteristic sound of an aircraft traveling faster than sound. "And that was it. I sat up there feeling kinda numb, but elated" (Yeager and Janos 1985, 130). Yeager took the X-1 to Mach 1.35 (890 mph) less than a month later.

The Cold War was heating up, and progress made in breaking the sound barrier was considered sensitive information. Unlike the mass media blitz that accompanied the first astronauts, Yeager would have to wait for his "fifteen minutes of fame." He received the Harmon Trophy (an international award presented annually to the world's outstanding aviator) in 1954 for his work with the X-1A, but it wasn't until Tom Wolfe's book *The Right Stuff* in 1979 (and the movie in 1983) that his notoriety spread beyond the narrow domain of the aviation enthusiast.

In January 1949 the X-1, with Yeager as pilot, achieved the only ground takeoff of the X-1 program. He reached just over 23,000 ft. before the limited propellant was exhausted. This was reportedly done because the FAI rules that governed speed records required that the aircraft be capable of taking off under its own power.

With these aircraft now operating at pressure altitudes above 63,000 ft. (the Armstrong limit), the Army Air Forces worked with the David Clark Company to produce the T-1, the first standardized mechanical pressure suit for use against the combined effects of depressurization and g-forces. Although uncomfortable when not inflated, it became almost unbearable when it was inflated, and work continued to find better pressure suits.

At the same time that the Army was contracting with Bell for the X-1, the Navy was working with Douglas Aircraft Company, to build three experimental aircraft, designated the D-558-2 Skyrocket, to investigate high-speed flight using the swept-wing concept advocated by Robert Jones at NACA's Langley Research Center. A sweep allows the wings to remain free of the shock wave generated by the nose of the aircraft, and the compressibility effects of supersonic airflow are delayed to higher Mach numbers.

In a March 1945 memo, Jones noted that he had recently made a theoretical analysis that indicated that a V-shaped wing traveling point foremost would be less affected by compressibility than other planforms. He went to theorize that if the angle of the V is kept small relative to the Mach angle, the lift and center of pressure remain the same at speeds both above and below the speed of sound.

His ideas received considerable encouragement when the spoils of war revealed the advances made by the German aircraft industry with respect to the advantages of a swept wing for high-speed flight.

The D-558-2 (or Dash-2, as it was often called) originally used a combination of jet and rocket power and had a 35-degree swept wing. The first Skyrocket was powered by a Westinghouse J-34-40 turbojet engine rated at 3,000 lb. of thrust and an LR8-RM-6 rocket engine—the Navy designation for the XLR-11 used in the X-1. The idea was that the jet engine would take the craft up to altitude, where the rocket would accelerate it into the transonic region. Operating out of Muroc's HSFRS, the three D-558-2s looked identical and had a length of 42 ft. and a wingspan of 25 ft. Fully loaded, the Skyrocket weighed almost 16,000 lb., with an empty weight of 9,500 lb.

The first flight at Muroc was in February 1948. Initially configured only for takeoff from a runway, it was soon learned that the J-34 was insufficient to power the heavily loaded Skyrocket off the ground. The technique eventually called for one or two chambers of the LR8 to fire during the takeoff roll to assist in getting airborne (JATO units were also used on occasion). In June 1949 the number 3 aircraft exceeded the speed of sound, using both the jet and rocket units. Gene May, the Douglas test pilot on that flight, noted that as the Mach meter moved past 1.0, the flight got glassy smooth— the smoothest flying he had ever known.

Following the initial testing, it was apparent that the amount of rocket fuel available for the high-speed portion of the flight program was not

The Douglas D-558-2 Skyrocket being dropped from the modified bomb bay of a Navy B-29 designated the P2B-1S. Courtesy of Department of Defense.

sufficient. When the safety issue of the long takeoff run was added, Hugh Dryden, NACA's director of research, recommended removing the J-34 to make room for larger propellant tanks for the rocket engine and air-launching the D-558-2 in a manner similar to the X-1.

The three D-558-2 airplanes flew 313 times, with the last flight occurring on December 20, 1956. The number 1 aircraft is now at the Planes of Fame Museum in Chino, California. The number 2, the first aircraft to fly Mach II, is on display at the National Air and Space Museum in Washington, D.C. The number 3 is displayed on a pedestal at Antelope Valley College, Lancaster, California.

Understanding the complex aerodynamics associated with high-speed flight proved a more difficult engineering feat than many had anticipated. After piercing the once-impenetrable sound barrier, another obstacle was encountered—control at high altitude.

Mach II: Deadly Competition

In 1952 Walt Williams, then the head of HSFRS, requested that NACA director Hugh Dryden approve a flight of the Skyrocket to Mach II. However, NACA had a policy of not setting speed or altitude records deliberately, although they had been setting unofficial records all along the research path. The Dash-2 had already achieved Mach 1.96 in August of 1953.

After test pilot Scott Crossfield sought and received approval from the Navy's Bureau of Aeronautics (the test program's sponsors), the NACA policy was relaxed and an attempt was scheduled to achieve Mach II. Because the Air Force was also planning a Mach II flight in their X-1A, an air of competition had been generated. Crossfield claims that he simply "dropped a hint" to the Navy about how great it would be if they could beat Yeager and the Air Force to Mach II. The Navy knew that the Air Force X-1A was a more capable plane and that it would exceed whatever speed the Skyrocket was able to reach, but the goal of being the first to Mach II, even if only for a few weeks, stirred the interservice rivalry.

In addition to adding nozzle extensions to increase the thrust by about 6 percent, the fuel (alcohol) was chilled so that more could be accommodated in the tank of the Skyrocket. The propellant tank regulators were positioned so that the pilot could adjust them to provide more pressure during the flight—increasing the thrust. The aircraft was carefully waxed to reduce drag, and a flight plan was devised to fly to 72,000 ft. and then push over into a slight dive. On November 20, 1953, Crossfield flew to Mach 2.005—1,291 mph. This was the fastest the Skyrocket would ever fly.

Three weeks after Crossfield became the first man to fly twice the speed of sound in the Navy Douglass D-558-2, Chuck Yeager prepared to better that speed record in the Air Force X-1A. Yeager had inherited the X-1A test program after the primary Bell pilot, Jean "Skip" Ziegler, was killed when the X-2 he was testing blew up in the belly of its carrier plane.

On December 12, 1953, the flight began with the B-50 (which had replaced the B-29 as the "Mother Ship") climbing to 32,000 ft. The X-1A, more than 4,000 lb. heavier than the X-1, hung in the cutout bomb bay. Yeager, in his T-1 pressure suit, waited up front with the B-50 flight crew until they passed through 13,000 ft. before he entered the tight confines of the X-1A cockpit with the help of the launch crew. With the canopy secured in place, Yeager spent some time going through the checklist to prepare the aircraft for the drop.

Following a short five-second countdown, the X-1A released from the B-50 at 32,000 ft. and at an indicated airspeed of 210 mph (about 340 mph true airspeed). Yeager lit the first three chambers and climbed to over 70,000 ft. before pushing over into level flight and lighting the fourth chamber. The airspeed rose steadily and easily slipped through the Mach II point, eventually reaching Mach 2.5 (1,650 mph).

Yeager recalled, "Up there with only a wisp of atmosphere, steering an airplane was like driving on slick ice." Yeager's memory of the incident in later years differs somewhat from the transcripts of the communications, but the essence is intact. "The Machmeter showed 2.4 when the nose began to yaw left." Yeager had been cautioned that the vertical stabilizer might not be able to keep the airplane aligned with the direction of flight at high Mach numbers. "I fed right rudder, but it had no effect" (Yeager and Janos 1985, 201).

With the fuel exhausted, Yeager encountered lateral instability that caused the plane to flip out of control. His first thought was that he had lost the tail since his control inputs did not seem to have any affect. Although belted securely, Yeager's helmet hit and cracked the canopy—he struggled to regain control as the cockpit depressurized. The faceplate on the helmet fogged over as the T-1 suit inflated to counter the loss of cabin pressure, and he had difficulty seeing the instruments and the horizon.

He moved the controls to a position that he hoped would result in a spin—if the tail were still attached. That would ensure the aircraft would decelerate into a slow-speed configuration, allowing the use of normal spin recovery technique. At 25,000 ft., he was in a spin and proceeded to recover from that condition to a normal flight attitude. He reported, "Down to 25,000 feet over the Tehachapi's. I don't know whether I can get back or not." In reply to a query from the ground he simply said, "I can't say much more. I gotta save myself" (Yeager and Janos 1985, 201).

Soon, at lower altitudes and with the adrenaline subsiding, his thought process started to clear. "Those guys [Bell engineers] were right [warning about controllability at high Mach numbers]. You won't have to run a structural demonstration on this damned thing. If I had [an ejection] seat you wouldn't still see me sittin in here" (Yeager and Janos 1985, 201). Like the X-1, there was no ejection seat. The X-1A had been built to withstand 18 gs. Yeager had proved their design limits were sufficient but had encountered a new phenomenon that would have to be explored to move rocket-powered aircraft further into space—inertia coupling.

Inertia Coupling

At high altitudes during high-speed flight, the heavy mass of the fuselage can create a moment of movement that cannot be controlled by the aerodynamic surfaces (wing and empennage—the tail surfaces) because of the low air density. This can cause the aircraft to become uncontrollable.

The flight-test environment at Edwards for the experimental rocket planes was every bit as hazardous as what the astronauts would face some ten years into the future—but there was no public acclaim or headlines that followed each flight. Yeager's speed record of Mach 2.5 stood for the next three years.

Exploring Flight at Mach 3

Like the X-1, the X-2 was conceived by Bell Aircraft in 1944 but with a swept-wing configuration. Little was known about the characteristics of the swept wing, which appeared to delay and mitigate the effects of supersonic compressibility on the wing. Bell's Design 37D was proposed as a successor to the X-1 with almost twice its speed.

While the X-2 Starbuster was to explore flight at speeds and altitudes far beyond anything achievable with the first-generation X aircraft, little was known of the characteristics of the swept wing. It was also recognized that an aircraft flying at Mach 3 would generate skin temperatures that would weaken traditional aluminum structures—yet another barrier to space flight. A new material called "K Monel" (a copper-nickel alloy) was used in constructing the X-2, along with an advanced, lightweight, heat-resistant, stainless steel alloy in those areas of the airframe where significant heat build-up would occur.

The X-2 was sleeker that the X-1 series and had a more powerful rocket engine with a variable thrust rating from 2,500 to 15,000 lb.—2.5 times the power of the X-1. The new Curtiss-Wright XLR25 two-chamber rocket engine was regeneratively cooled and throttleable: the pilot could vary the thrust of the second chamber (the shuttle engines would require the ability to vary their thrust). It employed the lightest and most powerful (for its size) turbo-pump system for delivering the propellants into the combustion chamber. Moreover, as it was the most advanced man-rated rocket engine developed at that time, it encountered many significant engineering problems.

Because of the speed capability of the X-2, Bell and the Air Force decided not to install an ejection seat but to use the entire nose as an escape capsule. In an emergency, the nose assembly would separate from the aircraft and a stabilizing parachute deployed. At a safe altitude and speed, the pilot could then open the canopy and bail out.

Thought was given to canceling the program becuase of serious engine problems encountered. The follow-on project, the X-15, was in the design stage, and there was some thought that it might actually fly before the X-2, which was then three years behind its projected schedule. However, NACA and the Air Force also realized that the X-15 itself would undoubtedly experience delay from its own set of problems, and so the X-2 was continued.

By early 1953 the first flight-worthy engine was delivered and installed, and a series of no-drop flight tests were conducted over Lake Ontario (adjacent to the Bell factory) in March 1953. One of these tests involved the emergency dump system for the X-2's propellants: the volatile liquid oxygen and alcohol combination. On May 12, while testing the propellant dump procedure, the X-2 exploded and fell from the B-50 into Lake Ontario, killing Skip Ziegler and a B-50 crewmember, Frank Wolko. The drop plane was damaged badly but managed to land safely. It was later determined that its main wing spar had been cracked, and the big plane was scrapped.

It was several years before the cause of the explosion was determined. Leather gaskets were used extensively in the X-2's power plant to seal the propellant plumbing joints. These were found to be highly unstable when saturated with liquid oxygen, and any significant shock could cause the gaskets to explode.

Capt. Frank Kendall "Pete" Everest finally completed the first powered X-2 flight in November 1955, igniting only the smaller 2,500 lb. thrust chamber. The maximum speed attained was Mach 0.95. The third powered flight in April used both engines and attained a speed of Mach 1.40 at an altitude of 50,000 ft. However, these flights were not intended to push the X-2 to its limits, only to explore the basic handling with power. Three more flights were completed that expanded the speed envelope to Mach 2.53.

In July 1956 Everest made his final X-2 flight and achieved a speed of Mach 2.87 at 68,000 ft. while gathering data on aerodynamic heating. Capt. Iven C. Kincheloe then flew the X-2 to an altitude of 126,200 ft. in September. That record stood until the advent of the X-15 program.

Milburn Apt made his first and only flight in the X-2 in September 1956. The X-2 performed to its maximum potential, and Apt flew a nearly flawless flight profile, allowing him to reach a speed of 2,094 mph (Mach 3.196). Apt had been cautioned about the inertia coupling control instabilities that Yeager had experienced in the X-1A, and the flight plan called for him to decelerate before attempting a turn back toward Edwards. Nevertheless, he unexpectedly initiated a sharp turn while still at a high speed. The X-2 tumbled uncontrollably. Apt separated the escape capsule but was unable to complete the bailout and was killed when the cockpit impacted the ground.

With the destruction of the second aircraft, the program came to a sudden end. The X-2 was an example of the problems encountered when pushing the state of the art. The X-2 contributed new construction techniques and advanced materials that were incorporated into the development of subsequent high-speed aircraft, such as the XB-70 bomber and the SR-71 Blackbird.

North American X-15: Winged Astronauts

Even before the X-2 made its first powered flight, its successor was in the planning stages. In 1954 NACA issued a requirement for a hypersonic, air-launched, piloted research vehicle with a maximum speed of Mach 6 and capable of attaining altitudes of more than 50 mi. (twice the performance of the yet to be flown X-2). North American Aviation Inc. was awarded the contract for this new rocket-powered airplane that would provide information on thermal heating, high-speed control and stability (inertia coupling), and atmospheric reentry. It was designated the X-15 (there had been numerous manned and unmanned X-craft in the 1950s that followed the X-1 and X-2 and had been given the intervening X designations).

A single-chamber liquid-propellant engine, the XLR-99 built by Reaction Motors, powered the X-15. Like the X-2's main chamber, the XLR-99 was throttleable and used liquid oxygen and anhydrous ammonia as propellants to generate 57,000 lb. of thrust—more power than the V-2 rocket. Because of development problems with the XLR-99 engine, the early flights were made with two Reaction Motors XLR-11-RM-5 engines, each rated at 8,000 lb. of thrust.

The vertical tail consisted of both an upper dorsal fin and lower ventral fin (a cruciform configuration) for added stability. The dorsal tail also contained air-brake surfaces. The ventral fin was jettisoned for landing, allowing two skids to be extended from the aft fuselage. It was later determined that the ventral fin was not needed, and it was not used after the seventieth flight. A conventional nose gear completed the tricycle arrangement. Like the X-2, the plane was incapable of conventional (runway) takeoff due in part to its unique landing gear.

The ability to climb into the lower fringes of outer space, where there is essentially no atmosphere, required that the conventional aerodynamic controls be augmented by twelve hydrogen peroxide thrusters—four in the wingtips and eight in the nose. This was the first use of thrusters to control an aircraft's attitude. (They had been tested on the X-1B, however.) Using

Three Axes of Control—the Reaction Control System

An aircraft requires the pilot to have control over its three axes of movement—pitch, roll, and yaw. This is achieved with movable control surfaces attached to the three primary lifting surfaces—wing, horizontal stabilizer, and vertical rudder. Because the X-15 flew beyond the point where the atmosphere could exert aerodynamic force on these surfaces, it required a reaction control system. Similar Soviet/Russian systems are termed attitude control systems. Using highly concentrated (90 percent) hydrogen peroxide, the resultant steam pressure is routed through nozzles to provide the required thrust to effect movement about the selected axis.

this reaction control system, or RCS, the X-15 could overcome inertia coupling and maintain a desired attitude during its high-speed flight at extreme altitudes.

With a length of 52 ft. and a wingspan of just 22 ft., the X-15 was truly as much rocket as it was airplane. Constructed primarily from titanium and stainless steel, the leading edges of the wings were covered with Inconel X nickel, an alloy that can withstand temperatures up to 1,200 °F (Heppenheimer 1984, 32). Three X-15s were built as a part of Project MX-1226, with the first being completed in September 1958.

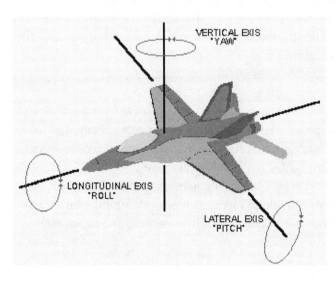

The three axes of control relative to a winged vehicle. Courtesy of Department of Defense.

All X-15 missions were flown from Edwards AFB, where they were dropped from a converted B-52 bomber at an altitude of 45,000 ft. and a speed of 500 mph. Unlike the previous manned X-craft, the pilot had to enter the plane while it was on the ground because it hung under the right wing. This meant that the pilot would spend several hours in the confines of the cockpit for a brief ten-minute flight.

Following the drop, the pilot flew a predetermined flight profile, depending on whether high speed or high altitude was the dominant objective. An inertial guidance system, similar to that used in an unmanned missile, aided the pilot in controlling and navigating the rocket plane.

The first glide test of X-15 was in June 1959. North American Aviation engineering test pilot Scott Crossfield put the plane through a series of pitch and bank maneuvers during the all-too-brief, three-minute flight. As he slowed for landing, the X-15 began a series of severe pitch oscillations caused by inadequate rate responses of the flight control system. Matching the pilot's control response with the fly-by-wire computer-controlled aerodynamic surfaces was a challenge to engineer. Crossfield's heavy breathing was heard on the air–ground communications as he fought to keep the X-15 from stalling, and only his extraordinary skill avoided loss of the plane and the pilot. Adjustments were made, and subsequent flights did not encounter the problem.

The first powered flight with the XLR-11s took place in September 1959, but it was not until the following year that the XLR-99 engine made its debut in November 1960. During an early ground-static firing of the XLR-99 with Crossfield in the cockpit, one X-15 was blown in half when a pressure regulator and a relief valve failed and pressurized the ammonia tank beyond its structural limit. Despite the dramatic explosion and resulting fire, Crossfield was not injured, and the plane was rebuilt.

Static Testing

To evaluate new designs, engines (and often complete airframes) are set up on test stands to be run without any intent of flying. This allows the use of more sophisticated diagnostic equipment to be used for monitoring such aspects as pressures, temperatures, and vibrations. Detailed inspections are then made of various components after the test firing has been completed—an option that the live firing of a rocket might not permit.

The North American X-15 being dropped from the wing of a B-52 over the desert of Southern California. Courtesy of NASA.

For the X-15 pilot, the thrill began with the drop itself, and the ignition of the XLR-99 that produced almost 2 gs of acceleration. The brief, ninety-second burn time accelerated the craft to Mach 6 and 4 gs. This was a busy time for the pilot, as he had to pitch up and maintain the climb attitude until the designated pushover point. The critical parameters of speed, altitude, and air loads had to be monitored to ensure the planned flight was flown correctly.

Following fuel exhaustion, the pilot performed the engine shutdown sequence while maintaining the appropriate pitch and heading parameters using the small reaction control system thrusters. If the test required a high-altitude ascent, the craft continued to coast upward for several minutes to its maximum altitude—exposing the pilot to weightlessness for brief periods. The pilot reoriented the X-15 for its reentry back into the atmosphere. Any significant deviation from the various speed, altitude, or time specifications could affect the quality of the test data being recorded. The pilot endured up to 5 gs during the reentry.

When the experimental phase of the flight ended, it was time to locate the runway and position the 15,000 lb. glider for landing on Rogers Dry Lake bed. The pilot could not be too low or too high on his approach—there was no power available for corrections. One or more chase planes were vectored to help the X-15 pilot during this critical final descent phase. Those pilots called out headings and altitudes and checked the aircraft from the outside to ensure there were no anomalies that posed a hazard. This routine was employed with the space shuttle landings two decades later.

A typical flight lasted less than 15 minutes and covered nearly 400 mi. The maximum speed achieved by the X-15 was 4,534 mph (Mach 6.72, while the highest altitude was 354,200 ft. It had taken forty-four years to progress from the Wright Brother's first powered flight to Yeager's Mach 1 flight in 1947. Mach 2 was exceeded six years later by Crossfield in the D-558-2, and Milburn Apt reached Mach 3 in the X-2 only three years after that. However, the next Mach numbers fell quickly, in 1961, when the X-15 flew through Mach 4 in March, Mach 5 in June, and Mach 6 in November.

The first flight into space (50 mi., as defined by the Air Force) was flown by Robert White in July 1962, while Joe Walker exceeded the Kármán line (62 mi.) in July 1962. Flights of the X-15 occurred almost weekly during this time. Neil Armstrong flew the X-15 seven times before transferring to the Gemini and Apollo Programs to become the first human to walk the Moon. He did not earn his astronaut wings in the X-15, as his highest flight only achieved 207,000 ft. (39 mi.).

A mishap occurred in November 1962 when an engine failure forced Jack McKay to make an emergency landing at Mud Lake, Nevada. However, the flaps failed to extend, and, with a touchdown speed of 290 mph, McKay misjudged the flare and a longitudinal oscillation caused the left main skid to collapse. The X-15-2 swerved and flipped over on its back. McKay sustained injuries but returned to flight status. The aircraft was substantially damaged, but the crash provided an opportunity to perform extensive modifications.

The resultant X-15A-2 was a test bed for development of a Mach 8, air-breathing, hypersonic ramjet engine (HRE) attached to the lower ventral fin. An additional 29 in. were added to the length of the fuselage between the existing tanks for the liquid hydrogen to power the HRE. Two large external fuel tanks added alongside the fuselage under the wings increased the burn time of the X-15s engine to attain Mach 8. The propellant in these tanks was consumed first, and then they were jettisoned at Mach 2. The gross weight of the X-15A-2 exceeded 51,000 lb., and it was a testament to the capabilities of the B-52's ability to carry such a load under its wing.

To withstand the added heating due to increased velocity, the entire aircraft surface was coated with an ablative-type insulator similar in concept to that used in the Mercury and Gemini spacecraft. The sprayed-on ablator

The X-15A-2 with the white ablative covering and the two external fuel tanks. Courtesy of NASA.

worked successfully but proved unrealistic because of the extensive preparation time and operational problems. The airframe had not been designed for this type of heat-resistant material, and it inhibited access to external electrical connections and maintenance panels.

To flight-test the modified aircraft, a mock-up HRE was attached for the first, and only, maximum-speed test of the X-15A-2 in October 1967. During this flight, the shock wave impinging off the mock-up HRE caused severe heating damage to the lower fuselage and almost caused the loss of the aircraft. The near catastrophic situation resulted from a lack of thorough analysis as to the effect of hypersonic flow on pylons (the mechanical support of the HRE).

The return from high-altitude flights were particularly hazardous because the vehicle had to reenter the denser layers of the atmosphere at the correct angle or face the possible disintegration of the craft from temperatures and air loads—as was the case with flight number 191. In November 1967 X-15-3 dropped from its B-52 at 45,000 ft. with test pilot Maj. Michael J. Adams, who climbed at full power to reach a peak altitude of 266,000 ft. in less than three minutes. During the descent, the X-15 began a slow yaw, and within thirty seconds the plane was at right angles to the flight path. There has been some conjecture that the attitude direction indicator had precessed 90 degrees. At 230,000 ft., the X-15 encountered rapidly increasing dynamic pressures, and at Mach 5 Adams radioed, "I'm in a spin, Pete"

(Mindell 2011, 60). (The ground communicator was another veteran X-15 pilot, Pete Knight.)

Adams fought to recover, using both the aerodynamic control surfaces and the reaction controls. He transitioned from the spin at 118,000 ft. but went into a Mach 4.7 dive at an angle of 45 degrees—inverted. Then a rapid pitching motion began, resulting in dynamic pressures that subjected the X-15 to 15 gs. At 65,000 ft. and a speed of Mach 3.93, the X-15 broke up. Mike Adams died, and the X-15 was destroyed.

It was decided at that point to end the X-15 program. Eight more planned missions were conducted before the final flight in November 1968. Of the 199 missions flown, 109 exceeded Mach 5, and 4 exceeded Mach 6. Thirteen pilots qualified for astronaut wings based on the 50 mi. American criteria of the "threshold of space," while two flights (both piloted by Joe Walker) exceeded the 62 mi. FAI definition. The surviving examples of the X-15 are housed at the Smithsonian Air and Space Museum (Washington, D.C.) and the U.S. Air Force Museum (Dayton, Ohio).

The two disciplines of aeronautics and astronautics were being merged as humans ventured further from the planet with wings and rocket engines, and a new term was coined by the Air Force in 1959—aerospace. The X-15 represented a new paradigm that would combine the power of the rocket with the maneuverability of the aircraft, ultimately to provide a reusable spacecraft that could fly back from space—the space shuttle.

6

COLD WAR ARMS RACE

As the spoils of World War II were examined following the end of that conflict, the U.S. military, which had been blindsided by the rapid advances in several areas of technology, looked closely at the V-2 rocket in particular. If the Nazi timeline of future weapons had any validity, the ability to send explosive warheads not just hundreds of miles but thousands of miles was in the not-too-distant future. An intercontinental ballistic missile (ICBM), with a 5,000 mi. range, when mated with the nuclear warhead, could prove to be the ultimate weapon—as well as a viable deterrent to future wars.

One noted scientist who, in 1946, considered the ICBM to be a project for several decades into the future—but not the present—was none other than Dr. Vannevar Bush. This was the man President Franklin Roosevelt chose in 1940 to chair the National Defense Resource Committee. This committee, which became the Office of Scientific Research and Development in 1941, was an umbrella organization for coordinating most military research activities during World War II. Bush had urged the president to move forward aggressively with the development of the atomic bomb in 1941. Now, just a few short years later, with respect to the development of an ICBM he declared, "I don't think anybody in the world knows how to do such a thing," and "I feel confident it will not be done for a very long period of time." One of the primary reasons for Bush's outlook was the inability of guidance systems to hit a target at that distance with any degree of accuracy.

The rocket would have to carry at least 10,000 lb. on a ballistic arc that would extend perhaps 600 mi. into space and fly 5,000 mi. to hit a target within a few miles. However, the technologies necessary to create the ultimate weapon were also the elements necessary to propel scientific equipment, or even humans, into space.

The Convair MX-774 assumed a shape similar to the German V-2 but was noticeably smaller (1948). Courtesy of Convair.

The Atlas: An Intercontinental Ballistic Missile

Despite tight postwar budgets, the Army proceeded cautiously to examine the technology. The Atlas rocket (which would launch America's first human into orbit) began life as a study contract issued by the Army Air Forces Technical Command in 1945 to Consolidated Vultee Aircraft Corporation (Convair). The MX-774 project was a rather modest 31 ft. technology demonstrator that had the outward appearance of the V-2. It did not possess the capabilities of an ICBM but was to provide the engineers with experience in this new field of rocketry (Launius and Jenkins 2002, 70).

The first flight articles were in final assembly when the budget realities of postwar America required difficult choices. On July 1, 1947, the MX-774 project was canceled. Convair was permitted to use unspent allocations to flight-test three missiles in 1948 from White Sands Proving Grounds in New Mexico. While none of the flights achieved the full-powered duration, the basic objectives of the tests were successful. Convair management kept the project alive as an in-house study.

In October 1950 the Rand Corporation completed a feasibility study begun a year earlier, which confirmed the military practicality of long-range rockets. A follow-on Air Force project in 1951, labeled MX-1593, was awarded to Convair. It became the first American ICBM known to the world as the Atlas (military designation SM-65). By 1953 Convair had completed the initial design studies for a vehicle that would deliver the specified 15,000 lb. warhead a distance of 5,000 mi. The design called for five engines producing 600,000 lb. of thrust in a unique "skirt" arrangement. All five were ignited at launch. Four fixed booster engines surrounded a larger, main gimbaled thrust chamber. A single set of propellant tanks fed all the engines. The configuration was referred to as a 1.5-stage missile. The rocket would stand almost 100 ft. tall with a diameter of 12 ft. (Launius and Jenkins 2002, 74).

The SM-65 was programmed as a ten-year development effort with three distinct phases that would minimize technological risk and result in operational deployment in 1963. With the Cold War heating up in light of announced developments with the hydrogen bomb, the Atlas project was reviewed by the Strategic Missiles Evaluation Committee in February 1954. They recommended that its development be accelerated.

H-bomb tests subsequently showed that the warhead for the Atlas could be made significantly smaller and lighter than expected. Based on the anticipated weight reduction, the five-engine SM-65 design was replaced by a smaller, three-engine configuration that retained the basic

---

Thrust Vector Control: Gimbaling the Engine

Early rockets used vanes in the exhaust to steer the vehicle much like the rudder steers a ship. However, this technique slowed the exhaust gases, resulting in a less efficient engine. By mounting the engine on a gimbal—a pivoted support that allows the engine to be move about a single axis—and by using two gimbals, both pitch and yaw can be stabilized. Roll control (the third axis) can be achieved in several ways. On the Atlas, small vernier engines mounted on the side provided that control.

one-and-a-half-stage concept. The final configuration of the Atlas would have a length reduced to 80 ft. with a diameter of 10 ft. The Atlas was publicly announced on December 16, 1954, and the program was given a high priority because of known Soviet progress.

This downsizing of the Atlas, while providing for a more economical ICBM, would subsequently leave the United States with less capability to launch large satellites when the space race began in 1957. One controversial aspect of the Atlas was its thin-walled, stainless steel propellant tanks. Using the enclosure of the pressure-stabilized propellant tanks as the outer body of the rocket provided the necessary structural rigidity and significantly reduced the weight to improve the mass fraction. However, this required that the propellant tanks always had to be pressurized even when empty—essentially acting like a balloon. This construction technique would present some problems when the Atlas was used as a launch vehicle for spacecraft. An advisory committee recommended the development of a second ICBM

In a television broadcast, President Eisenhower tries to calm the fears of the nation following the launch of *Sputnik I* (1957). He revealed that a warhead reentry vehicle had successfully recovered from its fiery plunge through the atmosphere. Courtesy of the Dwight D. Eisenhower Presidential Library.

Protecting Reentry Vehicles

It was understood that the kinetic energy of the incoming reentry vehicle was divided between that generated by the compression of the shock wave at the front of the nose cone and that generated by the air friction with the skin of the nose cone.

Heat generated by shock wave compression is somewhat removed from the cone itself as it lays outside the boundary layer, which served as a form of insulation. Sharp, pointed objects tend to permit the boundary layer to rest against the object, allowing the heat to be readily transmitted to it along with the friction-induced heat. By using a more rounded, "blunt" face to the object, the boundary layer is strengthened and somewhat distanced from the structure itself.

However, this is not the total answer, as the heat that transmitted to the nose cone still has to be handled by the structure. The first of two alternatives is to use a heat-sink material such as copper or beryllium, which could absorb the conducted heat and retain its structural integrity.

A second method of handling the induced heat is to use the ablation method. Here the outer layer of the nose cone is coated with a material that melts away and removes heat in the process. In effect, it is a method of a controlled burn of the outside of the nose cone to protect its content. Both of these methods would find their way into the design of manned spacecraft.

(the Titan) as a backup, with a structure of more conventional design, in case the Atlas's lightweight construction proved unsatisfactory.

While the problems of accelerating a missile to achieve intercontinental ranges appeared to be resolved, the ability to protect the warhead during its 15,000 mph reentry back into the atmosphere was still a major concern. No known materials could withstand temperatures of up to 12,000 °F. The Ames Research Laboratory in Sunnyvale, California, found a solution.

The Soviet R-7: Semerka

The development by the Soviet Union of an initial series of rockets in the late 1940s was based on the German V-2 technology. As with early rockets in the United States, these Soviet rockets also provided a firm foundation

---

Payload

The term "payload" relates to the cargo that a rocket is carrying. In the case of an ICBM, the payload is a nuclear warhead. When used as a satellite carrier, the payload is the spacecraft itself (Zaehringer and Whitfield 2004, 61).

---

on which to build projects that were more advanced. However, the passion of Sergei Korolev (as with Wernher von Braun) for space exploration focused beyond the ICBM. For him, these were only learning tools for the real job that lay ahead.

The Soviet designers also recognized the problem of igniting the second stage of a rocket in the vacuum of space. To avoid this problem, a scheme similar in concept to the Atlas was to group the stages in parallel. This configuration, which the Soviets called the "packet," offered several advantages. All the stages would ignite for liftoff, with the outer segments discarded along the trajectory as they depleted their fuel. The R-7 varied in absolute height depending on the payload but was typically 95 ft. tall when configured as an ICBM, with a diameter of just over 34 ft. across the width of the packet assembly. The Soviet's R-7 (NATO designation "Sapwood") weighed about 600,000 lb. at liftoff with a thrust of more than 900,000 lb. The Atlas was virtually half that weight, at 267,000 lb. with a thrust of 360,000 lb.

Weapon aside, the packet concept provided a rocket that Korolev felt sure would open the heavens to mankind. All of the technological challenges necessary to perfect an ICBM would have to be mastered to launch and recover satellites.

The initial design of the Soviet ICBM was well advanced when nuclear testing at the desert site of Semipalatinsk in October 1953 indicated that thermonuclear (hydrogen) bombs of significantly greater destructive power than fission weapons were feasible. The first hydrogen bombs, however, were massive in size, and initial specifications called for a warhead weight that might be as large as 15,000 lb. for the rocket now designated the R-7 (the seventh rocket designed by the Korolev team). While the breakthrough in size of a hydrogen warhead would also be achieved by the Soviets, the capabilities of the R-7 were not reduced, and this decision would serve them well in the future space race.

Combustion instabilities plagued the engineering efforts to construct large rocket engines that produced more than 100,000 lb. of thrust.

The concept of the packet (multiple stages surrounding a central core) and clustering (multiple chambers in each packet element) are apparent in this view of the base of the R-7. Courtesy of Energiya.

However, the R-7 design could not wait for Soviet technology to catch up with the requirement and schedule. The single thrust chamber (engine) originally envisioned for each module was replaced with a cluster of four chambers.

In May 1954 the Soviet government authorized development of the R-7 (official designation 8K71) ICBM. Following some additional changes, the design was frozen in March 1955.

Testing the ICBM: From Failure to Success

The first flight-worthy R-7 vehicle was launched on May 15, 1957. A fuel leak at liftoff caused a fire in the engine compartment of one of the booster

---

Clustering Rocket Engines

The requirement for higher-thrust rocket engines than the current technology makes available can be satisfied by grouping two or more thrust chambers into a single stage. This technique was done in most all first-stage boosters and in several upper stages.

modules and destroyed the rocket after ninety-eight seconds of flight. The second rocket was launched on July 12, 1957. This time failure came earlier as a short circuit in the control-system power supply resulted in a rapid roll that caused all four outer packets to tear away from the core after only thirty-three seconds of flight.

The first launch of America's ICBM, the Atlas, took place on June 11, 1957, from Cape Canaveral, one month after the first Soviet ICBM test. The missile rose for about ten seconds, then the brilliant white exhaust flame turned to yellow smoke as the engines shutdown prematurely. A problem in the fuel system had doomed the missile (Launius and Jenkins 2002, 77).

With the knowledge of the cause of the second R-7 failure, the third Russian missile was launched on August 21, 1957. This time all systems performed as expected, and the missile warhead flew the entire route, achieving an altitude of 600 mi. before reentering the atmosphere over the Kamchatka Peninsula and disintegrating. This last aspect had been anticipated, as the nose cone material was not yet perfected. The missile had flown a complete configuration test on only its third attempt, although it was far from being an operational ICBM.

The dilemma for the Russians now was if and how the success should be reported to the world. A saber can't be rattled unless the opposition knows of its existence. In a brief announcement on August 26, 1957, the TASS News Agency of the Soviets reported "a super long-distance intercontinental multistage ballistic rocket was successfully tested a few days ago" (Raibchikov 1971, 143). Much of the world was unimpressed because there were no pictures and no impartial observers to confirm this latest Soviet boast.

Korolev's team launched another R-7 on September 7, which was also successful. Those who were intimately connected with the development of the R-7 now adopted a more personal and affectionate name for their creation—it was called Semerka, "Old Number Seven." This moniker would remain with the rocket throughout its more than fifty-year lifetime.

Meanwhile, a second flight of the Atlas in September failed in a manner almost identical to the first. After much investigation, the problem appeared to be excessive heat from the engines being drawn up into the base of the rocket. This problem was resolved, and the third flight of a limited range of 500 mi. was finally accomplished in December 1957. However, this was not a fully configured missile. The first Atlas flight with a full range of over 6,000 mi. was finally achieved in November 1958. However, by then, significant events in the Soviet Union had already diminished the accomplishment.

If the Soviet announcement of August 26, 1957, was to be believed (and many chose not to), the Soviets were more than a year ahead of the American ICBM effort. The stage was set for launch of the first artificial Earth satellite—*Sputnik I.*

Contrasting the Contenders

The ability to use the R-7 as a satellite and a lunar launch platform at such an early stage in its development was a tribute to the perseverance of the entire Korolev team. The first R-7 ICBMs were placed on nuclear alert (operational) on October 31, 1959, when six were available for the next several years. They were phased out for simpler and less expensive missiles. It was never a viable ICBM, but it became an outstanding launch vehicle for the Soviet Union.

The Atlas D was the first operational version that went on to demonstrate 90 percent reliability with a 9,000 mi. range and a 1 mi. accuracy factor. During a flight to its maximum range, it reached a high point of 760 mi. and reentered the atmosphere at 16,000 mph. With a lighter payload (or an upper stage), it could place a manned spacecraft in orbit. The first squadron went on "combat alert" at Vandenberg AFB on October 31, 1959—the same day as the Soviet R-7.

The Atlas was used for only a few years as an ICBM before being replaced by smaller and more cost-effective missiles. However, the Atlas and R-7 served their countries for more than fifty years as satellite launch vehicles. Most of the 130 Atlas missiles manufactured as ICBMs were refurbished for

Satellite Numbering Schemes

At the start of the space age (1957 and the Sputniks), satellites with the same basic name (Sputnik, Pioneer, Vostok, etc.) were differentiated with roman numerals. This general trend continued with few exceptions through the Gemini program (1966). It was realized that the numbers could quickly become awkward, and the trend then became to use arabic numerals. This text will use the original designators to retain the historical significance.

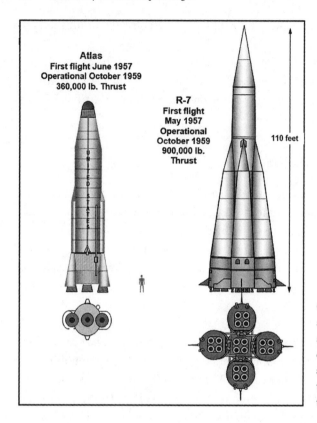

Comparison of Atlas and R-7 configurations shows the relative size of each. The R-7 has almost three times the thrust of Atlas. Courtesy of Ted Spitzmiller.

this purpose. An additional 350 missiles were manufactured specifically as satellite launch vehicles with a variety of configurations for a life that would finally end in 2005 (the Atlas launchers identified as Atlas IV and V are no longer part of the original lineage). Because of their configuration (all booster engines igniting simultaneously), the Atlas and the R-7 are often referred to as one-and-a-half-stage rockets.

Although the designation R-7 is used to describe the generic rocket, each of the many configurations that were to come had a more precise nomenclature. All rockets derived from the original ICBM are identified in this text as the R-7 to avoid any confusion on the reader's part. The R-7 is the only Russian vehicle that has been used for human flight, and it is still in production (2016) with more than one thousand having been launched.

The actual demonstrated orbital payload of the Atlas and the R-7 (without adding upper stages) was essentially the same. The largest satellite orbited by an Atlas was the 2,981 lb. Mercury spacecraft. The largest satellite orbited by the R-7 was the 2,919 lb. *Sputnik III*. The impact of technology was considerable.

7

THE RACE INTO SPACE

The Earth Satellite: First Concepts

The idea for an artificial Earth satellite with humans on board was first described by Hermann Oberth in his 1923 publication *Die rakete zu den planetenräumen* (The rocket into interplanetary space). Because the capability of radioing scientific data automatically (a process called telemetry) was not developed until shortly before World War II, the concept of an unmanned satellite was not defined in a scientific paper until the 1940s. The possible uses of low Earth orbits (within a few hundred miles of Earth's surface) for automated satellites as envisioned during this period included weather observation and communication platforms. Author Arthur C. Clarke was in the forefront of speculation, producing several articles during this period that culminated with his 1952 book, *The Exploration of Space*. His later science fiction work became the popular movie *2001: A Space Odyssey*.

In October 1945 a proposal was made by U.S. Navy lieutenant Robert Haviland and commander Harvey Hall to the Naval Research Laboratory (NRL) on the feasibility of using rockets to explore the upper atmosphere. This led directly to the involvement of the NRL in the early use of "sounding" rockets such as the small WAC Corporal and Aerobee and the larger V-2 and Navy-developed Viking. This upper-atmospheric research, occurring in the years immediately following World War II, saw scientific instruments lofted to altitudes of up to 100 mi. for a few minutes of readings over a specific point. It was recognized that a satellite in Earth orbit could continuously probe this region on a global basis—for weeks at a time—or longer. Several studies in the early 1950s advanced the concept of the unmanned satellite (in the United States and the Soviet Union) as a means to study the environment of space immediately above the Earth's atmosphere.

> ## Definition of a Satellite
>
> A satellite is an object that orbits another celestial body such as a planet. It may be a natural moon orbiting another body—our Moon is a satellite of the Earth. It can also be a device put into orbit around the Earth, or any other planetary body, to transmit data.

The U.S. Army Air Forces also understood the significance of artificial satellites for reconnaissance. It issued a request (classified secret) to the principal aviation companies for the design of an Earth-orbiting satellite in February 1946. The $1 million contract, issued to the Douglas Aircraft Company in July of that year, was transferred to the newly created Rand (Research and Development) Corporation. The resulting report concluded that a satellite launch vehicle capable of placing 500 lb. in low Earth orbit was possible with existing technology. The weight specified was considered the minimal useful size for a military application.

D. Griggs, in a Rand report, *Preliminary Design of an Experimental World-Circling Spaceship*, stated, "The achievement of a satellite craft by the United States would inflame the imagination of mankind, and would probably produce repercussions in the world comparable to the explosion of the atomic bomb" (2).

This last observation was particularly perceptive but was greeted with more than a bit of cynicism by many in the scientific and military community. For the two years that followed, there existed a loose alliance between the Army and the Navy as well as several serious moves to proceed with the project.

However, when Rear Adm. Leslie Stevens requested formal research and development funding for the project, it was declared that there was no military or scientific utility commensurate with the costs. However, it was not difficult for visionaries of this period to see that the technology of the German V-2, then being launched in the southern desert of New Mexico, was advancing the day when humans would move into outer space.

At the fourth International Astronomical Federation conference in 1953, Dr. S. Fred Singer of the University of Maryland presented the MOUSE—a minimum orbital unmanned satellite of Earth—a comprehensive analysis on the subject. A satellite with miniaturized components could provide data on cosmic rays, atmospheric density, and other items of interest to science. Using newly invented transistor electronics, whose thirst for battery

Orbital Mechanics

Orbital mechanics is the science (physics) that defines how a satellite can remain in space. Placing a satellite into orbit around the Earth requires accelerating it to a velocity corresponding with the selected altitude. For example, at an altitude of 100 mi. above Earth (the lowest altitude that permits a stable orbit because of the retarding force of the atmosphere—drag), a speed of approximately 17,500 mph is required. That satellite will take about ninety minutes to complete each revolution (Zaehringer 2004, 57).

With respect to the figure below, assume a rocket is launched on a ballistic trajectory from a site on Earth. Trajectory A represents a velocity of 6,000 mph while B is 10,000 mph and C is 15,000 mph. Note that the distance traveled increases as expected. Assume a fourth rocket (Trajectory D) achieves orbital velocity for the selected altitude. Its path would continue to curve (fall) around Earth but would not have an impact point.

Once a spacecraft achieves orbital velocity appropriate to the desired altitude, it shuts down its engine and is essentially in a free fall. The centrifugal force of the spacecraft balances the gravitational force of Earth, and the spacecraft and its contents (such as an astronaut) are essentially experiencing weightless conditions but are still well within the gravitational bonds of Earth.

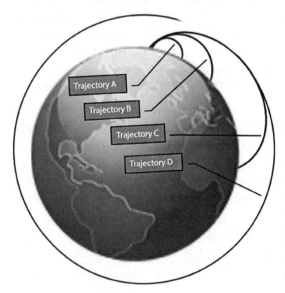

Achieving orbit is a function of the velocity of the vehicle parallel with Earth's surface. An illustration similar to this was drawn by Galileo. Courtesy of Ted Spitzmiller.

power was significantly less than that of large and heavy vacuum tubes, a satellite of perhaps 10 lb. could be constructed and orbited with a launch vehicle not much larger than the V-2. The international scientific community had periodically formed cooperative efforts to study various aspects of the Earth and its environment. With MOUSE, it was making a concerted effort to interest governments (specifically the United States) in investing in a satellite program.

Prodded by such proposals as Singer's MOUSE, America's National Science Foundation (NSF) in 1954 became involved with planning an International Geophysical Year (IGY)—a cooperative effort among fifty nations. Set to begin in July of 1957, the IGY was an eighteen-month period of scientific exploration of the Earth and the upper atmosphere (Brzezinski 2007, 92).

Science Fiction and Science Fact Converge

For Wernher von Braun, the years between 1946 and 1950, although filled with a variety of assignments in support of the U.S. Army (his new employer), did not provide an opportunity to move beyond the V-2 in developing a new and larger rocket. During this period, von Braun privately embarked on a study of sending a large expedition to Mars using the existing technology of the V-2. He expressed the scenario in the form of a novel but provided extensive notes in an appendix to support all of the technical assertions. *The Mars Project* was an insightful look at what could be achieved if (or when) mankind decided to travel beyond the gravitational bonds of Earth. The expedition he envisioned consisted of ten ships, each weighing 4,000 tn. and operated by seventy men. Making the trip in formation, the flotilla carried all the provisions to be self-sufficient for more than two and one-half years.

Following publication of *The Mars Project*, von Braun joined with Irish author Cornelius Ryan to write a series of articles on manned space flight for *Collier's* magazine, a popular monthly publication. They appeared as five installments in serial form between 1952 and 1954. To illustrate these articles, artist Chesley Bonestell Jr. was called upon to bring to life von Braun's vision. Coupled with Bonestell's own creative imagination, these interplanetary vehicles were revealed in realistic paintings that depicted places that humans had yet to visit and would fire the imagination of many.

What may have made the *Collier's* articles so enthralling to the nonprofessional was that von Braun was not simply addressing the hardware of fire and steel but the human element. While he recognized that some aspects of space exploration could be achieved with unmanned probes, he felt strongly that a human crew must be present to observe, correlate, and report. However, the simple presence of humans requires extensive life support systems to allow normal physiological functions in a temperature- and pressure-controlled environment. Additionally, the relatively frail human body might be adversely affected by acceleration during powered portions of the flight and by radiation and weightlessness during the months of travel to Mars.

As a stepping-stone to the Moon and Mars, von Braun proposed a manned space station in the shape of a large wheel, orbiting about Earth at an altitude of 1,075 mi. Because of the then-unknown effect of weightlessness on the human body, the wheel (some 300 ft. in diameter) would rotate slowly to generate its own gravity (one-fourth that of Earth) through centrifugal force. In the March 1952 issue of *Collier's*, von Braun optimistically estimated that it would require ten years and $4 billion to achieve the space station. He had the preliminary design of a massive, 265 ft., three-stage, 7,000 tn. rocket needed to carry 36 tn. of cargo. It was to be powered by fifty-one improved V-2 engines, each generating 500,000 lb. of thrust (ten times the power of the V-2).

Walt Disney, the man who made Mickey Mouse a household pet and who was an aviation enthusiast, read the *Collier's* articles with great intensity. In 1954 he was eager to round out a new weekly TV series called *The Wonderful World of Disney* with futuristic segments that would highlight *Man in Space*. In collaboration with von Braun, Disney's animators brought to life his imaginary creations. A younger generation was being acquainted with the prospects for the future.

All three of these endeavors (the Mars novel, the *Collier's* articles, and Disney) were viewed by von Braun as critical to getting his ideas of going into space before the public (and especially the unencumbered minds of the young) to more easily obtain government funding. The first of three Disney shows, *Man in Space*, was aired in March of 1955, and President Dwight D. Eisenhower is reported to have requested a copy. However, after some brief analysis in the media of the TV shows, there was no movement in Congress to appropriate money for such a venture.

The famous "Roswell incident," in which an alien spacecraft was claimed to have crashed in an isolated part of the New Mexico desert not far from the White Sands Proving Grounds in 1947, reignited the widespread speculation that we are not alone in the universe. Several movies in the early

1950s provided additional motivation to this end, including *The Thing from Another World* (1951).

The spacecraft envisioned by the serious futurists, such as von Braun, were also a part of the entertainment media's contribution to spurring the imagination. Two feature films of the era, *Destination Moon* (1950) and *The Conquest of Space* (1955) used realistic spacecraft and reasonably good special effects, and they presented some of the perils of space flight with a degree of technical accuracy. The public was slowly being drawn to the possibilities of space travel.

America's Satellite Plan: A Minimal Effort

In an effort to initiate a satellite program, the American Rocket Society worked with the National Science Foundation in planning the IGY. The National Science Foundation considered three satellite proposals: the Army's 15 lb. Project Orbiter using a modified Redstone, the NRL's 30 lb. Project Vanguard, and an Air Force proposal to use the yet-to-be-built Atlas. The latter, which could orbit several thousand pounds, was not expected to be available until 1960.

While von Braun was creating visions for the American people in the early 1950s, he was also finally creating a new rocket for the Army Ballistic Missile Agency at the revitalized Redstone Arsenal in Huntsville, Alabama. The Army had managed to retain the core of the German rocket team that developed the V-2 during World War II. Under the technical direction of von Braun, the team continued to operate organizationally in the same manner as they had in Germany. Each of the ten laboratories that represented the various aspects of missile development (engines, structures, fuels systems, etc.) was headed by one of the Paperclip (the code name of the program that brought them to the United States in 1945) technologists.

Design work of the 69 ft., 60,000 lb. Redstone (as officially named) was completed in 1952. Essentially an outgrowth of the V-2, it had a modest 200 mi. range and employed a single North American Rocketdyne liquid-fueled rocket engine of 75,000 lb. thrust. The first flight of the Redstone occurred in August 1953, and by 1955 the Chrysler Corporation began production. The first operational nuclear-tipped Redstone unit was deployed to West Germany in June 1958.

The follow-on to the Redstone was a missile with a 1,500 mi. range named Jupiter. To develop and test various Jupiter systems, the head of Army Ballistic Missile Agency, Gen. John B. Medaris, authorized von

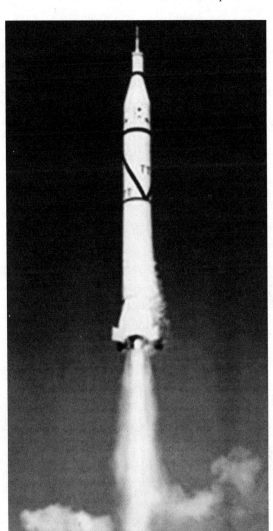

The proposed launch vehicle for Project Orbiter was the Army Jupiter-C. Note the cylindrical bucket at the top that contained the small solid-fuel rockets as a second and third stage. The pencil-like fourth stage was the satellite itself that protrudes from the bucket. Courtesy of Department of Defense.

Braun to modify a set of Redstone missiles with lengthened fuel tanks and up-rated engines. The first of these was designated Jupiter-A and outwardly looked identical to the Redstone. A second group of twelve, designated Jupiter-C, would prove the technology for protecting the warhead during heat of atmospheric reentry.

To achieve the high velocities needed to replicate the 10,000 mph speed of the reentry into the atmosphere, eleven small solid-fuel rockets sat atop the Redstone (as its second stage), arranged around the perimeter of a cylindrical aluminum "bucket" that was 30 in. in diameter and 4 ft. long. Each of these rockets was only 6 in. in diameter and 3 ft. long and produced 1,800 lb. of thrust. Nestled within these were an additional three rockets as a third stage. The nose cone to be tested was then set on top of the bucket.

This was exactly the configuration that von Braun proposed as a satellite launch vehicle for a project called Orbiter for the satellite effort then being discussed for the IGY. If a fourth stage were used in place of the nose cone, orbital velocity for a 15 lb. satellite could be achieved.

The Navy, as a part of its upper atmosphere research, had developed the rather modest, nonmilitary Viking sounding rocket, built by Martin Aircraft. The Navy proposed using the Viking as a basis for developing a new and extremely efficient three-stage vehicle defined by NRL project manager Milt Rosen. Rosen's wife suggested the name of the program, the rocket itself, and the satellite it was to launch: Vanguard, which is defined as "in the forefront."

The Vanguard program would have to endure all the teething problems of a major high-tech endeavor on a modest budget. It had to place into orbit a satellite capable of doing useful scientific work but weighing less than 30 lb. To accomplish its task, *Vanguard's* 44 in. diameter first stage

Dr. John Hagen, director of the Vanguard project, displays a model of the rocket and a full-size replica showing the internals of the satellite. Courtesy of Naval Research Lab.

was powered by the General Electric Company's X-405, a 27,000 lb. thrust engine—half the power of the fifteen-year-old V-2. The second stage was based on the Aerojet Corporation's Aerobee sounding rocket. A solid-fuel third stage, initially built by Grand Central Rocket Company, completed the 22,000 lb., 72 ft. tall rocket. On the surface, Vanguard was an impressive and elegant program whose initial schedule would orbit a satellite within thirty months (by the end of the IGY in 1958).

On July 29, 1955, White House Press Secretary James Hagerty officially announced that the United States would launch an Earth satellite as a part of the IGY. To von Braun's dismay, the Navy's Vanguard proposal was accepted—his Jupiter-C would not be used. There were four dissenting votes by the committee that had made the selection. One of them was the "Lone Eagle," Charles A. Lindbergh.

There were some who recognized that both Vanguard and Project Orbiter were entirely too modest. Dr. I. I. Rabi, a distinguished physicist and then chairman of the Office of Defense Mobilization Scientific Advisory Committee, alerted the Eisenhower administration in 1956 that the *Vanguard* satellite was "too small and that we should be making bolder plans for much larger satellites . . . in view of the *competition* we might face" (Killian 1977, 2; emphasis added). His advice was to prove prophetic. Here was perhaps the last opportunity for America to defuse a possible space race. However, no one who could do anything was listening. The CIA reported in the summer of 1957 that the Soviets had the capability. Nevertheless, complacency was dominant in America as October 4 drew near.

The world was told that the first Earth satellite would be American, it would be called *Vanguard*, and it would weigh 30 lb.

The Soviet Plan: Not to Be Second

In mid-1953 press coverage about the possibility of satellites, especially von Braun's *Collier's* series, prompted Sergei Korolev to ask Mikhail Tikhonravov to prepare a presentation for the Soviet Academy of Artillery Sciences that would support a request for an official Soviet space program (Lardier and Barensky 2010, 65). The information collected included a large number of clippings from the American and European scientific and popular press as well as tentative American plans for artificial satellites.

Accompanying the presentation was a detailed description of a proposed program complete with calculations to show that Soviet technology was capable of achieving these goals. Furthermore, by using the Soviet intercontinental ballistic missile (ICBM), then in the planning stages, a satellite could be significantly heavier than those envisioned by the Americans. With a total satellite weight capability of 3,000 lb., using the R-7 ICBM, it was anticipated that an animal container might be installed on later satellites. Follow-on aspects included piloted spacecraft, a space station, and flights to the Moon. The report was the equivalent of von Braun submitting his visionary articles to Congress rather than *Collier's*. The proposal concluded with a note that creation of an artificial satellite could have vast political significance (to get the backing of the Communist Party).

In February 1956 Soviet premier Nikita Khrushchev made a visit to review progress with the R-7. This was Khrushchev's first top-secret briefing on the direction being taken in the ballistic missile program. On viewing a mock-up of the R-7, he was overwhelmed by its size. He stood in silence as Korolev explained the U.S. satellite program and derided the U.S. launch vehicle—Vanguard. Korolev emphasized the capabilities of the R-7 and how it would position the Soviet Union as a leader in space exploration. Korolev also noted that the major costs of R-7 development were being borne by the military; the added costs of its use as a launch vehicle would be comparatively small. Khrushchev, still somewhat awestruck, simply nodded his approval: "if the main task [of developing the ICBM] doesn't suffer, do it." With that, the Soviet satellite program had been elevated to the highest political levels and had survived (Siddiqi 2003b, 150).

Using the Cyrillic alphabet (the basis of the Russian language), Objects А, Б, В, and Г were designations for different nuclear warheads planned for the R-7. With the advent of the Soviet satellite, the next letter Д (English D) represented the fifth object. Thus termed Object D, this 3,000 lb. satellite was far more sophisticated than anything contemplated as a first step into space by visionaries in the United States in 1956.

While the R-7 was hardly a practical ICBM because of its massive size, cost, and need for elaborate launch preparations, it would provide the Russians with a relatively reliable vehicle that could be used to launch large Earth satellites. The velocity needed to throw a 15,000 lb. warhead 5,000 mi. is about 15,000 mph. To achieve low Earth orbit (100 mi.), the R-7 could propel 3,000 lb. to a velocity of 17,500 mph.

Chief designer Sergei Korolev was driven by the desire to beat the Americans into space. He had been able to gauge the intellect and passion of Wernher von Braun through the *Collier's* articles and recognized that he faced a more than formidable opponent. Although he was disdainful of the capabilities of both the Jupiter-C and Vanguard, he did not want his

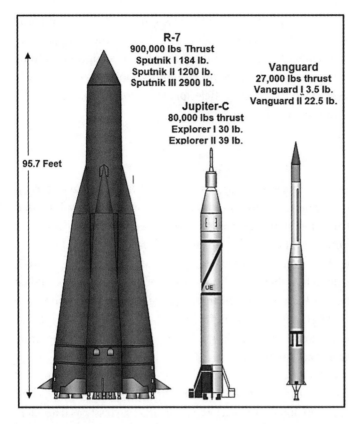

R-7
900,000 lbs Thrust
Sputnik I 184 lb.
Sputnik II 1200 lb.
Sputnik III 2900 lb.

Vanguard
27,000 lbs thrust
Vanguard I 3.5 lb.
Vanguard II 22.5 lb.

Jupiter-C
80,000 lbs thrust
Explorer I 30 lb.
Explorer II 39 lb.

95.7 Feet

The relative size of first Earth satellite launch vehicles (1957–58) illustrates the advantage that the R-7 gave to the Soviets. Courtesy of Ted Spitzmiller.

creation, no matter how superior it was to the Americans in size or complexity, to be *second* to orbit.

All of the satellite plans were dependent on the R-7 completing its first successful tests. In retrospect, Korolev had set an extremely optimistic schedule. Most large rockets developed to that date had required dozens of launches to achieve a reasonable reliability. The R-7 represented complexity an order-of-magnitude beyond anything that had left a launchpad.

It was not long before the development of Object D fell behind schedule and the possibility arose that the R-7 might not meet its anticipated performance levels; Korolev decided that a smaller satellite might be the answer to all of his known problems and perhaps some that he had yet to encounter. In late November of 1956 he requested permission to launch two smaller and less sophisticated satellites during the early summer of 1957, before the official start of the IGY in July, with the intent of heading off the American effort.

Therefore, a smaller Soviet satellite, now designated *Simple Satellite Number One* (*Prostreishiy Sputnik*, or PS-1), came into Korolev's plan

(Siddiqi 2003b, 155). This plain, polished 184 lb. sphere would contain only two radio transmitters, batteries sufficient for ten days, and temperature-measuring instruments. Within a month, the design was approved—to be launched only after one or two successful R-7 flights. Object D was rescheduled for April 1958. Two PS-1 flight article spheres were assembled, one for the launch and another as a backup.

Sputnik: Fellow Traveler

In keeping with the nature of a closed society and tight military secrecy, the Russians were not about to announce when, or what, they intended to launch. Instead, they made a series of generalized statements that were again taken by much of the world as groundless boasts.

In September 1957, following their second successful launch of the R-7, the Soviets indicated that they would launch a satellite "soon" that would be considerably heavier than the American *Vanguard*. Much of world opinion reflected on the statement as Russian propaganda. Aware of the effect that a failure would have on their credibility, the Soviets held secret the time, place, and vehicle that they would use.

On the night of October 4, 1957, at 10:28 p.m. Moscow time, the engines of an R-7 ignited and the 600,000 lb. booster lifted off the pad, propelled by a thrust of over 900,000 lb. The rocket performed flawlessly as it accelerated into the night sky. In less than eight minutes, the rocket had achieved orbital speed as it depleted the available fuel. The nose cone separated and the 23 in. sphere was released into space. As the satellite passed over the Kamchatka tracking station, its signals were received. However, Korolev cautioned that they should not celebrate until the satellite completed its first revolution. When it did, the ballistics experts determined that the satellite was in orbit with a perigee of 142 mi. and an apogee of 592 mi. The inclination of the orbit was 65.6 degrees, while the orbital period was

---

Orbital Inclination

The angle formed between Earth's equator and the plane of the spacecraft's orbit is the orbital inclination (see figure opposite). Once established at launch, changing the orbital inclination by even a few degrees requires high levels of energy.

Orbital Parameters

Due to a lack of precision in achieving a final velocity, parallel with Earth's surface, most satellites do not achieve perfectly circular orbits. Their altitude varies from a high point on one side of the orbit to a low point on the other side—essentially an elliptical shape. The low point is referred to as the perigee and the high point as the apogee (see figure below).

The lack of circularization is referred to "orbital eccentricity"—the parameter that determines the amount by which its orbit deviates from a perfect circle. An eccentricity value of 0 is a circular orbit, values between 0 and 1 form an elliptical orbit—1 is a parabolic escape orbit (minimum-energy escape trajectories), and greater than 1 is a hyperbola.

96 min. Korolev was quoted as saying, "I've been waiting all my life for this day!" (Hartford 1997, 129).

So it was, on a Friday evening, that the Russian news agency TASS startled the world with its announcement: "On October 4th, 1957, the first artificial Earth satellite was successfully launched by the USSR. The Soviet Union proposes to launch more artificial satellites in the course of the

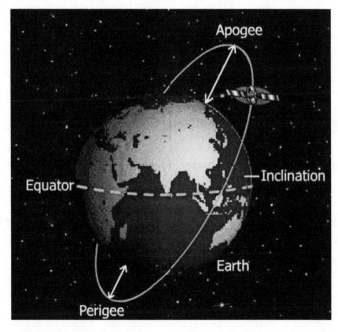

Orbital parameters of apogee, perigee, and inclination are illustrated. Courtesy of Ted Spitzmiller.

The first Earth satellite, the 23 in. diameter *Sputnik I*, was launched by the Soviet Union on October 4, 1957, to the surprise of the world. Courtesy of Energiya.

International Geophysical Year. The *Sputniks* will be larger and a broad range of scientific experiments will be carried on them" (Jordan 1957, 3). The first artificial Earth satellite was a 23 in. highly polished sphere called *Sputnik*, it was Russian, and it weighed an astounding 184 lb.

Initial reaction around the world reflected uncertainty. Sensing that the announcement held some significance but unsure of the implications, radio and television news stations around the world interrupted their regular programming to convey the terse thirty-second TASS news item. Then they went looking for an "expert" to tell them what was happening. By the next morning, the front page of the October 5 newspapers had large headlines the size reserved for major news events.

President Eisenhower's press secretary, Hagerty, declared the Russian announcement was "no surprise" and emphasized that the United States was "not in a race with the Soviets." Both of these statements were naive at best. To the average American and for many in the U.S. Congress, the significance of *Sputnik* was initially rather vague. Most had little appreciation of what it took to get a satellite into orbit.

What embarrassed the Americans was that they had been beaten to the record books for a very important entry. Moreover, they were to be beaten to many "firsts" in space flight over the next six years.

What angered the Americans was that those who were in control politically (essentially, the Eisenhower administration) allowed the Soviets to achieve this success when, as the story quickly unfolded, the United States could have launched a satellite (von Braun's Project Orbiter) more than a year earlier than the Soviets.

What shocked the Americans was the weight of this Earth satellite—nine times heavier than the long-anticipated American entry. It was convincing evidence that the Russians did have the power, the guidance, and the ability to launch ICBMs. This was seen as a threat to the very existence of the free world.

The president had carefully crafted military funding during his preceding term to avoid creating a large military-industrial complex funded by an inflated federal budget. Now, with the Soviet *Sputnik*, his entire military and economic strategy would unravel early in his second term.

As the opposing political party, the Democrats were quick to endorse the perception of the American public that there was a "gap" between the missile capabilities of the United States and those of the Soviet Union. This gap, they claimed, was caused by the inept handling of American priorities by the Republican administration (Oberg 1981, 35).

Lyndon Johnson, then the Democratic Senate majority leader, demanded a congressional investigation of the impact of *Sputnik* only a few days after its launch. This alleged "Missile Gap" was to cost the Republicans significant losses in the 1958 congressional races. As history was later to reveal, U.S. technology in virtually every critical defense area was ahead of the Soviets at that time, and six different long-range strategic missiles were being developed with the highest national priority. Nevertheless, technological image was paramount in 1957—like the garish tail fins of the cars rolling out of Detroit.

Eisenhower's failure was his lack of sensitivity to the impact of the first Earth satellite and the effect that satellite weight would have on the perception of technology in world opinion. His aversion to the German rocket scientists was tied to his firsthand knowledge of the Nazi concentration camps and the manner by which these "guest" scientists had been allowed to enter and pursue their careers in the United States. Could his anti-Nazi sentiment have played a role in this most fateful decision of his presidency?

The first edition of the October 5 issue of *Pravda*, the official Soviet newspaper, had only a few short paragraphs about the launch. With little information coming from their own official sources, *Pravda* (and TASS) simply parroted what the world was saying about the accomplishment. However, in subsequent issues, as world reaction expressed awe and amazement, *Pravda* used large bold headlines.

Literally overnight, Khrushchev achieved credibility with world leaders, allowing the influence of Communism to move with unprecedented assurance into many third world countries—particularly Cuba. Some political analysts have long felt that, had it not been for the prestige of *Sputnik*, Cuba's revolutionary leader at the time, Fidel Castro, might not have been lured into the Soviet sphere of influence; the Bay of Pigs debacle of 1961 and the Cuban missile crisis of 1962 all might have been avoided.

Sputnik II: Dogging America

While most of the engineers and technicians involved with the *Sputnik* launch enjoyed a brief holiday as they celebrated their success, Korolev and several of his key people were called to the Kremlin to accept the accolades from the premier himself. Almost from the start, Korolev was asked if something could be launched to celebrate the fortieth anniversary (on November 3) of the Russian Revolution of 1917. That date was less than four weeks away.

Korolev may have anticipated the request; he indicated that there was a chance that an even larger and more spectacular *Sputnik* might be prepared in time for that date. What he had in mind was not Object D, which was still lagging, but something more than just scientific equipment: something that would not only stir the imagination but also give the Americans another shock. Korolev proposed sending an animal as a passenger aboard the next *Sputnik*.

The passenger selected was a small female mongrel Husky named Laika, who had been retrieved from the streets of Moscow as a stray. She weighed about 8 lb., and the padded, pressurized cabin of the satellite allowed room for her to lie down or stand. The environmental system provided oxygen, food, and water, which was dispensed as a gelatin. Laika was fitted with a bag to collect waste and electrodes to monitor her vital signs. The total mass of the new satellite was 1,118 lb., six times heavier than *Sputnik I*! The launch occurred on the designated day, November 3, 1957.

The Soviets reported that the initial telemetry indicated Laika tolerated the weightless condition and ate and slept normally, and there were strong hints that Laika would return unharmed. However, there was never any intention to return the dog back to Earth, and when it became clear that

Sputnik II carried the dog Laika into orbit, bringing the possibility of human space flight a giant step closer. Courtesy of Energiya.

she was not to be brought back, the official Soviet statement was that she had been painlessly euthanized after several days of electronic observation of her condition.

However, Laika's demise was not as humane as the Soviets reported at the time. After reaching orbit, the thermal controls did not function as planned, and the interior temperatures quickly reached 120 °F. Because of the pressure to build *Sputnik II*, there was little time for testing. For almost the entire period of her flight, Laika suffered from these high temperatures and succumbed to heat exhaustion on the fourth day of the mission, on November 7. The dog's body burned up along with the satellite as it returned to Earth's atmosphere after 162 days in orbit. Laika's story remained a state secret for more than thirty-five years.

Laika's flight appeared to prove that humans could probably survive in space, and the excitement of this new era reached even higher levels. It was also apparent that a set of progressive milestones represented the ability of a nation to exhibit its scientific and technological prowess. These "firsts" became the focus of the space race. The orbiting of *Sputnik I* represented the first achievement and signaled the beginning of the race.

Weightlessness: Microgravity

Once a spacecraft achieves orbital velocity appropriate to the desired altitude, the engine is shut down and the satellite is essentially in a free fall. The same effect would occur in a simple ballistic trajectory. Because of the Earth's curvature and the high speed of the spacecraft, the spacecraft "falls" around the Earth. The centrifugal force of the spacecraft balances the gravitational force of Earth, and the spacecraft and its contents (an astronaut) is essentially weightless. The term microgravity is a more scientifically accurate term, as there is always some residual gravitation forces present.

If *Sputnik I* was an affront to the national pride of the United States, the launch of *Sputnik II* was overwhelming. Now the press and the experts openly talked of missions to the Moon and to Mars and of human spaceflight. Suddenly, within a period of thirty days, science fiction had become science fact.

The prestigious *New York Times* editorialized on November 10, 1957: "it must be hoped that the National Security Council . . . will not only be receptive to new ideas, but will take immediate steps to remedy deficiencies and put the U.S. again in a race that is not so much a race for arms or even prestige, but a race for survival."

The assumption that there were deficiencies in the U.S. program was quick to surface because the Russians were first, and *Sputnik I* weighed nine times that of the yet to be launched *Vanguard*. The use of the word "race" came quickly to the forefront in the lexicon of the press within the first twenty-four hours after the launch of *Sputnik I*. The Cold War arms race had evolved into the space race. The final word of the *Times* editorial, "survival," seemed to sum up the attitude of many Americans. The expression "we will bury you" that Khrushchev used in a speech earlier in the year was resurrected by the alarmists.

The first attempt to launch a fully configured Vanguard with a small 6 lb. test satellite on December 6, 1957, resulted in a massive explosion of the vehicle as it fell back onto the launchpad. America was humiliated in the eyes of the world. The Army had been given tentative approval to use its Jupiter-C, shortly after *Sputnik II* was launched and accomplished the

The first American attempt to orbit a satellite dies a fiery and humiliating death on the launchpad on December 6, 1957. Courtesy of Department of Defense.

successful orbiting of America's first satellite (*Explorer I*—the same name as the Army's aerostat of 1936) on January 31, 1958 (Killian 1977, 119).

NASA Established: Coordinating the Resources

As the surprise of *Sputnik* and the disappointment of the Vanguard launch failure settled into the nation's psyche in January of 1958, it was obvious that the United States was going to be involved in a protracted series of space programs to compete with the Soviet Union. President Eisenhower, Congress, and the emerging aerospace industry recognized the need for a concerted effort to address the startling display of apparently advanced Soviet technology. The disparate array of existing projects such as Vanguard and

Jupiter-C, coupled with those being presented for funding (a man-in-space effort and several proposals to send probes to the Moon) dictated that one organization should structure, direct, and fund them.

There were several proposals for such oversight that included a committee approach composed of civilian, military, and industry representatives. Eisenhower already had plans to create the Advanced Research Projects Agency (ARPA) to coordinate high-tech efforts of military space projects. However, as he envisioned America's long-term goals, he had strong preferences for a civilian-oriented space program, and he directed his science advisor, James Killian, to work with the Presidents' Science Advisory Committee to design such an organization (Killian 1977, 132). He did not want a cabinet-level "Department of Space," nor was he an advocate of even recognizing that the competition between the United States and the USSR constituted a "space race." Fortunately, Congress too was pushing for a "civilian space agency."

There was an immediate awareness by many that the basis for such an organization already existed. The National Advisory Committee for Aeronautics (NACA—always referred to by its letters and not as an acronym) was formed in 1915 to help direct research and development efforts in aviation. By the first week of March 1958, the Eisenhower administration had decided to transform NACA into a more capable organization that would perform at a higher level of planning and participation than that of an advisory committee—the National Aeronautics and Space Administration (NASA).

The NACA steering committee or "board" was composed of ten civil servants and seven individuals from the private sector. Eisenhower initially proposed that the new organization would be headed by a committee of eight civil servants and nine persons from the private sector. However, the Democrats interpreted the latter group as "industry"—politically, this approach implied "industry spending government money on industry"—a possible conflict of interest.

The decision to use a civilian-based space agency was a twofold ploy on Eisenhower's part. He recognized that the space domain was primarily associated with the role of the Air Force. However, he also understood that many of the technology gains being sought in space applied equally to civil and commercial use. To permit the Air Force to take the lead not only risked political censure in the eyes of the world but could also inhibit (or at least delay) the use of space for private enterprise. He was also well aware that what was accomplished in space in the next decade under a civil space agency could readily be transferred to military use—reconnaissance satellites being the exception.

The Department of Defense (and the Air Force in particular) was not pleased with Eisenhower's edict that NASA assume most of the space-based projects. However, Deputy Defense Secretary Donald Quarles supported the transition of NACA to become the new space agency, and legislation was introduced to Congress in April to make the change happen.

In early July of 1958 a House–Senate conference met to resolve legislative differences. It was decided that a single civil servant administrator would head the new organization, appointed and directed by the president and confirmed by the Senate to carry out the national space policy.

Eisenhower recommended Dr. T. Keith Glennan, the president of Case Institute of technology in Cleveland and a previous commissioner of the Atomic Energy Commission, for the job. However, Glennan would accept only if the current NACA head, Dr. Hugh Dryden, would serve in the number-two position. Glennan, no stranger to high-tech or to administration, recognized that Dryden, steeped in aeronautics, understood the coming space age more intuitively.

The Air Force had already made some progress in defining a manned spacecraft. This mission would transfer to NASA along with the first unmanned probes to the Moon initially assigned to the Advanced Research Projects Agency. However, the Air Force Dyna-Soar program (a more advanced manned spacecraft) was among those projects that would stay in the Department of Defense domain. It was also understood that much of the technology formulated by NASA would be immediately available to the military without it coming from their budget.

Eisenhower signed Public Law 85-568 (the bill creating NASA) in July 1958, to take effect that October. NASA was directed to downplay secrecy and to disseminate information to the maximum that security allowed. The National Aeronautics and Space Council was created to formulate space policy. This consisted of secretaries of defense and state, the NASA administrator, the Atomic Energy Commission chairman, and the president— who acted as council chairman.

8

DEVELOPING A MANNED SPACECRAFT

Even before *Sputnik II*, both the Soviets and the Americans had made tentative paper studies of a variety of possible paths for sending humans into space. Following the 1946 VR-190 proposal by Mikhail Tikhonravov, Sergei Korolev did some tentative work in 1956 that projected a manned suborbital flight to occur during 1964–67. Wernher von Braun's concepts from the *Collier's* articles continued to excite the imagination.

The U.S. Air Force, in February 1956, began examining the next step after the hypersonic X-15 rocket plane (only then beginning to take shape) that included a piloted hypersonic glider launched by a conventional rocket booster, similar in concept to the Eugen Sänger *Silverbird* of 1933. It also studied a possible manned ballistic rocket research program.

However, none of these paper projects received development funding. That would have to wait for the impetus of *Sputnik II*.

Critical Physiological Issues

By the end of the decade of the 1950s, all of the key elements necessary to put a human into space had been recognized and initial steps taken to address each. The stratospheric balloon flights of Project Manhigh, Excelsior, and Strato-Lab with the personages of Kittinger, Simons, McClure, Ross, and Lewis had validated the high-altitude pressure suit developed by both the Air Force and Navy. Thus, a spacesuit and a self-contained life-support system in the form of a space capsule, pioneered by Lovelace and Stapp, demonstrated that the human body could be sustained in the hostile environment of space.

There were still other unknowns however, such as the mysterious cosmic radiation emanating from space and X-rays from the sun. These had been measured by upper-atmospheric sounding rockets, but their effects on the body were not well known or understood. It was not known if the localized readings taken over specific points such as White Sands represented those that might be present at other latitudes around Earth.

The acceleration extremes (g-forces) endured by Stapp in the rocket-sled tests at Holloman AFB gave reasonable assurance that the human body could withstand the forces required to accelerate a rocket to orbital velocity in space and return. These were calculated as being 8 to 12 gs.

One of the few aspects of which there was little knowledge was weightlessness. Flying in an airplane performing parabolic arcs provides the crew with periods of microgravity of perhaps thirty seconds at a time—far too short to really understand the possible negative effect on various internal organs. How would the heart and cardiovascular system respond? What about bone and muscle conditioning? Will weightlessness confuse the sensory systems such as balance?

In addition, what about the fitness of the human subjects for space flight? Dr. Randy Lovelace had done pioneering work in this area, with the selection of Clifton McClure for *Manhigh III*, but had all the factors been adequately considered? For example, how important were the psychological dynamics? The balloon flights had lasted only a few days at most, and Earth was still a prominent feature 20 mi. below. How would humans react when they would be hundreds or thousands of miles from Mother Earth? How real was the breakaway syndrome of being psychologically isolated from Earth?

All of these aspects were about to come together.

Initial Proposals: Paper Projects Don't Fly

The exhaust plume of *Sputnik* had hardly cleared when the rush to catch up with the Russians began in earnest in dozens of military and corporate research centers in America. Combinations of existing boosters were lashed together (on paper) in an effort to find a possible arrangement that looked feasible as a "heavy-lift" rocket. However, it was one thing to play the game with paper and quite another to get funding to cut metal. With a dog on board *Sputnik II*, the reality of human spaceflight came into focus.

The NACA, which would become NASA within a year, convened a special committee on November 21, 1957 (just two weeks after *Sputnik II* was launched), in which critical aspects of human space flight were assigned to working groups. These included spacecraft systems, tracking, atmospheric reentry, and human factors. Each of the various NACA centers around the country participated, with the Virginia-based Langley Research Center

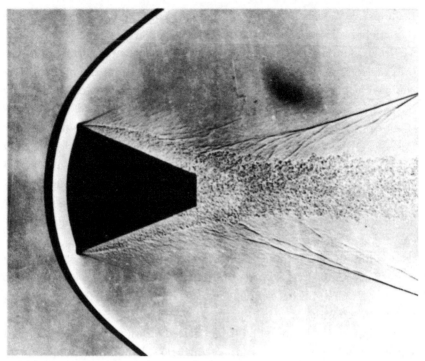

The blunt configuration was chosen for the American spacecraft because it would dissipate the reentry heat more effectively. Courtesy of NASA.

being represented by Maxime A. Faget, who headed the performance aerodynamics branch. By March 1958, Faget and his team had eliminated various approaches and had settled on the ballistic capsule as the path that would lead to the highest probability of success in the shortest time for the least money.

The Air Force initiated its own three-day seminar in January of 1958 at Wright-Patterson AFB in Ohio to examine possible projects and technologies to place a military man into space in the shortest period. The Advanced Research Projects Agency (ARPA), under Roy Johnson, assumed that human space flight was a responsibility of the Air Force, although no mandate had been issued by the Department of Defense to establish a need for such a project. A wide variety of spacecraft configurations were described, among them a 2,300 lb. ballistic capsule proposed by the McDonnell Aircraft Company, who inferred that a mission could be flown within twenty-four months from the start of the project. Optimism was rampant, and paper studies were still cheap.

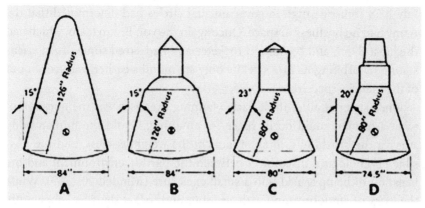

The design of the Mercury capsule went through several iterations before configuration D was selected. Courtesy of NASA.

The term "capsule" (rather than "spacecraft") was a holdover from the stratospheric ballooning days of the 1930s. That term would be used for Mercury, but the Russians used the word "spacecraft" because it projected a more advanced image. Beginning with the Gemini Program, NASA switched. The use of the word "CapCom" (capsule communicator) to denote the mission controller who talks to the astronauts is the last vestige of the term.

By mid-March 1958 the Air Force, based on its analysis, had gained approval for $133 million to initiate a Man-in-Space-Soonest (or MISS) Project (Wolfe 1979, 57). Its first objective: a simple ballistic ride into space with follow-on goals that included orbital flights of up to two weeks. Deeper in the paperwork were vague projections for circumlunar flights and lunar landings by 1965. The initial price tag for the Man-in-Space-Soonest Project was a shocking $1.5 billion.

The Air Force already had a "spaceship" in the fabrication stages in the form of the X-15. However, because it would not achieve orbit, it was not considered in the same realm as a spacecraft launched by a rocket—much to the consternation of the test pilots at Edwards AFB.

The Army Ballistic Missile Agency in Huntsville had its own idea of getting man into space with Project Adam, using the venerable Redstone booster for a suborbital flight of 115 mi. into space and 200 mi. down range. Dr. David Simons, who had flown the *Manhigh* high-altitude aerostat mission, had preliminary discussions with the von Braun team about the design of the capsule and the possibility of being its "passenger." However, by

July 1958 someone high in governmental circles had determined that the Army had no business in space. Once again the von Braun team, which had the most depth and breadth in rocketry and had saved some of America's stature by orbiting its first satellite only six months earlier, was frozen out of the man-in-space race.

One aspect of all of these early planning exercises was the question of what a human could do in space—if anything. Could the human body survive the crushing g-force of launch, the weightlessness, and the possible psychological disconnect between the Earthly environment and the isolation of being sealed into a small enclosure (Mindell 2011, 67). While the work of Dr. John Stapp with aerostats and rocket sleds had apparently answered many of these questions, there was a vocal group who insisted that the human element would simply be a passive passenger and not an active participant in the flight. Compounding the problem was the nebulous relationship between the automated aspects of a spacecraft, the bevy of ground-based flight controllers, and the spacecraft occupant—known as an astronaut.

Project Mercury: The Winged Messenger

With the pending reorganization of NACA into NASA set for October of 1958, a joint NASA-ARPA Manned Satellite Panel was formed in September. Administrator designee Dr. Keith Glennan had been given the specific mission of human space flight just a few weeks earlier by President Eisenhower. The basic layout established by Faget (and arrived at independently by McDonnell Aircraft) provided for a 12 ft. cone-shaped craft with a 7 ft. diameter heat shield on the broad end (Harland 2004, 11). The astronaut would lay reclined, and the capsule would reenter Earth's atmosphere backward so the g loads would remain in the same direction as at launch. Small retrorockets initiated reentry while a simple attitude control system of thrusters stabilized the craft about all three axes during retrofire and reentry.

To acclimate the astronauts to a vehicle that could present movements about all three axes simultaneously, a Multiple Axis Space Test Inertia Facility (MASTIF) was devised. MASTIF (also called the gimbal rig) was a nested set of three tubular steel cages inside of which the astronaut sat. Each of the cages represented one of the three degrees of axial freedom, and the cages could rotate independently or together. Pilots of conventional aircraft are often exposed to roll but rarely to high rates of pitch or yaw. With

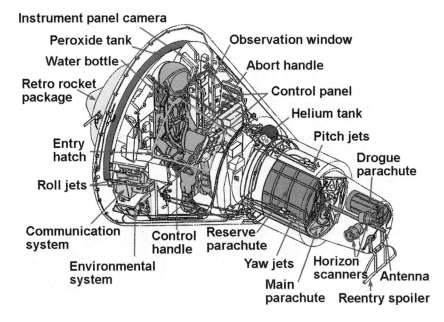

Instrument panel camera

Peroxide tank

Water bottle

Retro rocket package

Entry hatch

Roll jets

Communication system

Environmental system

Control handle

Reserve parachute

Observation window

Abort handle

Control panel

Helium tank

Pitch jets

Drogue parachute

Yaw jets

Main parachute

Horizon scanners

Antenna

Reentry spoiler

The final configuration of the spacecraft reveals the packaging of the various systems. Courtesy of NASA.

MASTIF however, significant rates of movement could be induced about all three axes at the same time. The objective was for the astronaut to use a hand controller connected to compressed air thrusters to damp out the movements to achieve a stable platform.

Alan Shepard was the first astronaut to experience the MASTIF; he was initially unable to bring it under control. Subsequent rides went better. It was not used for astronaut training after the Mercury Program was completed (Baker 1982, 47).

Three Axes of a Spacecraft

Because a ballistic type spacecraft (such as Mercury) has no aerodynamic surfaces and operates above the atmosphere, the reference to its three axes and control over each is somewhat arbitrary but is established as being the same as those of the rocket. The occupant obviously adjusts mentally to axes that are oriented the same as an aircraft, thus the pitch, roll, and yaw components are so aligned.

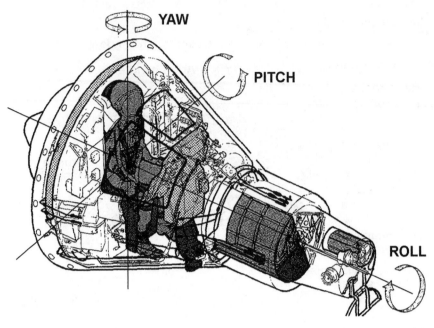

The three axes of the spacecraft represented the traditional aircraft orientation. Courtesy of NASA.

The various representatives on the Manned Satellite Panel were not unanimous on the configuration, but simple economics, time, and existing technology provided the greatest convincing arguments for the ballistic approach. The reentry dynamics were also somewhat vague, with some studies suggesting that, unless the capsule had extensive heat protection, the reentry angle had to be precise or the spacecraft might be incinerated.

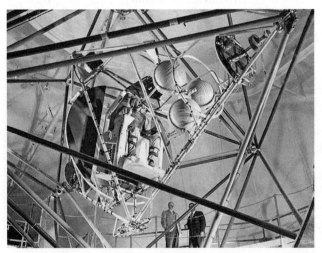

The Multiple Axis Space Test Inertia Facility (MASTIF) was constructed at the Lewis Research Center in Ohio (since renamed for astronaut John Glenn) to provide the astronauts with experience coping with multiple-axes orientation problems. Courtesy of NASA.

Atmospheric Reentry

When the perigee of a spacecraft is lowered to a critical value (slightly less than 100 mi.), it encounters increased numbers of air molecules, decelerating more rapidly, and returns to Earth within a few orbits. Because of the substantial amount of kinetic energy required to place it in orbit, this energy is dissipated during the reentry through the aerodynamic force of friction. This force results in the generation of extremely high temperatures on the structure of the spacecraft. The process is called "reentry," and the spacecraft will burn up unless it is protected from the heat.

Although the Atlas was the largest possible launch vehicle undergoing flight test at the time, its reliability and lifting capacity left much to be desired. Convair estimated that it could deliver 2,700 lb. to low Earth orbit, and early optimistic estimates indicated that a spacecraft could be built to that weight limitation.

By November of 1958 more than three dozen potential bidders attended a briefing in which key members of the manned spacecraft design team made presentations that described the systems and their rationale for selection. The objectives of the project were to orbit and recover a manned satellite, and to determine the human capacity to function in the environment of space (Kleinknecht 1963, 2).

As the project took on its own character, it had yet to be given a name. Most missiles of that era were named for Greek or Roman mythological figures, so when Abe Silverstein (who became head of NASA's Office of Space Flight Programs) proposed the name Mercury, it was readily accepted (Lawrence 2005, 40). The obvious oxymoron was that Mercury, the Winged Messenger of Greek mythology, had no wings in its NASA reincarnation, but it was the grandson of the Greek god Atlas!

Money: Not a Substitute for Time

McDonnell Aircraft Corporation had been an early proponent of a manned spaceflight program, and in October 1957 (immediately following *Sputnik I*), the company assigned several engineers to begin working on some of the problems to be resolved to put a man into space. It came as no surprise, then, when they were awarded an $18.3 million contract (with a $1.5 million fee) by NASA in February of 1959 for construction of twelve Mercury capsules.

While Faget's team had established the basic configuration, engineering reliable systems to conform to the weight restrictions became the critical aspect of McDonnell's responsibility. Some of the systems were an outgrowth of existing X-plane programs, such as the X-15—which was nearing the flight-test stage. However, these systems, such as the thrusters and environmental controls, required significant weight reduction. Even something as simple as the communication package had to be reengineered to fit the small space available.

The reentry process was initiated by a set of three small, solid-fuel retrorockets attached to the base of the heat shield. An independent circuit fired each, so it was almost a guarantee that one would ignite—and only one was needed to drop Mercury from orbit. Attitude control was important during retrofire to ensure that the correct thrust vector was established for maximum effect of the small retrorockets.

During the reentry itself, the reaction control system thrusters maintained the correct spacecraft attitude as the heat shield was designed for a precise angle for optimal weight savings. When the capsule descended into the more dense layers of the atmosphere and slowed to its terminal velocity at a subsonic speed, a drogue chute deployed to stabilize the craft, preventing any oscillations due to its unstable aerodynamic qualities. At 10,000 ft., the main conventional parachute opened, but its descent rate of 32 ft./s would result in a momentary forty-g impact.

To soften the touchdown, the heat shield detached from the main body of the spacecraft during the descent and extend down about 4 ft., revealing a fabric landing bag between the heat shield and the base of the spacecraft's pressure bulkhead. This bag had a series of holes that allowed air to be drawn in as the bag was extended by the weight of the heat shield. At touchdown, the air within the bag provided a compression effect as it escaped back out through the holes—softening the impact on the capsule itself. It was a simple if not elegant solution.

Because Earth's surface is predominantly covered by water, the craft had to survive a landing at sea in case of an early retrieval from orbit due to an emergency. This led to consideration of the ocean as the primary recovery area.

With the Atlas D being the only real option available as a launch vehicle, several modifications were made to it. The first was an abort-detection

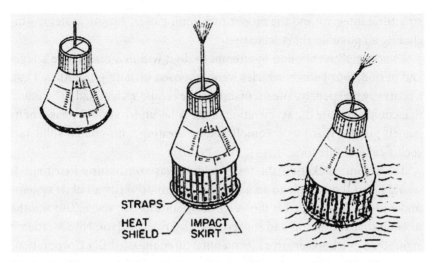

STRAPS

HEAT
SHIELD

IMPACT
SKIRT

The landing bag extended from the base of the capsule and served to cushion the impact.
Courtesy of NASA.

package that received input from a variety of sensors monitoring critical aspects of the vehicle's "health." The automated-abort-sensing and implementation system (commonly referred to as ASIS) allowed any one of the sensors to initiate an abort signal that would ignite an escape rocket and pull the spacecraft safely from the launch vehicle during the boost phase. An abort could be initiated by one of four sources—the astronaut, the flight director, the launch conductor, or an automated-abort sensor.

Because of the added structure that extended the height of the rocket almost another 15 ft., the load factor on the Atlas ultimately required that its skin thickness be increased. The added length also necessitated that the rate gyros (that sensed the flight path) be relocated to allow for the change in the control dynamics of the rocket.

On reaching orbit, the spacecraft separated from the Atlas by a set of small solid-fuel posigrade rockets, whose exhaust was shielded from the top of the Atlas propellant tank by a fiberglass faring. The term "posigrade" relates to a separation process whereby the spacecraft receives a slight increase or "positive" velocity as opposed to "retrograde" motion.

The idea of going straight to orbital flight with the capsule, without an intermediary step of a suborbital flight, was actually never contemplated. With the entire project open to world scrutiny, the risk of such a giant step was too much for those who held the reins of responsibility. Incremental testing, and lots of it, would precede an actual manned orbital attempt. The capsule itself was tested in a variety of unmanned modes to validate its structural integrity and the proper functioning of its various systems—including all possible abort scenarios.

To accomplish this, and to attempt to do it within a reasonable budget and period, four launch vehicles were proposed including the Atlas. First, a relatively inexpensive means of testing the escape system had to be found that could simulate the acceleration of the Atlas under various phases of its launch profile. This was accomplished by creating a fin-guided solid-fuel rocket called Little Joe.

The second rocket on the test schedule was von Braun's Redstone. It would boost the capsule to an altitude of 115 mi. to verify all of its systems under high g load as well as the weightless conditions of space. This was the mission that von Braun had suggested with Project Adam. NASA ordered eight Redstone boosters from its manufacturer, the Chrysler Corporation.

The third rocket was to have been von Braun's Jupiter intermediate-range ballistic missile to qualify the heat shield. However, the Army wanted as much money for its Jupiter as the Air Force wanted for an Atlas. The Jupiter rocket was canceled and the Atlas assumed its role—but with a price increased 32 percent to $3.3 million per missile.

Cost overruns began almost immediately in the Mercury Program. The McDonnell spacecraft contract itself grew over 30 percent within a few months to $41 million. The initial cost estimates were not the only optimistic aspect of the program that quickly fell by the wayside. The first schedule called for a manned Redstone flight by the end of April 1960 and the first orbital flight by the following September. A span of nineteen months was projected between the laying out of the plan and the first human orbital flight.

With respect to Project Mercury's prospects of orbiting a man before the Soviets, NASA deputy Hugh Dryden said that the race was lost before the space agency was founded. He noted that money is not a substitute for time.

Little Joe: Escape System

When America's first manned space program, Project Mercury, began in 1958, the Atlas missile on which the astronauts were to fly was quite unreliable, and there was no hesitancy in providing a means of escape should something go wrong during launch. If a malfunction were detected during the ascent to orbit, a small rocket would lift the spacecraft off the launch vehicle and propel it some distance away before the normal parachute recovery systems activated. This escape system had to function at high Mach numbers and extreme altitudes. Protecting the astronaut from the debris and flame of a catastrophic launchpad explosion as well as the wind-blast of a 3,000 mph abort at 20 mi. altitude meant that an ejection seat was ruled out.

Minimizing risk through careful design, the application of redundant systems, and the detection of a pending catastrophic event to allow an escape for the crew was referred to as "man-rating" a rocket. (The term "man," referred to species, not gender, and has been changed in our current politically correct society to "human-rating.") The result is a method that provides as much safety for the crew as is practical—though what an engineer might define as practical may differ from than that of a crewmember (Mindell 2011, 24).

The first aspect of an escape system is the network of sensors needed to determine if some aspect of the launch vehicle is not performing properly or is about to fail. Thus, a series of pressure and temperature sensors were devised along with accelerometers and rate gyros (sensing movement about the rocket's three axes) to determine when nominal conditions were being exceeded in various parts of the launch vehicle. If one or more critical parameters reflected an out-of-tolerance condition, the escape system activates.

Early intuitive designs had the escape rocket mounted under the Mercury spacecraft, but this was quickly changed to position the motor in front (a "tractor" arrangement) for several reasons. The first was that this "escape tower" jettisoned during the ascent, when it was no longer a factor in an abort scenario. This reduced the amount of weight the launch vehicle had to carry to orbit—allowing a greater spacecraft weight. Removing the escape rocket before entering orbit also reduced the amount of volatile propellants and pyrotechnics that could be a hazard to the spacecraft—and it would ultimately have to be discarded at some point in any case. Finally, there was an element of inherent stability in the tractor arrangement—no active guidance was necessary.

The Mercury launch escape system consisted of a 6 ft. long escape rocket that sat atop an 8 ft. triangular support structure. The exhaust of its 52,000 lb. thrust rocket was directed out of three nozzles canted away from the spacecraft.

Note that the spacecraft has a different weight at various times during its flight. At launch, with the escape tower, that weight is 4,257 lb. About

The Little Joe escape system test vehicle was a relatively inexpensive means of testing the various abort profiles. Courtesy of NASA.

1,200 lb. of that is the escape tower. If the escape rocket is not needed during the first 230 seconds of flight, then the ungainly structure is jettisoned by a smaller rocket that generated 800 lb. of thrust for 1.5 seconds.

To test the launch escape system, a low-cost launch vehicle was assembled using existing "off-the-shelf" solid-fuel rockets clustered to create the desired velocity conditions. At about $200,000 per launch, this approach was more economical than using an actual Atlas to explore the abort scenarios. The Little Joe abort test vehicle was 28 ft. in length and 80 in. in diameter. When the Mercury spacecraft and the escape tower were added, the total vehicle stood 50 ft. tall and weighed 27,900 lb. The thrust and burn time of Little Joe could be configured for a variety of tests to provide the appropriate aerodynamic and g-load conditions that the Atlas would present. The name Little Joe came about after one engineer noted that the initial configuration of the four rocket motors (as viewed end-on) resembled a "double deuce" dice roll called a "Little Joe." The name stuck!

To minimize cost, there was no active guidance in the rocket. The vehicle sat on a launchpad canted at a slight angle headed out over the Atlantic Ocean from Wallops Island, Virginia, where all the Little Joe tests took place. The first rockets in the cluster to be ignited burned for only 1.5 seconds, and provided almost 150,000 lb. of thrust. With a launch vehicle weight of about 28,000 lb., this provided an initial six-g acceleration to allow a set of fins to give aerodynamic "arrow" stability within seconds after launch.

Depending on the desired flight profile, the second group of either two or four motors then fired for durations of up to 38 seconds to accelerate the vehicle to the desired speed and altitude to emulate the flight profile of the Atlas. For simplicity, there was no staging; the spent rockets of the first group went along for the ride. Seven flight tests of the escape system were conducted over an eighteen-month period from August 1959 to April 1961. They performed their job, though not without an occasional misstep. On the first launch, the rocket took off 35 min. before it was scheduled because of a spurious electrical signal. The first Atlas launched in support of Project Mercury was used to test the integrity of the heat shield. It did not carry the abort system and was designated Big Joe.

Selecting the Astronauts: Finding the Right Stuff

As the hardware started to come together, the selection process of who would ride it into space took center stage. Originally, NASA's Space Task Group (STG) had thought that an "open" selection procedure would provide about 150 applicants, from which 36 would go through a winnowing down process, and a final 6 would be made offers to become astronauts. There was no shortage of volunteers from virtually every walk of life. However, shortly before the end of 1958, President Eisenhower decided that course was fraught with uncertainty, and he preferred to limit the application process to existing military test pilots. The reasoning was quite simple and pragmatic. These men had already proved their ability to operate in a high-risk, high-tech flight environment under pressure—and were already on the government payroll.

The basic requirements were the same as for the original STG selection criteria in that each candidate had to be between twenty-five and forty years of age, less than 6 ft. tall, in excellent physical condition, and hold a degree in science or engineering. To this was added the necessity of having completed one of the military test pilot schools and having logged 1,500 hr. of jet fighter time.

There was an interesting divergence of opinion over what many felt were excessive qualifications. Some viewed the astronaut as "spam in a can," simply doing what a monkey would have already accomplished. The Atlas launch vehicle was preprogrammed to get the spacecraft into orbit—so no input from the astronaut was required during this phase of the mission. As there were no aerodynamic control surfaces, the reaction control system positioned the spacecraft in various attitudes. An automated and preprogrammed flight plan could essentially "fly" the entire basic mission profile; thus, many considered the human component to be redundant (Mindell 2011, 121). Why, then, would a highly trained test pilot even consider the role?

Some in the Mercury Program management hierarchy initially looked on the astronauts as simply the human occupants of an automated satellite. Others, including the astronauts themselves, considered the occupant to be more than just a biological test specimen. Nevertheless, at this point in the program, to call the occupant a "pilot" was certainly stretching the characterization of the role since the astronaut would have no control of the path of the rocket, only the orientation of the satellite once it achieved orbit. Others felt that the human presence would add immeasurably to the utility of the mission with the ability to reason and make judgment evaluations. In fact, right from this first year of the space age, the debate began as to the human's usefulness in space as opposed to the cost and risk of getting him there.

Of the 508 military records provided to NASA by the Defense Department, 110 men were selected by a committee for further consideration. By the time the first 69 of these were reviewed, it was determined that there was no need to consider any more of the 110, as it was going to be a relatively easy task to find the 6 qualified candidates. Several were eliminated by the requirements such as height; 56 were given written and psychiatric examinations that reduced the group to 36, who were then asked to volunteer (Wolfe 1979, 59).

Four declined to be further considered, and it was at this point that the career aspect became a factor. What lay beyond Project Mercury was totally unknown. These military pilots had worked hard to position themselves in their careers. To be diverted for an unknown period into a project that had questionable use for their skills caused many of them to have second thoughts.

In groups of six, the thirty-two prospective astronauts—who were then considered "volunteers"—began the final and somewhat tortuous path to the ultimate selection. Each group visited the Lovelace Clinic in Albuquerque, New Mexico, for a weeklong series of intrusive physical and psychological tests (Wolfe 1979, 68). Dr. Randy Lovelace, a pioneer in aerospace medicine, was called on to create, as he had done with the Dr. Paul Stapp's aerostat occupants, a physical examination that would ensure that each candidate did not harbor any problem that could possibly interfere with the demanding flight that lay ahead. Several of these tests were widely criticized by the astronauts in their memoirs as being ludicrous by any standard. If there was one redeeming value to the Lovelace analysis, it may have been to determine if these physical and academic specimens of manhood had the right motivation to see the process through. Only one applicant failed to move on to the next selection step.

At the Wright Air Development Center in Dayton Ohio, another set of physical tests awaited the candidates. If the Lovelace ordeal stretched the limits of humility, this new series explored the extremes of heat, cold, isolation, noise, and physical endurance. Centrifuge tests required the men to endure high forces of gravity (g loads) while a spinning "Barany chair" determined the stability of their equilibrium. More in-depth psychological analysis was performed that included motivational and peer evaluation scrutiny.

The records of the final eighteen candidates were reviewed by the NASA committee, which passed its recommendations to the assistant director for manned satellites, Robert Gilruth. The legend that has grown up around the selection process is that the number could not be reduced to the desired six because there were just too many outstanding candidates. One account that may have some credibility is that the Air Force and Navy were equally represented with three each, and it was decided to add a Marine to the list to round out the service's participation. Oddly enough, the Marine selected (John Glenn) was the only one who had not completed a four-year degree.

The new NASA administrator, Glennan, approved the selection and the "Mercury Seven," as they were called, were introduced to the world at a press conference in Washington, D.C., on April 9, 1959. The spectacle of the conference took on a surreal twist from the moment that the presenter

During the introduction of the Mercury astronauts, a reporter asked who felt they would return safely from being rocketed into space, by a show of hands. The smiles reflect their thought process. Note that John Glenn has two hands in the air. *Left to right*: Deke Slayton, Alan Shepard, Wally Schirra, Gus Grissom, John Glenn, Gordon Cooper, and Scott Carpenter. Courtesy of NASA.

who was addressing the assembled press, said, "Gentlemen, these are the astronaut volunteers." Among the popping flashbulbs and ecstatic applause, the candidates themselves were astounded by the reaction (Wolfe 1979, 89). They had not expected this kind of immediate adulation—if any. After all, they had yet to do anything heroic—except endure Randy Lovelace's probing.

Perfecting the Spacecraft: Compromises

Each of the new astronauts were assigned to a part of the Mercury Program to follow the progress being made and to assist the engineers with any aspect that might involve the pilot—if, indeed, the occupant would deserve that title. Every week the astronauts would meet as a team to discuss the problems that each had uncovered and arrive at a recommendation to be passed on to the engineers. It was obvious from the start that a spacecraft designed by engineers for an environment that they had never experienced was going to be a challenge. Likewise, the desires of the astronauts themselves played a key role in such things as control and instrument placement and even window size and location. Their engineering knowledge and test pilot experience was invaluable in helping define system functionality and flight profiles.

One of the first systems to be defined was the life-support environment of the spacecraft. If the capsule were to hold ambient sea-level pressure of 14.7 lb./sq. in. (760 mm Hg) with the normal mixture of 20 percent oxygen and 78 percent nitrogen, the weight of the pressure vessel necessary to contain it would be too heavy for the Atlas. In addition, the expected leakage through welds and openings for electrical cabling would be significantly greater, and increased amounts of air would have to be provided to make up for the loss. A mixed-gas environment would also be more complex in achieving the proper ratio of gases.

It was determined that if the capsule's interior pressure was maintained at about one-third that of standard atmospheric pressure and used 100 percent oxygen, the weight and complexity of the system could be noticeably improved. The original design called for the capsule to contain an ambient atmosphere at launch, and as the vehicle climbed, the air within would be vented into space until the desired 5.5 lb./sq. in. was achieved, at which time the vent would be closed. During this period, oxygen would flow into the craft to enrich the mixture until the 100 percent level was reached. The astronaut, in a spacesuit with the visor down, would breathe 100 percent oxygen during launch.

If a fire occurred in the spacecraft while in space, the astronaut would close his faceplate, the spacesuit would pressurize, and the atmosphere in the craft would be vented into space creating a vacuum that would extinguish the fire from lack of oxygen.

However, during one test of the environmental control system (ECS), a McDonnell engineer lost consciousness from lack of oxygen. It was determined that there was a problem with the regulation of pressure between the spacecraft and the pressure suit. In an effort to avoid this problem, which could have been resolved, it was decided to fill the spacecraft with 100 percent oxygen prior to launch. It was recognized that this was a major hazard, as a fire occurring any time before the capsule got into space would not be controllable. However, the decision was made to proceed with the "easy" solution—time was of the essence. That decision would have major repercussions in the years to come.

Another decision was to place most of the systems within the small craft's pressurized hull to minimize the number of holes that would have to be sealed against loss of the cabin's environmental pressure, which would have to be continually made up by the limited supply of oxygen. This would prove to be an unfortunate choice with respect to the ability to maintain and modify systems in assembled spacecraft.

Astronaut Navy lieutenant commander Walter "Wally" Schirra, age thirty-six, was assigned to monitor the environmental control system and the space suit. Existing suit technology was called upon to handle protecting the astronaut, should the cockpit environmental control system fail at any point in the mission. The Navy's Mk IV, built by B. F. Goodrich, was adapted and most of the changes to the spacesuit involved mobility issues as the astronauts sought to be able to easily reach any switch or control.

The layout of the instrumentation and development of the spacecraft simulators was the responsibility of Marine lieutenant colonel John H. Glenn, age thirty-seven and the oldest. The youngest, Air Force captain Gordon "Gordo" Cooper, age thirty-two, was given the task of tracking the preparation of the Redstone booster. Air Force captain Donald "Deke" Slayton, age thirty-five, was assigned to monitor the Atlas.

Communications was another critical system. Because of the unknowns involving the astronaut's ability to function in space, it was desired to maintain virtually continuous communications. However, with the spacecraft being above the ionosphere, some radio communications technologies would not work. High-frequency and ultra-high-frequency communications bands were used that limited communication to "line-of-sight." This meant that a whole network of tracking stations had to be established around the world in several foreign countries. Even then, periods of time occurred when the capsule was out of touch. A basic rule was established that no more than ten minutes of "dead time" would be allowed between contacts, nor would any period of contact be less than four minutes.

Mexico, Nigeria, Spain, Australia, and the United Kingdom became key partners whose countries and territories were supplemented by the use of special tracking ships: *Rose Knot Victor* and *Coastal Sentry Quebec*. Navy lieutenant Malcolm Scott Carpenter, age thirty-three, and Navy lieutenant commander Alan B. Shepard, age thirty-five, monitored the global communications network and recovery operations, while Air Force captain Virgil "Gus" Grissom, age thirty-three, provided the liaison for the spacecraft electrical systems.

Two of the first changes instituted by the astronauts were the size and placement of the window and the escape hatch. The original design called for a relatively small porthole off to the side of the cockpit. However,

The Mercury spacecraft instrument panel had the overall layout of an aircraft with the reaction control system (RCS) attitude control set up for the right hand. Courtesy of NASA.

"pilots" need a large and unencumbered view of the horizon directly in front of them. This along with the fact that the hatch could not be opened quickly from the inside caused some tension between the astronauts and the engineers. Both of these features were changed to the satisfaction of the astronauts.

Thousands of changes to drawings, hundreds of minor changes to systems, and a few dozen major changes resulted in no two Mercury spacecraft being the same. This in itself caused much time to be lost. Arbitrating between the engineers and the Mercury astronauts, NASA's Faget and McDonnell's John Yardley were able not only to maintain a good working relationship but also to move the required changes through with a minimum of delay.

Creating Mission Control: The Flight Controllers

While the focus of spaceflight is typically on the astronauts and the impressive launch vehicles, there is also a vast array of supporting equipment and behind-the-scenes people. A precedent existed for this in the preceding programs that had flown men to the edge of space in the rocket planes (X-1 to X-15) and the less glamourous aerostats (*Manhigh*, *Excelsior*, and *Strato-Lab*). Through these programs many of the various aspects of flight test had been experienced, but putting a human in a ballistic rocket and sending him into orbit was an order of magnitude greater. The launch vehicles themselves likewise had pioneered the use of detailed preparation and checklists to help ensure a successful flight.

As NACA morphed into NASA, several key individuals became a part of Project Mercury. Robert Gilruth, who was the head of the NACA Pilotless Aircraft Research Division, was assigned to lead the STG in November 1958 (Swanson 2012, 66). His association with space technology was somewhat limited to the use of relatively small solid-fuel rockets to accelerate aircraft models to Mach 1 to observe their behavior. Most of these tests took place at Wallops Island, Virginia. Although not nearly as large and sophisticated as the Cape, the facility was close to the Langley Research Center and provided the required range safety consideration out over the Atlantic Ocean.

Gilruth selected other NACA staff, including Belizean-born Maxime Faget as chief designer and Chuck Mathews as chief of flight operations. At this point there were no flight operations as there was nothing to fly. That virtually all of these people were aviation oriented was also somewhat of an obstacle to overcome. Gilruth himself noted in later years that, before *Sputnik*, he had never thought of people flying in space. It wasn't until *Sputnik II* and its canine passenger, Laika, that it dawned on him that the Soviets were intent on flying a man in space (Swanson 2012, 64).

Walt Williams had been the lead engineer on the Bell X-1 project (the first to exceed the speed of sound) and at the time was chairman of the X-15 Flight Test Steering Committee. He had received a bachelor's degree in aeronautical engineering from Louisiana State in 1939 and joined the Mercury Program as the operations director in August 1959.

Christopher Columbus Kraft Jr., who graduated with a bachelor's degree in aeronautical engineering from Virginia Polytechnic Institute in 1944, worked for NACA at the Langley Research Center in Hampton, Virginia. Assigned to the Flight Research Division under Robert Gilruth, he was involved with aircraft stabilization and control issues. Kraft was invited by Gilruth to be one of the original thirty-six members of the STG assigned to Project Mercury Flight Operations Division under Chuck Mathews.

This group of people, referred to as Mercury Control, were tasked to direct the flight operations from the ground to assist the astronauts. Other early members of the team included the first flight controllers such as Glynn Lunney and Gene Kranz. Kraft noted that it was through these people that, "if something went wrong—when something went wrong—I wanted to be ready to identify the problem and then either correct it or control the outcome" (Kraft 2001, 100).

The number of major components to be synchronized included not only those in the rocket but also the spacecraft systems and its human occupant. Each of the primary systems was monitored at a console in the control room by an engineer with a cadre of supporting people in other locations. It was into this control center that the radar, telemetry, and communication flowed to coordinate the flight operations.

The team that coalesced consisted of individuals who became highly knowledgeable on various aspects of the fight. These would include the flight dynamics officer (FDO, pronounced "Fido"); the range safety officer (RSO); the environmental, electrical, and communications officer (EECOM), the reentry controller (RETRO), and a host of others. The flight director was simply known as "FLIGHT"—with Kraft becoming the first to hold that title.

The Mercury control room was modest by today's standards—being about 60 ft. square. The controllers sat at consoles facing a large map of the world on which were plotted the three orbit tracks of the spacecraft. Initially, a small model of the spacecraft, suspended by a wire, was moved manually based data from the tracking stations. While today's control room contains a plethora of computer generated video displays, the early Mercury flights conveyed critical data by mechanical plotters, meters, and status boards.

Kraft notes in his biography, "I saw a team of highly skilled engineers, each one an expert on a different piece of the Mercury capsule. We'd have a flow of accurate telemetry data so the experts could monitor their systems, see and even predict problems, and pass along instructions to the astronaut" (Kraft 2001, 92).

One member of the STG was the "voice of Mercury Control," Air Force lieutenant colonel John A. "Shorty" Powers. At just 5 ft. 6 in., Shorty, as

The Mercury control room with the individual consoles for each controller as seen from
the viewing area overlooking the facility. Courtesy of NASA.

he was known to everyone, was a dynamic personality whose abbreviated
height in no way reflected the impact he had on the press, and TV coverage in particular. His direct, no-nonsense style and his proclivity for finding the right words set the standard for future NASA public affairs officers
(Thompson 2004, 229). Unfortunately for Shorty, he also could tell tall stories that eventually got him in trouble with NASA management. He left the
program in 1963—but not before he became well known to the public.

Gene Kranz recalls that his initial assignment was to organize Mercury
Control's countdown script, which would eventually include a book of
"mission rules" (Kranz 2000, 17). His responsibility was to determine what
events had to take place—and in what order—for the entire mission to be
flown. The controllers followed the spacecraft through the launch phase,
into space, and back. Their responsibility did not end until the astronaut
was recovered.

Those who were charged with the actual launch of the rocket itself were
encased in the concrete firing room called the "blockhouse," only a few
hundred yards from the rocket's launchpad. Although many of the test

Orbital Track

Each orbit passes directly over a path or track that can be traced on a map of Earth (see figure below). Note that the ground track of a spacecraft as launched from Cape Canaveral into a 100 mi. near-circular orbit is inclined about 28.5 degrees to the equator. Its first orbital track is depicted as a solid line. As it completes the first orbit by crossing 80 degrees west longitude, it begins its second orbit shown with a dashed line. Note that the ground track between the two orbits is displaced by 1,500 mi. at the equator—illustrating the rotational effect of Earth over the ninety-minute period of the orbit.

engineers for the Redstone rocket had been part of the World War II German contingent, virtually all of the Atlas and Mercury controllers were young American and Canadian engineers.

The Mercury flight control room was somewhat more remote but still placed relatively close to the launch area to facilitate communicating the data from the vehicle and spacecraft to the controllers. When the program

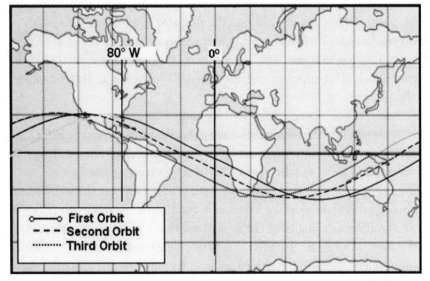

The track for the first three orbits of the Mercury spacecraft. The orbital inclination results in the satellite never passing over any part of Earth with a higher latitude. Courtesy of NASA.

moved on to the Gemini Program in 1965, these controllers moved to Houston.

At this point in the program (1961), however, the hub of the communications network was actually located at the Goddard Space Flight Center in Greenbelt, Maryland. This was where the only computer in the project was located to receive the radar and telemetry data and compute the spacecraft's orbit.

Putting together the countdown script for the subsystems that had to work in concert with each other as well as the growing ground support equipment was a challenge for Kranz—a former Air Force fighter pilot who had flown F-86s in the Korean War (Kranz 2000, 34). To assist in this endeavor, a spacecraft simulator was built to support integrating many of the elements. To help anticipate and evaluate the various possible anomalies, the systems engineers had to devise a series of scenarios that provided the controllers and astronauts with possible options when something did not perform as planned. Simulations were a way of life that enabled the various factions to discover problems and reengineer solutions. Hundreds of simulator and dress rehearsals preceded the actual launch. The training of the astronauts and the launch team was often an integrated operation (Kranz 2000, 262).

Because of the relentless countdown clock that indicated the time before launch, the emphasis was to identify any anomalies that might require the countdown to be stopped and the problem fixed before continuing— this was called a "hold." The exercise of real-time judgment had to be kept to a minimum and was typically reserved for the unexpected. Once the launch took place, problems required predefined abort scenarios or a quick work-around.

One of the early decisions was the question of who should talk directly to the astronaut. The astronauts themselves decided that only another astronaut would have the equivalent knowledge and perspective. Because it was not known what effects space flight might have on the human occupant, a constant monitoring of the astronaut's vital signs as well as communications was mandatory. At key tracking stations around the world, one of the nonflying astronauts would be stationed for the communications role. At others, an engineer served that function. The locations of these facilities was noted on the world map by a series of relatively small rings that represented the communications range for that station.

For the most part, the ground support was all civilian—NASA and its contractors. The Department of Defense was involved because the range facilities were run by the Air Force. Compared to the facilities that would evolve through Apollo and the shuttle programs, these early facilities were primitive.

The first fight that involved the new Mercury control staff was the unmanned Mercury Redstone-1. The inexperienced controllers received their baptism of fire when the countdown reached zero but the rocket failed to leave the launchpad.

Object K: Korabl

In February 1958, at the same time as an initial spacecraft proposal was being pursued by the Americans, Soviet chief designer Sergei Korolev directed a team headed by Mikhail Tikhonravov to concentrate on a manned spaceship. Like their American counterparts, time and existing technology would play a major role in their decisions. However, unlike the Americans, the Soviets had the R-7 (now with an upper stage), which provided a payload weight advantage by a factor of three. The upper stage that was added to the R-7 enabled it to orbit a 10,000 lb. spacecraft—three times the weight of the emerging Mercury capsule. On the other side of the coin, the Soviets needed that weight advantage because their limited technology inhibited the development of smaller and lighter equipment.

Within Tikhonravov's organization was a young man of thirty-two. As a sixteen-year-old, he had been captured by the Nazis during the war. Shot and left for dead, Konstantin Petrovich Feoktistov survived and went on to become the head for planning piloted space apparatus. He was the Soviet equivalent of NASA's Maxime Faget, and he would prove an important ingredient in Korolev's future plans for space exploration.

One of the first problems to be addressed by the Soviet team was spacecraft reentry. Several configurations were examined, including the cone shape being pursued by the Americans. However, the simplest shape that provided maximum volume with least surface area was a sphere. By selectively placing the center of gravity behind and below the reclined pilot, it was possible for the craft to assume a stable orientation during a ballistic reentry that avoided having to provide for attitude control during that critical period. However, the entire sphere had to be coated with an ablation material to protect it during the initial reentry stages before the self-orientation was completed. One of the drawbacks with a sphere was that it induced higher (but tolerable) g loads than other configurations.

Unlike Mercury, the Soviet craft used a sea-level atmosphere with 79

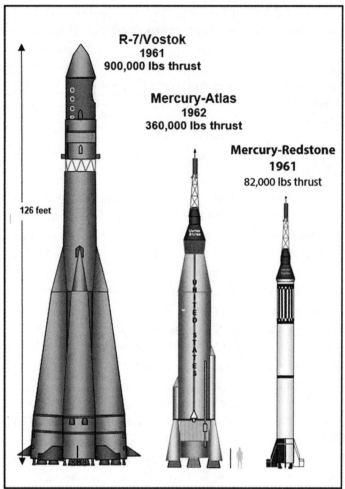

R-7/Vostok
1961
900,000 lbs thrust

Mercury-Atlas
1962
360,000 lbs thrust

Mercury-Redstone
1961
82,000 lbs thrust

126 feet

A comparison of the first American and Soviet manned launch vehicles. Courtesy of Ted Spitzmiller.

percent nitrogen and 21 percent oxygen. This required a thicker pressurized container and subjected the pilot to possible nitrogen embolism (the "bends") if the capsule experienced rapid decompression and the pressure suit (with 100 percent oxygen) had to be inflated. On the other side, however, the fire danger was greatly minimized. With respect to the space suit, some believed it was not necessary and should be omitted to save weight. This suggestion was accepted for later flights with the second-generation spacecraft, with disastrous consequences.

It was also decided that only a portion of the total satellite would return to Earth to simplify the reentry problem still further. A double truncated instrumentation cone attached to the descent apparatus provided for the

The Vostok spacecraft with the reentry sphere on one end (*left*) and the reaction control system and life-support equipment on the other. Courtesy of Energiya.

electronics and control system and the 3,500 lb. thrust storable liquid propellant "braking engine" that fired for 45 seconds to decelerate the satellite for return to Earth. While a solid-fuel system was simpler, the available thrust-to-weight ratio dictated a more efficient engine, which was discarded after retrofire. Like Mercury, orbital altitudes were selected to allow the spacecraft to reenter from natural orbital decay within the period of the available supply of life-support consumables (ten days for the Soviet craft) if retrofire failed to occur.

As with Mercury, a parachute slowed the spacecraft after reentry but did not prevent a momentary high-g impact on landing. To avoid any injury at touchdown, it was decided that the pilot would use an ejection seat to exit the spacecraft and complete the trip by parachute. The ejection seat also provided for emergency escape during portions of the powered ascent. A recovery from orbit within the Soviet Union on dry land permitted the entire flight envelope to be kept secret, although provisions were made for the pilot should an unplanned landing at sea occur.

The use of an ejection seat for emergency exit was not the first option explored. Korolev had wanted an escape tower topped by a solid-fuel rocket

---

Orbital Decay

Because there are still air molecules at extreme altitudes (100 mi.), there is some drag on the spacecraft. Each time it hits one of these molecules, energy is lost—slowing the spacecraft down and causing it to drop into a lower orbit. Because of its descent into a lower orbit, it encounters even more air molecules (higher density), and the increasing drag continues to lower the orbit. Thus, a spacecraft with an initial orbit of 100 mi. can expect a lifetime of only a few orbits, depending on the drag produced from its size (form and mass).

---

similar to Mercury, but the weight of the system was excessive. Instead, an ejection seat that provided for the first 40 seconds of flight was chosen. Should a malfunction occur after that period, the spacecraft would separate from the rocket and assume a normal reentry profile.

The crew cabin had three portholes, one of which had a special optical "viewer" that had precise lines engraved on it to permit manual orientation of the spacecraft to the horizon for retrofire. A miniature globe showed the pilot his current position and where he would land if the retrofire sequence were initiated at any particular time.

Again, like Mercury, the attitude control system was fully automated but provided a manual backup. However, unlike Mercury, the manual backup was seen as just that—a backup. The pilots chosen to ride the Soviet spacecraft were more passenger than pilot, while their American counterparts became an integral part of the spacecraft development process and the flight plan profile.

There were two primary reasons for this view. The first was that, as in America, there was a strong feeling that the psychological impact of being in space might prove an overwhelming experience that could incapacitate the pilot. The effects of high g-forces at launch followed by weightlessness and isolation were unknown. A second reason was that the spacecraft designers were not aviation people and did not know or understand the piloting environment. To them, the occupant was a passenger. This view hindered the development and utility of first- and second-generation Soviet spacecraft. Thus, the role of the cosmonaut reflected the social philosophy of the Communist state: tight control and little autonomy.

With the preliminary design complete, a meeting of the chief designers, held in November of 1958, reviewed the prospects for orbital and suborbital manned flight as well as the reconnaissance satellite. It did not take long to dismiss the intermediate step of suborbital flight—the Soviets moved directly to a manned orbiting satellite.

Moreover, as the resulting human space flight technology would transfer effectively to a reconnaissance satellite as soon as another "first" in space was achieved: the military was satisfied. However, not everyone was. There were voices in the Politburo who felt that the risk was too great and that, should a failure become public, it would do immeasurable damage to Soviet prestige. Nevertheless, Korolev's *Sputnik* aura was still preeminent, and his confidence—and that of the team—was high.

The spacecraft went through several names during its design phase that related to its origin as a reconnaissance satellite. However, when it emerged with its own identity, it was designated Object K, with the "K" being an abbreviation of Korabl—Russian for "ship." The initial prototype that appeared in the spring of 1959 had a total weight of 10,400 lb., of which the descent module represented about 6,000 lb. The craft had a diameter of 8 ft. and a length of 14 ft. The total height of the assembled R-7 was now a towering 123 ft.

The First Cosmonauts: A Different Perspective

The Soviets arrived at the same conclusion as their American counterparts: military pilots were the most logical choice for occupants of the new spacecraft that was starting to take shape in the summer of 1959. The fact that the Americans had already chosen seven military test pilots gave a renewed sense of urgency to the Soviet selection process. The criteria were similar in all respects but two: the Soviets were not looking for engineering test pilots. A bachelor's degree in engineering or science was not a prerequisite, nor was high performance jet time. Korolev's philosophy reflected the automated environment, and in this respect, he was looking for men who could execute a specific script laid down for the cosmonaut by the "controller." Men who could, and would, follow directions to the letter were selected (Mindell 2011, 89).

The records of several thousand potential candidates were reviewed, and those who appeared to possess the desired qualifications were interviewed and tested physically and academically. The purpose of the exams was not

First group of twenty Soviet Cosmonauts with Sergei Korolev. Yuri Gagarin is on Korolev's left. Courtesy of Energiya.

disclosed to the pilots at this point. By the end of 1959 a pool of candidates was available from which Korolev decided to select twenty. It was reported in later years that the number twenty was decided on simply to be larger than NASA's seven. The twenty reported to the new cosmonaut training center in Moscow, under the direction of Yevgeniy A. Karpov, in February 1960.

This first group of twenty (and their ages) consisted of Ivan N. Anikeyev (twenty-seven), Pavel I. Belyayev (thirty-four), Valentin V. Bondarenko (twenty-three), Valeriy F. Bykovskiy (twenty-five), Valentin I. Filatev (thirty), Yuri A. Gagarin (twenty-five), Viktor V. Gorbatko (twenty-five), Anatoliy Y. Kartashov (twenty-seven), Vladimir M. Komarov (thirty-two), Yevgeny Khrunov (twenty-five), Alexey A. Leonov (twenty-five), Grigoriy G. Nelyubov (twenty-five), Andriyan G. Nikolayev (thirty), Pavel R. Popovich (twenty-nine), Mars Z. Rafikov (twenty-six), Georgy S. Shonin (twenty-four), Gherman S. Titov (twenty-four), Valentin S. Varlamov (twenty-five), and Dmitri A. Zaikin (twenty-seven). The Soviet cosmonaut's average age (twenty-six) was almost ten years younger than their American counterpart.

The training regimen consisted of lectures on various subjects such as rocket fundamentals, geophysics, astronomy, and radio communication. A demanding physical conditioning program and intensive parachute training required that each candidate make forty jumps.

As the date for the launch moved closer, advanced training was provided for a subset of six candidates who were called (oddly enough) "The Vanguard Six." In some respects, this training program was similar to that of the Mercury astronauts. It was at about this period (the summer of 1960) that the cosmonauts (as they were to be known) finally came face-to-face with the chief designer, Korolev, and were introduced to the spacecraft they were to fly. Unlike the Mercury astronauts, the cosmonauts were not a part of the spacecraft development team.

9

THE FIRST ORBITAL FLIGHTS

Korabl-Sputnik: Unmanned Flight Tests

With the published (and much revised) Mercury schedule in the summer of 1960 showing a possible suborbital flight by December of that year, Korolev had his work cut out for him. It was vital that the Soviet spacecraft be in orbit before the Americans. If NASA's schedule had been optimistic, Korolev's was as well.

Many of the components for the spacecraft had already undergone tests using the R-5A rocket (essentially the Soviet equivalent of the Redstone) when the first unmanned version of *Object K* was launched into orbit at Baikonur Cosmodrome on May 15, 1960. This prototype did not have all of the systems and lacked the ablative coating since it was not to be recovered. The primary objective was a test of the attitude control system. A 195 mi. by 230 mi. orbit was achieved by the R-7 booster with a 65-degree inclination. The Soviets announced the successful launch and used the name Korabl-Sputnik 1, which loosely translated to "satellite ship." The TASS news agency referred to it as a spaceship, saying that it was a manned prototype; the Western press referred to it as *Sputnik IV*.

There was significant cause for concern in the United States, as the satellite was reported to weigh 10,005 lb.—three times the weight of *Sputnik III* (Object D), launched exactly two years earlier. The space race was now well into its third year, and the United States had yet to orbit anything heavier than the 1600 lb. Discoverer series reconnaissance satellites.

The flight plan called for Korabl-Sputnik 1 to remain aloft for up to four days while a variety of tests were run on the various systems. On the fourth day it was determined that there were some problems with the attitude control system, and when the retrorocket was fired, the spacecraft was pointed 180 degrees from that which was intended. This caused the spacecraft to be

Sergei Korolev with one of his space dogs. The Soviets favored the canine over the primate in their space medicine program. Courtesy of Energiya.

boosted into a higher orbit, where it stayed for five years before disintegrating on reentry.

A second Korabl-Sputnik was launched on July 28, with two dogs, Chayka and Lisichka, and more advanced attitude-control sensors. However, nineteen seconds into the launch, a fire developed in one of the R-7 s boosters causing the vehicle to be destroyed ten seconds later. As there was no escape system in these test vehicles—the dogs were killed.

A third Korabl-Sputnik (labeled Korabl-Sputnik 2 by the Soviets, thus not acknowledging the earlier failure) was launched on August 19 with two more dogs, Belka and Strelka, and a variety of mice, rats, insects, and plant seeds. This launch was successful, and the spacecraft attained an orbit of 191 mi. by 211 mi. For the first time, two TV cameras recorded the responses of the dogs in a weightless environment—startling the doctors observing them from the Earth. While the vital signs of the two animals were within tolerable limits, the two appeared lifeless. After several orbits they showed some movement, which appeared convulsive, and Belka vomited.

Telemetry from the spacecraft indicated that the primary attitude control system had again malfunctioned, but the backup (using the sun as a reference) was successful and the satellite reentered properly. At the designated altitude, the dogs were safely ejected and landed only a few miles from their intended point of touchdown—becoming the second object ever recovered from space (the American *Discoverer XIII* had accomplished that "first" only a week earlier). The flight of Korabl-Sputnik 2 was an important milestone in the spacecraft development because it essentially validated all of the critical systems.

Although both dogs were found to be in excellent condition, because of their alarming responses in space, it was recommended that the first human flight be limited to a single orbit. Two additional flights of the prototype were recommended before the more advanced manned version was to be scheduled. There was still time to beat the Americans. Then a tragedy occurred that delayed further testing for several weeks, but it had nothing to do with Korolev's manned spaceflight program.

OKB-586 chief designer Mikhail Yangel's R-16 was poised on a nearby launchpad on October 24 for the first test of this new storable-fuel ICBM. A propellant leak was discovered, and repairs were completed without draining the propellants or recycling the countdown. Almost two hundred technicians and managers including the chief of the Soviet strategic missile forces, Marshal Mitrofan Nedelin, were standing in the area immediately around the base of the rocket. As the countdown was set to continue at T-30 minutes, and the assembled group prepared to leave for the blockhouse and safer viewing points, the second stage of the missile unexpectedly roared to life. The destructive force of the exhaust ripped into the first stage, spilling its toxic and flammable content down upon the men below. An enormous fire ensued, and most of those in the area were incinerated including Marshal Nedelin.

Operations were brought to a standstill at the test facility while the carnage and destruction were analyzed and cleaned up. Fortunately, for Korolev, the Americans were having their own set of technical problems, and the December launch date for the first manned Mercury suborbital launch had long since slipped into 1961.

Korabl-Sputnik 3 soared into the sky on December 1, 1960, and achieved the precise low orbit planned for the manned launch (112 mi. by 155 mi.). On board were the dogs Pchelka and Mushka as well as improved biomedical sensors to record more accurately and completely the condition of the dogs while in the space environment. After a full day in orbit, the retrorocket fired, but the burn was less than planned. The new perigee was low enough that the spacecraft began its reentry after completing another orbit and a half—the initial calculation showed that it would land beyond Soviet territory.

Now an automated sequence of events aboard the spacecraft went into action that called for explosives to destroy the ship if onboard sensors determined it was not on its planned flight profile. The objective was to ensure that no useful components that could lead to determining the technology employed in the spacecraft systems could be recovered by the Americans. Thus, the spaceship was destroyed and its canine passengers killed. The Soviet news agency, TASS, simply announced that an incorrect reentry profile had resulted in the craft being destroyed on reentry—a rare admission of failure for the Communist regime.

The trouble with the braking engine (the Soviet term for retrorocket) was quickly isolated and remedied, and the last of the prototype spacecraft was erected within three weeks. The first flight of the R-7 with an improved RD-0109 upper-stage unit thundered aloft on December 22, 1960, with the dogs Kometa and Shutka. All went well until ignition of the new upper stage. A failure in the gas generator for the pump that supplied the propellants to the engine caused a premature shutdown. The emergency escape system was activated and the spacecraft separated from its carrier, arching more than 133 mi. into space and reentering 2,200 mi. downrange in one of the most inaccessible regions of Siberia.

Rescue forces were immediately sent but did not arrive in the vicinity until almost two days had elapsed. Moving through waist-deep snow, the technicians carefully examined the spacecraft, which still had its destruct charge set to detonate sixty hours after landing (to destroy any evidence), and it had been more than sixty hours since launch. They disarmed the charge and found the dogs in the spacecraft alive but undoubtedly very cold in the −40 °C temperatures. The ejection seat to which their enclosure was attached had failed to exit successfully due to a series of malfunctions. The escape system was redesigned. There was no mention of the aborted flight in the Soviet press.

With the two failures, it was recognized that the next flight of a fully man-rated and equipped spacecraft could not be made in the timeframe scheduled, and that two more unmanned flights were needed. The pressure on the Soviet team was now intense, as a successful American suborbital effort would dim their accomplishment. The first man-rated spacecraft was launched on March 9 (called Korabl-Sputnik 4 by the Soviet press). The flight was perfect in all respects. Korabl-Sputnik 5 went aloft on March 25, carrying the dog Zvezdochka (Starlet). It was another complete success.

At this point, there was some discussion whether the manned launch should be timed for the annual May Day celebration. Khrushchev actually preferred that it occur before or after May Day, as he was getting somewhat anxious that a failure resulting in loss of life would undermine the propaganda success that had thus far been achieved. However, the imminent launch of a manned Mercury spacecraft dictated that the attempt be accomplished as soon as possible—and a life had already been lost in the fledgling Soviet space program.

Cosmonaut-trainee Valentin V. Bondarenko had been undergoing a training session in an isolation chamber filled with 50 percent oxygen at reduced pressure to simulate the spacecraft environment. On the tenth day of the planned fifteen-day exercise, Bondarenko removed a biomedical sensor attached to his body and cleaned the skin with an alcohol wipe. He tossed the moist pad toward the garbage bag in the corner, but it landed short on a hot plate and immediately ignited. Bondarenko tried to extinguish the small flame, but it quickly grew and ignited his flight suit, engulfing him in flames that resulted in severe burns to virtually his entire body. He died within hours. The incident became a state secret that was not revealed until 1986.

Nevertheless, the stage was now set for the first manned Soviet launch.

Mercury: Beset with Problems

If the Soviets had encountered problems along the way with the development of their manned spacecraft, the Americans could well empathize. On the first test of the Mercury escape system using the Little Joe solid-fuel rocket in August of 1959 at Wallops Island, an errant electrical signal caused the escape rocket to fire thirty-five minutes before the flight was scheduled to launch (Kleinknecht 1963, 5). This test, designed to evaluate the escape system under maximum dynamic pressure (Max Q) was rescheduled with another spacecraft and modified circuitry. The spacecraft used in these tests were termed a "boilerplate" and were simply low-cost shape and center-of-gravity replicas, as there was no need for the internal systems of the production spacecraft.

A few weeks later, the first Atlas with a Mercury capsule launched from the Cape on September 9, 1959. The objective of the test called Big Joe was to prove the integrity of the heat shield. As the missile arched over in its

---

Maximum Dynamic Pressure: Max-Q

The point in flight where the vehicle is experiencing the maximum aerodynamic stress is based on the air density and velocity. The altitude at which this condition is encountered depends on the acceleration of the vehicle but is typically about 35,000 ft. for most rockets launched with the intent of orbiting a payload.

---

trajectory, the planned separation of the two booster engines at the point of booster engine cutoff failed to occur, and the skirt unit remained attached. The added weight did not allow the Atlas to achieve the desired velocity. Adding to the problems, the capsule did not separate at the planned time, and its thrusters attempted to wrestle the entire Atlas into a 180 degree attitude change to present the blunt end of the capsule forward to the line of flight for the reentry. This exercise in futility resulted in the thrusters running out of fuel. However, when the capsule did separate, it was able to withstand the unanticipated heating until its center of gravity performed a self-alignment with the atmosphere. Although not a complete success, the test was encouraging (Kleinknecht 1963, 6).

Little Joe 1A (a repeat of the August test) validated the Max Q requirement. Another Little Joe test in November provided for further low-altitude escape sequences, but the escape rocket fired late and the test objectives were not fully accomplished. The next Little Joe test in December carried a monkey safely through the abort sequence, and another in January of 1960 was also successful.

Delays in the production of the spacecraft continued to plague the test schedule. One aspect that had not been considered in the schedule was that the Mercury capsule was so small that only one or two technicians could work in the tight confines at any one time. In addition, the production workers still lacked the sensitivity for a "clean-room" environment. Small pieces of wiring, loose nuts, and other debris were left in the craft during their manufacture and modification. This could not be tolerated in the weightless environment of space, where anything not fastened down floated around and could lodge in critical places.

The first production Mercury capsule was delivered in April of 1960 and was fired in an "off the pad" abort test at Wallops Island in May. It was clear at this point that the December 1960 manned suborbital launch schedule would not be met.

The wreckage of the Mercury spacecraft recovered after the failed MA-1 launch. Courtesy of NASA.

However, the big test of a production spacecraft with an Atlas D, termed MA-1 (Mercury-Atlas) occurred in July of 1960. This capsule carried no escape tower and was to qualify the production spacecraft structure under maximum aerodynamic loads and after-body heating during the powered ascent. A heavy rain shower passed over the Cape and delayed the launch. At ignition, the booster rose cleanly into the sky, but as it began entering the area of maximum dynamic pressure at 40,000 ft., the sides of the Atlas buckled, and the entire vehicle was destroyed.

The failure was heavily covered by the press and played up in the aerospace media. Only four days earlier, the Apollo follow-on program had been proclaimed by NASA, but press announcements did not accomplish milestones in space exploration. The 1960 presidential race was in full swing, and the young Democratic candidate from Massachusetts, John Kennedy, remarked that if a man was to be placed in space before the end of the year, "his name will be Ivan" (an obvious reference to Russia's superior capability). Although Kennedy had no way of knowing, one of the Soviet cosmonauts then in training was named Ivan!

On Election Day, November 8, another test with the Little Joe was made to verify the escape system at the most critical portion of Atlas flight profile: where the g-forces were at their peak. However, early separation of the spacecraft occurred and the test objectives were not met.

The first test with an unmanned Mercury-Redstone (MR-1) occurred on November 21. The Redstone was touted as the most reliable missile in America's arsenal. Nevertheless, a failure in the launch sequence caused the Redstone to shut down just as it was lifting off, and it settled back on the pad. Fortunately, its structural integrity allowed it to remain upright, but a portion of the escape sequence initiated; while the capsule remained firmly attached to the Redstone, the tower itself shot 4,000 ft. into the air. Three seconds later the parachute deployment began from the nose of the capsule as it sat on the Redstone—amid the residual exhaust smoke. Since the self-destruct system of the booster was armed, it was not prudent for anyone to venture near to cut away the parachute and begin the clean-up process until the next morning, when the batteries that powered the explosives had exhausted their charge.

The damage to the Redstone was minimal and quickly repaired, and the source of the erroneous sequencing was located for both the booster and the escape system. On December 19, MR-1A (as the revised shot was designated) lifted off to execute a flawless flight. After one more successful test of the Mercury-Redstone, then a man would fly. That next test carried a chimpanzee named Ham (an acronym of Holloman AeroMedical). As the test pilots who derided the spam-in-a-can approach to human space flight had decreed two years earlier, man would fly only after a monkey had shown the way.

MR-2 launched on January 31, 1961, after an agonizing series of "holds" to the countdown, causing the poor chimp to be sequestered in the capsule

The first Mercury-Redstone (MR-1) experiences a shutdown on the pad just after ignition. Incorrect sequence of sensing data caused the escape tower to fire. Courtesy of NASA.

for more than four hours before ignition. As the booster climbed into the morning sky, the cabin inflow valve erroneously opened, literally sucking all the air from the capsule. The spacecraft environmental control system sensed the potentially lethal situation and inflated Ham's custom-made enclosure. To add to his discomfort, the trajectory was more vertical than the profile called for, and the liquid oxygen depleted at about the time the Redstone's engine was scheduled to shut down. The abort sensor recognized the low pressure in the liquid oxygen line and was quicker than the shutdown sequencer, resulting in the triggering of the escape system. Poor Ham had to endure another sudden high-g load as the spacecraft was lifted an additional thirty miles into space.

Because the escape system had activated, the retrorocket firing was not carried out, and the capsule descended at a faster velocity than planned, resulting in a 15 g load for Ham instead of the planned 12 gs. However, Ham's experiences were not over yet. As the capsule descended into the water, it landed at a slight angle, and the heavy beryllium heat shield that was hanging down to extend the flotation bag slammed into the side of the pressure vessel that was now providing buoyancy and shoved a bolt through the titanium bulkhead allowing water to begin entering the capsule. The heat shield was quickly torn away by the wave action, and the capsule tipped on its side allowing more water to enter through the open cabin pressure relief valve. Ham was in deep trouble, but the recovery forces were unaware of

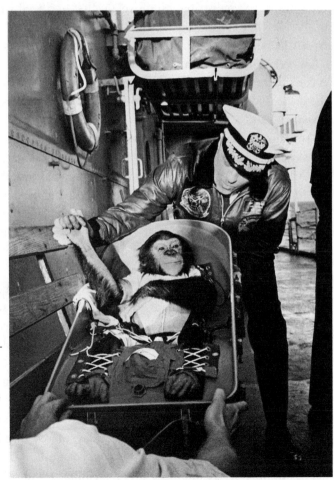

Ham is hauled aboard the recovery vessel, and when his enclosure was opened, he greeted the captain with a handshake. Courtesy of NASA.

his plight. As the capsule had overshot its intended landing area, it was not hoisted out of the Atlantic until almost two hours had elapsed. Fortunately, Ham survived the ordeal and lived a long and quiet life thereafter.

The straps on the landing bag were increased in number and size, and a fiberglass bulkhead was inserted between the heat shield and the titanium pressure bulkhead. However, Wernher von Braun was adamant that another test of the Mercury-Redstone was needed before he considered it "man-rated."

As the debate over the need for another Mercury-Redstone test before a manned flight began, the next Mercury-Atlas, MA-2 was scheduled for launch on February 21, 1961. This event was also contentious, because MA-1 had experienced a structural failure. The D model Atlas had undergone a weight reduction program, and its already thin skin had been shaved even more. The redistribution of weight that occurred with the addition of the Mercury capsule on top of the Atlas had caused the MA-1 failure. NASA had ordered the thicker-skinned Atlas, and the first arrived within a few weeks. The Atlas on hand at the Cape was the thin-skinned version. A controversial decision was made to place an 8 in. wide steel band around the critical location on the body of the existing Atlas instead of waiting (Swanson 2012, 73).

The launch occurred on schedule, and the rocket climbed into the morning sky. Everyone who was aware of the steel band was watching intently as the Atlas began to generate the "frozen lightning" contrail indicating that the Max Q region of aerodynamic stress was being encountered at 30,000 ft. MA-2 continued on its way—the fix had worked and the capsule, driven to 114 mi. above Earth, sped over 1,400 mi. downrange. The instrumentation revealed that it had endured a Max Q of 991 lb./sq. ft. and 15.9 gs of deceleration on its eighteen-minute flight.

As for the Mercury-Redstone, NASA was ready to fly an astronaut, but the ground rules for a manned space flight dictated that at least one full-scale rehearsal covering all the key elements had to be successful, and according to the von Braun team, the Redstone needed that one more test. However, the program had run out of production Mercury capsules, so an old boilerplate used in the Little Joe 1B flight was brought out and refurbished. On March 24, MR-BD, which incorporated some eight modifications to the Redstone, flew a near-perfect trajectory. The next MR flight would carry an astronaut. The following day, March 25, 1961, the Soviets flew their last unmanned test with Korabl-Sputnik 5. Even Hollywood could not have scripted a more exciting cliffhanger to see who would be first into space as April 1961 moved onto the calendar. Only a few in the Soviet Union really knew how close it might be.

Vostok: First Man in Space

It was 5:30 a.m., April 12, 1961, when cosmonauts Yuri A. Gagarin and Gherman S. Titov were awakened from a sound sleep at a small house in the secret Baikonur Cosmodrome deep in the Soviet Union. It was the same dwelling formerly occupied by Marshall Nedelin, before his untimely demise the previous October (Siddiqi 2003b, 272). Sensors placed surreptitiously beneath their mattresses to monitor movement confirmed the restfulness of their slumber to the doctors who were hovering over their every move. After a sparse breakfast of meat paste, marmalade, and coffee, they underwent a brief medical examination and then donned a light blue pressure suit covered by an orange jumpsuit.

The two, selected from the short list of six by a series of reviews, had to pass the final approval of both the chief designer, Sergei Korolev, and Soviet premier Nikita Khrushchev. Gagarin had only received the nod for being the first cosmonaut four days before and was told the following day. Titov was prepared to assume the primary role right up to the moment of launch, should anything inhibit Gagarin from completing the preparations.

Gagarin had been a favorite choice of the selection committee as his easy-going personality and intellect complimented his personal commitment and ability to focus on a task at hand. In peer votes for who they would like to see be the first in space (cast by the cosmonauts themselves), he had scored higher than any of the others. He had also satisfied the Communist Party since he came from a working-class family (having grown up on a collective farm), was a devout atheist, and his ethnic background was Russian.

Following a short ride in the transport van to the launchpad, the two were greeted by a bevy of officials including Korolev—who reportedly spent a restless night himself. He had concerns for the reliability of the upper stage of the rocket. At the base of the service structure to one side of the giant R-7, Gagarin and Titov parted, the former taking the elevator to the top level and the waiting spacecraft and the latter returning to the transport van to continue his standby role. The R-7 itself had been erected just two days earlier—on the tenth.

Gagarin was helped into the spacecraft and secured by 7:10 a.m. (T-120 minutes). All through the procedure, various technical and party officials

Cosmonaut Yuri Gagarin prepares for the first flight above the Kármán line (62 mi.) and into orbit. Courtesy of Energiya.

had been observant for any signs that the stress of the momentous occasion might be too much for the young twenty-six-year-old. However, Gagarin was quite composed considering the history-making event of which he was the center. The spacecraft hatch was closed at T-80 minutes, but incorrect seating caused it to be removed and laboriously reinstalled. The count continued, and at T-30 minutes, the technicians left the service tower, which was then lowered out of the way.

With 15 minutes to launch time, Gagarin put on his gloves as the spacecraft communicator in the control room reported to him that his pulse was 64 and his respiration was 24, good signs of his controlled physiological disposition. At T-5 minutes, Gagarin lowered and locked the helmet visor. Only he knew the thoughts crowding his young mind as the final seconds ticked off. Perhaps he had already reconciled the odds of completing the next 90 minutes successfully, as he sat atop a combination of thousands of parts that had only flown sixteen times in their present configuration, and only half of those flights had completed their mission.

As the countdown reached zero, the thirty-two rocket engines beneath him came to life and a half-million pounds of volatile propellants began to burn—hopefully in a controlled manner. Here was the first man to ride the fire into the cosmos, the first to experience the sound and vibration of sitting atop the most powerful creation of man, the first to sense the full measure of this new adventure.

The forces of gravity experienced by the cosmonaut during launch are in direct proportion to the current weight of the rocket relative to the thrust being generated. Thus, at liftoff the R-7 produced about 50 percent more thrust than its weight, and Gagarin experienced 1.5 gs. As fuel was consumed, the weight of the rocket decreased, and the g-forces increased. At T+119 seconds, the four boosters had devoured virtually all of their fuel, and the thrust to weight ratio pinned Gagarin to his seat with 8 gs.

Just then the four boosters shut down and separated, falling off to each side as if in a carefully choreographed, slow-motion ballet. The central core unit continued to thrust onward as the g-forces reduced back to 3, and the rocket arched over to become parallel to Earth's surface at over 100 mi. above the frozen wasteland of Siberia. The aerodynamic shroud that covered the spacecraft was released, and it too fell quickly behind the accelerating rocket, which now subjected its occupant to more than 5 gs. Gagarin experienced difficulty in enunciating words into the radio as the skin on his cheeks was drawn tightly backward, and his body weighed more than 1,000 lb. His heart rate had risen to 150 beats per minute.

At T+300 seconds, the core unit of the R-7 shut down, and the g-forces again reduced as the upper-stage RD-0109 came to life. In the control room at Baikonur, those who observed Korolev, reported that he was visibly shaking during the ascent as he agonized over the possibility of something going wrong during the eleven-minute powered period. However, the telemetry showed all three stages gave the expected performance.

Suddenly, at T+676 seconds, the acceleration ceased, and Gagarin perceived he was being thrown forward against the restraining seat harness. In reality, it was the suddenness of going from a high-g environment to weightlessness that provided the illusion. With a perigee of 110 mi. and an apogee of 188 mi., somewhat higher than planned but acceptable, man was in space for the first time.

Gagarin reported that he felt fine with no physical problems and was making an initial adaptation to his weightless environment. He was immediately drawn to the view from the small portholes—the muted colors of Earth contrasting with the bright white clouds a hundred miles below and the deep blue of the vast Pacific Ocean.

The ground controllers could hear from his reports that there appeared to be no psychological impairment. As in the Mercury Program, the Soviets had stationed tracking ships (*Sibir, Suchan, Sakhalin,* and *Chutkotka*) along the ocean path of the spacecraft for communication and telemetry reception.

Although the Soviets gave no advanced notice of the launch to the world, sealed press packets had been prepared for the Soviet news agency, TASS, and the state radio and TV news organizations. Once Gagarin reached orbit, they were telephoned and authorized to open the packets and make the flight public using the carefully scripted information provided. As was the case with many important announcements, it was preceded by the patriotic Soviet anthem, *How Spacious Is My Country*. In addition, another sealed packet contained the notice of a fatal malfunction if it occurred after the initial announcement was made. A third packet contained information relative to an emergency landing made outside the Soviet Union. The final revelation in the press packets was the name of the spacecraft—Vostok (the Russian word for "east"), which had been a closely guarded secret to that point. Soviet schoolchildren and factory workers were given the day off to celebrate.

The time passed quickly as Gagarin completed the rather simplistic chores assigned, and soon it was time to prepare for reentry. The mission was designed to be flown automatically by the spacecraft without any intervention by the cosmonaut. Nevertheless, should something go wrong with the automated sequence, the pilot had to be allowed to control the ship manually. In the event of an emergency, the last three digits of the combination to unlock the manual control system were in a sealed envelope that the cosmonaut could open (Siddiqi 2003b, 272).

The spacecraft's attitude was set to align the braking rocket opposite to the line of flight based on the position of the sun. The time of day for the flight was chosen specifically to allow the sun to be used for this purpose. At T+75 minutes, the retrorocket fired for 40 seconds. The spherical descent module was then supposed to separate from the instrument section, where it had been held in place by four metal straps. However, something went wrong. Gagarin felt a sharp jolt and observed that the craft began to rotate slowly around its longitudinal axis at about six revolutions per minute. He could see the continent of Africa periodically passing the window with each revolution. He reported through the voice channel that the braking rocket had functioned properly, and he should be landing in the designated zone.

Apparently, the metal straps that fastened the two segments of the spaceship together malfunctioned. Only some of the electrical cables between the two units were disconnected while the remaining connections held them together. It was conceivable that conflicting signals could have interfered with the subsequent reentry sequence. Gagarin perceived he had a potentially serious problem but retained his composure and even reported through his telegraph key the code "VN" indicating "all is well"—apparently believing there was no need to excite those on the ground with a potential problem over which they had no control. It may be fortuitous that the destruct package, which had accompanied all the unmanned missions, was not on board as it might have detected a faulty reentry and destroyed the spacecraft. The KGB representative on the flight planning committee had argued for its inclusion but was overruled.

Gagarin felt another bump about ten minutes after separation was to have occurred, and it is now assumed that the two units finally parted at this point, perhaps because of the building aerodynamic loads. It is also possible that, had the two segments not separated, the heat on the descent sphere could have exceeded critical values.

Gagarin now witnessed through his small openings to the world outside, a bright purple light that surrounded the spacecraft as the heat pulse created an ionized sheath of air, inhibiting the ability to communicate by radio. The flaming reentry had begun (Siddiqi 2003b, 277).

The cosmonaut felt the increase in temperature radiating through the hull and noted the sphere rotating about all three axes as the spacecraft entered the more dense layers of the atmosphere and the g-forces built. The oscillations gradually dampened as the aerodynamic forces and the center of gravity of the spaceship reached equilibrium. As the g-forces approached 10, Gagarin later reported that his vision blurred and he "grayed-out" for a few seconds (he had previously experienced as many as 13 gs in the centrifuge at the astronaut-training center).

A large contingent of more than two dozen aircraft and seven recovery teams were spread across the expanse of the Soviet Union, should the spacecraft fall outside the intended landing zone. At about 20,000 ft., the main parachutes opened and the hatch covering the ejection seat jettisoned. Within a few seconds, a small rocket charge ignited, and the seat shot away from the descent module. Two seconds later the seat restraints released, and Gagarin tumbled free to open his own parachute for the final phase of the landing.

It was a beautiful spring day, and the farmland of the Saratov region over which he was swinging gently in the parachute harness, southeast of Moscow and north of the Caspian Sea, was familiar territory to him. He landed softly in a field just one hour and forty-five minutes after launch. Yuri Gagarin had become the first man into space. The Soviet Union had once again beaten the Americans to another important milestone in the space race. The "backward" country continued to lead the United States more than three years after the first *Sputnik* had jolted the Americans into action.

America's Response: "We're All Asleep Down Here"

It was still dark on the morning of April 12 in the United States when the telephone rang on the nightstand next to where Lt. Col. John "Shorty" Powers was sleeping soundly in Florida. Answering the phone, he heard an excited New York reporter on the other end inquire if Powers had any comment on the launching of the first Russian cosmonaut. Powers, as the astronauts' public affairs officer, and others in the NASA hierarchy, had already prepared a statement for the expected announcement of a successful Russian manned flight into space. The preparation was a result of the United Press International report of a rumor that the Soviets were about to send a man into space, and the CIA's confirmation.

But being caught at 4:00 a.m. was just a little too much for the eighth astronaut and voice of Mission Control. "We're all asleep down here," he responded in reference to Florida's location south of New York. The reporter fittingly noted his comment, which was subsequently published with its double entendre, and it became an embarrassing icon of the position of the United States in its race with the Soviets (Thompson 2004, 282).

James Webb, President Kennedy's new NASA chief, appeared on nationwide television at 7:45 a.m. EST to congratulate the Soviets. He also wanted to reassure the nation that Project Mercury would achieve the same goal "soon." The following day Webb and several of his NASA team were called before the House Space Committee where strong disappointment was expressed that America had again come in second best.

At a presidential press conference, a reporter asked when the United States would catch up. Kennedy's stressed response was, "However tired anybody may be, and no one is more tired than I am, it is a fact that it is going to take some time. We are, I hope, going to go in other areas where we can be first, and which will bring perhaps more long-range benefits to mankind" (Bizony 2006, 25). This was a strong indication that Kennedy had been doing some soul searching.

The first of the thick-skinned Atlas missiles, 100-D, was launched two weeks later on April 25 in the first attempt to actually orbit an unmanned MA-3 Mercury capsule. The decision to orbit rather than perform another suborbital flight was made as a response to the Gagarin flight as a means of providing some lift for America's morale. However, the Atlas guidance system failed to pitch the rocket over into a trajectory to the southeast and instead continued its vertical flight. The destruct signal was sent, and another failure was etched into the American psyche. For the escape system engineers, the unplanned test resulted in a successful exercise of their systems. The Mercury capsule was pulled from the booster and descended safely under parachute. However, that was little consolation for the nation.

On April 28, a Little Joe rocket made its third attempt to simulate a Max Q escape, but one of the solid-fuel rockets in the cluster failed to ignite. However, the test conditions exposed the capsule to double the aerodynamic loads and three times the g load of a worst-case Atlas abort. The capsule escape system was considered "qualified."

Just as with the Soviet program, the previous February Project Mercury had moved three of its astronauts to a short list for the first flight. Alan Shepard, John Glenn, and Gus Grissom went through the final intensive training program with the actual capsule—production spacecraft No. 7. The original launch date of March 6 had been slipped to the end of April and then into early May. The press had begun the guessing game as to the name of the first American astronaut, when in fact Robert Gilruth had already told "the Seven" back in January that, barring any change in his physical or psychological profile, Shepard would make the flight with Glenn as the backup.

Unlike the anonymity of the Soviet cosmonauts, the identities of the American astronauts were well known to the world, as was the façade of their characters, thanks to a contract with *Life* magazine. The contract was in itself somewhat controversial as it initially provided a yearly $25,000 payment to each of the astronauts to allow their lives to be revealed to the world (Wolfe 1979, 110). A total of twenty-eight issues carried the stories, beginning in September 1959 and ending in 1963. The problem of subsequent payments and exclusive coverage became more involved with the second and then third group of astronauts, and subsequent groups did not have contracts with *Life*.

While the *Life* series appeared to convey the homogeneity of the astronauts, the depth of these individuals and their often superhuman egos would not come out for some time to come. Alan Shepard, for example, was a complex man with an inner drive to be the best. His quick wit with whomever he chose to befriend was often hidden by a veneer of cold steel to those he chose to shut out.

As the flight date drew closer, the press grew more demanding for information and access, and there was a point at which NASA felt that the pressure might be too much for not just the astronauts but the technical team as well. Because the Russians had been so secretive, there was some derision of the validity of their flight in the world press. Thus, NASA and the Kennedy administration felt obligated to open as much of the launch preparations as possible without interfering with the flight itself. Kennedy felt that the United States was leaving itself exposed, should a fatal flight occur, and, like Khrushchev a month earlier, was apprehensive about the negative propaganda that might result.

Mercury-Redstone MR-3: Alan Shepard

With the arrival of the first week of May, more than three hundred press representatives from virtually all of the major countries of the world descended on the Cape to cover the unfolding story. On May 2 Shepard went through the first stages of the countdown and had gone as far as suiting up, but the weather didn't cooperate and the flight was "scrubbed" (another word in the lexicon of the new space jargon that means canceled) two hours and twenty minutes before launch. It was at this point that NASA decided to reveal the identity of the astronaut. A second launch attempt on May 4 was also canceled.

Shepard was awakened at 1:30 a.m. on May 5 to begin the process of preparing for the flight once again. The medical people and NASA management observed Shepard carefully to see if he was showing sign of strain from the delays. However, Shepard was well prepared in all respects and had, as Gagarin undoubtedly had, reconciled the risks with the steps taken to mitigate them. While the concept of fear, as most would define it, was not a factor in the astronaut's thought process, the phrase "heightened awareness" was.

Shepard ate what became the traditional preflight breakfast: filet mignon wrapped in bacon with eggs and orange juice. The diet for the three days before a flight concentrated on high protein, low-residue foods.

The suit-up process then began with a brief physical followed by the attachment (or insertion, as the case may be) of the three primary biomedical sensors for temperature, respiration, and heart rate. It was almost 4:00 a.m. when Shepard emerged from his quarters at Hanger S to begin the van ride to the launchpad. America's first satellite, *Explorer I*, had been launched from the same pad. All of the other astronauts were engaged in some aspect of the mission, with Glenn having run through the spacecraft systems test in the early hours of predawn.

Shepard began the elevator ride to the top of the Redstone at 5:15 a.m. and was helped into the tight confines of the capsule by the technicians headed by Guenter Wendt (Wendt and Still 2001, 33). There he found a sign left by Glenn that read "No handball playing here"—a reference to Shepard's exceptionally good athletic prowess at the sport. Again, another tradition was being established, with humor and camaraderie playing an important role in preparing the astronaut for flight.

When the spacesuit sensors were plugged into the telemetry system, Shepard's heart reported eighty beats per minute. The Mercury cockpit was filled with a wide variety of switches and instrumentation—far more than the rather sparse Vostok. It reflected a cockpit designed for a pilot. The hatch was bolted shut by 6:00 a.m. and Shepard started breathing 100 percent oxygen to purge his system of nitrogen.

Low clouds accumulated over the Cape, and the countdown was held several times and then recycled back to T-35 minutes. For more than four hours, Shepard lay on his back waiting for an opening in the overcast. At one point he related that his bladder was in need of being relieved, and there was a flurry of discussions on what to do. Because the flight was only to be a quick fifteen-minute "up and down," no provision had been made for the pilot's waste products. Finally the decision was made to allow him simply to urinate in the space suit (Shepard et al. 1994, 107). Although this had some negative effect on the electrical biomedical sensors, it resolved the problem. With some irritation over several other glitches that caused added delays, Shepard at one point commented, "Why don't you fix your little problem and let's light this candle" (Thompson 2004, 296).

Since capsule No. 7 had been mated with Redstone No. 7, Shepard named his spacecraft *Freedom 7* to reflect the accumulated sevens that had come together in the project, and another precedent was set: each astronaut was allowed to name his spacecraft (Shepard et al. 1994, 105). Several of the changes the astronauts had requested (some would say demanded), such as the larger window in front of the pilot and the new lightweight quick-release hatch, were not in this version.

As the countdown neared zero, Shepard's heart rate increased to one hundred beats per minute, and people all across the nation, many of whom had delayed going to work or school that morning, leaned toward their TV sets . . . 3—2—1—Zero! The Redstone rose rapidly into the morning sky with Shepard providing a portion of his own narrative, "Ahh, roger, lift-off and the clock has started." This was an expression that the other astronauts repeated as reassurance to those on the ground that the astronaut had his act together and was coherent—similar to comments from the aerostats twenty-five years earlier. Timing of the events was dictated by the onboard clock synchronized with the planned schedule. And then, "This is *Freedom 7*, the fuel is go, 1.2-G, cabin 14 psi, oxygen is go."

At T+45 seconds, the area of Max Q presented Shepard with vibrations and buffeting that soon smoothed a bit as the outside sky quickly changed from blue to black. "Cabin pressure holding at 5.5" was an important call at just after T+60 seconds that his predecessor, Ham (the chimpanzee), had been unable to make; it had caused his space enclosure to pressurize. However, with Shepard, failure of the cabin vent valve to close would have initiated the abort sequence.

All continued to go as planned, and the Redstone shut down at T+142 seconds after imparting a velocity of 5,200 mph. The escape tower jettisoned, and the small posigrade rockets separated the spacecraft from the Redstone. Five seconds later the automatic stabilization control system turned the capsule around. Shepard then changed the spacecraft attitude with the manual system essentially to prove that man could perform critical tasks in the weightless environment. He then performed a quick evaluation of his ability to observe Earth and reported these to Mission Control.

The precious seconds ticked off as the spacecraft was realigned for retro-rocket firing. This was not necessary since Shepard had achieved less than a third of the speed needed to go into orbit, but the functioning of all the critical systems had to be tested and evaluated. Retrofire was completed, the pack jettisoned, and the spacecraft aligned for the correct reentry attitude. At about 38 mi. above Earth in the descent, Shepard noted the .05 g indication—the atmosphere was beginning to show its effect on the spacecraft. The g-forces increased to a maximum of 11 as the atmospheric friction slowed the descent to a subsonic velocity over the next few minutes. Suborbital spacecraft were equipped with a beryllium heat sink rather than the ablation type because they did not experience the extreme heat of an orbital reentry.

The drogue chute deployed on schedule at 20,000 ft., followed by the main chute at 12,000 ft. At 9:49 a.m., after a fifteen-minute and twenty-two-second flight, Shepard's *Freedom 7* splashed down into the Atlantic in sight of the recovery forces. Within fifteen minutes, he emerged from the recovery helicopter onto the deck of the aircraft carrier USS *Lake Champlain*. Shepard was in excellent physical and psychological condition, and except for a few minor glitches, the flight had been perfect. This was reflected in the news media that reported the expression "A-OK," which was used by the NASA press spokesman, Shorty Powers. It was assumed that Powers

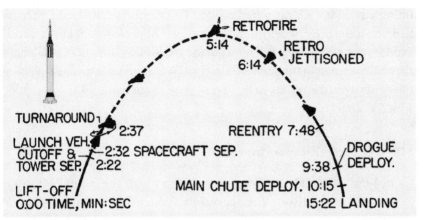

RETROFIRE
5:14
RETRO
6:14 JETTISONED

TURNAROUND
2:37
LAUNCH VEH.
CUTOFF & 2:32 SPACECRAFT SEP.
TOWER SEP. 2:22
REENTRY 7:48
DROGUE
9:38 DEPLOY.

LIFT-OFF
0:00 TIME, MIN:SEC
MAIN CHUTE DEPLOY. 10:15
15:22 LANDING

The suborbital trajectory of the Mercury-Redstone tested all of the critical systems.
Courtesy of NASA.

was relaying Shepard's exact words when in fact he was simply using his own jargon. However, "A-OK" had been indelibly impressed on the American lexicon of speech that day.

A few days later President Kennedy invited all the astronauts to the White House and awarded Shepard NASA's Distinguished Service Medal. There was a big parade down Pennsylvania Avenue. Not only was the nation excited about the flight but also the world responded most generously. Khrushchev was at first bewildered and then angered by the response. While the Soviets received gracious compliments on Yuri Gagarin's flight, there was not the same outpouring of emotion. But Shepard only went up and down—a short fifteen-minute flight, not around the world as Gagarin did—so why all the fuss?

The big distinction between Gagarin and Shepard was that, through the media, the world had become a part of the frustrating preparations as well as riding along with Shepard by way of Shorty Powers' real-time voice of Mission Control. They had suffered through the failures and postponements over the past two years. The American people (and to some extent the world) felt as though they had been a part of the drama. That was the difference.

However, the success that the Americans and the Soviets had achieved in their first flights was sorely tested in subsequent ventures as the capability and reliability of the various spacecraft systems were pitted against the unforgiving extremes of the environment of outer space.

As the space race moved into the summer of 1961, the Soviets still had a commanding lead in terms of spectacular accomplishments and weight lifting advantage. Gagarin had orbited once around Earth, and Shepard had flown the brief fifteen-minute suborbital hop. Neither flight had truly exposed the pilots to the full rigors of the space environment or challenged the ability of man to function effectively in space for more than ninety minutes. However, the results to that point were encouraging. This perception would change.

MR-4: Grissom Escapes Drowning

Capt. Virgil "Gus" Grissom was given the nod for the second Mercury-Redstone flight (MR-4). The spacecraft, No. 11, had several significant changes from Shepard's. The large window was now installed, as was the new quick-release hatch. The seventy bolts that secured the hatch each had a small hole drilled into them to permit the bolt to fracture when a thin explosive fuse that was wrapped around them was detonated. Activated by a plunger close to the pilot's right hand and requiring a force of 7 lb. to initiate, the new hatch saved 40 lb.

Grissom selected the name *Liberty Bell 7* as a patriotic symbol akin to the capsule's shape and the oneness of the seven astronauts. Grissom was the shortest of the astronauts and more reserved. His personality reflected a high degree of professionalism, and he was said to be the most sensitive of the astronauts and yet the most prone to use extreme profanity. Grissom entered the small spacecraft at 4:00 a.m. on July 21 for the planned 6:20 a.m. launch. Because of Shepard's experience, Grissom's spacesuit had been configured just the day before with a urine collection device.

A few short delays, including one for letting some clouds move through, resulted in ignition occurring at 7:20 a.m. "Ah Roger, this is *Liberty Bell 7*. The clock is operating," Grissom responded to the call of "liftoff" from Mission Control. Shepard, who was the CapCom for the flight, responded with, "Loud and clear, Jose, don't cry too much." A comedian of the day had created a nightclub routine that used a reluctant Spanish-accented astronaut to parody various aspects of a flight into space. It had been a great hit with the astronauts as they had gone through training and many of the lines had become a part of their communications vocabulary (Thompson 2004, 235).

The flight was textbook perfect, with a few changes in the procedures from Shepard's journey just ten weeks earlier. As the spacecraft splashed down into the Atlantic, Grissom began the process of disconnecting himself in preparation for egress—except for the oxygen hose inlet and the helmet communication lead. He began recording some of the instrument readings while the recovery helicopters positioned themselves for the pickup. Grissom indicated to the recovery team that he needed another five minutes before he was ready to exit. A few minutes later he called, "OK, latch on, then give me a call and I'll power-down and blow the hatch." Grissom had removed the cover from the hatch safety pin and had removed the pin itself. All that was needed was a firm punch on the plunger.

But before the helicopter could secure the capsule, Grissom heard a loud "bang," and the hatch disappeared revealing blue sky and green seawater pouring into the cockpit. Grissom removed his helmet and literally dove through the narrow opening into the ocean. The helicopter pilot and his recovery crew observed the premature hatch activation and saw that Grissom was apparently moving clear. They proceeded to snag the sinking spacecraft as Grissom looked on from a few yards away. When he saw the helicopter had hoisted the capsule from the water and the horse collar to pull him up being lowered, he swam toward it—when suddenly it started back up without him.

The helicopter pilot had observed an "over temperature" warning condition of the engine on the instrument panel. Decisions had to be made quickly now. The spacecraft had filled with water, and along with the impact bag, that was slowly draining its content, it represented about double the weight of a normal capsule at this point. The helicopter pilot was reluctant to release the valuable cargo, and as he pondered his next course of action, the helicopter sagged under the weight of the capsule toward the water, its landing gear awash.

The pilot released the spacecraft, advised the standby helicopter of the situation, and proceeded back to the carrier while the second helicopter moved in toward Grissom, whose plight had suddenly worsened. Grissom had failed to close the oxygen inlet valve on the suit, which normally provided buoyancy. The suit was filling with water, and Grissom could barely keep his head above the encroaching swells that were now breaking over and threatened to submerge him.

The second helicopter hovered over Grissom. With only seconds to spare before he sank beneath the surface, he slid into the recovery sling. In minutes, he was hoisted aboard the chopper and taken to the aircraft carrier USS *Randolph*, where his condition was determined to be excellent. The capsule sank in 15,000 ft. of water but was recovered—forty years later (Wendt and Still 2001, 202).

NASA convened an investigation board headed by astronaut Wally Schirra. There was no conclusive proof as to what had happened. Grissom

The second helicopter ultimately retrieves an exhausted Grissom following his MR-4 flight. Courtesy of Department of Defense.

affirms that he did not intentionally hit the plunger for the hatch release, but the stigma of having lost his spacecraft hung heavily around his neck. Unlike Shepard, he was not invited to the White House for presentation of his NASA award, and one of the stories that circulated was that he had panicked. There was never any indication of that type of behavior from his communications or biosensors. The experience embittered him (and his wife), turned him inward, and hardened him professionally. He continued in the astronaut corps and became respected for his resolve and engineering capability. Fate would deal him another blow before his career came to an untimely end.

To avoid a repetition of losing the spacecraft (and perhaps an astronaut), an auxiliary flotation collar was developed that was attached to the spacecraft by the helicopter-deployed swimmer teams. It not only stabilized the spacecraft in the water but it also provided a steady platform for the swimmers to assist the astronaut in the egress procedure.

The original plan was for each of the seven astronauts to fly a suborbital Redstone to give them a taste of rocket flight before the main event of going into orbit and experiencing the rigors of the Atlas. However, NASA management decided only a month after Shepard's flight to cancel MR-6 and await the results of MR-4 before deciding on MR-5. With the exception of the hatch problem, Grissom's flight had again proved all the capsule's systems. And while the lethal prospects of a hatch blowing while in space caused a detailed analysis to be made of the design and circuitry, NASA needed only a bit more prodding by the Russians to move on to orbital flight.

Vostok II: Titov Overcome by Space Sickness

From the outside, the Soviet space program appeared to have a well-orchestrated agenda; however, the actual flights proceeded on an ad hoc basis. Following Gagarin's flight, chief designer Sergei Korolev wanted to move directly to a twenty-four-hour mission. However, the medical team and the cosmonauts themselves, including Gagarin, who was now the commander of the cosmonaut group, preferred a less ambitious three-orbit flight. Nevertheless, Korolev took his argument to the State Committee for Defense Technology and forced the issue in his favor.

While twenty-five-year-old Maj. Gherman S. Titov was elated to be selected to fly the second mission, he was still disappointed that he had not been given the first. He was considered by most of his peers to be of superior intellect and more urbane in many ways than Gagarin. However, that same sophistication alienated him from his contemporaries and the professional staff. His individualism almost caused him to be dropped from consideration for the second flight, tentatively scheduled for August of 1961. However, the exact schedule of the flight was determined by other considerations.

At a meeting in mid-July with Khrushchev, Korolev indicated that the next flight would take place about mid-August. Khrushchev then told Korolev that, in fact, the flight would occur before August 10. Although Korolev was given no reason for this rather arbitrary requirement, Khrushchev had planned to isolate West Berlin behind a block wall, and construction of that wall was to begin on the August 13. Khrushchev was again looking to a "space spectacular" to strengthen his hand in dealing with the West (Siddiqi 2003b, 292).

The launch procedures moved ahead smoothly for Vostok II, and the call from the blockhouse of "She's off and running" came at 9:00 a.m. on August 6, 1961. The powered portion of the flight went as planned, but shortly after the spacecraft achieved its 110-mi. by 160-mi. orbit, Titov began to feel disoriented. He later recalled being in a "strange fog," unable to read his instruments or to orient himself with respect to Earth, which was visible from the porthole. He tried several head and body movements to shake off the symptoms that were like vertigo a pilot might experience following a series of maneuvers.

The spatial disorientation Titov was experiencing had been predicted by both Soviet and American medical specialists as an effect of weightlessness on the sensory structure for a person's balance. The lack of gravity on the small bones in the inner ear cause conflicting nerve impulses to be sent to the brain (the average person perceives only a slight sensation of weightlessness when in an elevator that begins a descent or on an amusement ride). Both the American and Soviet cosmonauts had undergone training to familiarize themselves with the weightless phenomena, but only periods of up to thirty seconds at a time could be achieved by putting an aircraft in a parabolic arc (a gradual pitch-over from a steep climb).

By the second orbit, Titov considered requesting a return to Earth, but at the time he was not communicating with Mission Control. When communication was reestablished, he decided to hold off. The biosensors relayed some indications to the medical staff on Earth that all was not well with Titov, and by the third orbit, they decided to make an inquiry. However, Titov was not about to give in to what he felt certain was a condition that would clear if only he gave it enough time. He reported, "Everything is in order" (Siddiqi 2003b, 293).

The television pictures beamed down to those on the ground were not of sufficient quality to perceive the anguish that Titov was experiencing, and they were available only during brief periods of the flight. At the start of the sixth orbit, Titov elected to eat a meal, per the schedule. However, after sampling some of the various gourmet foods prepared as a paste and squeezed from tubes, he became nauseous and vomited.

After a short rest period, he resumed the experimental schedule exercising attitude control for the first time with the manual thrusters. The seventh orbit started his sleep period, but since he was still fighting the spatial disorientation problem, his symptoms now included a severe headache that seemed, as he later reported, to permeate his eye sockets. At his next pass over Moscow he somewhat surprised the controllers with the comment, "Now I am going to lie down and sleep, you can think what you want, but I'm going to sleep." For the most part, his sleep period, which lasted more than six hours, was restful, but when he finally awoke and decided to resume his activity, he still felt poorly. He performed some head and eye movements as well as some experiments to determine his mental acuity and was surprised to find that he was more oriented than he had been at the start of the flight. However, an attempt to take in some fluids resulted in another bout of vomiting.

As Vostok II over flew selected third world countries, Titov transmitted greetings and read a statement extolling the virtues and technical superiority of the Communist state. As the exact schedule was not provided to these countries, few were actually able to hear the transmission. However, the Soviet consulate in each country ensured that the text of the prepared statement was made available to the local media.

By the twelfth orbit, Titov finally felt better, and his condition continued to improve as the flight moved to its termination at the start of the seventeenth orbit. Following retrofire, the separation of the descent module occurred with a loud "crack." However, unusual sounds and the movement of the spacecraft led Titov to believe that, like Gagarin's flight, the separation was not clean. During the early phases of reentry, the instrument section was again torn from the descent module and the remainder of the return was normal. Titov ejected and landed after twenty-five hours and eleven minutes in space (Siddiqi 2003b, 294).

Titov was truthful during the debriefing about his physical and mental state but was also elated about the opportunity to fly and his ability to "tough it out." There was a high level of anxiety in Korolev and the medical team about the possible causes and effects of what was called "space sickness." As the United States and Russia gained more experience with man-in-space, it was established that good physical conditioning and exposure to the weightless environment with numerous aircraft flights performing parabolic arcs might diminish the onset of space sickness. It was also concluded that the astronaut should avoid rapid head movements.

To certify the missions of both Gagarin and Titov as new world records, an application was made by the Soviets to the Fédération aéronautique internationale (FAI). However, the FAI requires the pilot to return from a

---

Experiencing Weightlessness

Astronauts are required to be exposed to, and train in, the zero-g environment before their first space flight. This is accomplished by flying parabolic arcs in an aircraft. In the early days, the Americans used a Convair C-131. The aircraft is put into a steep (35 degree) climb and then the pilot "pushes over" to match the gravitational acceleration of Earth. Periods of up to fifteen seconds of weightlessness can be induced, allowing the participants in the padded passenger compartment to float in zero-g, often causing nausea resulting in the aircraft's moniker—the Vomit Comet. At the end of the arc, the aircraft is in a 30-degree dive that requires a 2.5-g pull out. A typical flight will perform thirty to fifty of these maneuvers. Subsequently the Boeing KC-135 and then the Douglas C-9 (in 2003), assumed that role and could achieve up to thirty seconds of zero-g.

---

flight in his craft in order for the record to be valid. The Soviets were not about to reveal any aspect of the flight that might compromise security, but they were not going to miss holding the record for the first space flight because of a technicality. So Gagarin was forced to be creative with his responses to the press regarding those aspects of the flight—such as the use of the ejection seat—during its final phase. It was not until ten years later in 1971 that the truth came out, and by then no one cared. Another interesting anomaly in "squeezing the truth" occurred when the site of the launch had to be identified to the FAI. As the real location of Tyuratam was still a state secret, the name of the small obscure and far away town of Baikonur, some 230 miles to the northeast, was picked. That is how the Tyuratam launch site became known as Baikonur. As there were no photos released of the spaceship or the R-7, there was continued distrust around the world as to Soviet claims.

The fame that followed both Gagarin and Titov as they toured the world to spread the Communist gospel in the months following their epic flights was also the proverbial curse. The once straight-laced and easily embarrassed young Gagarin became a womanizer and liberally imbibed vodka. Both he and Titov were reprimanded by the Communist Party for their behavior, and these were their only flights into space. As for Gagarin, the Soviets desired to keep the first spaceman as a safeguarded national treasure; unfortunately, his life was cut short by an accident while flying a MiG-15 during a proficiency flight in 1968. For Titov, it may have been his personality that alienated him from his peers and the Communist ideology. There was, of course, no fame or recognition for the chief designer, Korolev, who was protected by anonymity lest the United States find out his identity and attempt to have him assassinated.

Final Preparations for Mercury-Atlas

With the daylong flight of Vostok-2, NASA recognized that the Soviets had already exceeded the basic capabilities of the Mercury spacecraft. The pressure was on to "get an American in orbit," and any tentative plans for another Redstone following Grissom's flight were discarded. The failure of MA-3 in April had put the orbital attempt behind schedule. MA-4, the second thick-skinned Atlas D, was erected at the Cape and was mated with capsule No. 8 recovered from the MA-3 flight. Then a significant problem with some of the transistors in both the capsule and the booster electronics required their demating, and several weeks were lost changing out the suspect components.

Atlas 88-D finally erupted to life on September 13, 1961, and quickly accelerated the spacecraft and its "mechanical astronaut" into orbit. This was the first time the Atlas had ever lifted that weight to orbital velocity and only the second time it had been used to orbit a payload without an upper stage (the first being Project Score almost three years earlier). The booster engine cutoff (BECO) occurred slightly earlier than programmed, as did the sustainer engine cutoff (SECO); however, the high levels of performance of the engines provided the required energy.

The spacecraft itself used an automatic sequencer to provide all of the control and timing of the various events. The "mechanical astronaut" was actually a series of devices to monitor the environmental controls to ensure that oxygen replacement, humidity, and communications functioned properly. A single orbit would essentially prove all of the systems in the hostile space environment without inducing a prolonged opportunity for problems to occur. While the most obvious tests centered on the spacecraft and the performance of the booster, the entire manned spacecraft tracking and data transmission network was exercised for the first time.

For NASA, it was critical that an "almost real-time" monitoring of the spacecraft and its "mechanical astronaut" take place to provide the highest levels of safety. All of the seven astronauts were involved in the MA-4 test, with four situated at tracking sites around the world and three working at the Cape itself. As the spacecraft flew over each of the stations, the data streaming down was retransmitted via terrestrial links back to the Goddard Space Flight Center in Greenbelt, Maryland, at data rates of one thousand bits-per-second (pitifully slow by today's standards).

All of the planned orbital events went off at the required intervals, and the spacecraft was soon back on Earth following its ninety-minute, 25,000 mi. journey. Of greatest interest for many was the fact that the ablative heat shield had survived the reentry; the other tests with Big Joe had been sub-orbital and had not induced the full duration of heat as this orbital test did. NASA was elated. The American public (and the rest of the world), while thankful, saw the test as simply another hurdle to catching up with the Russians and not a space spectacular. One more test separated America from a manned orbital flight, and it was still possible that an American would orbit the Earth in the same calendar year as a Russian.

Using a chimpanzee, MA-5 was a final dress rehearsal for a manned flight and was planned for three orbits. Atlas 93-D arrived at the Cape in early October, but its launch was delayed as ungraded electronics were installed. Spacecraft No. 9 had been on hand since February, and it was quickly mated for a launch scheduled for November 7. However, problems continued to plague the project, and it was not until November 29, 1961, that MA-5 sat fueled and ready for launch.

The occupant was a chimpanzee named Enos. Like Ham of MR-2 fame, Enos had gone through an intensive preparation period that consisted of more than 1,200 hours of training over a period of sixteen months. Chimps were used in America's space program because they had the ability to learn a variety of tasks that could be performed at critical times during a flight in order to determine the ability of a primate to make cognitive functions under the same conditions that the astronaut would face. Four distinct activities were planned for Enos. One required that he pull a lever exactly fifty times to be rewarded with banana pellets. It was observed during training that he quickly pulled the handle forty-five times before slowing and making the last five pulls more deliberately until the fiftieth, when he placed his hand under the dispenser in anticipation.

Those who worked with the chimps felt the animals enjoyed the challenge and tolerated the uncomfortable aspects such as the centrifuge rides in anticipation of the rewards that followed. Unlike humans, the chimps showed no physiological or emotional indications of anxiety even while being prepared for an uncomfortable test they had previously encountered.

The Atlas lifted into the morning sky and powered itself into orbit over the next five minutes. Enos proceeded through the series of planned exercises with his actions monitored by telemetry and recorded by an onboard 16 mm movie camera—one of four that filmed various aspects of the flight.

President Kennedy was in the middle of a press conference when a note was passed to him of the launch. His quick wit was displayed to the pleasure of all when he looked up after reading the note and announced, "The chimpanzee took off at eight past ten." He paused just slightly and then added, "And he reports that everything is perfect and is working well."

A variety of problems plagued the mission after the first orbit. The device designed to give a mild shock to the chimp malfunctioned and buzzed the slightly confused primate even when he was doing his tasks correctly. Enos took it all in stride and continued to perform. An inverter that converts direct current to alternating current for the electrical system showed indications of overheating, but there was a backup if the first one failed.

More critical was a drift observed with respect to the spacecraft's attitude; it was consuming excessive thruster fuel and could deplete the supply if all three orbits were flown. At the last minute, the decision was made to bring Enos back at the end of the second orbit. As the decision was being relayed to begin reentry, a farmer in Arizona ran his plow through the buried communication line. But that is why backup capabilities were provided, and within seconds an alternate link reestablished communications. MA-5 recovery was made in the Atlantic without mishap (Kleinknecht 1963, 12).

In the postflight analysis, all of the problems were carefully evaluated, and the most serious, the attitude control, was traced to a small metal shaving that blocked a thruster fuel line. In a humorous political cartoon that appeared in a newspaper the following day, a chimp in a spacesuit is depicted walking away from his spacecraft with the caption reading, "We're a little behind the Russians but a little ahead of the Americans." It was time for America to orbit a man.

"A Real Fireball out There": MA-6 and John Glenn

The schedule called for the launch of MA-6 on December 19, 1961, but with only one Atlas launchpad allocated to Project Mercury, the time required refurbishing the pad after MA-5, and erecting MA-6 would run the mission into the Christmas holidays. Mercury project manager Robert Gilruth, after consultation with key members of NASA and his management staff, felt that keeping the thousands of government employees and contractors working through Christmas was inappropriate. Moreover, if the past was any indicator of the future, there was no guarantee that the launch would come in calendar year 1961 in any case.

If a lack of broad-based technology slowed the Soviets, America's impediment was its need for a miniaturized high-tech environment and the challenges that resulted. Meticulous tracking of any change, no matter how small, required paperwork, approvals, and confirmations, and the number of changes that transpired for the Mercury spacecraft were considerable. Not only was a new type of flight vehicle being engineered but it also had to be reconfigured depending on the specific requirements of each mission. And the fact that only one technician could be working in the capsule at any one time meant that the installation of each change was time consuming.

Spacecraft No. 13 was built in May of 1960 and then reworked to the new specifications, to include the larger window and quick-opening hatch. When, in April of 1961, it was determined that it might be the first manned orbital mission then set for late that year, it went through another set of changes before finally being delivered to the Cape in August of 1961—after which two hundred more changes were made before it finally flew. Many of these changes were based on the early flight experiences while others were a result of more exhaustive testing and the desire to provide the safest, most reliable vehicle.

Marine lieutenant colonel John H. Glenn had been among the three finalists for the Mercury-Redstone flights selected the previous February and would have been the pilot for MR-5, had there been one. Therefore, the first orbital flight of Mercury officially became his mission as soon as Enos had returned from orbit. Glenn was the oldest astronaut, then age forty. He had a long and distinguished career beginning with his Naval Aviation Cadet Training in March of 1942. He flew fifty-nine combat missions in the Pacific Theater of World War II and another sixty-three during the Korean War, where he shot down three MiG-15s in the last nine days of that conflict. He returned to the United States to attend the U.S. Naval Test Pilot School at the Naval Air Test Center, Patuxent River, Maryland.

Glenn came into the public eye in July 1957, when he flew the Navy's new F8U Crusader fighter/reconnaissance jet from Los Angeles to New York in three hours and twenty-three minutes—the first transcontinental flight to average supersonic speed. He was well liked by his fellow astronauts although he sometimes rubbed them the wrong way because he was

John Glenn enters his *Friendship 7* spacecraft on February 20, 1962. The tape covers possible sharp edges that might cut the spacesuit during entry. Courtesy of NASA.

straight-laced and was known to preach about good moral character to his more lusty fellow astronauts—he was not the typical Marine.

The 4,250 lb. spacecraft, No. 13, christened *Friendship 7* by Glenn, was mated to Atlas 109D on January 3, 1962, to complete the MA-6 configuration. The first flight date of January 23 slipped to January 27 because of problems with the Atlas. After lying on his back for over five hours in the spacecraft, the shot was postponed on that date because of bad weather. During the recycling process, a propellant leak discovered in the Atlas caused the launch to slip another two weeks. Then still more problems and bad weather moved the launch date out another few days. Because of the need for acceptable weather at three primary locations—the Cape, the abort area, and the recovery area—weather played a major role in many American space flights.

Glenn was awakened at 2:20 a.m. on February 20, 1962. There had been speculation in the press that the lengthy delays might have put extreme psychological stress on Glenn and that the anticipation might be more than any man could bear. However, the press and the public had much to learn about the emotional makeup of the Mercury 7 astronauts. There was no problem on Glenn's part as he ate breakfast, had his vital signs examined, and suited up for the flight. Only the addition of the blood pressure cuff on his right arm differentiated Glenn's preparation from that of Grissom and Shepard.

When Glenn arrived at the launchpad, a problem with the launch vehicle delayed by one hour his elevator ride to the top of the giant gantry for the squeeze into the spacecraft. Then, with the various connections made between the spacesuit and the spacecraft, the "closeout" procedures were completed. However, a broken bolt, one of the seventy that fastened the hatch, had to be replaced—more delay.

Finally, at 9:47 a.m., the missile came to life when the countdown reached zero! From launchpad 14, the great silver rocket, with its band of pure white frost defining the location of the liquid oxygen tank, began to rise, balanced on its orange-sheathed white-hot flame. Leaving its colorful backdrop of the red service tower and surrounding green shrubs, it accelerated into the blue morning sky. "Roger, the clock is operating. We're underway," Glenn announced.

"Godspeed John Glenn," replied the voice of CapCom Scott Carpenter.

Riding the Atlas was a new experience, as it was unlike the Redstone; indeed, it was unlike the Russian R-7. Glenn recalled that he could feel the small corrections that the powerful engines induced as they moved (gimbaled) to balance the 250,000 lb. rocket. He described it as being at the end of a long swaying pole. However, this pole was also vibrating as it surged into the atmosphere, building g-forces on Glenn and creating an aerodynamic shock wave as it pushed the air out of its way. "A little bumpy along about here," Glenn reported in the understated language of a test pilot. Through Max Q, thirty seconds into the flight, the vibrations smoothed a bit.

At T+129 seconds, booster engine cutoff occurred and the two booster engines slid away as the single sustainer continued to power the spacecraft into the sky, which had now turned from a bright blue through dark blue and quickly into the blackness of space. The g-forces dropped from six back to three with the change in thrust at staging, and then built again as the rocket consumed it propellants at an astonishing rate.

The escape tower appeared to jettison early as Glenn reported, "I saw the smoke go by the window," but that was a visual queue from some other unnamed event, and the tower finally left on schedule and was plainly visible. "There, the tower went right then! Have the tower in sight way out." This was an important event because it indicated that a malfunction at this point was handled by simply separating the spacecraft from the Atlas. It was also

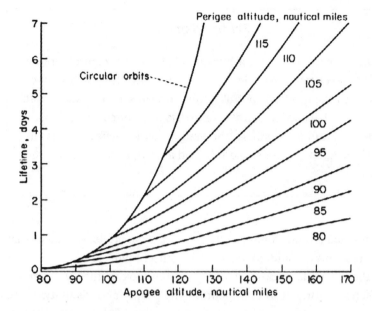

Perigee altitude, nautical miles

Apogee altitude, nautical miles

Approximate orbital lifetime in days (vertical axis) based on apogee (horizontal axis). Glenn's parameters indicate about an eighteen-hour (twelve orbit) duration. Courtesy of NASA.

a critical event because the weight of the tower had to be shed in order for the Atlas to reach orbital speed.

Glenn continued to report to Mission Control not only his observations of the launch phenomena but also readouts from the various systems. At T+280 seconds Mission Control responded, "Roger, twenty seconds to SECO [sustainer engine cutoff]"—it was as much a wish that all that was now needed to achieve a milestone was within their grasp. Sustainer engine cutoff was an indisputable event as the 7.7 gs and the sound of the engine suddenly disappeared. Glenn then commented, "Roger, the capsule is turning around and I can see the booster doing turnarounds a couple hundred [yards] behind me. It was beautiful!" The experience of sight and sound has to be one of the most enthralling that a human has ever witnessed.

Mission Control now related another important item. "Roger Seven, you have a go, at least seven orbits." This particular communiqué has been often misinterpreted as the intended length of the mission. This was not the case. It had always been planned for three orbits. The statement was to advise Glenn that the computer at Goddard Space Flight Center had analyzed the velocity vectors imparted by the Atlas, concluded that the low 87 mi. by 141

204 · The History of Human Space Flight

Fly-by-Wire

In previous aircraft systems, the movement of the flight control surfaces was by a direct mechanical linkage. This was also the primary mode of actuating the reaction control system thrusters in early spacecraft. However, as a backup, electrical wires to the thruster-control solenoids provided a redundant path. As the ease of routing wires was less intrusive and lighter, newer spacecraft (and aircraft) became completely "fly-by-wire."

---

mi. orbit was stable, and would result in at least seven orbits—ensuring that a premature reentry would not occur (Kleinknecht 1963, 12).

From the Bermuda tracking station, Grissom took up the CapCom duties as Glenn continued to comment on the beauty and visibility. More instrument readouts followed as the capsule maintained its rearward facing, 34-degree pitch-down attitude with the automated control system until a series of telemetry readings to Goddard confirmed the integrity of all the spacecraft systems. Over Africa, Glenn switched to fly-by-wire attitude control and reported that he had no problem positioning the spacecraft using outside references, especially when the pitch down was increased to about 60 degrees.

Moving out across the Indian Ocean, and now communicating through one of the tracking ships, he moved into his first sunset. Glenn again was in awe of the visual impact of his high-flying perspective: "The sunset was beautiful. It went down very rapidly," and he then added, "I have no trouble seeing the night horizon," and speculated that moonshine from behind him was probably illuminating that portion of Earth.

Halfway around Earth from the start of his journey, he continued to make observations while out of range of the tracking ship, but they were recorded by an onboard tape unit: "Friendship Seven broadcasting in the blind, making observations on night outside . . . I can identify Aries and Triangulum (constellations of stars)."

Over Australia Glenn observed the lights of the city of Perth, as its residents turned on as many lights as possible in their early morning hours for his viewing. He asked CapCom Gordon Cooper at the Muchea Tracking Station to thank the people of Perth. Then his first sunrise in space evoked the comment, "Oh the Sun is coming up behind me in the periscope, a brilliant, brilliant red." He then proceeded to describe thousands of brilliant luminescent particles (that became known as "Glenn's fireflies") that seemed to swirl around the spacecraft but quickly disappeared when full sunlight bathed *Friendship 7*.

Glenn now encountered the first of several problems. The spacecraft was not holding its position relative to the yaw axis (left and right), and he concluded the low power (1 lb.) left thruster was not functioning. There were two sets of thrusters for each axis, and the low-powered units were used for small adjustments. However, with the left thruster out, the automated control system used the larger thruster when the yaw passed the 20 degree tolerance point. Glenn switched back to the manual "fly-by-wire" system so that he would not consume so much hydrogen peroxide attitude-control fuel.

At about this time, two hours into the flight, Goddard Space Flight Center noted a telemetry code of "Segment 51," an indication that the landing bag had deployed. As the bag was contained between the heat shield and the bulkhead of the capsule, it could only mean that the heat shield had unlatched—if the Segment 51 indication was correct. The retrorocket package with its three metal straps was probably still in place over the heat shield and secured directly to the bulkhead of the spacecraft. However, when that unit was released (after retrofire) the heat shield would separate from the bulkhead—if it was loose. If this occurred, it would undoubtedly be torn away during the reentry and the spacecraft would burn up from the incinerating 1,600 °C temperatures.

Not wanting to alarm the astronaut, Mission Control proceeded to make oblique queries as to the status lights that Glenn was observing and on occasion asked if he heard any noises when the capsule attitude was changed by the thrusters. Back at Mission Control, queries were made to determine if the Segment 51 signal was valid and possible alternatives if it was. The validity of the signal could not be confirmed, and they thought that it was probably erroneous. However, Flight Director Chris Kraft could not make that assumption, and a contingency plan was examined to leave the retrorocket package in place after it fired. It was felt that the straps would hold the heat shield in place until enough aerodynamic force to keep it there after the straps for the retropack melted through.

However, there had been no evaluation made as to what the effect of having the retropack in place might do to the heat shield because of an uneven heat pulse generated by the module. Moreover, this was only an option if all three of the retro's solid-fuel rockets fired. If one failed to fire, then the reentry heat would eventually ignite it, and its exhaust might burn through the heat shield or cause the spacecraft to assume an attitude that allowed the heat pulse to impinge on the side of the spacecraft.

Although Glenn had responded to the landing bag light queries, Mission Control had not confided in him their worst thoughts. With all of the rapid communications going on, Canton Tracking Station assumed that Glenn had been made aware of the situation and finally let the cat out of the bag. "Friendship Seven, this is Canton. We also have no indication that your landing bag might be deployed. Over." Glen immediately responded, "Did someone report landing bag could be down? Over." "Negative," reported Canton, "we had a request to monitor this and to ask if you heard any flapping." Now some of the other communications that Glenn had fielded began to make sense to him, but he thought perhaps the "fireflies" he had reported had started the queries, and he again gave more description of them.

Over Hawaii, as the time for retrofire approached, Glenn was told of the Segment 51 indication and given some more checks to make. They came back negative. There was no confirmation that the bag had deployed. Glenn put *Friendship 7* into the retrofire attitude (flying backward with the capsule pitched nose down by 34 degrees), and the three solid-fuel rockets fired as planned. Now Flight Director Kraft made his decision, and it was communicated by Texas CapCom to Glenn. "We are recommending that you leave the retro-package on through the entire reentry." Now Glenn wanted the complete story. "What is the reason for this? Do you have any reason? Over." However, all he got was, "Not at this time; this was the judgment of Cape Flight" (Glenn 1999, 271).

With only about four minutes until the ionization sheath of reentry blocked radio communications, Glenn realized any clarification had to be swift. It was: as the capsule came into range of the Cape CapCom, the answer was, "we are not sure whether or not your landing bag has deployed. We feel it is possible to reenter with the retro-package on." There was no time for Glenn to reflect on the engineering aspects of a statement that began with "we feel" (Kranz 2000, 71). The Cape advised Glenn to perform some of the reentry tasks manually, such as the periscope retraction, to bypass the retro jettison event. The first of these occurred with the detection of the upper fringes of the atmosphere when the .05 g sensor light flashed on.

Now another problem confronted him—only 15 percent of his attitude-control fuel remained, and the most critical part of the mission was ahead of him. The pitch attitude of *Friendship 7* had to be held to within 10 degrees during reentry. With the realization that there might be a serious problem with the heat shield, Glenn elected to concentrate on those things that he could control and to move those that he couldn't out of his conscious thoughts.

As the g-forces and ominous sounds began to build, the super-heated ionized air choked off communications with Mission Control. Glenn was on his own now, and he could see the orange glow surrounding his capsule plainly through the window. A part of the energy that the Atlas rocket had generated to get him into orbit at 17,500 mph now had to be dissipated by friction with the upper atmosphere for his return. He watched as molten pieces of what he hoped were the retropack began flashing past, and he could feel the heat building on his back—which was less than a foot from the searing temperature of the heat shield. He also kept making his observations into the tape recorder with one comment that seemed to sum up his feelings of the moment. "That's a real fireball out there!"

As the glow outside the window subsided, Glenn thought he had made it through the toughest part. But just then the fuel supply became exhausted, allowing the capsule to begin to oscillate. He debated deploying the drogue chute to help stabilize the craft as he had now decelerated to less than 700 mph. However, the automatic sequencer beat him to it, and the main chute soon followed. He established communication with the recovery forces and again repeated his observations of the reentry: "My condition is good, but that was a real fireball, boy. I had great chunks of that retropack breaking off all the way through."

The landing was uneventful, and within minutes the U.S. Navy destroyer USS *Noah* had retrieved *Friendship 7*. Glenn was eager to exit the hot confines of the spacecraft as it sat on the deck of the ship. Rather than take the long way out through the egress hatch at the top, Glenn advised the crew to stand clear, and he blew the hatch. In striking the plunger to initiate the explosive charge, Glenn bruised the palm of his hand and cut his knuckles. This had also happened during training, and it was a clear indication that Grissom (whose hands showed no injury) had not used the plunger during his ill-fated landing in *Liberty Bell 7*.

America was ecstatic with the results of the MA-6 mission. The entire country had come to a virtual standstill during the four-hour- and fifty-five-minute flight as many stayed home from school and work to keep up with the events as they unfolded. Even though there was no live television broadcast of the spacecraft in orbit or of the splashdown, the running commentary of Col. Shorty Powers and the "talking heads" of enthusiastic

A comparison of the Vostok and Mercury spacecraft. Note the use of a shroud (dotted line) for Vostok to protect sensitive components (such as antennas) during ascent. With Mercury, all systems were imbedded and hardened within the aerodynamic structure. Courtesy of Ted Spitzmiller.

newscasters such as Walter Cronkite kept the nation and the world appraised of each aspect of the flight.

During the debriefing, it was clear that Glenn was not pleased that Mission Control had not kept him informed of the heat shield problem. Nevertheless, when the various factors were considered, such as the truly unknown environment in which Glenn was flying, the decisions made by Mission Control appeared reasonable. Still, Glenn and the other astronauts made it clear that, from that point on, the pilot in command of the spacecraft must be given all of the facts related to the flight so that a balanced evaluation could be made. As for the Segment 51 light, it turned out to be an erroneous indication that resulted from a loose rotary switch. The thruster problem was caused from a small metal shaving, just as it had with Enos in MA-5. Better clean-room facilities and a more cautious work ethic were emphasized.

Glenn's flight had clearly demonstrated that humans could function in space. He established detailed and accurate observations, cognitive evaluation, and judgment. While Titov had experienced space sickness and had been somewhat incapacitated for more than half of his mission, he too had prevailed emotionally until his physiology had stabilized.

John Glenn, accompanied by the other six astronauts, took part in a big ticker-tape parade down Broadway in the heart of New York. The cheers effectively drowned out the skeptics and those who still felt that a massive manned space program was the wrong response to the Russian threat. However, the fact remained that both the United States and the Soviets had encountered potentially serious problems in their space flights. That neither had lost an astronaut was fortuitous. America moved ahead cautiously, but the Soviets needed to continue to demonstrate a commanding lead.

Glenn returned briefly to the space program thirty-six years later at the age of seventy-eight. He participated in an experiment on board *Discovery* (STS-95) related to the physiology of the human aging process. He orbited the Earth 134 times and experienced only 3 gs on launch and reentry.

10

EXPLORING THE UNKNOWN

With Glenn's flight completed, albeit ten months after Gagarin's, the United States breathed a bit easier. The Soviets had not given any indication that had Titov suffered disorientation—space sickness. Despite the careful scripting of each flight, both countries would learn that there were many unknowns yet to be encountered.

MA-7: "I'm out of manual fuel . . ."—Scott Carpenter

At the same time Glenn was named to MA-6 in November 1961, Deke Slayton was assigned to MA-7. However, Slayton had displayed a mild idiopathic atrial fibrillation during a routine examination—essentially an irregularity in the heart's ability to cope with stress. NASA had sent Slayton to several specialists, and the results had always come back the same; he was physically fit to fly. Nevertheless, NASA continued to elicit medical opinions until finally one summary concluded that so long as there were other astronauts equally qualified who did not have any problem, Slayton should not fly.

Slayton did not accept the decision gracefully. However, realizing he was up against a power structure that could not be moved, he resigned himself to performing a ground-based job and adapt. He was assigned the role of chief astronaut—the man who would select, manage, and discipline the astronaut corps for more than ten years. Known as a tough, decisive taskmaster, his opportunity to fly in space would eventually come, but for now his slot for the next orbital mission passed to Navy lieutenant Malcolm Scott Carpenter. Carpenter had the least jet time of any of the astronauts because he was primarily a multiengine patrol-bomber pilot since completing naval flight training in April 1951.

Although MA-7 was to be an identical flight profile to MA-6, Carpenter was to expand the role of man-in-space, and to that end, the schedule of scientific and engineering experiments was impressive. America, despite the desperation of the space race, was pressing to do science in space. The only major change in the spacecraft configuration was the deletion of the Earth-path indicator since Glenn indicated that it was relatively easy to determine position from outside references, and the communications network had worked well.

Spacecraft No. 18, mated with Atlas 107D, was named *Aurora 7* by Carpenter for the light that this mission would shine on the dawn of a new age. The launch, originally set for May 15, 1962, was delayed by problems with the booster, the spacecraft attitude control system, and modifications to the parachute deployment mechanism until May 24. Unlike previous Mercury-Atlas flights, the countdown went smoothly, and an early-morning liftoff occurred at 7:45 a.m.

Carpenter was not as expressive as Glenn had been and even more understated as he passed through Max Q. As each booster had its own personality, it is possible that Carpenter was given a smoother ride than Glenn. On arrival in orbit, Carpenter noted that the horizon sensor was off by about 20 degrees—a significant amount (Carpenter and Stoever 2002, 257). He also had to adjust the suit temperature continually, which ran on the high side. With these distractions, Carpenter set about accomplishing each of the tasks on the flight plan, and by the end of the first orbit, he had kept up with the schedule.

The Russians had by this time given some indications that Titov had experienced some space sickness, but Carpenter felt completely at home in this new and strange environment, just as Glenn had been. Some attribute this to the extensive weightless training the American astronauts received and the fact that the Mercury capsule had a smaller interior so the astronaut felt more an integral part of the ship rather than a passenger. Carpenter even tried to induce vertigo by performing sharp head movements. His visual acuity appeared to improve in space, perhaps because of the weightless condition, and he reported being able to see the dust trail from trucks driving along unimproved roads, and the wake of ships at sea.

Carpenter was able to observe Glenn's "fireflies," and this took more of his time than he realized. He determined that, if he hit the inside of the spacecraft, he could generate the luminous particles and that the hydrogen peroxide thrusters also generated them. "I can rap the hatch and stir off hundreds of them," he reported. He soon found himself behind schedule with all of the activities (some of which were poorly planned by the experimenters). A few times he accidentally bumped the manual control system causing it and the automatic system both to be active, which was depleting the thruster fuel more quickly than anticipated.

By the time he was ready for retrofire, there were several uncompleted tasks left to prepare the spacecraft for reentry. Moreover, when the retro-rockets failed to fire by the automatic sequencer, Carpenter's manual response placed the ignition about three seconds late. This in itself was not a major problem. However, the fact that the spacecraft was not properly aligned (being about 25 degrees off from the proper azimuth) meant that the retros were pointed to the left of the flight path. The combination of all of these factors put his splashdown almost 300 mi. beyond the intended point where the recovery forces were located.

As he passed over the West Coast of the United States, he ominously reported to CapCom Alan Shepard, "I'm out of manual fuel, Al" (Carpenter and Stoever 2002, 286). Carpenter was potentially in big trouble. He allowed the spacecraft to drift and waited for the .05-g indication before attempting to align the ship to the proper reentry attitude. He realized that at any moment the fuel supply for the automated system, which he was using in the fly-by-wire mode, would also become exhausted. However, by the time that occurred, the spacecraft had driven itself into the denser layers of the atmosphere and had achieved a degree of stabilization. As the spaceship became subsonic, it oscillated, as Glenn's had done, and Carpenter elected to deploy the drogue chute early.

The last communication he had heard from the Cape was that he would be landing long and recovery forces were an hour away. After splashdown, Carpenter elected to egress through the top hatch to avoid any possibility of losing his capsule as Grissom had. He inflated the life raft and proceeded to climb in. He noted, as his feet began to get wet, that he too had neglected to shut the suit inlet hose valve, but he quickly remedied the problem. He lay in the raft for over two hours before being retrieved by helicopter.

While his flight was essentially a complete success, Carpenter's performance came into question. That he had been given the most demanding experimental schedule—which, with few exceptions, he had handled quite well—did not seem to impress NASA management. That he had run out of fuel during reentry and had landed long caused considerable distress to those whose heads would roll (if he had not been successfully recovered) seemed to be the primary focus. His independent spirit also played a role, as his attitude appeared to be too cavalier to suit the more conservative elements in Mission Control (Kranz 2000, 91).

Carpenter never again flew in space. He recognized the warning signs and subsequently took a leave of absence from NASA to participate in the Navy's Man-in-the-Sea Project as an aquanaut in the SEALAB II program in the summer of 1965.

Vostok-3 and Vostok-4: First Group Flight

Following Titov's flight, the emphasis for the available R-7s and Vostok spacecraft (of which eighteen were in various stages of construction) was switched to the Zenit reconnaissance satellite. However, Korolev continued to remain focused on human space flight. He envisioned that each flight must be a significant step beyond the previous one—Titov's experiences notwithstanding. Within a month after completing Titov's flight, Korolev began planning a group flight of three Vostoks launched on successive days, with the longest remaining in space for at least three days.

However, Korolev continued to be restrained by others who felt that two ships were all that the current tracking and recovery facilities could effectively handle and that two days should be the limit because of the unknowns revealed by Titov's space sickness. Moreover, although Korolev was

The Vostok spacecraft as it appears in a museum, attached to the upper stage. Courtesy of Energiya.

eager to launch as soon as the Zenit schedule freed up the launch facilities, the spectacular success of Glenn's flight in February of 1962 suddenly put pressure from above to accelerate the next Vostok mission. The enthusiasm shown by the world's media toward the openness of the United States' efforts put the Soviets in a defensive position. The edict to launch by the middle of March 1962 was unrealistic, but the process was begun.

Two Zenit satellites had been planned before the next Vostok booster could be erected on the launchpad, and these were delayed by a series of problems. Zenit failures in December 1961 and again in January 1962 put that program behind schedule. The first was finally in orbit on April 26, 1962, but the next, on June 1, experienced a catastrophic failure that damaged the launchpad, which required more than a month of rework before it was again operational.

The objective of the group flight, as it evolved, was for the two spacecraft to be launched in such a manner as to pass close to each other. Because Vostok had no onboard propulsion, they could not actively rendezvous. However, even the possibility of the two cosmonauts being able to observe each other's ship would be a monumental achievement. The length of the mission was also still cloudy as Korolev pushed for three to four days, but it was officially to be a two-day flight at that point.

The cosmonauts were chosen less than two weeks before launch and included Capt. Andriyan G. Nikolayev and Maj. Pavel R. Popovich, who were both then thirty-two years old. Nikolayev was launched first in Vostok III on August 11 at 11:30 a.m. into an orbit of 113 mi. by 147 mi. and an inclination of 64.98 degrees. The medical team at Tyuratam closely monitored his condition, and the world was informed of the ongoing mission after he completed the first orbit. Nikolayev reported no problems, and on a subsequent orbit, the flight plan called for him to unstrap himself so he could float freely within the cabin (something that the smaller Mercury capsule did not allow). In an effort to blunt some of the criticism of Soviet secrecy, live TV pictures of Nikolayev were routed for broadcast within the USSR. There was no capability to provide live coverage across the world at that time.

As soon as the rocket had left the pad, the launch crew began its restoration, and the Vostok-4 booster rolled out. Popovich was launched at 11:02 a.m. on August 12 into a 112 mi. by 148 mi. orbit with an inclination of 64.95 degrees. The precision with which Korolev's team was able to launch and the orbital insertion parameters achieved were impressive. The world was not told the nature of the mission nor that rendezvous was out of the question. The Soviets were not about to clarify their intentions. The media assumed that the two ships would rendezvous, which gave the Soviets another dramatic and technological first (Zimmerman 1998, 101). Korolev's first deputy, Vasily Mishin, noted in later years: "With all the secrecy, we didn't tell the whole truth, the Western experts, who hadn't figured it out, thought that our Vostok was already equipped with orbital approach equipment. As they say, a sleight of hand isn't any kind of fraud. It was more like our competitors deceived themselves all by their lonesome. Of course, we didn't shatter their illusions." (Siddiqi 2003b, 361). The two ships never came closer than three miles and then gradually drifted farther apart.

Because Nikolayev's physical condition was not showing any signs of space sickness, the decision was finally made to allow him to go for the third day, and after the completion of that, there was more disagreement over the fourth day, which he finally was allowed to complete. Khrushchev requested a fifth day, but several problems, such as a steadily decreasing temperature in the spacecraft, finally prompted the recovery on August 15, after three days and twenty-two hours. Popovich followed just six minutes later, after two days and twenty-two hours. Korolev had again upstaged the United States.

MA-8: "She's riding beautiful"—Wally Schirra

The Mercury spacecraft had been engineered to get an American into orbit—but three orbits, back in 1959 when Mercury was being designed, was a big step. However, Titov's daylong, seventeen-orbit flight had upped the ante, and NASA carefully reviewed the Mercury design to see if it could provide for more extended operations. Simply another three-orbit flight would prove little and would be viewed by the world as evidence that America's technology was still woefully behind the Russians. Thus, MA-8 sought to expand the endurance capabilities until the next generation spacecraft was available.

The obvious problem in increasing the time in space was the availability of the consumables. Most obvious was the attitude-control fuel—the hydrogen peroxide had been totally depleted by both Glenn and Carpenter. However, the environmental supplies of oxygen, water for the cooling system, and lithium hydroxide for removing the carbon dioxide from the air were also limited. Finally, the batteries for the electrical systems could not be readily extended, especially since the Atlas was already at the limit of its ability to boost the 3,000 lb. craft into orbit; the physical confines of the capsule left virtually no room for any appreciable expansion.

With respect to the hydrogen peroxide, there were several approaches to making it last longer. First was more discipline in its use and the structuring of the flight plan and the experiments. Drifting was an option for extended periods. The tolerance that the ship could be allowed to drift (called the deadband range) when a desired attitude was commanded was expanded from ±3 degrees to ±6 degrees. The electrical system had several components that could be powered down when not in use, and it was determined that this technique itself doubled the battery life.

The biggest problem with the oxygen was the leakage rate of the spacecraft itself, which was replenished from the two cylinders. The initial design requirement of 1,000 cc/min in the pressurized mode (in space) was reduced to 600 cc/min., and 460 cc/min was actually achieved.

Despite all the effort, the weight of spacecraft No. 16 increased by almost 50 lb. as it was mated to Atlas 113D.

Navy lieutenant commander Wally Schirra had been selected more than a year earlier for the flight and had served as the backup for Carpenter. He could be terse and outspoken, but, like Grissom, he held a strong belief in good engineering principles. A Naval Academy graduate, class of 1945, he flew ninety missions during the Korean War and received the Distinguished Flying Cross. He was perhaps the most well-prepared astronaut to date, as he had spent countless hours in the simulators as well as the spacecraft itself, going through the various activities. He was determined to fly the perfect mission.

Aiming for a September launch, the Americans were once again upstaged by Korolev with the group flight of Vostok-3/4 and its four-day mission. A series of slippages finally saw *Sigma 7*, as Schirra had named the capsule in honor of the sum total of the American effort, take to the sky on October 3, 1962. "Ah, she's riding beautiful, Deke," he commented to CapCom as he "slipped the surly bonds of Earth."

The booster engines did not perform well, and the sustainer had to fire for a longer period to make up for the deficiency—using fuel that the boosters had not burned. Sustainer engine cutoff occurred fifteen seconds later than in the previous missions. The velocity vector was sufficient to place Schirra into an 87 mi. by 153 mi. orbit (Kleinknecht 1963, 13). Other than a problem with the suit temperature (similar to what Carpenter experienced), the flight proceeded smoothly as Schirra demonstrated discipline over the use of thruster fuel and powered down unnecessary equipment to conserve the electrical power.

Schirra was the most vocal of the astronauts, continually chatting away into the recorder when he was out of range of a tracking station. Even with the addition of three more tracking ships to the Mercury communications network, the ten-minute limit of being out of communication range was expanded to thirty minutes since the track of the craft for the period of six orbits covered a wider swath of the Earth. Because of the extended flight, the splashdown shifted to the Pacific, and Schirra reentered with more than 50 percent of his reaction control system fuel remaining. As he descended under parachute, the excess fuel was vented because it represented a hazard to recovery personnel.

He landed less than four miles from the primary recovery ship to the cheers of the crew. He elected to remain in the craft as it was hauled aboard the aircraft carrier USS *Kearsarge*, and he then blew the hatch. He received the same bruise to his hand as Glenn had, again vindicating Grissom. He had flown the perfect mission, thanks to the pioneering efforts of Shepard, Grissom, Glenn, and Carpenter and the tireless work by NASA and its contractors.

MA-9: Lots of Problems—Gordon Cooper

With Schirra landing with an excess of consumables, the planning for the last Mercury flight, MA-9, swung into high gear. It was to be Air Force captain Gordon "Gordo" Cooper's turn, the youngest and last of the seven who had yet to fly (save for the grounded Deke Slayton). While Cooper was not as outspoken with his spiritual beliefs as Glenn, he was perhaps the most philosophical of the astronauts. He named his spacecraft *Faith 7*, "as being symbolic of my firm belief in the entire Mercury team, in the spacecraft which had performed so well before, and in God."

The flight plan was quite aggressive for the small craft as it called for more than a day in orbit—thirty-three hours. One orbit more than Titov's, the mission reflected the growing confidence of the NASA Manned Spacecraft Center that had recently moved into the new facilities built in Houston, Texas. The flight also required two shifts of controllers for the first time.

The heavy, 75 lb. periscope was removed as the previous astronauts felt its value was not worth the weight. The rate stabilization and control system, one of the four attitude-control modes, was also removed, saving another

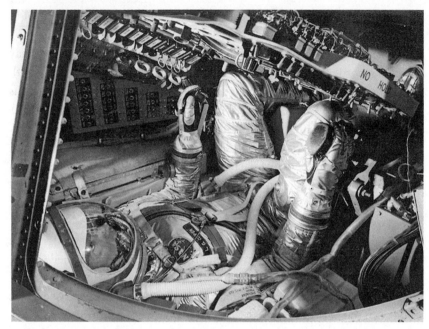

Astronaut Gordon Cooper in the Mercury capsule shows the restricted space. Courtesy of NASA.

12 lb. In their place, two of the 1,500 watt-hour batteries were replaced with 3,000 watt-hour batteries, and another auxiliary tank with 16 lb. more hydrogen peroxide was added. The oxygen supply was increased, along with water for environmental cooling and for drinking. The thrusters were replaced with a more efficient and lower-powered set. In all, there were more than 183 changes to the spacecraft to make it suitable for the extended mission. Yet Cooper's capsule weighed only 4 lb. more than Schirra's.

Atlas 130D with spacecraft no. 20 in place was on pad 14 by April 22. Then a series of delays occurred, including a scrubbed flight on May 14 after Cooper had spent four hours on his back in the capsule—a result of problems not directly associated with the Atlas or the Mercury spacecraft. At 8:04 a.m. on May 15, 1963, the last Mercury flight lifted off, and five minutes later sustainer engine cutoff occurred. As had happened with previous flights, the temperature of the interior of the spacecraft reflected the heat generated on its exterior by the acceleration through the lower layers of the atmosphere. Cooper's suit was unable to compensate for the 118 °F cabin temperature. However, after several orbits and some patience and sweat by

Cooper, the interior temperature fell to 95 °F while the suit reflected a more comfortable 70 °F.

Cooper observed Glenn's "fireflies" with each sunrise, and transmitted the first TV pictures of himself at two frames per second. With lots of time available, Cooper also spent considerable time evaluating the topography of Earth and, like Carpenter, claimed to be able to see extremely small details, including the smoke from a steam locomotive as he passed over India.

With a more organized flight plan, Cooper was also able to conserve fuel and power down unneeded equipment; by the fourteenth orbit, all was going exceptionally well. In marked contrast to the propaganda broadcasts made by the cosmonauts and the anti-American rhetoric that emanated from them, Cooper transmitted a short prayer: "Father, thank you for the success we have had in this flight. Thank you for the privilege of being in this position, to be up in this wondrous place, seeing all the many startling, wondrous things You have created. Help, guide and direct all of us that we may shape our lives to be good, that we may be much better Christians" (Kleinknecht 1963, 415).

Within a few more orbits, Cooper was going to need some of that divine guidance as the .05-g light came on. This was supposed to indicate that the spacecraft had sensed the upper limits of the atmosphere. Mission Control confirmed the orbit was stable and that it was an erroneous indication. More investigation revealed the signal had initiated some reentry activities (as it should, if it had been valid) and that several events had to be performed manually to resequence the process. At that point, it was found that both gyroscopes and the horizon scanner were without power. The attitude control for the reentry had to be performed manually.

On the twenty-first orbit, the automatic stabilization and control main 250 v-amp inverter blew a fuse. This unit converts the direct current of the batteries to alternating current for some of the electronics. Switching to the backup, Cooper discovered that whatever had killed the primary had also taken out the secondary unit. This essentially incapacitated all automated sequences.

Cooper now reported that the carbon dioxide level was building up. Because he had exhibited such calm demeanor toward all of these problems, Mission Control advised him to take a Dexedrine tablet to arouse his awareness level. However, Cooper's response simply reflected a man who was unflappable in the face of trouble—one of the reasons they had chosen him as an astronaut!

Cooper's Mercury flight was the first to return solely using the manual proportional mode with the astronaut controlling all aspects of the reentry, including the firing of the retrorockets. When the parachute blossomed over the blue Pacific, *Faith 7* was just 1 mi. from the intended splashdown point.

Although there were two more capsules and Atlas boosters available, NASA decided there was little to be gained from another flight, and the effort and dollars should be put into the next generation spacecraft—Gemini.

Vostok-5 and Vostok-6: Lady Cosmonaut Valentina Tereshkova

With the success of Titov's flight and before planning for the first group flights of Vostok-3 and Vostok-4 had been formalized, Lt. Gen. Nikolai Kamanin, chief of the cosmonauts, put forth the notion of sending a woman into space. While he was met with dissent, his stated objectives supported the current Soviet space philosophy in that eventually a woman would fly in space, and it must not be an American. The flight of a woman under the Communist banner strengthened the Communist party among women of the world. After a selection process that involved combing the parachute and flying clubs for possible candidates, five women were selected for cosmonaut training by April of 1962 (this was reduced to four when one dropped out). The candidates were an engineer, Irina Solovyova; a mathematician and programmer, Valentina Ponomaryova; a textile worker, Valentina Tereshkova; a teacher Zhanna Yorkina; and a shorthand secretary, Tatyana Kuznetsova.

After the Nikolayev and Popovich group flight, another plan come together by March of 1963. A male cosmonaut would launch for a long-duration (eight-day) flight. A few days after his launch, the woman would orbit for two to three days. Despite the rhetoric of the opportunities for women in the Soviet Union, it was generally understood that this would probably be the one and only flight for a woman, and that the rest of the team would not have another opportunity. Thus, the competition was intense among the four remaining women.

It is interesting to note the criteria that General Kamanin used in making the selection:

Ponomareva [Ponomaryova] has the most thorough theoretical preparation and is more talented than the others . . . but she needs lots of reform. She is arrogant, self-centered, exaggerates her abilities and does not stay away from drinking, smoking and taking walks [a euphemism for promiscuous activity]. Solovyena [Solovyova] is the most objective of all, more physically and morally sturdy but she is . . . insufficiently active in social work [not an ardent Communist]. Tereshkova . . . is active in society [being a member of the Young Communist League] . . . is especially well in appearance. Yerkina [Yorkina] is prepared less than well . . . We must first send Tereshkova into space flight . . . she is a Gagarin in a skirt. (Siddiqi 2003b, 362)

During a visit of the female cosmonauts to Tyuratam to witness the launch of Vostok-3/4, Tereshkova had caught the eye of Korolev and became his favorite choice as well. Although Ponomaryova had cultivated a loyal following, in the end Tereshkova was selected and Solovyova was suited up on launch day as her backup.

None of the first four Vostoks had included any significant scientific experiments or observations, but this changed with the selection of Sr. Lt. Valeriy F. Bykovskiy and his long-duration flight that included a series of photographic experiments.

A variety of problems confronted preparation for the mission, including the fabrication of a spacesuit for a female. Unlike the smoothness of the countdown that characterized other launches, the R-7 for Bykovskiy's Vostok-5 had many difficulties. Ranging from radios to gyroscopes, the problems brought Korolev and his deputies to harsh words.

As the delays moved the launch into mid-June, the estimated shelf life of the assembled Vostok spacecraft became an issue. Unless flown by August, it became unflightworthy. Just when it appeared that the launch would take place on June 11, a significant solar flare caused a high level of radiation in the upper atmosphere—another delay of a few days. As the countdown proceeded toward ignition on the afternoon of June 14, 1963, numerous small problems continued to plague the rocket. When a power cable umbilical failed to eject minutes before the launch, frustration and impatience dominated Korolev's thinking. He decided to proceed; the cable was ripped from the missile as it rose into the sky.

The R-7 produced less than nominal performance and Vostok-5's low orbit of 109 mi. by 139 mi. made the likelihood of a full eight-day mission questionable. Nevertheless, Bykovskiy proceeded to work through the lengthy flight plan that included a brief period of manually orienting the spacecraft as well as much photography. At one point a film canister jammed in a camera, and he discovered that another canister had no film at all.

Tereshkova's prelaunch preparations proceeded more smoothly than Bykovskiy's, and she followed him in Vostok-6 three days later on June 16, 1963. Her orbital inclination, while within .02 degrees of Vostok-5, was intentionally launched out-of-plane so that there was no possibility of a rendezvous, even if they had been capable of orbital maneuvers. They did come within several miles of each other on a few occasions, but otherwise it was somewhat anticlimactic compared to the Vostok III/IV flight. The big news for the world, of course, was that the Soviets had orbited a woman; another "first" and a propaganda triumph had been recorded for the Russians. In another anomaly of Soviet thinking, the TASS announcement indicated that Tereshkova was a civilian, when in fact she had been given a commission in the Soviet Air Force as a junior lieutenant for the purpose of the flight.

Knowing that the Americans had developed sophisticated methods of intercepting communications, the Russians prepared a series of coded phrases for the cosmonauts to use if there were problems. The term "feeling excellent" indicated just that—all was well. However, the expression "feeling well" was cause for concern, as some aspect might require early termination of the flight. The most ominous phrase "feeling satisfactory" required immediate termination at the first opportunity.

Live television of Tereshkova highlighted the flight and thrilled the Russian citizenry. However, she experienced space sickness within the first few orbits. Consideration was given to bringing her down early, but she asked to continue the flight with the knowledge that Titov's equilibrium had returned to a more normal state after the first twelve hours. Two days into the flight there were strong indications that she was still unable to perform adequately when she failed to adjust manually the attitude of the spacecraft. With the help of Gagarin and Titov radioing up instructions, she was finally able to accomplish the basic maneuvers necessary if the automatic system failed and she had to position the spacecraft.

Tereshkova was brought back to Earth after three full days. During retrofire and reentry, she did not communicate at all. However, she had logged more time in space than all six of the Mercury astronauts. Bykovskiy's reentry was more event-filled. His spherical reentry module again failed to

The first female in space, cosmonaut Valentina Tereshkova in Vostok VI. Courtesy of Energiya.

separate cleanly from the instrument compartment in a situation similar to that of Gagarin and Titov. He had spent almost five days in space.

During the postflight debriefing, Tereshkova noted of her incapacitation that "removing the film was very difficult. I didn't conduct any biological experiments—I was unable to reach objects." Yet she also added, "Weightlessness did not arouse any unpleasant sensations. . . . I threw up once but that was because of the food, not to any vestibular disorder" (Siddiqi 2003b, 371). A critical review of the flight, which was kept secret for decades, revealed that virtually all concerned, including Kamanin and Korolev, considered her performance inadequate. There has been some speculation that this judgment was possibly influenced by gender bias.

Although the group of female astronauts continued in training in anticipation of another flight in the coming Soyuz multiman spacecraft, none would fly. Another twenty years elapsed before the Soviets would send a woman into space—and that to upstage the Americans who scheduled a female for a shuttle spacewalk.

The completion of the Vostok-5/6 group flight effectively closed out the Vostok manned flight operations. Korolev wanted to turn his attention to the development of the new super booster and a more advanced spacecraft capable of orbital rendezvous—and his health was failing. Russia, at the end of the sixth year of the space race, still appeared to have a commanding lead over the Americans. As 1963 ended, so did both the Vostok and Mercury programs, after each sent six flights into space. Both countries had demonstrated that humans could function in space and that the spacecraft and its systems had exhibited a resiliency and reserve capability to overcome serious problems.

11

COMMITMENT TO THE MOON

A Change in Political Leadership

The presidential race of 1960 had three key issues, two of which were interconnected: the Cold War and the space race. The Republican candidate, the incumbent vice president, Richard M. Nixon, had a strong record of anti-Communism, but the indifferent endorsement given him by the outgoing president Dwight Eisenhower, had hurt his standing. Being a part of the Eisenhower administration, which many believed had "allowed" the Russians to achieve the first spectacular successes in space, continued to put Nixon on the defensive during the campaign.

The Democratic challenger, Massachusetts senator John F. Kennedy, hammered the elusive "missile gap" (the alleged disparity between the American rocket capability and that of the Soviet's) at every opportunity. The election was one of the closest in history, and most analysts credit the missile gap issue as costing the Republicans the election.

The new American president, John F. Kennedy, was sworn in on January 20, 1961. Amid what many considered the darkest period of the Cold War, he set the nation on notice during his inaugural address when he said the American people should "ask not what your country can do for you—ask what you can do for your country." It was a powerful speech that set the tone for his brief but dynamic and event-filled presidency. It was also a foreshadowing of a commitment he made, less than four months later, to overcome the Soviet lead in the space race.

Searching for an Alternative

The first NASA chief, Dr. T. Keith Glennan, had been the primary architect of the early space program that Kennedy inherited. Glennan's goal was the Moon, and under his direction, a preliminary plan for man-in-space had taken form. Without a firm understanding of the lunar environment (including its surface) and without a clear requirement of the technologies needed, a lunar exploration agenda had been mapped out by 1961. Glennan formed the Goett Committee (named for its chairman, Dr. Harry Goett, the director of NASA's Goddard Space Flight Center) in May 1959. Its objective was to identify the key steps in a human space flight schedule. Within two months, it defined a manned circumlunar landing as a follow-on to America's first man-in-space effort, Project Mercury (Harland 2004, 12).

However, Eisenhower, in a move to emphasize his feelings that a space race was not the proper course of action, reduced NASA's budget request for fiscal year 1961. This left the fate and direction of America's space program up to the new administration with the admonition that "further tests and experiments will be necessary to establish if there are any valid scientific reasons for extending manned space flight beyond the Mercury program" (Baker 1982, 58). It is interesting to note the use of the phrase "valid scientific reasons."

Even before taking office, Kennedy realized that the issue of the Soviet lead in space exploration had to be high on his priority list. Although his campaign had promised a "new frontier," Kennedy himself was not enamored with the issue of space flight and of the Mercury program in particular. Nevertheless, he recognized that in dealing with the Soviet Union, only a position of unqualified strength held sway with the leaders in the Kremlin. He was keenly aware that opinions in many nations uncommitted to either communism or capitalism (thus prime targets for Soviet expansion) believed that the Soviet Union had moved to a dominant position—and no one likes to be on the losing side.

Now that he sat in the Oval Office, Kennedy had the responsibility to deal with these problems and to make decisions that affected the future of the United States—indeed, the world—possibly for decades to come. Perhaps the most pressing concept that Kennedy had to wrestle with was just how important was the "conquest of space." The visionaries of the day were saying, without equivocation, that "the nation that controls space controls the world" in the same way as aviation had proved a dominant military factor in World War II and the postwar period. Moreover, they contended that it was imperative that this "control" be established within the next decade. There was little doubt in the minds of anyone in the free world that control of "outer space" by the Soviets represented a threat to democracy and personal liberty worldwide.

James Webb accepts the position as NASA administrator in a meeting with President Kennedy in the Oval Office in 1961. Courtesy of National Air & Space Museum.

To manage the growing NASA bureaucracy, Kennedy sought a man with special capabilities. It was felt that the first NASA administrator, Keith Glennan, appointed by Eisenhower, did not have the right political connections—aside from being a Republican. What was needed was someone who could not only organize and budget a project that had few equals in history but who also had the political savvy to deal with the Congress. More than a dozen persons were approached, who flatly rejected the job.

When James Webb was initially asked, he likewise declined. Here was a man in his mid-fifties who had performed successfully a variety of roles as a lawyer, Marine, and corporate executive. He had served his country in the Department of the Treasury, as director of the Bureau of the Budget, and in the State Department. His North Carolina background added to his persona, with a southern drawl that could soften his otherwise strong presence. He recognized that an enormous amount of political pressure existed, and he endeavored to be his own man. Could the program be run without interference? With pressure applied by several influential people including Vice President Lyndon Johnson, he finally accepted—but only after president Kennedy asked him personally and guaranteed his "independence."

Next to Glennan, no one person shaped NASA's formidable years, or created its image and influence, more than James Webb. It has been said that it was only through his strong leadership that all of the elements needed to achieve the Kennedy goal could be welded into a cohesive and successful program.

To help assess the space race situation, Kennedy had formed an Ad Hoc Committee on Space, chaired by his future science advisor and former Massachusetts Institute of Technology president Dr. Jerome Wiesner. Wiesner was not in favor of a human space program, which he felt was too risky and costly. Like Eisenhower, he preferred an unmanned program that emphasized science.

The day after Gagarin's flight, the president met with Wiesner; the new NASA director, James Webb; and his deputy, Dr. Hugh Dryden. Webb and Dryden reviewed the status of America's space program and the question of how to compete with the Soviets. They indicated that the Soviets would orbit the first multimanned ship, establish a space station, and probably circumnavigate the Moon. Kennedy was insistent: "Is there any place we can catch them? What can we do? Can we leapfrog?" Dryden felt that a crash program, on the scale of the atomic bomb project of World War II, might land a man on the Moon in ten years—the only possible "space spectacular" that might be achieved before the Russians.

It was a technological gamble that could cost $20 billion or more. Kennedy was obviously expressing sticker shock when he replied, "Can't you fellows invent some other race here on Earth that will do some good?" His comment reflected his inner thought that man's venture into space was simply a grandstand act of little scientific importance. Nevertheless, the imagination of the world had been captured by the possibility of a flight to the Moon—and Kennedy knew it. The meeting ended with the president imploring that they provide a winning solution, and he added, "There's nothing more important." If Eisenhower had been blind-sided by *Sputnik*, Kennedy was trapped by the space race—a race in which the United States did not have the option of withdrawing. Moreover, if that were so, then Kennedy had only one solution available to him—to win.

Kennedy sent a "Memorandum for the Vice President" dated April 20, 1961, that summarized the key questions and asked for "an overall survey of where we stand in space." He ended the memo with the request, "I would appreciate a report on this at the earliest possible moment." The questions that Kennedy posed were then answered point by point in a responding memo a few days later.

Then, three days after Shepard's flight (May 8, 1961), following his award of NASA's Distinguished Service Medal at the White House, Kennedy was talking with Robert Gilruth and George Low. Both were advocating a circumlunar flight. Kennedy queried further about why a lunar landing was not being contemplated. Gilruth was somewhat surprised that the president seemed receptive, but cautioned him that the effort was "probably an order-of-magnitude bigger challenge than a circumlunar flight." Still, the president asked what it would take. Gilruth responded, "Sufficient time, presidential support, and a Congressional mandate." When Kennedy pressed further as to how much time, Gilruth and Low compared thoughts and suggested, "Ten years" (Kraft 2001, 143).

The two memos represented the sum total of the dilemma and, along with these conversations, provided Kennedy with all that he really needed to make his decision.

"Land a man safely on the Moon"

In assessing the impact of major space "firsts," the Soviets had now achieved the first three: (1) the first satellite, (2) the first unmanned Moon probe, and (3) the first man in space. However, Kennedy pondered if the fourth major first, landing a man on the Moon, could be achieved before the Soviets and what importance the world would place on that accomplishment. In addition, to be truly effective, this fourth accomplishment had to demonstrate a commanding lead or its impact might not be significant. It was understood that colonization of the Moon was expected as a possible immediate follow-on as well as expeditions to Mars.

Overall, a vigorous space program could be a significant shot in the arm for economic and technological growth in additional to its political implications. Despite future possible stimulants, the head of the Bureau of the Budget was opposed to an all-out program because of its drain on the federal resources.

The success of astronaut Alan Shepard's suborbital MR-3 flight on May 5, 1961, touched off a degree of euphoria in the country. Coupled with the political dilemma that Kennedy faced, his decision was not so much a surprise as it was a powerful visionary message delivered with a style that made the pronouncement a landmark in American history.

On May 25, 1961, Kennedy delivered a special message to Congress on "urgent national needs." He stated: "If we are to win the battle that is now going on around the world between freedom and tyranny, the dramatic

President John F. Kennedy delivering his challenge to Congress to compete with the Russians in a race to the Moon. Courtesy of National Air & Space Museum.

achievements in space which occurred in recent weeks should have made clear to us all, as did the *Sputnik* in 1957, the impact of this adventure on the minds of men everywhere, who are attempting to make a determination from which road they should take." There was no doubt Kennedy believed that, whatever his personal feelings about space, the duel between the United States and the Soviet Union compelled America to take a leadership role.

He then made the fateful commitment: "First, I believe that this nation should commit itself to achieving the goal, before this decade is out, of landing a man on the Moon and returning him safely to the Earth. No single space project in this period will be more impressive to mankind, or more important for the long-range exploration of space; and none will be so difficult or expensive to accomplish."

He also made it clear in the speech that it was up to Congress and the American people to validate his rationale and the road he was mapping out. While some saw his careful wording as an indication of his lack of full commitment, others perceived it as a way of getting Congress and the people to "buy in" to the commitment, to make them full partners in the decision. Whatever the case, it accomplished its goal. A careful reading of the challenge did not say that the United States would be first; only that an American would walk on the Moon within the next eight and one-half years—but the meaning was obvious. As it had at Pearl Harbor, the sleeping giant had been awakened and ultimately emerged the victor.

If Kennedy still harbored a reluctance to proceed down the path he so eloquently defined, it was not evident to the electorate. With the goal clearly established, Congress approved the funding, and progress began to happen—quickly. All this enthusiasm had been generated while an American had only fifteen minutes in space.

However, the organization that sought to plan and control this daunting project—NASA—had to grow from its role as an "advisory committee" of only a few years previous to a complex structure capable of overseeing the activities of hundreds of contractors and tens of thousands of individuals—and billions of dollars.

In the Soviet Union, there was little attention given to Kennedy's speech; after all, the Americans were good at speeches but poor on delivering their promises. In particular, Sergei Korolev, architect of the Soviet space program, had no direct reaction, and there was no assessment of a plan to ensure that the first man on the Moon was Russian. In the years that immediately followed Kennedy's challenge, several proposals were advanced from the Soviet chief designer to achieve a manned circumlunar flight. However, none was embraced by the Soviet leadership until it was too late to compete effectively.

President Kennedy continued to emphasize the importance of his commitment when he recognized John Glenn's accomplishment on becoming the first American to orbit Earth the following February (1962). "This is a new ocean, and I believe America must sail upon it."

In a highly publicized speech at Rice University in September 1962 Kennedy said,

> The exploration of space will go ahead, whether we join in it or not. And it is one of the great adventures of all time, and no nation which expects to be the leader of other nations can expect to stay behind in this space race. We mean to lead it, for the eyes of the world now look into space, to the Moon and to the planets beyond, and we have vowed that we shall not see it governed by a hostile flag of conquest, but by a banner of freedom and peace.

Then he added, "We choose to go to the Moon! We choose to go to the Moon in this decade and do the other things, not because they are easy, but because they are hard." Few could have believed that he did not have a strong personal dedication.

A year after Kennedy made the commitment, NASA chief Webb told Kennedy that there were still some in the scientific community who expressed doubt that it was possible to send men to the Moon. Kennedy responded that the Moon landing is NASA's top priority. "This is, whether we like it or not, a race. . . . Everything we do [in space] ought to be tied into getting to the Moon ahead of the Russians. . . . Except for defense, [it is] the top priority of the United States government. . . . Otherwise, we shouldn't be spending this kind of money," to which he surprisingly added, "because I'm not that interested in space."

12

RENDEZVOUS IS THE KEY

Even before the first Vostok or Mercury spacecraft took men into orbit, the visionaries in both the Soviet Union and the United States were considering the next step. The initial planning had to be somewhat tentative because the effects of space flight on the ability of humans to function effectively in this unknown environment were yet to be discovered. The obvious requirements for advancing humanity's reach into the cosmos were evident; the technology and techniques necessary to join two or more craft in orbit were needed to assemble a space station or to send men to the Moon and planets.

This capability required levels of satellite tracking and computing that were essentially unknown in the early 1960s. Likewise, the ability to restart an upper stage reliably in the vacuum of space was still being worked out. The terms that moved to the forefront in the ever-widening vocabulary of space were "rendezvous" and "docking." Rendezvous is the ability for two satellites, launched at different times, to be brought alongside each other in space. Docking is the process of physically joining the two satellites.

Orbital Union: The Hohmann Transfer

The basic physics involved in bringing two satellites into "union" to permit rendezvous and docking was defined by the German engineer Walter Hohmann in a 1925 publication *Die Erreichbarkeit der Himmelskörper* (The Attainability of the Celestial Bodies). The Hohmann Transfer represents the most energy-efficient path (but not necessarily the shortest or fastest) for a spacecraft to transfer from one orbit to another (Zaehringer and Whitfield 2004, 61). To accomplish the Hohmann Transfer, guidance and tracking procedures had to be developed and the supporting radar systems and computer hardware built and software written.

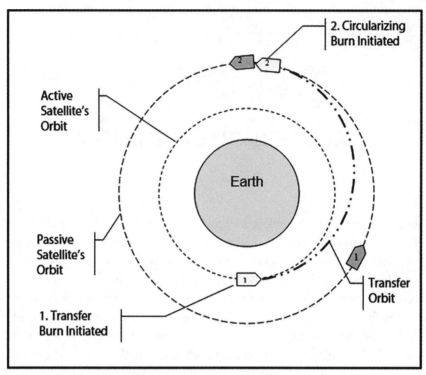

Hohmann Transfer Rendezvous Technique. This method is also used to transfer from one planetary orbit to another. Courtesy of Ted Spitzmiller.

Rendezvous in Space

A rendezvous between two satellites requires the two orbits have the same inclination (the path of one satellite must be directly above or below the other). Because the lower satellite has a shorter orbital period, it will pass beneath the higher satellite numerous times over the course of multiple orbits. In practice, the lower satellite is usually the "active" interceptor while the existing satellite plays a passive role.

The transfer requires the creation of a new elliptical orbit for the active satellite to make its apogee match the orbital altitude of the passive satellite at the point where the two orbits intersect (see figure above). At the proper time, the active satellite is powered into this transfer path, and halfway through this new elliptical orbit, the active satellite intercepts the passive satellite's altitude and executes a burn to match the orbital parameters of the passive target (Harland 2004, 91).

Soyuz Development: A Challenge to Soviet Technology

The primary mission of a second-generation spacecraft (as envisioned by Soviet chief designer Sergei Korolev) was for a manned circumlunar flight—although the Communist Central Committee had not yet approved a lunar program through 1962. However, Korolev also had to accommodate the military to ensure funding, so their requirements were always paramount in the planning documents. For both missions, the new ship had to be capable of rendezvous and docking to provide for orbital assembly of large space stations or refueling for lunar transfer missions. Several concepts were explored and heatedly debated among the various chief designers and the military before a final configuration was established.

The primary designers were again Mikhail Tikhonravov and Konstantin Feoktistov, of Korolev's team, the architects of the first Soviet manned spacecraft—Vostok. They proposed a craft that provided a ballistic reentry but had an offset center of gravity allowing some corrections in the descent to permit limited selection of the landing footprint and to provide lower-g reentry profiles. A wide range of possibilities was considered, including an upgrade to the Vostok. Although winged vehicles had been explored, it was obvious that these were far too ambitious for the time schedule and the state-of-the-art to be considered. There were occasional differences of opinion between Korolev and his staff that made him periodically rethink and revise his plans. This caused him to recycle some of his ideas and resulted in delays in the implementation (Siddiqi 2003b, 463).

Another factor that continually moved him to reevaluate the technical direction was the reports coming to him by way of the Western press, which frequently published optimistic schedules for the various projects being developed in the United States. While the "openness" of the American program was often an advantage to Korolev (allowing him to upstage the American effort), it also caused him to make premature decisions that haunted the Soviet program in the long run.

The final second-generation proposal, code-named Soyuz (a Russian word for "union"), was submitted by Korolev in mid-1962. It consisted of three components: a three-man reentry module, a cylindrical instrumentation and equipment module (on one end of the reentry module), and a spherical living quarters and docking adapter on the other end. All modules were enclosed in an aerodynamic shroud that was jettisoned before achieving orbital velocity.

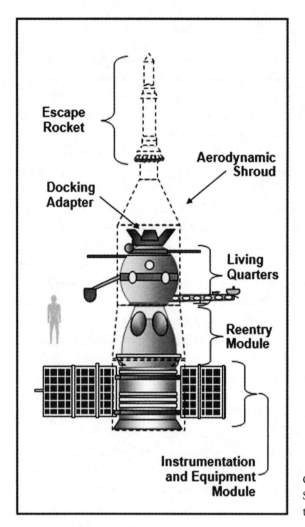

Escape
Rocket

Aerodynamic
Shroud

Docking
Adapter

Living
Quarters

Reentry
Module

Instrumentation
and Equipment
Module

Configuration of the
Soyuz spacecraft. Cour-
tesy of Ted Spitzmiller.

The reentry module now had a dish-shaped heat shield similar to the Mercury capsule to minimize the weight of the ablative material; as a result, the reentry module now required an attitude control system unlike the passive off-set center of gravity that stabilized Vostok during its reentry. The rendezvous procedure between two satellites was completely automated, although provisions were made for the crew to intervene. The fact that the docking adapter was not visible from the reentry module (from which the spacecraft was controlled) meant that electronic alignment of the two ships and television cameras played a critical role in the final technique.

Ejection seats were no longer an option for either emergency escape or the final descent to land. Thus, a solid-fuel, tractor-type, emergency escape rocket (similar to that of the Mercury capsule) was configured to pull the spacecraft from the booster in the event of a malfunction. Because the ship was again recovered on dry land, some means of slowing its descent at touchdown was required to avoid a bone-crushing g load on impact. This meant the inclusion of another solid-fuel retrorocket suspended from the parachute assembly and triggered by a feeler probe that sensed when the craft was several feet above the surface to soften the impact with the ground (Siddiqi 2003b, 469).

While Vostok had weighed in at 10,400 lb., Soyuz tipped the scales at 15,000 lb. This required elongating the upper stage of the R-7 to provide for more fuel for Earth-orbital flights and a new, bigger booster (still several years from being tested) for the manned circumlunar flights. The initial schedule called for the first flight of the Soyuz in early 1965.

Mercury II: An Interim Solution

The Goett Committee convened in 1959–60 to outline America's plan to regain the technical lead in space exploration. One of its recommendations was the development of a second-generation spacecraft following Mercury. It would carry a three-man crew and serve as the basis for a possible (but at the time unapproved) manned lunar program. In July of 1960 the project was formally named Apollo (Evens 2009, 406). Although work proceeded on the Apollo spacecraft, its role and schedule was still in doubt as the Kennedy Moon challenge had yet to be issued. It was also recognized that Apollo (like Soyuz) represented a quantum leap in many areas of systems technologies that had yet to be defined and mastered. Apollo also required a million-pound thrust booster that was still years from being available.

As Project Mercury moved toward its first manned flight in early 1961, the McDonnell Aircraft Company, in concert with NASA's Space Task Group (primarily Gilruth and spacecraft designer Maxime Faget), performed a study of an enlarged version called Mercury II. The McDonnell Company was hopeful of continuing its role in providing spacecraft to NASA (Harland 2004, 21).

Because Project Mercury had been established on a crash basis and within the limits of the Atlas booster, this new round of planning was predicated on the availability of the Atlas Centaur with its liquid hydrogen second stage. It allowed double the weight of the spacecraft and provided more capability. Several approaches were examined. The first was simply to expand some of the consumables so that the spacecraft could perform a one-day (eighteen-orbit) mission (this was eventually accomplished by upgrading the basic Mercury spacecraft). A second approach involved some structural changes to move many of the systems outside of the pressure hull so that they could be more easily accessed for problem isolation and engineering change modification. These latter operations had been consuming a significant amount of time and had put the basic Mercury program well behind its initial schedule. The most ambitious of the possible configurations was a two-man capsule more than twice as heavy as Mercury.

With the advent of the Kennedy challenge, the design of Apollo was mired in the complexities of its possible lunar mission, with many of its subsystems representing unknowns. What was needed was an interim spacecraft that bridged the gap both technologically and chronologically between the anticipated end of the original Mercury program in 1963 and the projected start of Apollo in 1965. As the various possibilities were explored, a version that could be the basis for a manned circumlunar flight was also suggested, but that put the whole existence of Apollo in question and introduced more complexities. The Mercury II proposal that was finally approved was to build a scaled-up version of the original Mercury. A set of attached segments on the aft end housed a hypergolic propulsion unit to provide orbital maneuvering and rendezvous. Provisioned for long-duration flights of up to two weeks, it also had features to allow the crew to accomplish extravehicular activities—a spacewalk.

The Atlas Centaur launch vehicle, on which the increased weight could be accommodated, was another obstacle as its availability continued to slip. Its high-tech nature also made its ability to launch with precise timing for the needs of rendezvous problematical. If Mercury II was to become a NASA program, another launch vehicle had to be found.

Hypergolic Propellants

The use of a traditional oxidizer, such as oxygen, and a fuel like kerosene requires some form of ignition to begin the combustion process—a part of the problem with igniting a second stage in space. To counter this and provide for a simpler ignition sequence, two propellants that would ignite on contact may be used—hypergolic fuels.

Titan II: A More Capable Launch Vehicle

During the early years of the Atlas project, a recommendation was made from several sources that a second intercontinental ballistic missile be developed not only as a backup but also to provide a heavier lift and longer-range capability. The Glenn L. Martin Company, which had just begun work on the Vanguard rocket, was chosen in September 1955 to produce WS 107A-2—the Titan missile. At the time, it was envisioned that only one intercontinental ballistic missile (either Atlas or Titan) would actually become operational—the first one to complete its development program and become available for deployment. With a first stage powered by a set of Aerojet General LR-87 engines of 150,000 lb. thrust each and a second stage hosting a single LR-91 of 80,000 lb. thrust, the Titan was a true two-stage missile (Harland 2004, 19).

Titan flight-testing began in February 1959, almost two years after Atlas. The first four flights were successful, but then a string of failures set the program back. With the Cold War and Khrushchev's threats still placing a burden on military preparedness, it was decided to deploy both missiles, with fifty-four Titans being in place by 1962. However, Titan was powered by liquid oxygen and kerosene, a combination that was not conducive to quick response and that could not be held in a high state of readiness for a prolonged period.

The impetus for a more powerful version of the Titan was born of the concept of multiple reentry vehicles and the need for more flexible reaction time in the launch sequence—enter Titan II. With lighter-weight warheads and extreme accuracy (arriving within 1 mi. of the target after a flight of more than 6,000 mi.) being achieved, it became more economical for each missile to carry several warheads and have them target objectives within a few hundred miles of each other with a single launch. This thwarted possible anti–missile defense systems by presenting multiple threats. To accomplish the improved performance, the 8 ft. diameter of the second stage was widened to match the 10 ft. first stage, and the total length went from 98 ft. to 104 ft. A second major change was to use storable (but less energetic) hypergolic propellants so that the missile could remain fueled for an extensive period and achieve a quick launch.

The combination of A-50 hydrazine and N204 nitrogen tetroxide was teamed with a new set of engines that produced more thrust than the Titan I, to make up for its lack efficiency as compared to the use of the favored cryogenic oxidizer—liquid oxygen. The first stage now had 430,000

Storable Propellants

Although liquid oxygen, a low-temperature (-297 °F) cryogenic oxidizer, has a superior specific impulse (I_{sp}) to other chemicals, its low temperature means that it can't be retained in the tanks of the rocket for extended periods of time because it literally boils off and must be periodically replenished. It is not a good choice for strategic military missiles that have to be maintained in a high state of readiness for a launch within minutes of a notification. Thus, "storable propellants," although lower in I_{sp}, are preferred for such military applications—and for spacecraft that will not use the propellant for days or weeks after launch.

lb. thrust, and the second, 100,000 lb. This quicker launch capability and greater weight lifting of the new Titan II also proved to be the selling point for the Mercury II application.

The proposed allocation of a dozen Titan IIs to NASA was also fraught with political bickering between NASA and the military, who were moving forward with another version—Titan III. NASA wanted Martin to enlarge the propellant tanks of the Titan, but this would affect the military program. After much debate, NASA finally settled for the standard Titan II with some special upgrades for the process of "man-rating" the booster (Launius and Jenkins 2002, 162).

Evolution of Gemini

As the planning for Mercury II became more defined in scope and schedule, it was separated from the Mercury program in the first days of 1962 and given a new name: Gemini (a constellation of stars represented as "The Twins"). It required independent funding (apart from Project Mercury) for ten manned flights to develop the technologies to take America to the Moon. It ultimately cost almost $1 billion—more than ten times the original Mercury budget (which ended after spending $350 million). The Gemini Program had five primary objectives: (1) to develop spacecraft systems for long-duration Earth-orbital flights of up to fourteen days, (2) to determine man's ability to function during these extended missions, (3) to demonstrate rendezvous and docking with a target vehicle, (4) to develop improved launch procedures and techniques for rendezvous, and (5) to

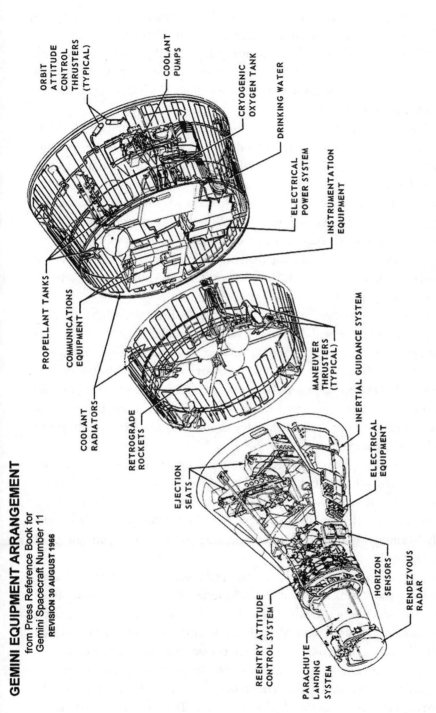

GEMINI EQUIPMENT ARRANGEMENT

from Press Reference Book for
Gemini Spacecraft Number 11
REVISION 30 AUGUST 1966

ORBIT ATTITUDE CONTROL THRUSTERS (TYPICAL)

COOLANT PUMPS

CRYOGENIC OXYGEN TANK

DRINKING WATER

ELECTRICAL POWER SYSTEM

INSTRUMENTATION EQUIPMENT

PROPELLANT TANKS

COMMUNICATIONS EQUIPMENT

COOLANT RADIATORS

RETROGRADE ROCKETS

MANEUVER THRUSTERS (TYPICAL)

INERTIAL GUIDANCE SYSTEM

EJECTION SEATS

ELECTRICAL EQUIPMENT

REENTRY ATTITUDE CONTROL SYSTEM

PARACHUTE LANDING SYSTEM

HORIZON SENSORS

RENDEZVOUS RADAR

The three primary Gemini components—the spacecraft (*left*), retrograde module (*center*) and OAMS module (*right*). Courtesy of NASA.

move the primary recovery mode to land rather than water. The original schedule called for completion of all objectives by October 1965.

The McDonnell Aircraft Company (again the prime contractor, as it had been for Mercury) moved swiftly not only to establish a modular approach to building the spacecraft but also to refine their newly developed project management methods.

The new spacecraft consisted of three segments, the first being the 12 ft. long scaled-up Mercury crew module that provided for two astronauts (hence the appropriate reference to the twins of Gemini). Virtually identical in basic shape to Mercury, it had a maximum diameter of 7.5 ft. at the base of the heat shield (compared to Mercury's 6.2 ft. base). The second segment, forming a part of the widening flange to interface to the 10 ft. diameter Titan upper stage, contained four 1,200 lb. thrust solid-fuel retro-rockets. The third segment contained the on-orbit electrical power, and the orbital attitude and maneuvering system (OAMS—pronounced "ohms") and other equipment.

The OAMS had four functions: (1) to provide thrust to enable the space-craft to rendezvous with the target vehicle, (2) to control the attitude of the spacecraft, (3) to separate the spacecraft from the second stage of the launch vehicle, and (4) to provide abort capability at altitudes above 300,000 ft. OAMS comprised sixteen thrust chambers: eight 25 lb. attitude thrusters for pitch, yaw, and roll, and eight 100 lb. thrusters to maneuver the spacecraft fore and aft, up and down, and left and right with respect to its longitudinal axis.

Because the crew module required attitude control (pitch, yaw, and roll) on its return to Earth, it also needed two redundant sets of 25 lb. thrusters and fuel for an independent reaction control system, which was controlled by the computerized guidance system that allowed the spacecraft to use its offset center of gravity to change its landing point by taking advantage of its positive lift-to-drag ratio.

Electrical power for the spacecraft presented a new set of problems. Mercury and Vostok had used sets of batteries. Batteries would be a part of the crew module to provide power during reentry. However, the mission duration for Gemini extended up to two weeks, and there was no way that much battery power could be allocated considering the weight limitations. Given the high levels of electrical power for the many new systems, includ-ing the radar needed for the rendezvous operation, and given the weight restrictions, a new solution had to be found. The obvious second choice was solar power. However, with the expected maneuvering that Gemini

Fuel Cells

Working on the reverse principle of electrolysis (in which electrical power is used to separate water into its two primary elements of oxygen and hydrogen), the fuel cell combines hydrogen and oxygen to produce electrical energy with water as a by-product—and heat. The process involves a simple chemical reaction of hydrogen with a catalyst to free electrons.

would perform, the structure of the solar panels would have to be quite robust or capable of retracting during the powered phases. This was deemed unacceptable. What was needed was a revolutionary capability to generate electricity. Some research had been done on an electrical generation device known as the "fuel cell."

Fuel cells had never been used in practical applications in the early 1960s, and they required the spacecraft to carry the cryogenic liquid hydrogen (-423 °F) in addition to liquid oxygen. The latter was already a component needed to supply the breathing oxygen for the crew. Extensive work was required to make this power source (located in Gemini's third adapter) a viable option. General Electric, whose fuel-cell design used ion-exchange membranes rather than gas-diffusion electrodes, was chosen to develop two units for each spacecraft.

The pure oxygen cabin environment was retained for Gemini along with the lithium hydroxide for removal of carbon dioxide. An omen of future consequences of this volatile atmosphere occurred in September of 1962 with a fire in a simulated space cabin at the U.S. Air Force School of Aerospace Medicine, Brooks Air Force Base, Texas. The experiment was designed to provide information on possible effects of the crew breathing pure oxygen in an environment of 5 psi for a duration of fourteen days. One of the two participants (non-astronauts) was severely burned. Had the fire occurred in a sea-level environment of 14 psi, the event would have been fatal.

One of the main problems with the Mercury spacecraft was the inability of the environmental control system to manage the heat generated by the astronaut, the electronic equipment, and the friction during the initial passage of the spacecraft through the atmosphere on its ascent into orbit. Gemini relocated many of the electrical systems outside the pressure hull, but it would still create more than three times the heat of Mercury. This problem was addressed by a heat exchange system located on the skin of the adapter section that provided an ample area for cooling.

Many of the systems were segmented into two independent (or semi-independent) parts, including the battery power, environment, and attitude control, as these capabilities had to be provided to the spacecraft for reentry following its separation from the adapters. A new set of mission rules dictated that, should the spacecraft have to switch to any of these reentry systems during a mission, then the mission had to be terminated. Those Mercury flights that ended with depletion of the attitude control fuel during reentry could not be tolerated with Gemini.

With the move to the Titan II and its hypergolic propellants, the emergency escape system during launch was revisited. Originally, it was envisioned that Gemini use an escape tower powered by a solid-fuel rocket, as had been the case with Mercury. However, a catastrophic explosion of the Titan did not present as lethal a fireball in size or temperature as the Atlas Centaur (the alternative launch vehicle) would. This allowed the use of ejection seats for low-altitude abort profiles up to 60,000 ft. (including pad aborts). Retrograde rockets in the forward adapter (for reentry) were used to separate the spacecraft from the booster from 60,000 ft. to normal second-stage burn completion. This approach also allowed use of the seats after the capsule returned to the lower levels of the atmosphere following high altitude aborts. Likewise, following reentry, if there were any malfunction with the landing system, the crew could elect to eject.

To achieve the required ejection timing, the hatches had to open in less than one-third of a second. The system allowed the astronauts to make the abort decision but, as with Mercury, provided a ground-activated backup. It was also determined that if either pilot elected to eject, the system ejected both—using rocket-propelled seats set at a slight angle to each other to ensure no trajectory conflict between the seats during the process. While the ejection method had been carefully thought out and engineered, the astronauts viewed the system with suspicion.

The recovery system for the spacecraft itself used a single 80 ft. parachute for a water landing. However, unlike Mercury, which landed on its back, Gemini transitioned its longitudinal axis during its final descent to a more horizontal position of 35 degrees relative to the horizon. This allowed the capsule to impact with the crew in an upright position and the spacecraft to float like a boat. This position was also more desirable during crew egress in that the large crew access doors opened out and allowed some

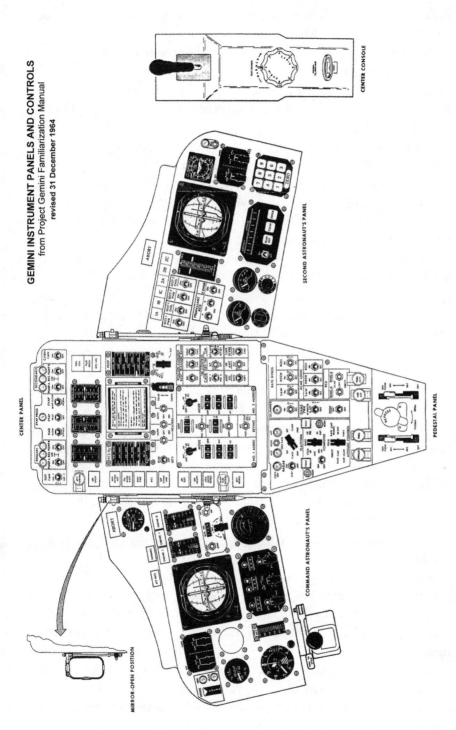

GEMINI INSTRUMENT PANELS AND CONTROLS
from Project Gemini Familiarization Manual
revised 31 December 1964

CENTER CONSOLE

CENTER PANEL

MIRROR-OPEN POSITION

COMMAND ASTRONAUT'S PANEL

SECOND ASTRONAUT'S PANEL

PEDESTAL PANEL

The Gemini instrument panel had a layout similar to a typical high-performance aircraft. Courtesy of NASA.

protection from wave action during the process by "splash curtains" that could be raised when the hatches were opened.

Two horizon sensors (a primary and backup) were part of the spacecraft's guidance and control system (GCS). They identified Earth's infrared horizon to provide reference for aligning the inertial guidance platform and to provide commands to the control system for the spacecraft's attitude in pitch and roll axes.

In developing the rendezvous capability, several approaches were employed. An optical method used a flashing light on the target vehicle and visual alignment of the spacecraft to perform the final rendezvous and docking. A completely automated system used radar, a computer, and a gyroscopically stabilized platform to allow the spacecraft and the target vehicle to electronically lock on to each other at distances up to 500 mi. and maneuver automatically from that point on. The automated method was the most efficient in terms of both time and fuel used; it was also the most complex. Because of the possibility of failure of one or more components in the fully automated system, a series of integrated procedures were produced to allow the completion of the rendezvous using a combination of the two methods. Gemini allowed experimenting with several rendezvous techniques.

With respect to the docking requirement, NASA planners decided to use a target vehicle that was more than just a satellite to which Gemini could link itself. A fully stabilized unit including a restartable engine was employed to provide experience in coordinating the activities of two semi-autonomous vehicles, the mechanical linkage, the electrical interfaces, and the control transfer techniques. Next to the rendezvous capability itself, the linking with another "live" rocket that contained propellant was the most daring of the Gemini objectives.

The Lockheed Agena (launched by the Atlas) was selected as the target satellite because it possessed all of the required capabilities and was in the process of achieving a high-level of reliability. It was the basis for America's satellite reconnaissance program. Eleven Atlas-Agena vehicles were procured as rendezvous targets with modifications that included radar and visual navigation and tracking aids as well as an external rendezvous-docking unit.

The rendezvous radar, situated in the forward section of Gemini, located and tracked the target vehicle during rendezvous maneuvers. A transponder beacon (a receiver/transmitter) located in the Agena replied to an "interrogation" signal from the Gemini radar. A digital computer in Gemini integrated all of the data (position, distance, velocity, etc.) and computed the required burn of the OAMS system. It provided an incremental velocity indicator (which visually displayed changes in spacecraft velocity), a manual data insertion unit (a keypad), and a display to read out the computer solutions.

The weight of all three segments of Gemini eventually grew to 8,355 lb. in orbit with a total length of 18 ft. Following separation of the various adapters during the reentry process, the spacecraft weighed 4,840 lb. at splashdown. It was a capable space vehicle in all respects.

As an extension of the goal to determine the human ability to function during long-duration missions, the possibility of one of the astronauts leaving the spacecraft while it was in orbit was discussed in March 1961. The event was termed an "extravehicular activity" or EVA. While this originally did not take a high profile, it was seen as an opportunity to develop a spacesuit capable of providing more mobility than simply an emergency response to cabin depressurization. The ability of the spacecraft hatch to be repeatedly opened and to reseal effectively in weightless and vacuum conditions would prove techniques needed on the lunar surface with Apollo. By February 1963 extravehicular activity was a full operational objective of Gemini.

With the manned space flight program now expanded to Gemini, the need for more astronauts led to recruiting nine candidates who were selected in September 1962: Neil A. Armstrong, Frank Borman, Charles Conrad, James A. Lovell Jr., James A. McDivitt, Elliot M. See Jr., Thomas P. Stafford, Edward H. White II, and John W. Young.

As the program progressed into 1963, it was evident that both the initial schedule and the price had been overly optimistic. Budget overruns caused the program to exceed the $1 billion level, and there was an almost frantic effort to stem the rising costs. The number of ground and flight tests for a variety of systems, including the Titan II booster and Agena target vehicles, were reduced. There was serious talk of eliminating the Agena altogether since the Apollo could accomplish these tasks. However, the Apollo and its Saturn booster cost significantly more, and its use would extend that schedule to accommodate the required orbital experiments. Therefore, the original objectives of Gemini remained as the effort to minimize costs became almost an obsession.

The first orbital, unmanned Gemini-Titan flight slipped a full year from 1963 to the morning of April 8, 1964. It was not an operational spacecraft but simply a test of its structural integrity during the launch phase and to

NASA's second group of astronauts had a wide spectrum of success and failure. Back row (*left to right*): Elliot See (died flying into St. Louis), James McDivitt, James Lovell (first circumlunar flight, survived Apollo 13), Edward White (first American EVA, died in the Apollo fire), and Thomas Stafford. Front row (*left to right*): Charles Conrad, Frank Borman (first circumlunar flight), Neil Armstrong (first to walk the Moon), and John Young (first shuttle flight). Courtesy of NASA.

verify that the modified Titan could loft it into orbit. As such, GT-1, as it was called, had the normal weight, center of gravity, and moment of inertia. It was not to be recovered and so therefore had no ablative heat shield. In fact, four large holes in the aft end of the spacecraft ensured its destruction when it reentered the atmosphere. GT-1 achieved an orbit of 100 mi. by 200 mi. At the end of three orbits, four hours and fifty minutes after launch, the first Gemini mission was completed, although it actually stayed up for nearly four days, allowing the worldwide tracking stations, controlled from Goddard Space Flight Center in Maryland, to follow the vehicle by radar. It reentered over the South Atlantic following its sixty-fourth orbit.

The next Gemini-Titan (GT-2) was an unmanned, suborbital flight to qualify the heat shield. Originally scheduled for August 1964, a series of storms, including three hurricanes, overtook the Cape and forced continual postponements and subsequent retests of systems to ensure that lightning had not damaged any of the electrical circuits. McDonnell was having significant problems with spacecraft No. 2 and probably would not have met any of the earlier schedules. The second flight was set for mid-November 1964. The third, and perhaps the first manned flight with GT-3, was scheduled for the end of January 1965.

While NASA and its contractors labored to get the second unmanned Gemini aloft, a new Soviet announcement once again upstaged the American effort. On October 12, 1964, the Soviet Union orbited Voskhod I. By all accounts, it was not just the Soviet's second-generation spacecraft; it was

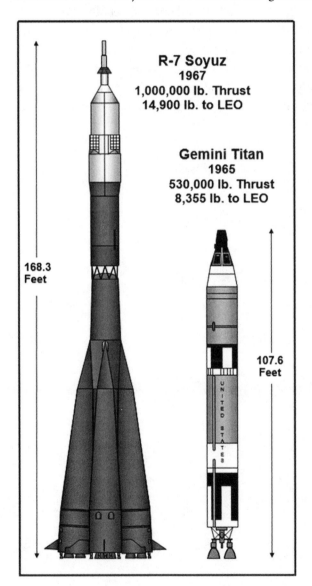

R-7 Soyuz
1967
1,000,000 lb. Thrust
14,900 lb. to LEO

Gemini Titan
1965
530,000 lb. Thrust
8,355 lb. to LEO

168.3 Feet

107.6 Feet

Second-generation manned spacecraft launch vehicles showing their comparative sizes. Courtesy of Ted Spitzmiller.

in a class with Apollo, with a crew of three cosmonauts. The crew flew in a "shirtsleeve" environment (flight suits rather than space suits), and all remained in the spacecraft to a landing on terra firma. The implication was that the Soviets had made the spacecraft environment so safe that there was no need for the backup protection of the space suit.

Could it be that the Soviets had been able to move directly to a lunar-capable vehicle so quickly? All indications seemed to point in that direction. The chief designer had once again been able to "pull one more rabbit from his hat" (as one of his colleagues recalled) and out-engineer the Americans, who were still months from flying its interim two-man Gemini and several years from its three-man Apollo. However, there was more to the story than the terse TASS announcements of another Soviet "first." While the world press was heralding the "advanced second-generation" Voskhod, the Soviets were breathing a sigh of relief, reveling in their sleight-of-hand, but concerned over their indeterminate future goals.

13

CONCEIVING A MOON ROCKET

Of all the surprises that the Russians had sprung upon the United States in the fall of 1957, the weight of their Sputniks was the most worrisome. It gave not only clear evidence that they had developed an intercontinental ballistic missile (ICBM, the R-7) but that its schedule was ahead of America's Atlas program. Therefore, it was natural that one of the first objectives for the United States was to develop new boosters of greater power. The magic number, in terms of thrust, was 1 million lb. Estimates at the time placed the Russian booster at about 500,000 lb. when in fact it was closer to 1 million. It was their conservative engineering and resulting greater structural weight that prevented even larger payloads in the early Sputniks. However, development of upper stages for the R-7 revealed a satellite launch capability of 10,000 lb. by 1960.

If America had been caught napping by Sputnik, it was not from lack of vision by those in key technology positions. Even before the Atlas started taking shape in the Convair manufacturing facility, the Air Force began studies for larger rocket engines. A 1955 contract with the Rocketdyne Division of North American Aviation examined the feasibility of a single-chamber engine with a thrust of up to 400,000 lb., which was designated the E-1, and an even larger engine of 1 million lb. was considered possible. Although the Air Force had no specific role for such large engines, it was prudent to continue to see where the state of the art could be directed.

A November 1956 meeting of the USAF Scientific Advisory Board recommended engines of 5 million lb. of thrust. However, it was recognized that, should the need occur for such power, a better interim solution was to cluster (group) smaller engines. A design competition for a single-chamber engine of 1 million lb. of thrust resulted in a contract to Rocketdyne in June 1958 for such an engine designated F-1.

Saturn I: 1 Million Pounds of Thrust

With the launching of Sputnik I in October of 1957, Wernher von Braun's Army Ballistic Missile Agency (ABMA) team intensified its possible options for building a large booster, looking to the Rocketdyne E-1 that promised almost 5000,000 lb. of thrust and, ultimately, the F-1 with almost three times that power. However, these were still paper projects. The need to move quickly to catch up with the Russians required that existing technology be employed. Thus, while the E-1 and F-1 offered promising futures, the "future" for the von Braun team was already upon them.

The Advanced Research Projects Agency (ARPA), formed in early 1958 to coordinate Department of Defense space activities in response to the Sputnik challenge, began discussions with ABMA to examine their heavy-lift concepts. The agency encouraged von Braun to consider clustering existing engines to shorten the development cycle. The obvious choice of an engine was the Rocketdyne LR-89 (LOX/RP-1 propellants—rocket propellant-1 or refined petroleum-1, a highly refined form of kerosene outwardly similar to jet fuel, used as rocket fuel) that powered the Jupiter intermediate-range ballistic missile (IRBM) (as well as the booster stage for the Atlas). It was the largest U.S. engine that had any track record.

However, to achieve the required thrust, at least eight of these engines had to be used, and this number of engines presented many engineering issues. The anticipated savings of time and money loomed as an enticing motivation. Given the initial designation of Juno V, the project's funding was approved in August 1958 with the goal of achieving the first static test within eighteen months, and, in keeping with the optimistic outlook of the times, the first flight test was anticipated by September 1960 (Launius and Jenkins 2002, 302).

For the first time, the von Braun team had been turned loose on a project of greater potential than any since the V-2, almost twenty-five years earlier. To provide a strong differentiation from its origin, the name of the vehicle was changed from Juno V to Saturn. Two objectives were immediately established—to simplify the LR-89 to improve reliability and performance, and to minimize the time necessary to assemble the propellant tank structure.

With the formation of NASA in October of 1958, it was obvious that it was just a matter of time before von Braun's ABMA team transferred to that new organization. However, it wasn't until March of 1960 that the move

The Efficiency of Propellants—Specific Impulse

Propellants with higher theoretical exhaust velocities have a higher potential specific impulse (I_{sp})—which is the measure of the efficiency of the rocket's propellants. It represents the thrust generated per pound of propellant consumed per second (Zaehringer and Whitfield 2004, 62). The energy produced by a rocket engine may be expressed as a relationship between the mass of the propellants consumed (m), the exhaust velocity (C) created, and the resulting thrust (T) generated by the engine. Essentially Newton's Second Law, more commonly expressed as F = ma, may be restated as T = mC. It illustrates that more thrust (T) is generated when higher exhaust velocities (C) are achieved, or when more propellant mass (m) is consumed per unit time.

Exhaust velocities can range from 2,000 ft./s for simple solid-fuels such as gunpowder to 8,000 ft./s for a mixture of liquid oxygen and gasoline. The highest exhaust velocities of 12,000 ft./s (or more) are available from chemical fuels such as liquid oxygen and liquid hydrogen.

Measuring I_{sp} is similar to the efficiency of a car in miles per gallon. Modern liquid propellants have specific impulses of about 300 lb./s, and high-energy liquid propellant combinations (liquid oxygen and liquid hydrogen) typically have an I_{sp} of up to 450 lb./s The theoretical maximum for chemical fuels is about 500 lb./s.

of the team from the Army to NASA was completed—with the bulk of the ABMA facilities becoming the George C. Marshall Space Flight Center (MSFC). Some believe the choice of the name was President Eisenhower's way of mitigating the role of the German team by overlaying the name of the only professional soldier to win the Nobel Peace Prize—and the man who was a primary architect in the defeat of Nazi Germany and its postwar reconstruction.

Although the risks of using highly volatile liquid hydrogen (LH2) as a fuel had yet to be mastered, the decision was made in the spring of 1960 to go with LH2-powered upper stages for Saturn. With the Pratt & Whitney RL-10 being the only LH2 engine then under development, it was readily determined that at least six of these engines were needed (90,000 lb. of thrust) as a first step in providing a potentially useful upper stage for Saturn. This new LH2 stage was given the designation of S-IV. The designation reflected it as the fourth stage of the proposed Nova—a rocket of 10 million lb. of thrust that von Braun thought necessary to take man to the Moon. The von Braun team, now fully a part of NASA's MSFC, understood that engineering that number of questionable engines was a challenge of even greater magnitude than their eight-engine booster.

As the various future configurations of Saturn began to proliferate, and as the massive Nova rocket took shape on the drawing boards, it was equally obvious that a new LH2 engine of 200,000 lb. was needed, and that it too would be clustered. The dependence on LH2 to provide a heavy-lift program now became vital. While science and the military continued to be used as the dominant reason for America's move into space, Eisenhower's Scientific Advisory Committee stated that "at present the most impelling reason for our effort has been the international political situation which demands that we demonstrate our technological capabilities if we are to maintain our position of leadership" (Baker 1982, 58). The report emphasized the importance of LH2 development.

Of all the possible configurations put forth for Saturn, two moved forward to the flight-test stage. The first was designated Saturn I, with the six RL-10 engines as a second stage; the second, called Saturn IB, used the new LH2 engine (designated the J-2) as its second stage (Lawrie 2005, 50).

As noted, the engine chosen for the Saturn I first stage was based on the Rocketdyne LR-89, which was redesignated as the H-1. Its 150,000 lb. of thrust was upgraded over several years to 165,000, to 188,000, and finally to 205,000 lb. To facilitate the first-stage structure, the tank assembly of the Jupiter intermediate-range ballistic missile was used as a central core around which eight Redstone tanks were assembled.

The Saturn I had the ability to orbit 22,000 lb., while the enhanced Saturn IB, with its LH2 second stage, delivered 35,000 lb. to low Earth orbit (LEO). In July of 1962 the Saturn IB configuration (called the C-1B at that time) was chosen by NASA to provide the Earth orbital testing of the Apollo hardware.

One of the first advances made (because of clustering) was that of automated checkout of the various rocket systems. In the past, manual readings and verification of instrumentation and switch settings preceded the various steps in the countdown. However, with so many items composing this new rocket, techniques for sequencing through the various checklists had to be automated. Large banks of electromechanical stepping switches provided the early concept of programmed instructions that soon give way to digital computers (Bilstein 2003, 238).

The problem of two or more NASA organizations handling different aspects of the hardware came to a head in 1960—the responsibilities of the booster development team and the spacecraft team were in conflict. The new MSFC (headed by von Braun) was charged with developing the booster while the Space Task Group was given the spacecraft itself. To ensure that both groups were operating as equals, the Space Task Group was renamed the Manned Spacecraft Center, with Robert Gilruth as its director. This essentially elevated the spacecraft development to the same organizational level as the rocket itself.

Static testing continued throughout 1960, and it was not until October 1961 that the first Saturn I (SA-1 denoted the first Saturn-Apollo test) took flight with inert upper stages. Over the next two years that followed, the Saturn I program proceeded at what some considered a snail's pace compared to its first eighteen months. As von Braun had predicted, the LH2 upper-stage development had consumed more time than its optimistic supporters had projected. The first of the Block II vehicles, SA-5, with the LH2 S-IV upper stage and eight aerodynamic fins to aid flight stabilization, flew in January 1964.

The Block II also used the uprated 188,000 lb. H-1 engines that now gave the Saturn a total of 1.5 million lb. of thrust. The propellant tanks were lengthened and had a simplified propellant interchange system to ensure that little propellant was left unburned in any of the nine tanks. Over the test period of the ten Saturn I vehicles, the propellant utilization increased from 96.1 percent to 99.3 percent. The hold-down time during engine start and stabilization was reduced from 3.6 seconds to 3.1 seconds, providing another performance improvement.

Launch Vehicle "Hold Down"

The United States typically employed launchpads for most of its rockets that allowed the engines to be tested several days prior to launch. Called flight readiness firings, the tests ensured that the vehicle was physically secured so that it could not take flight prematurely. During the actual launch sequence, this capability also allowed the engines to build to full thrust, and various temperatures and pressures were verified, before committing the vehicle to flight by releasing the hold-down arms.

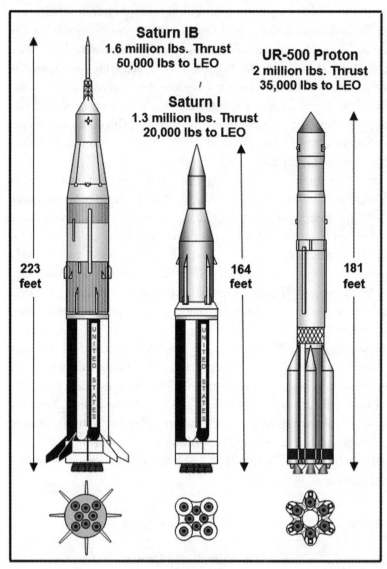

Evolution of the Saturn I (the addition of fins to the first stage) and a comparison with the Soviet UR-500 Proton. Courtesy of Ted Spitzmiller.

With a launch every six months, it was not until SA-6 in May 1964 that an unmanned, boilerplate Apollo spacecraft payload was finally orbited using a live six-engine S-IV LH2 upper stage.

What was extraordinary about the entire Saturn I test-flight regime was that not one catastrophic failure had been experienced. Virtually all other large rockets had undergone the ignominious fate of a fiery explosion early in its flight, often accounting for 50 percent of the first dozen or more. Thorough ground testing and the ability to simulate and understand the flight environment paid enormous dividends.

Taming Liquid Hydrogen: Developing the J-2

The development of a significantly larger LH2 engine for use as an upper stage began even before the smaller 15,000 lb. thrust Pratt & Whitney RL-10 (originally developed for the Centaur rocket) had been test fired. Studies by NASA in 1959 showed that not only was LH2 a key ingredient but that its full importance was revealed in engines of 150,000 lb. to 200,000 lb. of thrust—and clustered for even greater power levels. Pratt & Whitney's pioneering expertise in LH2 notwithstanding, the Rocketdyne Division of North American Aviation, an early participant in LOX/RP-1 experimentation, was awarded a contract in June of 1960 to proceed with the development of such an engine—designated the J-2. It was felt that Pratt & Whitney had its hands full with the RL-10, which at the time was a critical ingredient in the Centaur program. Within eighteen months, Rocketdyne was test firing an LH2 thrust chamber, and by October 1962 full thrust tests of a complete J-2 engine were accomplished for periods of up to 250 seconds.

Pratt & Whitney provided Rocketdyne with data on the innovative engineering concepts it had developed for the RL-10—an important legacy of information (Launius and Jenkins 2002, 340). The first production J-2 engines were delivered in April 1964. By the end of that year, full-duration static tests of up to 410 seconds were achieved—although each engine was required to have a usable life of 3,750 seconds to provide for a high level of reliability. Computers performed the prefiring checkout, propellant loading, and ignition sequence.

In the end, the J-2 went through a series of upgrades that improved performance, with the J-2 providing 230,000 lb. of thrust and a I_{sp} of 421 lb./s. The development of this engine was one of the most significant engineering feats of the Apollo program.

The first flight of a Saturn IB with the new J-2 engine serving as the S-IVB second stage occurred in February 1966. The launch vehicle was designated AS-201 (Apollo Saturn—the word "Apollo" now preceding the word "Saturn"). It carried a more complete but yet unmanned Apollo command and service module through the entire suborbital launch and reentry sequence, which lasted a total of thirty-four minutes as it arched over 1,000 mi. into space to impact 5,000 mi. downrange.

AS-203 followed (numerically out of sequence) in July 1966 to flight-test the second-stage restart process, although a restart was not actually attempted. AS-202, launched in August 1966, performed another suborbital test of the command module heat shield using a "skipping" profile to expose it to high levels of thermal stress. The spacecraft was recovered near Wake Island in the Pacific after traveling three-quarters of the way around the world. The Saturn IB, now 224 ft. tall, and the Apollo spacecraft were ready for their first manned space flight. However, tragic events would intervene in the progress of the American space program.

Selecting a Path to the Moon

As the Saturn I took shape and its upper stages were defined, the commitment by President Kennedy in May 1961 to land a man on the Moon within the decade required that one of several possible paths to the Moon be chosen (Lawrence 2005, 41). The specific configuration and size of the rockets developed after Saturn I was determined by this mode. By the spring of 1962 work had progressed as far as it could on the lunar program without this decision being made. There were seven variations of achieving a manned lunar landing, which could be categorized into three primary modes. Depending on which mode was chosen, one or more of a series of increasingly more powerful boosters be developed, ranging in designation from C-1 through C-8. (All of them were paper proposals at this point except the C-1, which was the Saturn I.) Each increment in the numerical designator represented an increase of 1.5 million lb. of thrust.

The first and most obvious path (or mode) was direct ascent. Its basis is the popular notion that a rocket sent its occupants from Earth directly to the Moon, as was generally depicted in the science fiction movies of the day. The entire spacecraft landed on the lunar surface and returned, perhaps leaving a portion of its structure there, such as the descent engine and landing gear, and weighed an estimated 140,000 lb. This required a Class V vehicle such as the proposed C-8 Nova—a 500 ft. tall beast with at least eight F-1 engines to provide 12 million lb. of thrust. This method did not require any form of rendezvous and docking (which had yet to be demonstrated) and was considered to present the least technological risk as it involved a single multistage rocket.

Development of this rocket, called Nova, was initially backed by MSFC (the von Braun team). However, the energy that it required demanded the use of LH2 upper stages, which had yet to prove themselves. The sheer size

> ## Parking Orbit
>
> When low Earth orbit was considered as a staging site for fights to the Moon and to the planets, the term "parking orbit" was defined to establish that mode. This term is used for an orbit that positions the vehicle for the next stage of a flight to higher orbital altitudes or to the Moon or planets. Its use allows a wider descretion in launch timing.

was also of concern since fabrication and movement of such a large, heavy structure presented significant problems. The projected (and typically optimistic) schedule for Nova showed its first launch in the autumn of 1967.

The cost of Nova ($200 million in 1961 dollars) per launch sent shock waves through NASA management. Using the existing concepts of flight-testing, it was estimated that perhaps as many as fifty launches of the smaller Saturn and the proposed Nova might be required to prove and man-rate the spacecraft and the rocket.

A second mode to reach the Moon used a somewhat smaller launch vehicle of the Saturn C-4 type (5 million lb. thrust) and a LEO to assemble or refuel a translunar stage. It then proceeded to the Moon from its parking orbit. This method (termed Earth orbit rendezvous—EOR) had significant technological risk as it required several rockets to launch, rendezvous, and dock. Von Braun had originally envisioned this method in his *Collier's* magazine articles some ten years earlier, when a large space station was seen as the jumping-off point for lunar and planetary missions. The price and the time schedule again made heads shake.

There were also variations of both the direct ascent and EOR that involved the placing of equipment on the Moon in advance of the astronauts. This too required multiple launches and the ability to accurately preselect and land next to these supply depots.

A third path to the Moon resulted from an evaluation of the roles of each piece of hardware and the energy needed to transport that hardware to only those points in space where their function was needed. For example, the primary role of the Apollo "mother ship," as it was sometimes called, was to transport the crew through the atmosphere (up and back) as well as provide the crew radiation shielding during the two- to three-day trip to the Moon. However, there was no need to take all the weight of the thick heat shield and all three crewmembers to the surface of the Moon. The mother ship could be left in lunar orbit while one or two crewmembers descended

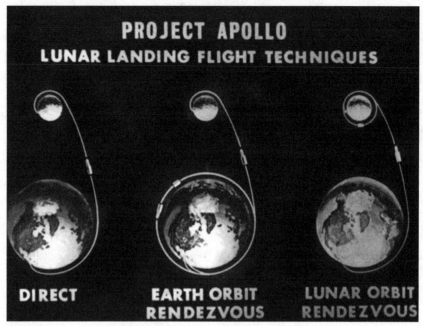

PROJECT APOLLO
LUNAR LANDING FLIGHT TECHNIQUES

DIRECT   **EARTH ORBIT RENDEZVOUS**   **LUNAR ORBIT RENDEZVOUS**

Three paths to the Moon, but only one held the promise of achieving the Kennedy goal. Courtesy of NASA.

to the surface in a lighter lunar lander. Likewise, the ascent back from the lunar surface could be done with only a portion of the lander, leaving the descent engine, its empty propellant tanks, and the landing legs (which were no longer needed) on the surface. The Chance Vought Company had done a significant amount of in-house research on this in hopes of finding a place in the lunar-landing program. They estimated that a lander of this type might weigh only 10,000 lb.—far less than a 140,000 lb. direct-ascent lander.

This latter mode, termed "lunar orbit rendezvous" (LOR), had many attractive benefits: one rocket (smaller than Nova), a shorter development period, and less cost. LOR reduced the total weight accelerated to escape velocity from 100 tons to 45 tons. However, as conceived in 1961, it represented the highest risk with respect to the need to develop rendezvous and docking. It also meant that many critical activities during the mission be performed a quarter-million miles from Earth, where the round-trip time to send and receive electronic signals took nearly three seconds—an eternity if a real-time, life-threatening condition had to be evaluated during lunar landing, takeoff, or rendezvous.

NASA management continued to block EOR and the most promising method, LOR, in preference to direct ascent. Gilruth felt that the need to rendezvous (LOR in particular) compromised mission reliability and flight safety. He believed that EOR and LOR were being advocated primarily because they appeared to promise the least cost of achieving Kennedy's goal, not because they represented the safest path to the Moon.

John Houbolt, an engineer at NASA's Langley Research Center, had thoroughly evaluated LOR and was firmly convinced that it was the only path that offered a lunar landing within the decade. The other methods of direct ascent and EOR required excessively complex hardware. In a rare and bold move borne of frustration, Houbolt wrote directly to NASA associate administrator Robert Seamans in November 1960, detailing the pros and cons of all the possible paths—with LOR being the preferred choice.

Because of the Houbolt letter, Milt Rosen, director of launch vehicles and propulsion (and a direct-ascent advocate), was tasked by Seamans to prepare a report on the various boosters and paths to lunar landing. Working with a group of representatives from the major NASA centers, the report ambiguously concluded that only by using rendezvous techniques was there be a reasonable chance of meeting the Kennedy goal. However, the feasibility of rendezvous would not be demonstrated for several more years (December 1965)—too late to redirect efforts to another mode (such as direct ascent) if rendezvous proved difficult to master. There were also proposals for a costly parallel development as a hedge against failure of one

NASA Engineer John Houbolt pushed the LOR concept aggressively. Courtesy of NASA.

method. The report, submitted in late 1960, stated that LOR presented the highest technical risk, and the group essentially recommended the direct ascent and development of the C-8 Nova.

The logic within the report, however, made more converts (Gilruth and Max Faget in particular) to LOR, and now it was Rosen's turn to see the light. The decision had become a critical issue, and Gilruth and Rosen often debated the merits. On one occasion, Gilruth made a concluding point that centered on the massive size of a direct-ascent lander and the power and complexity that it required. He contrasted it with an LOR lander with less than 10 percent of the weight but more maneuverability and abort options. The basic argument was sound, and Rosen now stood in agreement on LOR.

However, von Braun was not yet in the LOR camp. He had vacillated between direct ascent and EOR. Joe Shea, then deputy director of the Office of Manned Space Flight, had been tasked by Gilruth to resolve the problem. Following an intense meeting, with Shea, von Braun realized that the issue needed to be resolved and that LOR was as good a choice as any. If the other centers and NASA management felt LOR was the way to go, he concurred and commented, "We are already losing time in our overall program as a result of lacking a mode decision" (Neufeld 2007, 376). He also noted that the rocket the size of C-8 Nova presented difficult fabrication and transportation issues in addition to the technology itself. It was also about this time (1962) that the von Braun team, in its continuing effort to construct "paper" boosters, discovered that a five-engine F-1 configuration (known then as the C-5) might just provide the required power for LOR with a single launch.

The decision was made; America would place its prestige and the lunar-landing commitment on lunar orbit rendezvous—a mode that required building a rocket five times as powerful as the Saturn I. Called the Saturn C-5, its name was later revised to simply Saturn V.

Origins of the Soviet Lunar Program

The desire to send men to the Moon and to Mars had always been a part of Korolev's personal plan that had to be subordinate to the interests of the military and the realities and expediencies of the economic and political climate. In fact, Korolev felt that a manned Mars mission was a higher priority than a lunar landing (Siddiqi 2003b, 333). The Heavy Interplanetary Ship project (designated TMK), which he proposed in the early 1960s, provided preliminary planning for a three-man crew to assemble an interplanetary mission in EOR using multiple launches of a large booster. It was envisioned that by 1967 a Mars landing could be achieved. These goals, however, seemed to ignore the many problems of long-duration human flight that included the requirement for nuclear reactors and electric ion engines. Moving his vision into the "official" arena had to wait for proper timing.

Early in 1960, while at the peak of his prestige with the Soviet power structure (although some felt his influence was on the downward trend), Korolev and his key engineering staff created what was termed the "Big Space Plan." Among the projects was a 3.5 million lb. thrust booster with nuclear upper stages to provide the capability of sending men to the Moon and Mars using EOR.

By mid-1960 the Communist Central Committee had essentially approved the essence of Korolev's plan with a decree entitled *On the Creation of Powerful Carrier-Rockets, Satellites, Space Ships and Mastery of the Cosmic Space 1960–67*. Among the specifications was the requirement to launch payloads of up to 200,000 lb. to LEO. The decree allowed Korolev's organization to begin designing a rocket known as the N1 with the ability to send 100,000 lb. to LEO and the N2 with a 160,000 lb. capability. The initial schedules for these projects were just as overly optimistic (1962 and 1967, respectively) as their American counterparts. The euphoria of the post-Sputnik period generated a plethora of visions from virtually all of the chief designers, most of which never became a reality. With few exceptions, most were canceled by political maneuvering or lack of funding before achieving their goals.

As reports of America's progress showed concerted and successful results (specifically with the Saturn I), a meeting was held with many of the top chief designers and scientific leadership in April 1963 to assess the possible implications of the Kennedy challenge issued almost two years earlier. Korolev reported that the N1 was capable of accomplishing the lunar landing. About this time, Korolev's emphasis shifted from a Mars landing to a lunar landing as he recognized that the political leadership was concerned by the American commitment (Siddiqi 2003b, 401).

Toward the end of July 1963 a letter from the British astronomer Sir Bernard Lovell to NASA's deputy administrator, Hugh Dryden, set off an interesting debate in the American media. The letter stated that Lovell had met personally with several top-level Soviet academicians (including USSR Academy of Sciences president Mstislav Keldysh) who informed him that there were no current plans for a Soviet manned lunar program. The concept of a manned space program had been initially criticized by some in the American scientific community when the Mercury program was begun in 1958. Each major new program (Gemini and Apollo) again received that same denunciation as a waste of taxpayer money.

The Lovell letter brought that debate to the forefront once again. The position of those opposing the lunar-landing plan in particular was that unmanned probes could do as well for significantly less money. It was recognized by many in the government, as well as NASA, that one of the prime reasons for the expensive human space program was the competition with the Soviets—not science. If, indeed, there was no Soviet program to send men to the Moon, then the United States was spending billions racing against itself.

Khrushchev added to the debate by publicly proclaiming in October 1963, "At the present time we do not plan flights of cosmonauts to the Moon by 1970" (Siddiqi 2003b, 400). The implications of his statement reinforced the notion that there was no space race. Although the Apollo lunar program continued, the events at the close of the decade of the 1960s led many to speculate that indeed there had been no intent on the part of the Soviets to reach the Moon before the Americans. Keldysh and Khrushchev knew that the Soviet objective was to beat the Americans to the Moon. By refusing to acknowledge their goal, the Soviets could not be "beaten" politically.

Kennedy seemed to reassess his own commitment in a September 1963 speech, when he called for a joint effort with the Soviets to conquer the Moon together. Khrushchev seemed to open the Iron Curtain a crack a few weeks later when, in another statement to the press, he said (with regard to Kennedy's offer), "It would be useful if the USSR and the United States pooled their efforts in exploring outer space . . . specifically for arranging a joint flight to the Moon."

Here was a golden moment for both Kennedy (who had never liked the Moon race option) and Khrushchev (who, while wanting to beat the Americans, could not really afford the price tag) to reach a new accord. With the assassination of Kennedy three weeks later and the Soviet military recalcitrance to reveal any aspect of their technology, the opportunity passed.

However, the reality of a Soviet lunar program was assured during this same period with Korolev's latest proposal for a comprehensive five-point effort that included a manned circumlunar flight, an unmanned lunar rover, and a manned landing by 1968. The mode for each was to be EOR.

The plan called for the manned Soyuz (called product 7K), along with a 9K translunar injection stage and a series of 11K tankers to rendezvous in Earth orbit to refuel the 9K for its flight to the Moon with 7K as its payload. The advantage of EOR was the ability to assemble a large (200 ton) structure that provided a high degree of reliability and redundancy. The success that the Americans were achieving with their big boosters (the first flights of the Saturn I and the live static firings of the F-1 engine) had finally moved the almost insolvent Soviet government to approve a manned lunar-landing plan—but the final decree was slow in coming.

Early in 1964 Gen. Nikolai Petrovich Kamanin, the cosmonaut chief, summarized the effort in his diary: "The Central Committee is approving a plan for sending an expedition to the Moon in 1968–1970. The N1 rocket . . . will be used for this purpose. The mass of all the systems . . . will comprise about 200 tons. The plan is still only on paper, while the Americans already have done much for carrying out flights to the Moon." Kamanin headed the cosmonaut office from 1960 until 1970. He was a vocal proponent of the manned space program and kept a series of diaries that have revealed much about the Soviet space program.

A meeting with Khrushchev in March 1964, in which Korolev apparently received his lunar program approval, suggests that Khrushchev was somewhat ambiguous in his own desires. He wanted the Soviets to remain in a commanding lead in the space program, but at the same time, he lamented the costs and the sacrifices that he was forcing on the peasants who made up the vast majority of the populace. Even with Khrushchev's reluctant approval, it was still several months before a complete plan and funding was forthcoming.

Finally, more than three years after Kennedy's challenge, the Soviets had an approved and funded program for sending men to the Moon. However, the competition was not simply between the USSR and the USA. Korolev also had to compete with Mikhail Yangel, another chief designer who had taken over Korolev's ICBM role. He had a manned lunar proposal, the R-56—which ultimately withered for lack of official approval. Another program was also receiving funding—chief designer Vladimir Chelomey's manned circumlunar effort with the UR-500.

Chelomey's UR-500 Proton

With chief designer Sergei Korolev's R-7 ICBM falling from favor as a weapon by 1960, the door had been opened for other chief designers to pursue their projects. Emerging to prominent visibility was Vladimir Chelomey, who had Soviet premier Khrushchev's son, Sergei, as a member of his design group. In addition to the UR-200 ICBM that used storable propellants and was thus more flexible and responsive to the military needs, Chelomey also had plans for a space booster of greater capability than Korolev's R-7. As expected, the design was also touted to the military as a heavy-lift ICBM called the GR-2. Chelomey's proposal was approved in April 1962, with the first flight scheduled in 1965. With funding always restricted in the hard economic times that surrounded this era, Chelomey's venture diverted precious rubles from Korolev's projects.

Using the designation UR-500, the heavy-lift ICBM (and space booster), as it emerged in its 1963 design, had many of the structural attributes similar to Saturn. Six RD-253 engines of 330,000 lb. of thrust each provided for a total of 2 million lb. The second stage used a single 540,000 lb. thrust engine for 212 seconds (Siddiqi 2003b, 303).

Although the configuration was visually similar to the R-7, the surrounding tanks did not "stage" but remained a part of the booster throughout the first-stage flight, as with the Saturn I.

Chelomey used the UR-500 as a basis for a manned circumlunar project designated "Lunar Ship 1" (LK-1). It included a manned spacecraft that Sergei Khrushchev noted was "similar in outward appearance to the American Gemini," and the program received approval in an official decree in May 1964. Carrying only one cosmonaut, Chelomey's LK-1 was scheduled to make its historic flight in time for the fiftieth anniversary of the Russian revolution—November 1967.

Critical analysis of Korolev's Soyuz and Chelomey's LK-1 presents an interesting schism in view of the American lunar-landing program. Under what rationale did the Soviet's believe that a manned circumlunar program would compete with Apollo? Did the Soviets discount America's abilities at this point? The Soviet leadership—in particular, Premier Khrushchev—seemed to regard America's efforts with disdain. While the Soviet space program gave the impression of a carefully planned and vigorous effort, its ability to vie with the Soviet military needs always left it on the short end of funding—save for the creative manipulations of Korolev and Chelomey.

Although Chelomey envisioned the UR-500 as a means to a manned circumlunar flight, he received the requisite military support from the Soviet Air Force, which had been beaten out of one project after another by the Soviet Strategic Missile Forces. They looked to the UR-500 as the basis for competing with the American Dyna-Soar project. This was another indicator that the openness of American missile developments played a strong role in directing the Soviet efforts. Likewise, the cancellation of Dyna-Soar in late 1963 eventually redirected Soviet military interest away from the UR-500.

Despite all the political intrigue that surrounded the various projects being developed by the chief designers, the UR-500 found its way to the launchpad in July 1965. The initial two-stage version placed an 18,000 lb. scientific satellite in orbit on its first test. The Soviet press played up the weight, which, when the final stage was included, amounted to 26,000 lb. in orbit. The American Saturn S-IB (with its LH2 S-IVB second stage) had exceeded that a week earlier with the AS-203 flight that placed a total of 58,000 lb. in orbit.

The Soviet satellite was named Proton I, and the launcher was identified with that same nomenclature by the Soviet press, although the Chelomey team had been calling the big rocket Gerkules (Hercules). While the booster performed its first flight flawlessly, a malfunction in the payload resulted in no radio contact with the onboard experiments.

Three more tests of the UR-500 over the next eighteen months resulted in two successes. Here was a rocket that would play an important role as a large space booster and would endure for over fifty years. It was also a threat to beat the Americans to the first manned circumlunar flight.

The N1: Korolev's Giant

Valentin Glushko, who had been the premier engine provider, had moved away from large LOX-based chambers because of combustion instabilities and the desire of the military for storable fuels. It was Glushko's engines that propelled Chelomey's 2 million lb. thrust UR-500. A confrontation between Glushko and Korolev effectively divided the Soviet program into two opposing camps (storable versus cryogenic propellants). Korolev prevailed, but only marginally, with his selection of LOX for the first two stages of the new generation super booster designated the N1. Formal approval came in September 1962 from the Council of Ministers and the Central Committee with the objective to "ensure the leading position of the Soviet Union in the exploration of space." Specifically, the new booster should initially be capable of placing 150,000 lb. in LEO. This could be increased to 200,000 lb. with the availability of LH2 in the upper stages. The first launch was scheduled for 1965 (Siddiqi 2003b, 483).

Whereas the Saturn V was being configured specifically for the lunar

The N1 first stage ultimately had thirty engines and was assembled in a horizontal position. Courtesy of Energiya.

mission (and the LOR mode in particular), the N1 initially was not optimized for any specific role. This may have been another political maneuver to ensure that the military supported the project. The development of these super boosters again pointed up the fundamental differences between the American and Soviet programs. With Eisenhower's insistence that the basic exploration of space be carried out as a civilian activity under the auspices of NASA, the battle for funding occurred in the House and Senate, allowing the military and civilian programs to be more clearly delineated.

For the Soviet military, the N1 provided the ability to orbit large space stations that could be used for reconnaissance or for command and control, directing orbital weapons back toward targets on the Earth. From a scientific perspective, the N1 offered the opportunity to explore Mars, which was still seen by Korolev as a more desirable goal than the Moon.

With the help of Nikolai Dmitriyevich Kuznetsov, a leading designer of aircraft engines, by the end of 1962 the basic configuration of the N1 was established. The first stage consisted of twenty-four engines of 360,000 lb. thrust each, in a circular arrangement, providing a total of over 8 million lb. of thrust. The second stage employed eight 390,000 lb. thrust engines with over 3 million lb. of thrust. The second-stage engines were similar to the first stage except they were optimized for the vacuum conditions of space and thus achieved higher thrust levels.

Technology aside, a major obstacle in building the N1 (or any rocket of that size) is the ability to move such a large structure from its manufacturing facility to the static test stand and on to the launch site itself. No transport capability was feasible. As a result, the N1 first stage was never static-tested before its first flight. This was unprecedented.

The United States also faced that problem but built all the Saturn V stages as complete units and transported the first and second stages by barge, using a manufacturing facility (Michoud, in Mississippi) that was colocated with a navigable waterway. The Kennedy Space Center provided the receiving terminus on the Atlantic.

Whether it was the financial situation, technical considerations, or just the strange politics of the Soviet space program, EOR was abandoned for LOR by mid-1964, and the configuration of the N1 had to change to accommodate the heavier lifting requirement.

Although committed in spirit to LH2 upper stages, the resources of the design bureaus were such that progress was slow and never provided an LH2 engine for decades. By late 1965 it was realized that the third and fourth stages had to be LOX-kerosene, at least for the initial versions of the N1. To make up for the lack of high-energy upper stages, six more engines were added in a smaller concentric ring within the first-stage complement of twenty-four, bringing the total to thirty—and providing almost 30 percent more thrust than the Saturn V.

The N1 booster that finally took shape had a takeoff thrust of 10 million lb. The diameter at the base of the rocket was 55 ft., compared to the 33 ft.

The N1 being raised from the horizontal to the vertical firing position was an impressive sight. Courtesy of Energiya.

of the Saturn V. Pitch and yaw of the first and second stages were controlled by reducing power on one side of the thrust circle as none of the engines were gimbaled. Roll control was achieved by four 15,000 lb. thrust vernier engines.

However, the performance calculations for the N1 initially showed only 175,000 lb. could be lifted to LEO compared to 250,000 lb. for the Saturn V. The impact of the LH2 development program in the United States was the significant difference. The huge N1 rocket was approximately 345 ft. tall.

Saturn V Takes Shape

To achieve the LOR path to the Moon, the Saturn V required the ability to send almost 100,000 lb. to escape velocity. Fifty pounds of booster weight was required to accelerate each pound of payload to that speed. To accomplish this task, the first stage generated 7.5 million lb. of thrust with five clustered F-1 engines. The emphasis in development of the F-1 focused on reliability and the desire to remain within the state of the art whenever possible. The F-1 burned 2,200 lb. of fuel and 4,400 lb. of oxidizer each second under a pressure of 1,200 lb. to achieve the rated thrust of 1.5 million lb. Designated the S-1C, the construction of the first stage (33 ft. in diameter and 75 ft. in length) was contracted to Boeing. The first stage emerged with four large fins at its base—aerodynamic stability of the fins eased the work of the control unit in gimbaling the huge F-1 engines during most of the

The Saturn second stage with its five clustered J-2 LH engines represented an exceptionally efficient unit. Courtesy of NASA.

flight and provided a more stable platform during staging. The S-1C stage was static fired for the first time in 1965 (Lawrie 2005, 13).

A cluster of five J-2s provided 1 million lb. of thrust for the Saturn V second stage and was designated as the S-II. It had the same 33 ft. diameter of the first stage. The tank structure accounted for only 3 percent of the overall weight of the fully fueled assembly—a significant engineering feat by Douglas Aircraft. The third stage was the S-IVB powered by a single J-2 engine.

Assembling the giant rocket posed yet another set of engineering problems. Traditionally, American rockets had been erected "on the pad"— hoisted from a trailer-transporter to the vertical position, one stage at a time by a crane atop a gantry that enclosed the rocket. The various electrical, hydraulic, and cryogenic connections were then made to the ground support equipment, and the gantry rolled back before launch to leave the rocket standing free.

However, this method posed many problems for a rocket as large as the Saturn V. There were to be two launchpads, and this required building two massive gantries—literally buildings that have to be movable. In addition, the proximity of the launchpad to the gantry meant exposure to damage, should the rocket fail during the initial phase of launch.

The answer to the problem came from the 1928 science fiction movie *Girl in the Moon* that was created by the famous German director Fritz Lang. In the movie, the massive rocket is assembled in a large building and then transported to the launchpad on a set of rails. The movie's technical advisor (rocket pioneer, and von Braun's mentor in the early years, Hermann Oberth) had devised this concept. The arrangement of having a remote assembly building and a transporter to then move the rocket to the actual firing site was adopted for the Saturn V.

A single 525 ft. tall vehicle assembly building (VAB) was constructed and became, at the time, the largest enclosed building in the world. It had more volume than three Empire State Buildings—130 million cu. ft. of space covering 8 ac. With four "high bays," each with a 456 ft. vertically sliding door, four completely assembled Saturn Vs could be accommodated (Evens 2009, 410).

The rocket was assembled on one of three mobile launchpads (MLP) that were 25 ft. high, 160 ft. long, and 135 ft. wide. Each had their own 400 ft. umbilical tower that provided connections for fuel and oxidizer, ancillary gases (such as nitrogen), ground electrical power, and communications links—on big swing arms. One large opening in the base of the

Umbilical Connections

As the name implies, like an unborn child connected to its mother, the rocket is connected to various ground support equipment prior to flight through a series of electrical, pneumatic and cryogenic umbilicals. These supply control signals, monitoring test points, and replenishment of consumables before liftoff. At launch, all of these connections have to be "severed" to allow the vehicle to assume an autonomous state. Film of a launch often shows the elaborate umbilical towers that provide the connections and the rapid disconnect that follows a launch commit.

platform accommodated the flames of the five F-1 engine exhausts. Each MLP weighed 10 million lb.

The mobile pad was set on six 22 ft. high steel pedestals when in the vehicle assembly building or at the launch site. When on the firing site, four additional extensible columns were used to stiffen the MLP against rebound loads, should engine cutoff occur right after launch commit.

To move the MLP with the Saturn V to the firing site, another gigantic machine was created—the crawler-transporter. This vehicle could move these objects that weighed over 12 million lb. It traveled at 1 mph for the 5 mi. trip from the vehicle assembly building to the launch site. Yet the ride was gentle enough to place the MLP to within a fraction of an inch and keep its surface level so its towering cargo stayed perfectly vertical throughout the six-hour ride.

Automated testing of a wide variety of the rocket's systems, such as the monitoring of temperatures and pressures during flight, became another critical aspect of being able to move the project along and to achieve a successful flight-test program. More than 2,500 test points were monitored on the Apollo spacecraft, service module, and lunar module, while the Saturn V had more than 5,000 additional points. The emphasis on automated testing was predicated on the complexity of the rocket itself as well as on the lives of the three astronauts. Human error and the relatively slow speed at which a technician could read and interpret a test point (coupled with the volume of information) would dramatically increase the time required to prepare and launch the rocket. Finally, by using an automated system, the critical data could be electronically (and immediately) shared among several test conductors for evaluation.

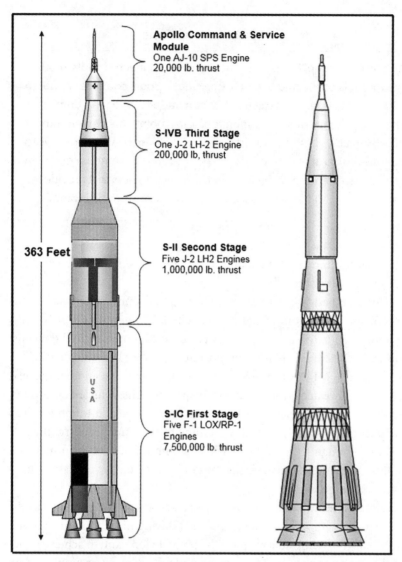

Apollo Command & Service Module
One AJ-10 SPS Engine
20,000 lb. thrust

S-IVB Third Stage
One J-2 LH-2 Engine
200,000 lb, thrust

S-II Second Stage
Five J-2 LH2 Engines
1,000,000 lb. thrust

363 Feet

S-IC First Stage
Five F-1 LOX/RP-1 Engines
7,500,000 lb. thrust

The American Saturn V and Soviet N1 represented the largest boosters that would fly in the 1967–72 period until the advent of the Space Launch System, which has yet to take flight. Courtesy of Ted Spitzmiller.

Guidance and control of the Saturn V was managed by the instrument unit, which occupied the upper 3 ft. of the S-IVB stage. This unit was essentially an inertial system that used a set of gyroscopes, accelerometers, a digital computer, and a variety of other components to monitor and control the giant on its journey. A direct lineage of the system could be found in the Jupiter, Redstone, and even the crude stepping switches of the V-2. An

IBM computer provided triple redundancy with critical functions presenting their data for comparison. If one set of inputs disagreed with the other two, the majority ruled, and the third set of results was disregarded.

When the new director of the Office of Manned Space Flight, George Mueller, reviewed his domain in September 1963, two project management factors stood out: time and money (quality was not considered a variable). The schedule for flight-testing of the Saturn V reflected the step-by-step approach that had been the traditional manner for proving new rockets in this high-risk environment. The plan called for several tests of the S-1C first stage before a "live" (fueled) S-II second stage was incorporated for a series of tests on it. Only after proving the second stage would a live third stage be added. In all, it was estimated that at least ten flights were needed to man-rate the Saturn I—and another ten for Saturn V. But if the Saturn testing continued to employ this conservative method, not only would it be virtually impossible to meet the lunar-landing goal by the end of the decade but the cost of the number of huge rockets required (which was approaching $200 million per launch) was prohibitive.

Mueller issued a proposal early in November 1963 in which he stated his belief that an accelerated schedule could be achieved that resulted in flying the first manned Saturn IB after only its third flight. More surprisingly, he proposed the same three-flight goal for the mammoth Saturn V. The concept was known as "all-up testing"—all of the stages were live beginning with the first test. The pressures of time and money could be mitigated, he felt, by the use of new and innovative testing techniques. These procedures used automated sequences enhanced by computers to achieve a high degree of assurance in comprehensive ground testing—before the rocket lifted into the air for the first time.

All-up testing was the final critical process to allow the United States to achieve the Kennedy goal. It figured prominently for both nations' efforts; one achieved world acclaim while the other hid its failures for years to come.

14

GEMINI CAPTURES THE LEAD

Voskhod I: Korolev's Last Rabbit

If America had established an optimistic schedule for its Gemini and Apollo spacecraft, Korolev likewise had visions of the near impossible for his three-man Soyuz—the follow-on to Vostok. He had originally established a time-table that called for its first flight in May 1963. However, the level of effort demanded of his design bureau was far beyond what his limited staff and budget could sustain. Work on the Molniya communications satellites and Zenit reconnaissance satellites, along with the unmanned probes to Mars and Venus, not to mention Korolev's commitment to advanced weapons to replace the R-7, stretched his staff's engineering ability to the limit.

A program to launch twelve more Vostoks, which had been approved in December 1962, was canceled when it was recognized that they had limited propaganda value. The new Soyuz had been delayed by many technical problems and the inability of the Soviets to manage the multitude of programs. Moreover, like the Americans, it was realized that some interim solution had to be defined to not only keep Soviet men in space but to ensure that the perceived lead in technology was maintained over the Americans. Tentative goals of higher altitudes and longer durations, while important, did not represent advancing technology in the eyes of the world.

Korolev's manned circumlunar proposal, accepted by the government in December 1963 and pushed by the optimistic schedules that NASA had published for Gemini, required that Soyuz fly before the end of 1964. As the early months of 1964 passed, it became clear that Soyuz would not fly in the same timeframe as Gemini. At this point the history of the Soviet program becomes somewhat nebulous. Korolev reportedly told his first deputy, Vasily Mishin, that Premier Khrushchev himself had directed Korolev to find some way of placing three men in orbit before the planned Gemini missions. However, Khrushchev's son writes that his father was not that aware of the technology aspects of Gemini. Sergei Khrushchev believed that it was Korolev himself who initiated the plan but did not want to be seen as advocating such a reckless and radical solution to their dilemma as was about to be produced.

Adding to the confusion, cosmonaut chief Gen. Nikolai Kamanin reported in his memoirs that "it was the first time I had seen Korolev in complete bewilderment . . . and could not see a clear path on how to re-equip the ship [Vostok] for three[cosmonauts] in such a short period of time." Thus, while it is unclear who was responsible for the decision, Korolev went forward with a re-layout of the basic Vostok to accommodate three men (Siddiqi 2003b, 384).

The project was given a new name, Voskhod (Sunrise) to disassociated it from the Vostok in every possible way in order to ensure that the West perceived it as a true second-generation effort. The official decree authorizing the project was signed in March 1964, and the first flight was expected the following August. Simply flying the first three-man ship was not enough. The announced plans of Gemini to introduce the extravehicular activity (EVA)—a spacewalk—prompted a second part of this ad hoc program; a Soviet cosmonaut would also leave a Voskhod (probably the second flight) and venture out into the hostile environment protected only in his space suit (Newkirk 1990, 35).

Even the booster had to be the upgraded R-7 used for the Zenit satellites because of the projected 1,300 lb. increase in the weight of the spacecraft. While the Soviets had often paced their program based on what the Americans had scheduled, this is the first real evidence that America was now driving the race and that their technology was being used as the measure of the Soviet's own progress. Perhaps the most important aspect of Gemini, the ability to rendezvous and dock, was not present in Voskhod. With the flight of America's 1.3 million lb. thrust Saturn SA-6 in May of 1964, which carried an unmanned boilerplate of the Apollo spacecraft, Korolev may have become anxious.

As Korolev's engineers pondered the problem, it became clear to them that the project represented an almost suicidal venture. Because of the weight and space restrictions, there was no room for the cosmonauts to wear pressure suits—a true shirtsleeve environment. Perhaps the most precarious design change was deleting the ejection seat, so there was no escape from the ship for a catastrophic failure. This also meant that during the return, unlike Vostok, the crew had to ride the descent module to the ground.

The impact was to be softened by a small solid-fuel rocket suspended from the parachute and fired just before touchdown.

Because of the tight budgets and timeframe, test articles were difficult to procure, but one test in particular was critical. Korolev authorized that Gherman Titov's Vostok II craft (which had been placed in a museum) should be equipped with the recovery parachutes and landing rocket to prove the design and technique. Unfortunately, the test did not go well, and the priceless artifact was destroyed. It was obvious that everything had to work perfectly or the three-crew members would surely die.

The Vostok life-support system, which normally sustained one cosmonaut for ten days, only provided enough consumables for the expanded crew for about three days. If the retrorockets failed, the crew would most likely perish before the orbit decayed naturally to reentry, therefore a backup retrorocket was added to the configuration. The Soviets could hide launch failures, but they definitely wanted to be able to announce a successful flight while the cosmonauts were in orbit; thus, the guaranteed safe return was pivotal. They had learned their lessons well from the public acclaim of Project Mercury and wanted to provide as much "openness" for the maximum propaganda value.

As Korolev listened to his designers lament the possible consequences of the redesign, he offered a "carrot" to motivate their innovation: one of the three cosmonauts would be a member of the design team. While this might seem a bit odd in that the prospects for a successful flight were significantly narrowed by the improvisation that was taking place, the opportunity to fly in space was overwhelming. A regular cosmonaut would command the mission, with the remaining slot would be allocated to a medical doctor (Siddiqi 2003b, 414).

The selection of the crew was a demanding and politically charged effort. Korolev had no authorization to make the flight offer to his staff. Other influential authorities had seen Voskhod as the opportunity to place a scientist in that third seat. After much bickering, several engineers were selected and underwent physicals to determine their suitability.

Konstantin Feoktistov, who had been in the forefront of the engineering work on Vostok and now was a key figure in its rebirth as Voskhod, was eager to get the opportunity to take his ride into space. He apparently completed the physical successfully (and actually was the only member of the engineering staff to do so). His selection was later challenged with the comments that ulcers, poor vision, a deformed spine, and missing fingers were disqualifying. He was also not the most socially acceptable—as he had a history of being difficult to communicate with and uncompromising. In support of Feoktistov was the fact that he knew more about the workings of Voskhod than any other person.

The command position was reduced to three candidates, one of whom had a Jewish mother. This was deemed unacceptable to the Communist hierarchy. The candidate supported by Korolev was Vladimir Komarov. He had been grounded by a heart anomaly similar to that which was keeping the American astronaut Deke Slayton from a flight slot. Doctors finally relented to Korolev's insistence, and Komarov was cleared for the flight.

While it had been agreed that the one slot go to a medical doctor, the Soviet Air Force had been insistent that he be a military man. Four candidates were selected for training, one of whom was not only the youngest (at twenty-six) but also a civilian: Boris Yegorov. Compounding the crew selection and the final testing of the spacecraft and the R-7 was the deteriorating physical and mental health of the chief designer, Korolev. His inability to control his emotional outbursts was costing him credibility and support even among his closest colleagues.

The actual training program was the shortest on record for either the Soviet or the American space programs, lasting less than four months. In the end, the final crew selection of Komarov, Feoktistov, and Yegorov was not made until a few weeks before the scheduled launch, which was postponed from mid-September to early October to allow for the retest of the recovery system that had destroyed Titov's museum piece.

The planned duration for the first Voskhod flight was a full day (eighteen orbits), and there were a few improvements to the basic Vostok systems. The spacecraft orientation method used a new "ion" technique that allowed the ship to sense the horizon and maintain a selected attitude while in the nighttime portion of its orbit. An improved video system transmitted twenty-five frames per second instead of ten, to smooth out the movements of the crewmembers to those viewing their activities from Earth.

The launch of a full dress-rehearsal unmanned test satellite, announced as Kosmos-47 to mask its intent, occurred on October 6, and all the systems functioned properly. It was returned to Earth successfully after the full one-day mission.

There was considerable pressure on the entire launch crew as the countdown proceeded for Voskhod I on October 12, 1965. Korolev, in particular, was showing the strain, and he was observed to be visibly trembling during the initial launch phase—as he had during Gagarin's flight. He knew that for the first twenty-seven seconds of flight there was no possible survival

The three Voskhod cosmonauts had to squeeze into the basic volume of the Vostok spacecraft, which left no room for space suits. Courtesy of Energiya.

of the crew if there were a major malfunction during that period. However, the R-7 and its upgraded upper stage performed admirably, and Voskhod I was soon in orbit, with the surprise announcement to the world by way of the Soviet news agency TASS.

Despite the short training period, all three crewmembers had reasonably useful work to perform, although by the second and third orbit both Feoktistov and Yegorov were beginning to show signs of space sickness—that debilitating, nauseous feeling that was an extension of vertigo. Significantly, both were able to complete most of their scheduled assignments, and the flight ended successfully with a landing after slightly more than twenty-four hours in orbit.

Everyone involved with the flight that had knowledge of the actual planning and spacecraft configuration breathed a sigh of relief—everyone except Soviet premier Nikita Khrushchev, who, during the course of the flight, had been deposed from power. While he might have been at the pinnacle of his leadership in some respects, he had alienated too many within the party structure—and two men in particular. Aleksey Kosygin and Leonid Brezhnev had orchestrated a coup that toppled Khrushchev after his reign of ten years. The centralized power between the Communist

Party and the government that Khrushchev had declared was about to be diffused by collective leadership between Kosygin and Brezhnev.

This transformation at the top also saw a change in the fortunes of the various design bureaus and the stature of several chief designers. Each of the primary figures—Korolev, Glushko, Yangel, and Chelomey—sought to move closer to the new power structure, and change was inevitable. While Voskhod was given high marks in the world press and more aspersions were cast upon the "failing" Americans, the flight itself did not achieve the preeminence it might have, had the headlines of the world not been dominated by Khrushchev's relatively peaceful but undemocratic removal.

Voskhod II: A Walk in Space

As for the second part of the Voskhod objectives that provided for a space-walk (code-named Vykhod—"exit"), they were still in full swing. Korolev had hoped to launch Voskhod II in time for the November celebration of the 1917 revolution, but there were too many problems yet to resolve. Essentially, the spacecraft was similar to Voskhod I but had only a two-man crew. The third couch was deleted; that space was needed for the cosmonaut to position himself for entry into an expandable, double-walled rubber air lock that was attached to one of the access hatches.

Unlike Gemini, in which the entire spacecraft was depressurized for the EVA, the Voskhod cabin could not be exposed to the full vacuum of space—its instrumentation and marginal life-support systems were not designed for that environment. Thus, a collapsible, thermally insulated air lock with a hatch at both ends was devised. Folded tight against the spacecraft sphere during the ascent, the 28 in. air-lock package unfolded in space (like an accordion) to its full length of 8 ft. and was pressurized to 40 percent of sea level. The cabin pressure had to be lowered to this value, necessitating both crewmembers wear pressure suits and breathe 100 percent oxygen (Siddiqi 2003b, 414).

Crew selection was simplified over the first Voskhod in that both cosmonauts were members of the final group of four who had not yet flown (and were still physically qualified). The training program was more intense than for previous flights, as it was understood that the unknowns of a spacewalk could present significantly more exertion than simply riding the spacecraft into orbit and back. It was also recognized that space sickness could be life threatening during EVA; if the cosmonaut vomited, it would obscure the faceplate and clog the environmental system, possibly asphyxiating the

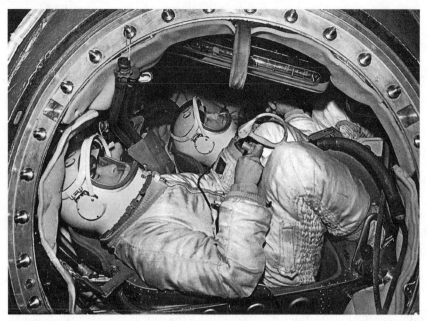

Cosmonauts Pavel Belyayev and Alexey Leonov in full pressure suits had little room for movement or ejection seats. Courtesy of Energiya.

spacewalker. All four candidates underwent extensive zero-g training in a TU-104 transport aircraft (similar to the Boeing KC-135 *Vomit Comet* that was used by NASA to acclimate astronauts to weightlessness).

A test of all spacecraft systems was to be accomplished with the launch of Kosmos-57 on February 22, 1965. It was an identically equipped, unmanned ship launched under the nebulous banner of the Kosmos series to disguise its role. The mission was fully automated. Once in orbit, the air lock was extended and the hatches activated in the same sequence as required for the manned mission. At the completion of the simulated EVA, the air lock was jettisoned.

However, the spacecraft was not positioned properly for retrofire and it entered a different orbit. This was sensed by the automatic destruct mechanism, and the ship was destroyed. Coming after a series of failures that had delayed the program, the presence of the Soviet secret police was soon obvious because they looked intently for a "subversive" among the test personnel. It was later revealed that erroneous commands from the ground caused the problem, and it was felt that, with one exception, the flight fully qualified the modifications.

The exception was the revised hatch ring (onto which the air lock was fitted) that remained on the spherical descent module. Because it had not returned to Earth, it could not be examined to ensure it did not present any problems during reentry. The workaround to verify the hatch, however, proved relatively simple. Since the Zenit reconnaissance satellites used the same descent module, authorization was given to attach the hatch ring to the next Zenit flight (identified to the West as Kosmos-59), which occurred on March 7. Recovery of the descent sphere was made on March 15, which showed no problems.

The EVA Voskhod II mission was set for March 18, and its scheduling was none too soon. The first Gemini was set to launch on March 23 (although the first Gemini EVA was not accomplished until the second manned flight in June). Soviet mission planners reported in later years that they had felt pressured by the pending Gemini program.

The hatch ring and inflatable airlock being attached to the Voskhod II spacecraft. Courtesy of Energiya.

Pavel Belyayev and Alexey Leonov were chosen as the prime crew for the flight that was now officially designated as Voskhod II (assuming it achieved orbit). Belyayev, who was in command the flight, was the oldest cosmonaut of the original twenty (then being thirty-nine years of age) and the only one with a college education. His flight opportunity had been delayed by an injury received during a parachute exercise. Leonov, age thirty, was to perform the EVA. Perhaps one of the most articulate of the original cosmonauts, he was also an accomplished painter and captured his EVA experiences in that form following the flight. He also authored several books in his later years that, after the fall of the Soviet Union, provided more insight into the Soviet space history.

The launch of the one-day Voskhod II flight began on the cold morning of March 18, fortunately without mishap, as again there was no way of escaping should the booster malfunction. Korolev was his usual apprehensive self. He had written to his wife only a few weeks earlier about the importance of the safety of the crew: "God grant us the strength and the wisdom to always live up to this motto and to never experience the opposite" (Siddiqi 2003b, 454). It was an interesting reference to the deity, officially banned from Soviet culture by the Communist Party.

Once in an orbit that ranged from 108 mi. to 311 mi., Belyayev activated the air-lock expansion while Leonov prepared his suit. The pressure in the cabin was lowered to the value established and verified in the air lock, about 300 mm of mercury (Hg)—40 percent of the 760 mm sea-level pressure. Leonov entered the air lock and connected the 16 ft. tether to the suit as Belyayev closed the hatchway behind him. It was a tight fit for the cosmonaut and his bulky spacesuit, with its life-support backpack. Although the air lock had an internal diameter of 40 in., the hatchway was only 25 in. wide. All of the actions involving the extension of the air lock and the opening and closing of both the inner and outer hatches were under the control of the Belyayev. The pressure in the air lock was lowered to 200 mm/Hg as Leonov verified the integrity of his suit. With the air lock finally depressurized to zero, the outer hatch was opened at about ninety minutes after liftoff as the spacecraft passed over Soviet territory. Several 16 mm film cameras provided for the historical footage, while a video camera recorded images transmitted to the ground and to a monitor for Belyayev, so he could be aware of Leonov's relative position and situation.

Leonov slowly emerged from the air lock into the expanse of outer space and tentatively held on to the edge of the hatch. His first observation was

Leonov's spacewalk, as revealed by a frame of movie film. Courtesy of Energiya.

that he could "see the Caucuses"—the range of mountains that run across the heart of eastern Russia. He then let go of the spacecraft and the film shows him facing the camera and drifting slowly away. He later recalled, "It was an extraordinary sensation. I had never felt like it before. I was free above the planet Earth and I saw it—saw it was rotating majestically below me" (Siddiqi 2003b, 455).

Science fiction films of the 1950s, such as *Destination Moon*, had predicted and depicted the spacewalk. Later films, such as *2001: A Space Odyssey*, added more drama to the event. Here, for the first time, a human was floating above the Earth in outer space. Traveling almost 18,000 mph with nothing between him and a most hostile environment but the thin veneer of the space suit, Leonov waved to the camera. He made an effort to take a picture of the spacecraft from his unique vantage point but was unable to activate the shutter. As the spacecraft and Leonov continued their orbital track, the Pacific soon appeared under them, and it was time to end the ten-minute excursion.

As Leonov started to reposition himself for a feet-first entry back into the air lock, he realized for the first time that the spacesuit had expanded to the point where both his feet and hands no longer extended into the boots and gloves of the suit. It was as though he had no hands and feet, just blunt limbs that were almost useless. Even the limited digital dexterity the suit had afforded was now gone. To make the situation even more critical, the suit had ballooned so that it was too wide to fit back through the hatch. He was beginning to feel the physical effect of the forced exertion as his heart rate now reached 143 beats per minute, and his respiration was twice normal, while his body temperature exceeded 100 °F.

Leonov recognized the solution was to decrease the pressure in the suit to the lowest level available, about 200 mm. This done, he was able to enter the air lock headfirst and then turn himself around so that he returned to the cabin in the required feet-first position. Given the dimensions of the air lock, this was an almost impossible task. He was near exhaustion when the increasing sweat and moisture condensed on his faceplate, obscuring his vision. He elected to open the faceplate before pressure in the air lock was stabilized (against mission rules). He rested for a minute before opening the inner hatch. The EVA had lasted just twenty-four minutes (Siddiqi 2003b, 456).

Subsequent EVAs encountered similar situations in both the American and Soviet programs. Movement in space was difficult and had to be carefully planned, and footholds and handholds had to be provided; otherwise, there was nothing to provide leverage for moving the body.

For the new leaders in the Kremlin, Kosygin and Brezhnev, being aware of the launch and watching the video of the EVA as it happened was a new and exciting experience. With Leonov now safely back in the spacecraft, the inner hatch was closed and the air lock released. The two cosmonauts breathed a sigh of relief—however, more danger lay ahead. Almost immediately, it was discovered that the inner hatch had not sealed properly. With the air lock gone, the cosmonauts could not reopen the hatch and attempt to remedy the problem. The life-sustaining atmosphere of the spacecraft was slowly venting into space, and the life-support system was replacing the air with an enriched oxygen content that reached almost 50 percent, increasing the risk of fire. Compounding the problem was a rotation of about two revolutions per minute imparted to the spacecraft when the air lock was jettisoned. It was decided not to address the rotation as the limited attitude thruster fuel needed to be conserved for the reentry positioning.

By the thirteenth orbit, the cabin air supply had been depleted to about one-third of its capacity, but the pressure appeared to stabilize at about two-thirds of the sea-level value (500 mm/Hg), although the oxygen content still presented an explosive mixture. The cosmonauts proceeded with some simple experiments as they waited for the reentry procedure to begin on the seventeenth orbit. However, at the appointed time, Belyayev reported that retrofire did not take place. Apparently the attitude had not been correct, and the automatic sequencer had detected it and prevented the engine from firing.

It was reported that there was confusion among the ground control crew, with Korolev taking command and assessing the situation. It was apparent that a manual retrofire would have to take place on the next orbit. The procedure, although relatively simple in concept, required Belyayev to lie across both seats of the spacecraft. This allowed him to get a visual alignment through the porthole of an etched line to establish the correct attitude for retrofire. Leonov had to lie under the couch, out of the way, yet holding Belyayev from floating around, while this was taking place. Then both cosmonauts returned to their seats before retrofire commenced so that the center of gravity would be in the correct position.

The reseating took the better part of a minute and caused retrofire to occur late. As had been the case with most of the Vostok missions, the instrument section failed to fully separate from the spherical descent module. This caused a higher g load, but the two pieces eventually separated.

Because of the delay, an overshoot of almost 2,000 mi. put Voskhod II down in dense forest and several feet of snow, and the report of a safe landing was delayed. It was several hours before a rescue helicopter spotted the red parachute, but it was unable to find a clearing to land. Warm clothes and food were dropped to the pair, and they were forced to spend a cold night in the spacecraft while wolves prowled its perimeter.

The next day a rescue team was able to ski into the area, but because of confusion over who had jurisdiction for the recovery effort, Belyayev and Leonov, along with the rescue team, spent the second night in the wilderness before finally being airlifted out after having chopped a clearing for a helicopter to set down (Siddiqi 2003b, 459).

The harrowing ordeal, both in space and on the ground, was not revealed for many years. In fact, Leonov, on several occasions during speaking tours over the weeks that followed the flight, noted how pleasant and easy his spacewalk had been, and Belyayev indicated that the Voskhod had the capability to change orbit (it could not).

Although there were plans for at least two more Voskhod flights, poor planning and indecision resulted in the Soviets not orbiting a man for more than two years. The Soviets touted their five hundred–plus hours in space

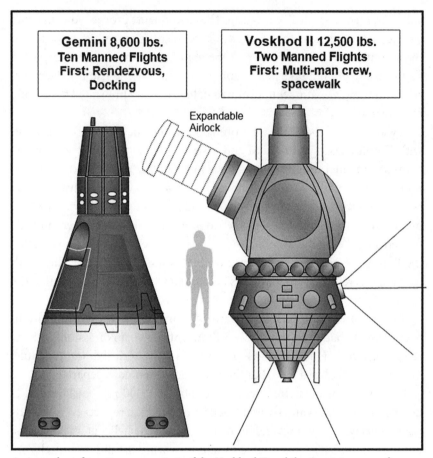

Gemini 8,600 lbs.
Ten Manned Flights
First: Rendezvous,
Docking

Voskhod II 12,500 lbs.
Two Manned Flights
First: Multi-man crew,
spacewalk

Expandable
Airlock

A size and configuration comparison of the Voskhod II and the Gemini spacecraft.
Courtesy of Ted Spitzmiller.

compared to the Americans' fifty-four hours and the fact that they had now sent eleven cosmonauts into orbit compared to America's four (they dismissed Shepard's and Grissom's flights).

The Voskhod effort, designed solely to upstage Gemini, delayed the introduction of the Soyuz by at least a year, perhaps sealing the fate of the Soviet program and ensuring that, barring any significant problems to the American effort, the Soviets would not be the first to the Moon. First Deputy Mishin, in a comment made twenty-five years later, indicated that the Voskhod program had not contributed to Soviet space advances and in fact had been "simply a waste of time."

Gemini III: Space Twins in Orbit

Attempts to launch the second unmanned Gemini-Titan 2 (GT-2) during the first week of December 1964 were frustrated by a series of problems. The launch was recycled (a term meaning that the countdown had to be restarted), and it finally left the pad on January 19, 1965. The flight was a ballistic, suborbital trajectory to subject the heat shield to a calculated overload to ensure it would survive all of the likely reentry profiles. After just two Gemini-Titan tests, the man-rated booster and its spacecraft were set for the start of the manned flight program.

However, the GT-2 was far more than simply a new spacecraft and booster. In the months following the last Mercury flight, a new NASA facility had sprung up near Houston, Texas, and this hub of the Manned Spacecraft Center was now operational. Control of the manned missions was transferred from Cape Kennedy to the Lyndon B. Johnson Space Center, or "Houston" (as it was blandly referred to on the communications channels), when the rocket cleared the umbilical tower. The Cape had undergone its own name change, from Canaveral to Kennedy following the assassination of the president in 1963, and the NASA facilities there were known as the Kennedy Space Center. The Cape itself reverted to its historic name of Canaveral in the years to come, to preserve that landmark.

Control of a manned space flight could only be effective if the data that reflected the condition of the spacecraft could be quickly and accurately transmitted to the Goddard Space Flight Center in Maryland—which remained the nerve center of the electronics network. Here information was filtered and routed through a greatly expanded telemetry system that now included eighty-nine ground stations and ships around the world. Some of the information was processed by a set of new IBM 7094 computers (in today's world, they are called supercomputers), working in tandem, before being sent on to Houston. Data rates increased dramatically, and the format ultimately went from a hodgepodge of analog signals to exclusively digital by 1966. Each spacecraft sent information regarding 270 aspects of various temperatures, pressures, and states of electronic and mechanical relationships (compared to the 90 channels of data for Mercury).

A third group of fourteen new astronauts was selected in October 1963 with the emphasis moving from test-piloting to engineering and scientific skills, although each astronaut was required to be a pilot. To this point, only John Glenn had retired from the astronaut corps to pursue his political

The IBM 7094 supercomputer in the mid-1960s at Johnson Space Center composed the Gemini real-time computer complex. Two of these computers operating in tandem had less computing power and storage capacity than a modern cell phone. Courtesy of NASA.

ambitions. Some, like Deke Slayton and Alan Shepard, were grounded because of medical conditions.

Virgil "Gus" Grissom had been named to command GT-3, the first manned Gemini flight, perhaps as a result of his personal dedication resulting from the bad press he received when his Mercury capsule sank during the recovery operation. No one could deny that his engineering expertise and drive set the standard for ensuring that this first Gemini flight was a success. Because of its tight confines and the fact that Grissom, being the smallest of the astronauts, was the only one who really fit into it, the spacecraft was occasionally referred to by the other astronauts as the "Gusmobile." As for a spacecraft name, Grissom selected *Molly Brown*, from the title of a 1960 Broadway musical and 1964 film—*The Unsinkable Molly Brown*—which hinted at his unfortunate episode with his Mercury capsule. NASA did not like it, and Grissom somewhat lightheartedly suggested an alternate name, *Titanic*. Because of the situation, *Molly Brown* was the last spacecraft to carry a name other than its official designation for the next four years (Harland 2004, 34).

Grissom's right-seat pilot for the mission was the rookie John Young. It was decided not to use the term "copilot." The traditional left-seat positon

NASA's third group of astronauts: Back row, *left to right*: Michael Collins, Walter Cunningham, Donn F. Eisele, Theodore C. Freeman, Richard F. Gordon Jr., Russell L. Schweickart, David R. Scott, Clifton C. Williams Jr.; front row, *left to right*: Edwin "Buzz" Eugene Aldrin Jr., William A. Anders, Charles A. Bassett II, Alan L. Bean, Eugene A. Cernan, Roger B. Chaffee. Four would not live to fly in space (Freeman, Williams, Bassett, and Chaffee). Courtesy of NASA.

was now defined as the "command pilot" while the right seat was simply termed "pilot." Neither astronaut was particularly talkative, avoiding the press and tending to business. They suited each other and made a good team for the tight confines of Gemini. The astronauts had pressed NASA management to make the flight an open-ended affair with a target of perhaps four days. However, NASA was still in the conservative mode, and three orbits were sufficient to prove the basic systems. One aspect that was a strong indicator of how smoothly things were starting to fall into place in the American space program was moving the flight forward from April into March.

Thus, it was just three days after Belyayev and Leonov had emerged from the dense forests of northern Russia that the 340,000 lb. GT-3 rocket came to life on the morning of March 23, 1965. At T+151 seconds, the first stage had exhausted its propellants and was shed as the second stage continued to accelerate the spaceship until T+335 seconds, when the orbital velocity of 17,600 mph was reached. Twenty seconds later explosive bolts separated *Molly Brown* from the spent second stage, and Grissom fired the orbital attitude and maneuvering system (OAMS) unit for the first time to move the spacecraft comfortably ahead of the now tumbling second stage, which was also in orbit.

Astronauts Gus Grissom and John Young are seated in Gemini III as the hatches are about to be closed. Courtesy of NASA.

At the end of the first orbit, Grissom again fired the OAMS for seventy-four seconds to slow the spacecraft and lower its apogee by 34 mi. The historic event caused a few smiles across the faces of the controllers hunched over their consoles at Mission Control Center. America had achieved a major "first" in space—a piloted satellite had changed its own orbit. Two more maneuvers were accomplished, with the last one lowering the perigee so that Gemini III would reenter the atmosphere even if the retrorockets failed to fire. The OAMS unit was then jettisoned, and the retrorocket sequence began. Both astronauts later reported that the "kick" from the 1,200 lb. thrust units was more than they had expected. The retrorocket adapter unit was likewise jettisoned, and Gemini III began its plunge back into the atmosphere.

As the spacecraft began to heat from the atmospheric friction, an ionized sheath of plasma enveloped it, cutting off communications with the ground. Grissom flew the spacecraft using computerized directions for pitch and roll. This in turn used the lift from the offset center of gravity to provide for some control over where it landed. It became immediately apparent, as the plot boards at Mission Control showed, that the spacecraft was not generating the predicted lift and landed some fifty miles short of its designated landing area.

The next surprise came when the spacecraft, descending on its parachute, transitioned from its vertical orientation to the almost horizontal position. The pitch forward was more intense than either astronaut expected, and, perhaps because they had not snugged the restraining belts, they were thrown forward into the windows. Grissom's helmet faceplate was badly broken while Young's was deeply scratched (Young 2012, 82).

During the mission review, virtually everyone was satisfied with the performance of the spacecraft and its systems. Only the fuel cells, which continued to exhibit problems in their development, were not on board. GT-3 flew with only batteries, as did GT-4, because its expected duration of four days was within the capabilities of the battery systems.

The technology embodied within Gemini was indeed a generation ahead of its Soviet contemporary, Soyuz, which did not fly for another two years. Moreover, when Soyuz did fly, it required several years of upgrading to perform to the level of Gemini. Despite the weight-lifting disparity of the Titan booster (10,000 lb. as opposed to the R-7's 14,000 lb.), miniaturization of spacecraft systems and its advanced engineering allowed the Americans to fly a truly capable machine. The only negative aspect of the flight was a corned beef sandwich smuggled aboard by Wally Schirra as another running gag among the astronauts. It brought the ire of management and a stern reprimand that "unauthorized" equipment or food would not be tolerated.

Gemini IV: America's First Walk in Space

Planning for GT-4 was in its advanced stages when GT-3 was hoisted on board the aircraft carrier USS *Intrepid*. However, the tentative EVA was not a sure thing. Several items still needed to be qualified. The suit itself (called the G4C) was a modified version of G3C used on GT-3. It provided for more air circulation and had two nylon micrometeoroid protection layers and two additional faceplates to reduce the intensity of the sun by 88 percent. Even politics played a role. There was some reluctance to perform the EVA at this early date as it might seem that the United States was reacting to Leonov's spacewalk from Voskhod II, when in fact the reverse was the case. All of the required tests and validations were completed by mid-May, and the EVA remained in the flight plan (Harland 2004, 50).

The biggest difference in the American spacewalk was that astronaut Ed White, who flew right seat in GT-4, had a handheld maneuvering unit to help him move in space. The small device, powered by two small oxygen cylinders, allowed measured thrust based on how hard White squeezed the trigger.

The biggest unknown in the GT-4 mission was the four-day duration itself—fully three times the flight of Cooper's thirty-four-hour MA-6. What data the Soviets had published on the effects of long-duration flights (Vostok-5 had the longest, with just under five days) revealed that the body appeared to exhibit mineral loss in the bone as well as possible cardiac problems and the unpredictable disorientation of space sickness. The medical teams kept a close watch on the biosensors as well as reports from the astronauts themselves. To help mitigate the symptoms, each astronaut used a bungee to exercise frequently (as much as could be done in the tight confines of the cockpit) during the flight to stimulate the metabolic actions of the body.

The duration of GT-4 also required the first real meal planning, which consisted of reconstituted freeze-dried and dehydrated foods to provide about 2,500 calories a day. What goes in must come out, and GT-4 was the first U.S. flight to deal with solid human waste. A plastic bag, which contained a biological neutralizing agent and an adhesive opening, was placed on the buttocks for defecation. After the event, the bag was sealed and the content "kneaded" to mix the neutralizer with the feces before storage. How easy this whole process would be in the weightlessness of space and the small volume of the spacecraft was yet to be discovered.

After its launch from the Cape, GT-4 was the first flight where control of the mission passed to the Manned Spacecraft Center in Houston (later named the Johnson Space Center). It was also the first time that a scheduled round-the-clock Mission Control was manned by three different sets of controllers known as the "red, white, and blue" teams. Likewise, a flight director, whose prominence became more visible to the public as the missions advanced in complexity and duration, headed each team. Many Americans came to know the names of Chris Kraft, Gene Kranz, and Glynn Lunney by way of the media reporting of Jules Bergman (ABC), Walter Cronkite (CBS), and Roy Neal (NBC).

GT-4 was the first launch telecast to Europe live by way of the *Early Bird* communications satellite placed in orbit the preceding year. Moreover, the "voice of Gemini Control" was now assumed by NASA's public affairs officer, Paul Haney, while Lt. Col. John "Shorty" Powers, who had provided the "voice of Mercury Control" faded into obscurity.

Two Air Force rookies were selected for the crew of GT-4; Capt. James McDivitt commanded the mission while Capt. Ed White flew right seat and perform the EVA. The countdown and liftoff was flawless, and within minutes the astronauts were in orbit and observing the second stage tumbling slowly behind them at a distance estimated to be about 200 yd. As noted on the previous flight, it appeared that it might be an easy task to move closer to the Titan's second stage to exercise the maneuverability of the OAMS system, and this became the first objective of their schedule.

However, the relative movement of two objects in slightly different orbits was more complex than it intuitively appeared. With each firing of the OAMS, the elusive second stage actually seemed to move away from them. NASA concluded that there were "strange forces" at work. It was determined that, even at these close distances, they had to use the computer to make the appropriate orbital changes. The chase ended because too much OAMS fuel was being used (Harland 2004, 52). It was time to move on to the EVA.

Ed White prepared his suit by going through the checklist that included attaching the 15 ft. tether that also provided the oxygen supply. McDivitt established the proper attitude and began the depressurization of the spacecraft. This first increment was to reduce the cabin's ambient pressure of 250 mm/Hg (one-third of sea-level pressure) to about 195 mm/Hg. If there were any leaks in White's suit, it would cause the cabin pressure to increase— there was no change. The next step was the complete spacecraft depressurization and opening of the hatch, which took place as the crew came into communication with the Hawaii tracking station.

White mounted a 16 mm camera to record the activities (there was no room or weight allocation for a TV camera). Twelve minutes after the hatch opened, White stated, "O.K. I'm separating from the spacecraft." As he floated out, he aligned the maneuvering unit with his center of gravity and pulled the trigger. The response was immediate and positive. He experimented with various power settings and thrust durations for the next few minutes. In the 1955 science fiction movie *Destination Moon*, one crewmember used a tank of compressed gas to maneuver in space to rescue another; science fiction had become reality in the short space of ten years.

With the maneuvering unit fuel exhausted, White described the sensations: "CapCom it was very easy to maneuver with the gun . . . this is the

Astronaut Ed White uses the maneuvering unit to move around in space during his EVA. Courtesy of NASA.

greatest experience . . . There is absolutely no disorientation associated with it." As White moved back to the spacecraft, the situation changed. Now he had to use the tether to move himself, and he found that his tugs and pulls sent him off at odd angles. Management in Houston became a little uneasy, and CapCom relayed that anxiety with the message: "The flight director says get back in!" (Harland 2004, 62). This was one of the few times that a direct order was issued from the ground-based flight director to a crew. There was some thought that perhaps White had become mesmerized by the experience, but that was not the case; it was just difficult to get himself back to the hatch.

White began the hatch-closure process but quickly determined that he could not exert any pressure on the hatch lever in the weightless environment. McDivitt had to hold White down, and it took three hands to complete the closure. In the process, White's heart rate, which had risen to 155 beats per minutes during the EVA, jumped up to 180. Now the Americans had learned what the Soviets had come to understand—movement in space required planning and handholds.

During the debrief following the flight, there was some discussion between McDivitt and White as to the usefulness of the maneuvering unit. White enjoyed the experience but quickly ran out of propellant, forcing him to pull on his tether to continue maneuvers. McDivitt recalled the gun as being "utterly useless" as it required precise aim through the user's center of mass in order to move in the desired direction without inducing unwanted rotation.

The two astronauts proceeded to move through the reminder of the mission without any further problems, although it became obvious that staggered sleep periods did not work out well since any communications or thruster activity tended to wake the sleeping partner.

A failure of the onboard computer required that the spacecraft fly a ballistic path during its reentry, resulting in slightly higher g loads. Other than some loss of bone mass and depleted blood plasma volume, the two astronauts came through the mission in good physical condition.

Of some significance at this time was the announcement of a fourth group of six astronauts in June 1965—in which the primary consideration had moved from being a pilot and engineer to being a scientist. The selection was made from a list of four hundred candidates supplied by the National Academy of Sciences.

Gemini V: Fuel Cell Problems

Planning moved forward for Gemini V, the first spacecraft to use the troublesome fuel cells needed to generate electricity for the eight-day flight. The flight's duration was determined to coincide with the time required to fly to the Moon, spend a day on the surface, and return. The significance was an important milestone in the upcoming Apollo program.

The Gemini V schedule also represented an interesting departure from the previous routine. In the past, hardware had determined the schedule. With the Gemini and Titan coming together with smoother precision and timing than the former Mercury-Atlas, now the astronauts slowed the schedule. Because of the inability of the Gemini V crew to get sufficient training in the simulators, Chief Astronaut Deke Slayton requested a two-week delay to the standard two-month launch cycle.

In addition to the long-duration flight, Gemini V astronauts Gordon Cooper and rookie Pete Conrad were to practice the first attempts at orbital rendezvous using a radar evaluation pod (REP) carried along in the equipment bay and released in orbit. In an effort to reduce cost, the REP simulated the Agena's radar response. Gemini would change its orbit and return to the REP to simulate the rendezvous technique using its onboard radar and a new computer. The range and rate data from the radar was analyzed by the computer (which held the relative positional data of the spacecraft) and then provided the appropriate spacecraft pitch and yaw attitude and the OAMS burn time to accomplish a rendezvous. A series of other scientific experiments and observations were scheduled to keep the duo occupied for the eight-day mission.

A problem with the fuel cells and then a thunderstorm canceled the launch at the T-10-minute mark on August 19, 1965. Recycled to Saturday, August 21, the launch occurred precisely at the scheduled 9:00 a.m. time. The ride itself was a bit disturbing to the two astronauts as the Titan exhibited some significant pogo effects (lateral oscillations). It was quickly determined that an incorrect control parameter in the guidance system had allowed the oscillations.

Within a half-hour of achieving orbit, the pressure in the fuel cell oxygen tank dropped. At first it was not critical since the fuel cells actually operated better at the lower limits. When that lower limit was exceeded, Conrad attempted to increase the pressure by electrically heating the oxygen to get it to produce more gas. There was no response. The heating unit had failed, and the pressure continued to drop. Without the proper pressure, the cell did not develop its rated electrical power (Harland 2004, 77).

The planned release of the REP took place on the second orbit, and the Gemini radar successfully interrogated its radar transponder. However, before any rendezvous maneuvers could take place, the oxygen pressure fell to the point where it was producing little electrical power. At that point, activities with the REP were canceled, and there was talk of returning the astronauts early. There was no real concern for safety because the reentry battery power, which had not yet been activated, was sufficient for their return, should that be necessary, and the batteries themselves provided power for approximately thirteen hours.

An effort was now made to save the mission. Some believed that the fuel cells might continue to operate at the low pressures and, with enough oxygen "boil-off," could return to normal operation without the heaters. The mission controllers worked through the electrical systems to turn off all unnecessary equipment—at the very least to reach the sixth orbit contingency recovery area.

As the oxygen pressure continued to fall, it reached a level from which it was not expected to make any recovery, and Cooper and Conrad prepared for a reentry. Mission Control was still determined to save the long-duration aspect of the mission, and when the pressure finally appeared to stabilize, Gemini V was given a "go" for completing at least a day in orbit. The fuel cell was the critical part to subsequent Gemini and Apollo missions. NASA was dedicated to finding a solution to the problem if possible—especially since the fuel cells were located in a part of the spacecraft that is not returned for examination. This was a perfect example of the role that Mission Control could play in having ground-based resources to augment the astronaut's needs when problems arose and real-time flight planning was needed.

Over the next several orbits, ambient heat provided a minute but perceptible increase in the oxygen pressure and, by the beginning of the second day, the mood had changed to one of guarded optimism. The REP module's batteries ran down, and that important aspect of the mission was no longer possible. The Soviet news agency TASS accused the United States of jeopardizing the lives of the astronauts for the sake of beating the Soviet's long-duration, five-day space flight.

Politics aside, astronaut Edwin "Buzz" Aldrin (who had yet to fly) had done his doctoral thesis on orbital rendezvous, so he worked with Mission Control to establish a series of OAMS maneuvers to intercept an imaginary satellite, now that the REP module was no longer available. These exercises worked effectively, but a live rendezvous had not actually been accomplished. The fuel cells continued their slow rebound, and the mission continued on a day-to-day basis with the pressure returning to about one-third of its intended value and consuming about one-third of its oxygen. The thrill of flying in space wore thin for the astronauts in an environment of empty wrappers and equipment floating around in the small cockpit, accompanied by unpleasant odors that were too intense to be removed by the activated-charcoal filters.

When the last day of the mission finally "dawned" (a figure of speech when dealing with a satellite in low-Earth orbit), both astronauts were eager for the return to Earth. The medical team was apprehensive about their ability to withstand the high g-forces after such a long period in space. A series of data-entry errors again resulted in the spacecraft missing the recovery area by an embarrassing 100 mi. However, within ninety minutes, Cooper and Conrad were plucked from the ocean, and a quick physical showed no more than the expected relatively low levels of metabolic deterioration.

Gemini V represented another turning point in the space race. Its long duration (finally exceeding the Soviet record), coupled with rendezvous capability and use of the temperamental fuel cells, set a new level of technology. It was almost a decade before the Soviets exceeded the Americans in space flight duration, and by then it would no longer be of major consequence.

Spirit of '76: Courage on Pad 19

Astronauts Wally Schirra and rookie Tom Stafford were assigned to Gemini VI; it was to be the first to rendezvous with a live Agena. As such, it was a short, two-day flight and did not have fuel cells on board. As the two lay on their backs during the GT-6 countdown on October 25, 1965, the Atlas-Agena roared off a nearby pad. After the Agena came around the Earth following its first orbit, GT-6 would launch for its rendezvous using the coelliptic method of the Hohmann Transfer.

All went well until five minutes into the flight of the Atlas rocket when the Agena D separated and initiated its firing sequence. Then all telemetry was lost. It is believed the Agena was destroyed in an explosion. The GT-6 flight was postponed. Not only would it be some time before a review board could determine the cause of the failure but there was also no Agena immediately available for a backup.

NASA was in a quandary as to how to proceed with the Gemini program. Then a surprising suggestion was made: unstack Gemini VI from Pad 19 and launch Gemini VII on its planned two-week flight. Then, immediately restack Gemini VI and launch while Gemini VII is still in orbit, and have the two rendezvous. The plan was called "Spirit of '76"—reflecting the reversed launch order as well as a play on the patriotic aspect with the slogan. It would require a change in how data from the spacecraft was returned to Houston (Harland 2004, 102). Although the Soviets had flown multiple manned missions, NASA had not contemplated having two manned spacecraft in space at the same time until Apollo, when the command module and the lunar module presented a similar problem.

With the alternative being a two-month delay in proving the ability to effect orbital rendezvous and the possibility of being upstaged once again by the Soviets, the pressure was on to see if the dual flight plan could really be accomplished. One of the few hardware changes was the installation of the Agena radar transponder in Gemini VII. When it was established that the concept was sound and doable, the change in the program was sent to President Lyndon Johnson to allow him to make the plan public, which he did on October 28. Gemini VI was unstacked the next day.

Station Keeping

While rendezvous was an obvious milestone on the road to the Moon, the ability of two spacecraft simply to fly along next to each other—station keeping—was a critical requirement. It was not known how difficult this might be, given the lack of knowledge and experience in bringing two satellites in close proximity. Previous attempts to station keep with the Titan's second stage had proved difficult. Thus, the success of Gemini VII was important.

The launch of Gemini VII occurred on December 4, 1965. Its objective was to acquire valuable data on the effect of longer flights on the human physiology. The spacesuit is uncomfortable to wear, even in the weightless condition of space where there are no "pressure points" on the body. To ease the burden of discomfort, rookies Frank Borman and Jim Lovell wore new lightweight spacesuits that were capable of being removed, even in the tight confines of the Gemini cockpit—but mission rules required that one astronaut must be suited at all times.

Following orbital insertion, Gemini VII was able to perform "station keeping" with the expended second stage, using newly developed procedures that allowed the spacecraft and the second stage to move to within 50 ft. of each other—another first (Borman and Serling 1988, 133). Following that exercise, the astronauts raised their orbit and began a more leisurely pace of scientific and engineering observation and experimentation for the next fourteen days. Activity on the ground, however, was more intense, and Pad 19 was ready to receive GT-6 the day after the GT-7 launch.

The erection of GT-6 was smooth and without any significant problems, and the countdown proceeded to a launch on December 12. As the count reached the magic "zero" mark, the turbopump began its characteristic start-up whine, and the flame emerged from beneath the rocket. Almost immediately, the two Titan engines shut down. Mission Control made the first call indicating the situation: "We have a shutdown, Gemini VI." As there had apparently been some motion in the booster, a call from the spacecraft indicated, "My clock has started." Flight Director Chris Kraft countered with "No lift-off, no lift-off." The astronauts themselves could not see what was happening, and there was anticipation that one or both would pull the D-ring and eject them clear of what threatened to be a fireball

(Harland 2004, 119). Their discipline held, and Schirra later commented that his distrust of the ejection system was such that it would have taken more than a premature shutdown before he would commit himself to using the "escape of last resort."

The astronauts were now sitting on over 300,000 lb. of volatile toxic chemicals, and some programmed sequential postlaunch events had begun with the start of the mission clock onboard the spacecraft. Mission Control and the blockhouse team immediately moved into action to "safe" the rocket. Guenter Wendt, the NASA technician who had secured all the previous astronauts in their spacecraft, positioned himself and his team to begin a hasty egress of Schirra and Stafford as soon as the booster had been declared safe. More than an hour and a half elapsed before they descended from the spacecraft. The thoughtful and somewhat courageous actions of the astronauts in not ejecting saved the mission. Had they ejected, the spacecraft would have been destroyed.

First indications were that one or more sensors in the Titan booster had detected a slight decay in thrust after engine start, but it took a full day of searching before a dust cap was discovered in a virtually inaccessible area of the turbopump gas generators. It should have been removed during the booster assembly process at the factory. Was there time to salvage the rendezvous mission? Following an assessment of the situation, it was determined that GT-6 could be recycled for three days hence—December 15.

The launch occurred on the designated day at the specified time. Gemini VI-A (the suffix recognizing the second effort) was in orbit some 1,250 mi. behind Gemini VII, but in a slightly lower track (100 mi. by 170 mi.) to allow the astronauts to catch their target. The first maneuver, a slight phase change to raise the perigee, occurred two hours and eighteen minutes into the flight, when the two spacecraft were about 450 mi. apart. This was followed by a plane adjustment of .007 degrees by yawing the spacecraft 90 degrees to the orbital track and firing the OAMS for forty seconds.

Several small corrections now put the two spacecraft in a position where the Gemini VI radar could lock onto the transponder carried in Gemini VII, which it did at 271 mi. The computer called for another burn at three hours and forty-seven minutes into the flight, using the aft thrusters for fifty-four seconds. The computer now switched to the final rendezvous mode. As they crossed the African continent, the final burn was made while both spacecraft were still in darkness. The radar called the distance as less than a quarter of a mile and closing at about 6 ft./s. The spacecraft

Gemini VII as seen from Gemini VI following the first successful rendezvous in orbit. Courtesy of NASA.

continued to move closer until they were separated by only 100 ft.; it was five hours and fifty-six minutes into the flight. Schirra canceled the closing rate with another small OAMS burn, and the two spacecraft began station keeping only yards apart. The first rendezvous in space had been achieved (Borman and Serling 1988, 143).

There was a lot of celebrating in Mission Control as the astronauts traded comments about the long beards on the occupants of Gemini VII. The crews took turns maneuvering about one another—station keeping. They noted on the rear of each craft the trailing adhesive tape used to seal the separation area between the spacecraft and the booster to keep dust and debris from entering the small opening while it sat on the launchpad. Following the maneuvers, Gemini VI used its OAMS to separate the two orbits to ensure the two craft would not inadvertently run in to each other during the astronauts' sleep period.

The following day, as Gemini VI prepared for reentry, astronaut Stafford gave a report that had Flight Director Chris Craft sitting upright: "This is

Gemini VI. We have an object, looks like a satellite moving from north to south, up in a polar orbit. He's in a very low trajectory . . . looks like he may be going to reenter very soon." By now, all of Mission Control had their attention riveted on the report. "Stand by . . . it looks like he is trying to signal us." At this point Schirra played *Jingle Bells* on his harmonica, and Stafford added the bells—it was little more than a week until Christmas. It was another "gotcha" by the playful astronauts (Stafford and Cassutt 2002, 75).

The return of Gemini VI went well. With the corrected lift-to-drag ratio now programmed into the computer, the spacecraft landed within 8 mi. of the recovery forces. It had been a short and successful flight for the crew of Gemini VI.

But the spirits of the Gemini VII crew, temporarily buoyed by the rendezvous and banter with Gemini VI, sank a bit as they returned to their tedious and somewhat uncomfortable mission, which had two days to completion. It was obvious that long-duration space flights were as much a psychological as physical problem, particularly when living in tightly confined spaces was factored into the equation. Toward the end of day thirteen, the task of the astronauts became one of cleaning up the spacecraft, putting their spacesuits back on once again, and preparing for their return to Earth, which was uneventful.

The year 1965 saw the end of Soviet dominance in space. While it could not be recognized at the time, the overall capabilities exhibited by the two sparing nations started to show significant differences. With successful lunar and planetary probes as well as flying the new second-generation two-man spacecraft, America's vast technological infrastructure was steadily showing its supremacy. The initial five-manned Gemini spacecraft, launched over a period of just eight months, had effectively achieved two of the three primary objectives (EVA and rendezvous). The third and critical docking exercise was yet to be accomplished, along with the testing of different rendezvous modes and EVA equipment. These were demonstrated in the coming year, while the Soviet's second-generation spacecraft, Soyuz, was still two years from its first flight.

Gemini VIII: Docking Achieved

Gemini VIII, with rookies Neil Armstrong and David Scott, sought to achieve the elusive docking milestone and test a sophisticated human maneuvering unit with an ambitious EVA. It was scheduled for a mid-March

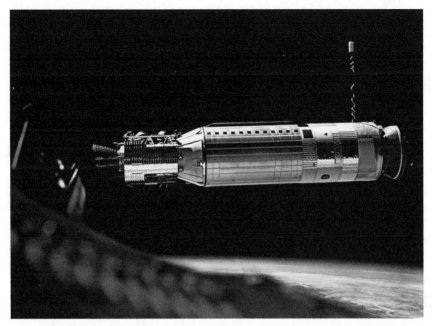

The Agena satellite as seen from the Gemini following rendezvous. Courtesy of NASA.

1966 launch. The various simulators were shared with the crew for the following Gemini IX mission. Astronauts Elliot See and Charles Bassett flew up from Houston to St. Louis to begin their work with the docking simulator since their mission involved the first restart of the Agena. As their T-38 approached Lambert Field where the McDonnell plant was located, poor weather dictated an instrument approach to minimums. The T-38, a small, two-seat, twin-engine, high-performance jet trainer, was used by NASA for the personal transportation and flight proficiency of the astronauts.

As their aircraft broke out of the low overcast, See recognized that they were not in a good position to continue the approach to touchdown. Inexplicably, he executed a sharp right turn at low altitude, apparently to perform a circle-to-land maneuver instead of climbing straight out as the missed approach procedure called for. With the afterburners roaring, an indication of See's recognition of the critical situation, the aircraft clipped the building in which their spacecraft was completing its final assembly and crashed, killing both See and Bassett (Evens 2009, 306). Astronauts Tom Stafford and Eugene Cernan, who had accompanied See and Bassett in a second T-38, landed safely and eventually were named as the prime crew for Gemini IX. The hazards of being an astronaut were not limited to the space missions, as demonstrated again within a year.

The Docking of Two Satellites

The ability to join two spacecraft and their critical electrical connectors was an obvious requirement for the lunar program—and for follow-on space station access. While the math and physics worked to bring the object into rendezvous, the engineering necessary to join the two was formidable.

A set of hydraulic dampers on the Agena acted as shock absorbers to soak up the impact of the Gemini. Small electric motors came to life and pulled the Gemini into what is referred to as a "hard dock" configuration. This allowed the two spacecraft to achieve a relatively rigid assembly—and for the electrical connectors to be lined up and plug compatible.

The Atlas-Agena target for Gemini VIII successfully achieved orbit on March 16. Astronauts Armstrong and Scott followed 101 minutes later on board GT-8. Using the basic coelliptic rendezvous technique, the Gemini twins (Armstrong and Scott) were station keeping a few yards from the Agena six hours later and preparing to perform the historic first docking. The target docking adapter (TDA), located in the nose of the Agena, consisted of a cone into which the nose of the Gemini was inserted.

Following an intense scrutiny of the Agena systems to verify that all was well, Armstrong lined up with the TDA and moved Gemini into it at a speed of about 8 in./s. The spacecraft were still on the night side of Earth, with a variety of position and status lights illuminated on the Agena, when the two spacecraft made contact and the mooring latches engaged.

Armstrong announced "Flight, we are docked" to confirm the telemetry indications that Mission Control was seeing on the ground. The last of the three critical aspects of rendezvous had been achieved—rendezvous, station keeping, and now docking. The ability to link two independent spacecraft (docking) was the last unproven aspect—the cornerstone of the lunar orbit rendezvous technique that America had gambled would take them to the Moon "before this decade is out."

The various systems of the two spacecraft were now electrically integrated to allow the astronauts to issue a variety of commands. Either the Gemini spacecraft or the Agena could control the attitude of the docked assemblies. Looking out the windows, the astronauts could see a set of illuminated instruments positioned on the Agena to provide direct readings of various conditions. All seemed to go smoothly.

Within thirty minutes of the docking, a slow yawing (left turning) of the assembly was noted. Scott decided to switch off the Agena's attitude control only to find that the yaw and roll moments to the vehicle increased. For some reason both astronauts felt that the anomaly was being input by the Agena and thought that their efforts to disengage the Agena control had failed. Soon the movement was so severe that they realized they had to disconnect from the Agena for fear the forces being experienced would tear the two assemblies apart (Hansen 2005, 260).

They were fortunate that the undocking process was quickly and effectively achieved—without ramming the two together. However, with the Gemini spacecraft now freed from the Agena, its roll and yaw rate increased rapidly, making it obvious that the problem was with their spacecraft and not the Agena. The roll rate of the Gemini and its occupants reached 360 deg./s. As a last resort, they systematically shut down as much of the OAMS system as possible and activated the reentry attitude control. Their wild ride was finally brought under control (Harland 2004, 157).

It was conceivable that the motion rates Armstrong and Scott had been subjected to may have significantly impaired their ability to cope with the problem. They were fortunate to have recovered from an apparent electrical short in the number eight thruster as soon as they had. As the event had occurred when the spacecraft was out of direct communication with Mission Control, the appearance of two separate craft to the tracking ship *Coastal Sentry Quebec*, on station in the Pacific, presented a puzzle that discussions with the crew soon resolved. With the activation of the reentry control system, the mission rules dictated that the flight be terminated at the earliest opportunity, and reentry and recovery (by secondary resources) went smoothly.

Over the next several weeks, a review of the incident caused the recovered spacecraft (and those yet to be launched) to be carefully examined. To virtually everyone's surprise, there were several areas where electrical shorts were possible. A further inspection of the yet-to-be-launched Gemini IX also revealed many areas that could present problems. It was obvious to NASA and McDonnell Aircraft that the quality of workmanship and design may have been sacrificed to meet deadlines, and quality had to be improved to avoid a future disaster. Within the year, similar poor design and workmanship by another contractor had devastating consequences.

Gemini IX: The Angry Alligator

The Gemini IX mission with the backup crew of Tom Stafford and Eugene Cernan (who replaced See and Bassett) prepared for their launch on May 17, 1965. However, for the second time, the target Agena failed to achieve orbit (the Gemini VI target had been the first). On this occasion, the problem was with the Atlas. The GT-9 launch was scrubbed, and an alternative target for rendezvous was prepared.

Several months earlier McDonnell Aircraft had been tasked by NASA to prepare a simple docking collar for launch by the Atlas as it was recognized that the Agena could be a long-term problem. This less expensive stand-in ($6 million versus $9 million) did not provide any functions other than attitude control and the ability for the Gemini spacecraft to dock with the assembly. The satellite was called the Augmented Target Docking Adapter (ATDA), and NASA turned to it to allow the remainder of the Gemini IX experiments, particularly a complex EVA, to be flown. With three more missions planned, Gemini IX did not need to fire up an Agena.

The backup plan was put into action, and an Atlas with the 2,200 lb. ATDA was launched successfully on June 1. Now another problem clouded the mission. Telemetry indicated a problem with the ATDA's protective shroud (nose cone). Subsequently, it was decided to launch Gemini as there was no way to confirm that the specific problem and the primary rendezvous procedures could be accomplished.

Stafford and Cernan followed the launch of the ATDA two days later, on June 3. Stafford, having flown on Gemini VI, became the first person to have a second space flight. The rendezvous used a slightly different technique that required only three orbits (termed M=3) instead of four. On joining up with the ATDA, it was obvious that there was a problem with the shroud; the rear stainless steel band was still in place although the forward portion had separated. Stafford characterized the appearance to Mission Control as looking like an "angry alligator." There was some thought of dislodging it by bumping it with Gemini or cutting it during an EVA, but it was decided that the risk was too high for either option (Cernan 1999, 122).

Three rendezvous maneuvers were accomplished during the first day, and then Gemini moved to another orbit to ensure there was no conflict with the ATDA during the sleep period. The second day was devoted to the planned EVA, the second in the U.S. program, following Ed White's spacewalk by one year. The primary objective was to test an Astronaut Maneuvering Unit (AMU) that allowed Cernan to become a separate satellite

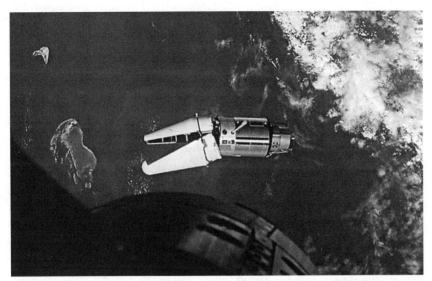

The ATDA failed to shed its nose cone and appeared to the crew of Gemini IX as an "angry alligator." Courtesy of NASA.

himself. The first part of the EVA went well since it was composed of relatively simple tasks that did not involve the AMU, which was stored in the back of the equipment adapter.

Following an initial thirty-minute work period, Cernan took a planned rest and then proceeded to the equipment bay to get into the AMU. He quickly discovered that the plan for strapping on the backpack AMU was flawed. A new material called Velcro had been used extensively to provide the astronaut with the ability to get some leverage, but this proved insufficient. Without adequate handholds, it was almost impossible to prepare the AMU or position himself to attach to it. As his pulse soared to 180 beats per minute and respiration accelerated, the small chest pack that provided the spacesuit environment was unable to keep the water vapor and temperature within the established tolerances, and his faceplate started to fog over (Cernan 1999, 136).

At about ninety minutes into the EVA, Cernan and Stafford, with the concurrence of Mission Control, decided that it was futile to continue the exercise. Cernan returned to the spacecraft in what appeared to be an almost completely failed mission. Following another sleep period, Gemini IX was brought back to Earth. If there was any redeeming value to the mission, it was that the new reentry program put the spacecraft down within a mile of the recovery ships.

Examination of the photos taken of the "angry alligator" ATDA showed that the two electrical leads that should have been connected to activate the separation of the shroud were taped neatly to the side of the ATDA. It was reported that the technician who was responsible to connect them had been suddenly called away to attend to his wife who was giving birth, and the person assigned to close out the prelaunch procedures didn't find them on his checklist and assumed that another tech would make the appropriate connections (Harland 2004, 189).

Cernan was the first astronaut to attempt to do any work during an EVA, and his experience was a revelation to NASA. The feedback he provided allowed subsequent EVAs to be better planned. His EVA resulted in him losing 13.5 lb. of body weight. The problems encountered led to the need for building the neutral buoyancy facility at the George C. Marshall Space Flight Center in Huntsville, Alabama. This large swimming pool–like laboratory allowed astronauts to work in an environment that closely simulates the weightless conditions in space.

Gemini X: Firing up the Agena

Sometimes failure breeds success. With the knowledge Cernan gained about movement during an EVA, the next missions were better prepared—Gemini X would be extraordinary.

Six weeks later the next scheduled Atlas-Agena successfully achieved orbit, followed by Gemini X with John Young and Mike Collins 101 minutes later. A late afternoon schedule was dictated by a backup plan, should the target vehicle, the GATV 5005 not achieve orbit. Agena 5003, from the aborted Gemini VIII the preceding March, had been moved to a higher orbit and was still available, should it be needed. The initial rendezvous (M=4) with Agena 5005 did not go well due to a variety of factors, including the inability to get good navigation sightings. Fortunately, the spacecraft had 50 percent more OAMS fuel than previous flights. Performance enhancements to the Titan and the availability of space in the equipment adapter allowed for considerable growth potential. The two spacecraft were docked within six hours of liftoff and ready to perform the last major objective of the Gemini program—a burn of the Agena engine to boost the astronauts into a higher orbit.

With the Gemini firmly latched into the docking collar on the front end of the Agena, the astronauts were facing backward, looking at the Agena. It was the first time this configuration had ever been attempted because of the normal constraints of g-forces. Facing backward, Collins and Young experienced only about one negative-g (sometimes called "eyeballs out") because the weight of the combined spacecraft was about 14,000 lb. and the primary propulsion system (PPS) of the Agena was 15,000 lb. of thrust (Collins 1974, 211).

Although their view of the Agena was somewhat obscured by the docking collar, John Young vividly recalled the events: "At first the sensation I got was that there was a 'pop' and then a big explosion. . . . We were thrown forward [restrained by their harnesses]. Fire and sparks started coming out of the back of that rascal. The light was something fierce" (Young 2012, 93). Their new orbit took them almost twice as high as any human had ever ventured, 476 mi. above Earth, where its curvature was more pronounced.

Following a nine-hour sleep period, the PPS fired again, lowering the orbit to permit a later rendezvous with Gemini VIII's Agena 5003. Again, Young was impressed with the burn. "It may be only one-G but it's the biggest one-G we ever saw." Mike Collins performed a "stand-up EVA," which did not involve actually leaving the spacecraft; he simply opened the hatch and floated only a foot or so above his seat to perform some experiments. The EVA was cut short when both astronauts experienced a burning in their eyes that was later traced to the lithium hydroxide that removed the carbon dioxide from their environment. The problem immediately cleared when the hatch was closed and they turned off the spacesuit environment system.

The next objective was to rendezvous with Agena 5003. They undocked from their Agena and executed several burns of the OAMS that soon put them station keeping with the now inert target. Collins then proceeded to perform a major spacewalk in which he maneuvered over to the Agena and retrieved a micrometeorite package that had now been in space for three months (Harland 2004, 217). A handheld maneuvering unit, similar to what Ed White had used, aided him. He still experienced considerable physical effort to complete the seventy-two-minute EVA. Following a final sleep period, Gemini X prepared for a reentry that dropped them within 3 mi. of their primary recovery ship.

At this point in America's program, preparation for the first Apollo flight (which had now slipped into early 1967) was consuming more of the time of both the astronauts and Mission Control. There was some talk of canceling the last two Gemini flights. These flights explored several areas that Apollo would have to spend time on; one in particular was an M=1 rendezvous. This scenario simulated the lunar module liftoff from the Moon and the

Astronaut John Young is hoisted on board the recovery helicopter. Courtesy of NASA.

one-orbit rendezvous with the Apollo command and service module. Any techniques that could be accomplished with Gemini would shorten the time to the lunar landing, so the last two flights were approved.

Gemini XI: M=1 Rendezvous

Liftoff of Gemini XI with Pete Conrad and Richard Gordon occurred on September 12, 1966, for a first orbit M=1 rendezvous with its Gemini Agena target vehicle, Agena 5306. The launch had to occur within seconds of its scheduled time to perform the maneuver successfully. One hour and thirty-four minutes later, Gemini XI was docked with its Agena. The ability to rendezvous and dock in such a short period surprised even a few in

Earth's Radiation Belts

Often referred to as the Van Allen belts, Earth's radiation belts are intense areas of charged particles (primarily electrons and protons) held in place by Earth's magnetic field. There are two permanent belts—the lower beginning at about 600 mi. above Earth. Most of these particles are a result of the solar wind, but some are due to cosmic rays. Solar activity can periodically create additional belts. Although the intensity varies, they are considered a hazard to both astronauts and electronic equipment.

They were discovered by experiments on board the early U.S. satellite *Explorer III*. They are named for the pioneering astrophysicist Dr. James Van Allen, who created the instrumentation and analyzed the results.

NASA, who had their doubts about this most difficult rendezvous. Launch, phasing, plane, apogee, and terminal-phase initiation tasks all had to be accomplished within a one-hundred-minute span.

The mission was a combination of scientific experiments and observations, several EVA activities, and the use of the Agena engine to achieve a higher apogee (853 mi.). The new apogee was positioned to avoid the Van Allen radiation belt. Conrad and Gordon exhibited the same thrill with the Agena ignition and the view from high above the Earth. "You can't believe it," exclaimed Conrad. "I can see all the way from one end, around the top . . . about 150 degrees. . . . The water really stands out and everything looks blue. . . . There is no loss of color and details are extremely good." At one point, when a rest period occurred during an EVA, both astronauts nodded off while Gordon floated in space (Harland 2004, 242).

Like the previous EVAs, however, Gordon also experienced the inability to position himself effectively for the various activities, despite more hand-holds, and one EVA session was ended early due to his fatigued condition. Several important experiments were conducted using a 100 ft. tether between the two spacecraft following an undocking.

Positioning the heavier Agena between Earth and the spacecraft, the effect of the gravity gradient (as had already been demonstrated with some unmanned satellites) confirmed that the heavier object remained closer to Earth, stabilizing the assembly. A second task involved rotating the two spacecraft around each other to create artificial gravity. Although the

A tether was rigged between Gemini XI and the Agena to perform an experiment with centrifugal force and gravity gradient. Courtesy of NASA.

rotation was only once every six minutes, and the astronauts indicated they could feel no gravity force, unsecured items in the cabin floated away from the rotational center. The reentry on the forty-fourth orbit was conducted solely by the autopilot for the first time and resulted in a splashdown only 3 mi. from the recovery force.

Gemini XII: All Objectives Accomplished

The flight plan for Gemini XII, which had been constantly revised since astronauts Jim Lovell and Buzz Aldrin began training more than five months before liftoff, was still in a state of flux when Gemini XI returned to Earth. While all of the primary objectives of the Gemini program had been achieved, the inability to perform an effective EVA still eluded NASA.

The Air Force had yet to strap an astronaut into its AMU and was anxious to fly it again. However, NASA was adamant that the basic ability of the crew to do effective work during a spacewalk had to be addressed. To accomplish this, it was decided to make the EVA the focus of an experiment using a set of "busy boxes" that Aldrin used to perform a variety of mechanical tasks that included threading and applying torque to bolts, moving switches, and plugging electrical connectors. These types of tasks were critical to building structures in space, as visualized by such luminaries as Wernher von Braun, Arthur Clarke, and Sergei Korolev. To aid in preparing the astronaut for EVA, NASA had by this time installed the Neutral Buoyancy Laboratory in Houston. By immersing suited astronauts under water, their movements closely replicated the weightless environment of space. This proved an invaluable tool for decades to come (Harland 2004, 250).

What emerged from the water-training simulations were forty-four handholds, foot restrains, and tethers that could be positioned to hold an astronaut in a specific place while activities were being performed, without having to use a hand to maintain that position. On November 11, 1966, as Lovell and Aldrin walked to the Gemini spacecraft from the elevator that had brought them to the top of the giant Titan rocket, each wore a sign on his back with a single word that expressed the finality of the mission—"The" on one and "End" on the other.

Astronaut Buzz Aldrin training for his Gemini XII EVA in the Neutral Buoyancy Laboratory as scuba divers observe for safety. Courtesy of NASA.

The launch of both the Atlas-Agena and the Titan were flawless, and Gemini XII began an M=3, coelliptic rendezvous. The radar failed when the two ships were still 70 mi. apart. This proved not to be a problem since Aldrin (often called Dr. Rendezvous by his cohorts at NASA) used a sextant and navigation charts to complete the task (Aldrin and McConnell 1989, 154).

An anomaly in the thrust-chamber pressure of the Agena PPS during its initial burn into orbit indicated a possible problem, and it was decided not to restart the Agena for a planned orbital change to a higher apogee. This canceled activity allowed another event (which had been previously considered) to take place: the observance of a solar eclipse from space. Little more than seven hours into the mission, Lovell positioned the spacecraft into a slightly modified orbit so that it would fly through the shadow of the Moon (its umbra) as it moved between Earth and the sun. It was a short, seven-second event that was captured on film for later analysis by astronomers on Earth (Harland 2004, 257). A two-hour stand-up EVA was then conducted, followed by a sleep period.

The next day problems with a fuel cell and an OAMS thruster caused some concern, but these had workarounds, and the long EVA with "busy box" tasks began. Using frequent rest periods, Aldrin's respiration remained around 20 per minute and his heart rate around 100. A repeat of the tether experiment of Gemini XI was conducted with a 100 ft. tether linking the two spacecraft. The EVA ended after two hours—humans could effectively work in space. Gemini XII returned after 94 hours which, when added to the previous human flight, gave the United States a commanding 1,993 hours in space compared to the Soviet's 507.

The Gemini program was a resounding success. It had demonstrated that humans could fly reliably for periods of up to two weeks in the hostile environment of space and to perform useful work there. It had allowed the various techniques of rendezvous, station keeping, and docking to be performed. Moreover, Gemini had shown that, with redundant systems and flexible flight plans, equipment failure could have a high degree of tolerance (Harland 2004, 271). Finally, Gemini had revealed that the technological depth of the American program had overtaken the Soviets. But it had also engendered, in the United States, a degree of complacency with respect to the high-risk aspect of their spacecraft design—the pure-oxygen environment. Within months, this aspect almost crippled the American lunar program.

15

PROJECT APOLLO

Defining the Command and Service Modules

As envisioned in 1959, Apollo was a three-man spacecraft initially developed as a follow-on to Project Mercury. The Mercury spacecraft and its systems had been constrained by the lifting ability of the Atlas and the compressed time element to compete with the Soviets. Although Apollo initially did not have these limitations, it had no booster until the Saturn IB became available. With a 13 ft. diameter, the 10.5 ft. high, cone-shaped craft was considerably wider than the diameter of either the Atlas or Titan. The basic thinking was that Apollo would serve as the basis for whatever missions followed Mercury. The Gemini spacecraft (unplanned at that time) ultimately filled the gap between Mercury and Apollo. It provided an interim capability to develop rendezvous and docking techniques.

When President Kennedy established the lunar-landing goal in May of 1961, Apollo became the primary spacecraft for that mission. In that role, the reentry protection for the return to Earth required considerably more coverage around the entire spacecraft—not just the heat shield. The spacecraft dipped into the upper layers of the atmosphere on its return to dissipate the high energy levels generated by the 25,000 mph speed. Using an offset center of gravity to achieve a degree of lift, it skipped like a flat stone across a lake. To some extent, this meant that upper portions of the cone were exposed to the heat pulse.

The heat shield, a phenolic epoxy resin ablative covering, varied in thickness from 2.5 in. at the base to .5 in. on the upper portion of the conical cabin. Provisions were made for openings for the thrusters, antenna, hatches, and four windows, which were a special high-temperature resistant glass. As a result of the need to envelop virtually the entire outer portion in ablative covering, equipment access could not be achieved through

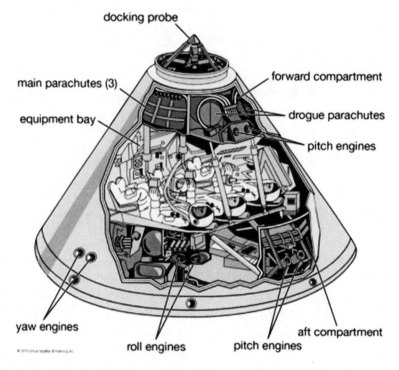

The basic layout of the Apollo command module. Courtesy of NASA.

external openings for maintenance and modification prior to launch, as had been done with Gemini, but would have to revert to internal access, similar to Mercury. This configuration, as it had with Mercury, delayed the ability to install, test, modify, or replace components.

The first design called for a two-gas environment for the cabin (nitrogen and oxygen) at one-half atmosphere (362 mm/Hg) because of the unknown physiological effect of breathing pure oxygen for durations of up to two weeks. Subsequent ground-based experiments showed no significant problems, and, because the two-gas system weighed at least 35 lb. more, the pure oxygen environment of Mercury and Gemini was adopted (at one-third of sea level atmospheric of pressure—258 mm/Hg).

Because the essential design of the spacecraft was conceived before the lunar orbit rendezvous (LOR) mode had been established, there had been many changes along the way. However, the basic cone-shaped reentry vehicle remained a fundamental part of the design. With LOR established as the mode to the Moon, and with the Saturn V as the vehicle, the weight of the command module was set at 12,000 lb., the service module at 52,000 lb., and the lunar lander 22,000 lb. This total had to be kept within the Saturn V's ability to accelerate 88,000 lb. to escape velocity. All through the development period, weight was a contentious factor as tradeoffs became critical. The basic structure of aluminum honeycomb was bonded between aluminum sheets.

Apollo used an attached service module configured to support specific assignments (for durations of up to two weeks) and added structures for possible lunar-landing stages as well. The service module, similar in concept to the Gemini adapters, contained the electrical power, attitude control, and an orbital maneuvering engine. It was discarded before reentry and burned up in the atmosphere.

The diameter of the cylindrical service module matched the outer 13 ft. cone of the spacecraft and was 15 ft. long plus an additional 6 ft. for the service propulsion system (SPS) engine's protruding nozzle. The SPS had a thrust of 20,000 lb. and could be restarted as many as fifty times for periods as short as one-half second or as long as twelve minutes. Its job was to (1) provide midcourse corrections to the lunar trajectory, (2) slow the

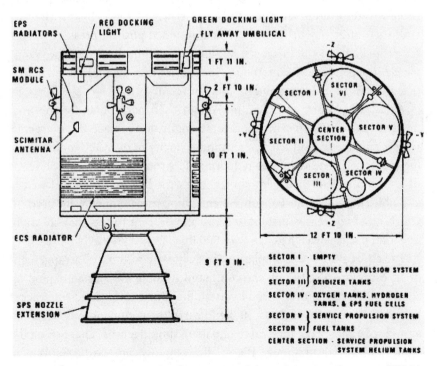

The various systems within the Apollo service module configuration. Courtesy of NASA.

spacecraft so that it entered lunar orbit, and (3) accelerate the spacecraft out of lunar orbit and back toward Earth. Each of these tasks was mission critical.

Failure to restart would put the astronauts' lives in peril. The SPS engine had to fire every time it was commanded to perform. The choice of hypergolic propellants (those that ignite on contact) was for obvious reasons, one being the length of the mission—typically eight days. Even though cryogenic propellants offered superior performance, their still-uncertain nature, in addition to their propensity to "boil-off," virtually eliminated any thought of their use. Thus, a mixture of 50 percent unsymmetrical dimethylhydrazine diluted with 50 percent hydrazine was the fuel, and nitrogen tetroxide was the oxidizer. Each propellant tank contained a sump tank at the outlet with a retention screen to ensure that there was always an initial flow to start the engine. Acceleration settled the fluids within the tanks for continued supply in the weightless condition.

The reaction control system (RCS) provided attitude control, supplying 28 lb. of thrust for as little as twelve milliseconds using hypergolic propellants. The problem of settling the RCS propellants in their tanks prior to firing was resolved in an innovative manner with a Teflon bladder in each tank surrounded by helium gas, which provided positive pressure to feed the propellants to the thrusters. Similar to Gemini, there were two completely independent systems, one in the service module for mission activities, and one in the command module for reentry.

As with Mercury and Gemini, Apollo used parachutes for the final descent back to Earth. This took some engineering, since the spacecraft weighed 12,000 lb. and fell at a terminal velocity of over 300 mph following reentry. That velocity was reduced to less than 200 mph by a series of drogue chutes and then pilot chutes. With the limited space and weight available, there was no backup system; thus, three chutes were used, of which any two were sufficient to arrest the descent to less than 20 mph. The total package weighed less than 500 lb.

The abort system was similar to Mercury, with a "tractor" arrangement of an escape rocket to pull the command module clear of an exploding Saturn V. Weighing over 8,000 lb. itself, the four-nozzle, solid-fuel rocket provided 150,000 lb. of thrust. It could carry the command module to an altitude of almost 1 mi. (and off to one side), should an abort happen on the launchpad. As with Mercury, the Apollo escape system used a test vehicle (called Little Joe II) to verify its ability to save the command module in the event of a catastrophic event.

The decision to award the contract for building Apollo to North American Aviation in 1962 was somewhat complex. Martin Aircraft submitted a technically superior bid and scored the highest points in virtually all areas of the selection criteria. North American Aviation's bid displaced the Martin bid when the point allocation was unexpectedly redistributed to place a higher emphasis on cost. However, when LOR mode was selected, North American Aviation realized that the lunar module would not be a part of the contract. It would not be a North American Aviation product that took the astronauts to the surface of the Moon.

There was some political maneuvering to allow North American Aviation to build the entire assembly. The influence extended into the Kennedy White House through the efforts of the presidential science advisor, Jerome Wiesner. Had it not been for Kennedy's preoccupation with the Cuban missile crisis, the decision might have been revisited. The lunar module contract ultimately went to the Grumman Aircraft Engineering Corporation and was the first spacecraft designed purely to operate in the environment of outer space and on the lunar surface in one-sixth Earth's gravity. There were no provisions for the craft to be able to reenter the Earth's atmosphere, as that was the job of the Apollo spacecraft.

Engineering the Lunar Module

The contractual negotiations during this era showed the close and unique relationship between NASA and suppliers such as Grumman. Although Grumman won the lunar model (LM) contract based on their design, the spacecraft that landed men on the Moon seven years later bore only a passing resemblance to the initial proposal (Kelly 2001, 53). The final design was the result of collaboration between Grumman and NASA engineers.

These pioneers, many of whom had been with Grumman since World War II, had to make the transition from designing and building navy fighters to high-tech space vehicles—an interesting and challenging move for most. As Thomas Kelly, a life-long employee of Grumman and the LM chief engineer, notes, "There were no precedents for what we were doing" (Kelly 2001, 53).

The LM was a unique part of the Apollo project in that it had to perform many of the basic functions of rest of the vehicle—and some that had never been attempted before. It had to land vertically, provide its own launchpad, depart vertically, employ a RCS about all three axes, achieve lunar orbit, and perform rendezvous and docking—all while providing life support

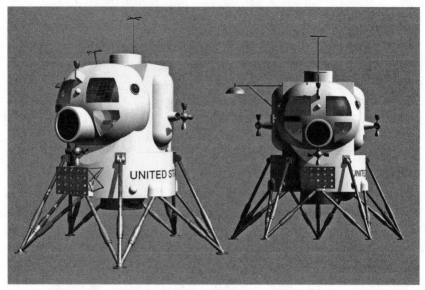

An early configuration for the lunar excursion module (as it was then called). Note that it had five landing legs. Courtesy of NASA.

for two crewmembers. And this had to be accomplished under extreme weight-control limits.

The module had originally been known as the lunar excursion module (LEM) but by 1966, the term "excursion" was dropped to avoid confusion in the minds of legislators who were then debating the merits of a follow-on to the Apollo Program. The LM (as it was renamed) did not have the ability to perform excursions remote from the lander itself, as some believed the earlier name implied.

Because the concept of LOR implied taking only what was needed as far as it was needed, the LM started life as a two-stage spacecraft. The first stage was the descent stage, which contained a 10,000 lb. thrust descent propulsion system (DPS) engine that fired twice. The first time it fired, the craft would drop out of lunar orbit toward the lunar surface, and the second time it fired, it would slow the LM to a soft touchdown on the Moon.

This second firing demanded much from the engine. It required a throttle to allow the computer, or the LM pilot, to change the thrust setting. This permitted the craft to approach the surface at a slow rate of descent and to compensate for the constantly changing weight of the craft as it burned off fuel during the descent (it consumed more than half its weight during the landing).

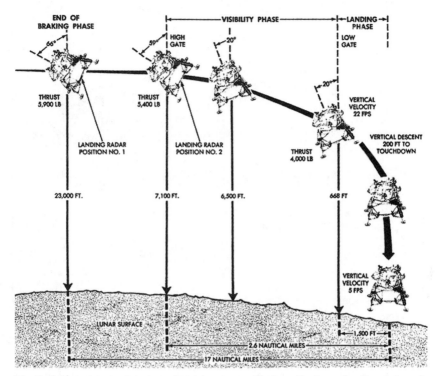

The final few minutes of a nominal LM descent profile from "high gate" to "touchdown."
Courtesy of NASA.

Initially, the Rocketdyne division of North American Aviation was considered as the contractor for the DPS because of their vast experience. However, the Space Technology Laboratories (later renamed TRW Inc.), who demonstrated a high degree of innovative and reliable engineering, ultimately received the contract.

In addition to containing the descent engine, the LM also had to provide the legs on which the craft sat on the lunar surface and provide a stable platform from which the ascent stage could be launched. The initial design called for five legs, but this was soon changed to four. The legs had to fold up during launch from Earth so that the LM fit within the S-IVB stage forward area.

Following the firing of the S-IVB to place the vehicle on a path to the Moon, the command and service module (CSM) separated from it and turned 180 degrees to point back toward it. It was critical that the S-IVB's attitude control system hold the assembly steady so that the Apollo spacecraft could approach and dock with the LM that was held fast to it. Following the

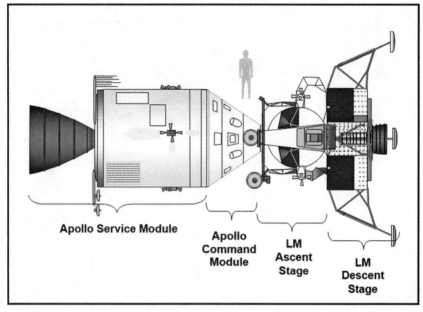

Apollo spacecraft, service module, and lunar module. Courtesy of Ted Spitzmiller.

accomplishment of a hard dock, the LM was released by the S-IVB, as it was now a part of the set of modules (Apollo, CSM, and LM) headed toward the Moon.

The pads on each leg of the LM were constructed with a crushable aluminum honeycomb material, and each leg had a shock-absorbing capability. Even the exhaust nozzle of the DPS had to be frangible (fragments when deformed) and able to be compressed up to 30 in. because of the possibility of the LM landing at a high descent rate or setting down on a bolder. If this occurred, the frangible exhaust nozzle would prevent the DPS from being shoved up into the ascent structure.

The LM electrical power and environmental systems were capable of spending up to twenty-four hours on the surface. Initially the fuel cell was considered for electrical power, but, as the problems with this innovative technology made its use problematical, it was decided to use battery power alone. The fuel cell ultimately proved itself on board Gemini, but the decision to use batteries had already been made.

The second-stage ascent propulsion system (APS) launched the craft off the Moon and into orbit for rendezvous with the CSM (that had remained in orbit around the Moon). This scenario was never fully tested until it was used in a manned flight. The ability of the spacecraft to contain all of the necessary attributes—to provide its own launchpad, countdown sequencing, and guidance into orbit around the Moon—was a formidable task. Unlike the DPS, the APS (built by Bell Aircraft Corporation), employed a fixed-thrust engine. Because of its single-purpose role, it did not have to be gimbaled, as was the DPS. The RCS provided the appropriate pitch, roll, and yaw during launch from the lunar surface into orbit around the Moon. It did have a slight bias to its mounting, being angled 1.5 degrees toward the forward leg—the leg that pointed to the azimuth of flight.

The compactness of the entire LM is illustrated by fact that the APS engine compartment enclosure protruded up into the cabin area and provided a seat on which the astronauts could sit while on the Moon. The DPS, APS, and RCS engines all used nitrogen tetroxide as an oxidizer and a blend of unsymmetrical dimethylhydrazine as a fuel in separate tanks. The RCS propellants could be cross-fed from the APS tanks if needed as a backup safety feature.

Initially two docking ports where considered, one on top and another on the front of the LM. However, it was quickly realized that only a single overhead port was needed for docking, and the second one was converted to a hatch to allow the astronauts to exit onto the lunar surface. This also necessitated overhead windows through which the astronauts looked over their heads during the docking process. This turned out to be a "pain-in-the-neck" process.

Weight was a critical consideration, and pounds were shaved from every possible system and structure. Even the astronauts had to give up their couches. They were suspended by a harness in a standing position, as there were no high-g powered phases in the one-sixth g lunar gravity. This also meant that they spent their sleep period in a hammock (Kelly 2001, 63).

The initial specifications called for the LM to weigh 24,000 lb., but that figure had jumped to 32,000 lb. by the time it flew. The increase was accommodated only because of the continual improvements to the Saturn V. Because the CSM configuration was frozen several years before the LM, the majority of the increased weight-lifting ability was allocated to the LM. Likewise, cost increases pushed the LM from an initial estimate of $400 million to a final figure of $2 billion (the CSM ended up at $3.5 billion).

With respect to the composition of the lunar surface, there were many uncertain variables. The possibility of the LM tipping over at touchdown produced a litany of conceivable conditions that had to be anticipated. The allowable sideways movement at touchdown, the angular surface the LM

Early Robotic Lunar Exploration

America's efforts to send probes to the Moon met with a string of failures beginning with the Pioneer program (1958–59). The follow-on Ranger series also met with little success from 1961 to 1964, until Ranger 7 through 9 finally sent back dramatic video of the surface prior to impact (1964–65).

The next program, named Surveyor, set down softly on the Moon in 1966 and transmitted the first detailed, close-up video images of the surface. Before the program ended in 1968, five of the seven missions landed successfully at various lunar locales.

The Lunar Orbiter program (1966–67) mapped the entire surface of the moon with what was then high-resolution video with five successive flights.

---

was contacting, and possible sliding action all had to be considered and mitigated (Kelly 2001, 66).

The efforts of the Ranger and Surveyor lunar probes had been intended to support the design of the LM. However, the accelerated timeline for the decisions that had to be made regarding the LM's configuration was such that virtually no feedback from these projects was incorporated into the lander. What was of value, however, was that Surveyor confirmed the assumptions made regarding the lunar surface.

Pushing the weight-reduction program while trying to accommodate all of the required complex propulsion, environment, and navigation systems caused innumerable delays. It was fortunate that the other two modules, the CSM, could be tested without the LM. The weight penalty of possible redundant systems caused considerable engineering analysis. Thoughts of possible in-flight maintenance (such as electronic component replacement) were not considered because of the perceived decreased reliability from such a model.

Designing a spacecraft for a mission that had never before been flown, and that was executed for the first time with humans on board, was a daunting challenge. Something as simple knowing as how much room was needed for the astronauts to put on and take off their spacesuits actually determined the cockpit dimensions—and the actual suits had yet to be fabricated.

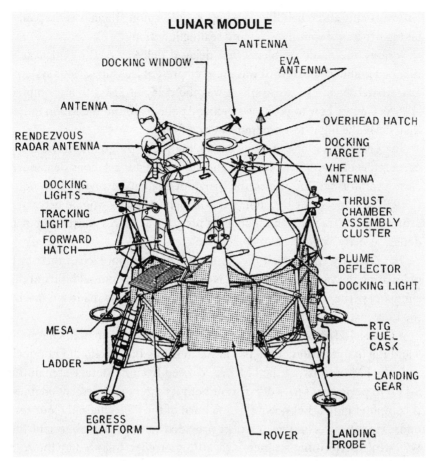

The final configuration of the LM showing most of the critical systems layout. Courtesy of NASA.

The thorough and thoughtful interchange that took place regarding fuel cells versus batteries involved issues of complexity, reliability, and weight. Tens of thousands of drawings were required. In the days before computers were integrated into the engineering, drafting, and manufacturing process, it is a marvel that the program progressed as fast as it did (Kelly 2001, 81).

The critical weight-reduction program and the painstaking process to track failures and to respond to them to ensure extremely high reliability were two of the top concerns. A crisis at Grumman occurred when the first unmanned flight-test article (LM-1) was emphatically rejected by NASA after it had essentially passed all of the test criteria at Grumman's Bethpage, Long Island, plant. The extent that Grumman and its suppliers went to ensure quality and reliability were impressive. Among them was the painstaking process of routing wires and sealing junctions.

Among the many problems that delayed delivery of the critical first manned flight article (LM-3) was the ever-present leaks in various systems exacerbated by stress corrosion. It was the delay of LM-3 that prompted NASA's George Low to propose sending Apollo 8 around the Moon rather than delay the flight-test program waiting for it.

The LM was just one of many exotic, high-tech programs whose management, economics, cost control, and subcontractor relations demanded close monitoring. The use of the program evaluation review technique (PERT) developed for the submarine-launched Polaris missile became a critical part of tracking and managing progress as NASA started to pinch dollars as early as 1965 (Gray 1992, 158).

The intensity with which the Apollo astronauts approached their task of learning the various LM systems is evident with the untold hours in the simulators. The sims were made by Singer-Link, a name made famous by the Link instrument flight trainers of World War II.

A Saturn IB (AS-204), on January 22, 1968, placed the unmanned Apollo 5 and the first iteration of the LM in Earth orbit. The Apollo did not dock with the LM but instead the LM was released from its storage area in the S-IVB to perform a series of burns of both its ascent and descent engines. A computer glitch shut down the first burn of the DPS after only four seconds. However, a contingency plan provided for two short burns of the DPS while performing a series of throttling profiles, followed by the APS exercising a simulated descent abort.

Another programming problem quickly depleted the attitude-control fuel. Because of the ability to cross-feed the fuel from the APS to the RCS, a stabilized second burn of the APS was achieved until the cross-feed was closed and the ascent module proceeded to tumble. By then, virtually all of the primary flight objectives had been satisfied. The mission was declared a success, and a second unmanned test of the LM was removed from the schedule.

Testing the Escape System: Little Joe II

The method of testing the launch escape system (LES) of Mercury was brought forward to the third-generation Apollo Program. However, the simple escape tower of Mercury grew into a rather complex structure for

Apollo as the size and destructive power of an errant 6 million lb. Saturn V required a greater escape distance.

The Apollo LES weighed 8,200 lb.—almost as much as the command module's 11,000 lb.—for a combined weighed 19,200 lb. (compared to 4,300 lb. for Mercury). Although the Apollo system was fundamentally similar to the one flown on Mercury in that it was a tractor rocket mounted forward of the spacecraft, that is where the similarity ended.

Near the top of the LES tower, two small canard wings unfolded eleven seconds into the abort sequence. These provided drag to orient the spacecraft blunt end forward, in preparation for parachute deployment. Because the exhaust plume from the small jettison rocket had left a residue on the Mercury spacecraft windows and because the blast effects and heat from the larger Apollo escape rocket could damage the command module, it was decided to shroud it with a "boost protective cover." This cover was physically connected to, and jettisoned with, the escape motor tower.

The main escape motor weighed 4,700 lb. and provided 155,000 lb. thrust for eight seconds. It would pull the entire assembly about a mile from the launch vehicle, should an abort occur on, or close to, the ground. A second motor in the tower, a small 2,800 lb. thrust pitch-control rocket, fired for one-half second perpendicular to the tower, pushing the spacecraft into an arching trajectory out of the path of the booster and possible debris and fireball.

To test the Apollo LES, a larger rocket than Mercury's Little Joe was required, which was named Little Joe II, a finned structure into which six to nine solid-fuel rockets could be mounted so that several abort scenarios could be configured similar to its namesake. It was a massive rocket that stood as tall as the Atlas booster that had taken the Mercury flights to orbit—85 ft. It had four different configurations, one of which weighed 139,000 lb. and generated up to 736,000 lb. of thrust—twice that of the Atlas.

The first launch in August of 1963 was without a command module to determine the basic flight characteristics. Six tests were then conducted at White Sands Missile Range in New Mexico between May 1964 and January 1966. There were three primary scenarios: the pad abort test, in which the Little Joe II was not required; a maximum dynamic pressure (Max Q) test at 45,000 ft., in which the LES pulled the command module away from an accelerating Little Joe II; and a high-altitude abort test.

Except for the last flight, the command module used for these tests was

The Little Joe II
rocket allows the
Apollo escape
system to function
under a variety of
worst-case scenarios.
Courtesy of NASA.

a "boilerplate." The first boilerplate test with Little Joe II initiated the LES in the transonic region at an altitude of about 15,000 ft. During the parachute deployment, a riser on one of the three recovery chutes failed. The remaining two returned the spacecraft safely, thus inadvertently validating the redundant role of three chutes. The last Little Joe II used a complete Block I Apollo CSM 002 to observe the functioning of the environmental system during the flight.

Following the second Little Joe II launch, the somewhat erratic unguided flight path of the vehicle became an issue. It was decided to install a simple attitude control system using hydropneumatically operated elevons on each of the four fins. A set of small reaction-control thrusters were also installed to assist with more precise pitch and yaw for the high-altitude tests.

A subsequent test failed when one of the new movable vanes on a fin stuck, causing the vehicle to roll rapidly. The abort-sensing system recognized the problem and activated the escape system at about 13,000 ft.

None of the escape systems on Mercury, Gemini, or Apollo, were ever called upon to perform during a manned flight. However, MA-3, an unmanned Atlas test designed to orbit the first Mercury spacecraft, strayed from its intended path and was destroyed by a signal from the ground. Its escape system worked as planned. The spacecraft was rocketed away to a safe distance and the parachute recovery was successful. The Soviets adopted the tractor-escape system for their Soyuz spacecraft. It has been used twice in the past forty years to save the crew.

Roles and Responsibilities: Man Verses Machine

As the Apollo mission plan unfolded, a contentious debate about how much each participant (man and machine) could or should play in the first lunar-landing expedition. The problem was the division of responsibility of automation in the spacecraft cockpit.

The robotic functions of an autopilot are an obvious challenge to the pilot's control authority. Its design functions assume some degree of the flight operations responsibility as the speed at which events transpire increase dramatically and human reaction time becomes an issue. The development of rocket-powered research aircraft of the 1950s (as exemplified by the X-15) illustrated the inability of the human senses and judgment to respond with appropriate physical reactive processes. This is especially true when dealing with a spacecraft that has no inherent aerodynamic stability.

Perhaps more profound are the implications of the pilot-in-command concept built on the legendary model of "the right stuff" that is threatened by automation. The antagonistic nature of the issue surfaced with the sudden advent of the space age and the prospect of man-into-space. The two diverse spacecraft design philosophies that saw the human element either as a relatively passive passenger or as a proactive pilot were hotly debated during the Mercury Program. Early astronaut candidates had to overcome the derisive spam-in-a-can comments by their peers and even overcome their own self-doubts as to how much they would get to "fly" the spacecraft.

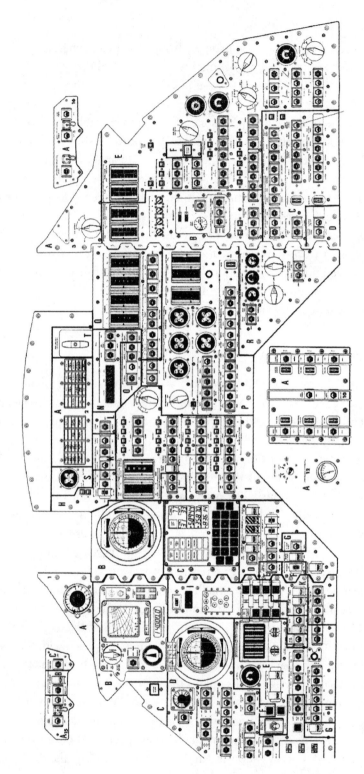

The complexity of the Apollo instrument panel, which spread before the three astronauts who flew the spacecraft. Courtesy of NASA.

Computers with the fundamental tasks of guidance and control also challenged the role of the crew and the interface used. During the 1960s early methods of programming and the advent of the word "software" came into general use. The initial program encoding was created as a primitive form of storing binary data on magnetic cores with inhibit and sense lines (firmware).

Multiple issues converged in the trilogy of the roles of the crew, Earth-based mission controllers, and spacecraft systems. Engineers had to proceed down various paths just far enough into some of the technologies to validate why certain functions and philosophies should be discarded or embraced during the engineering of the Apollo systems.

The lunar module, which had to "fly" to the surface of the Moon, was a more complex problem than the command module. It ultimately used a form of "fly-by-wire"—the pilot's inputs to the onboard control system specified the desired pitch attitude and vertical speed during the lunar landings—which the computer then maintained. To achieve the desired level of interaction, a simple and flexible interface to a series of programs was developed. Each program guided the LM through a specific part of the descent. Although the landings were described as being "manually controlled," the computer, in fact, flew them. The astronauts were at the mercy of a computer for their survival. With each mission, the actions and observations of the crew—as well as the sometimes misleading reports of the media—created erroneous impressions among the public as to the effectiveness of man over machine.

Many of today's military and some civil aircraft are fly-by-wire. There have been instances in which the man and machine disagreed over what function to perform at a given point—with disastrous consequences. The astronauts faced this dilemma.

Following President Kennedy's 1961 commitment to go to the Moon, the first contract awarded was to MIT for the guidance and control computer functions of the spacecraft. MIT and the automated systems were a critical part of the lunar program.

That the Apollo Program succeeded in meeting many challenges despite an unrealistic deadline of landing a man on the Moon before the end of the 1960s decade—not to mention politics, bureaucracy, mismanagement, and technical problems—is a testament to the motivation, skill, and integrity of so many of the participants.

16

DEATH STALKS THE ASTRONAUTS

January 1967 arrived with the United States enjoying the recent conclusion of its highly successful two-man Gemini project and the imminent launch of the first three-man Apollo spacecraft. Success with the Mariner planetary probes and the Surveyor lunar lander gave strong indications that a change in leadership in the space race may have taken place. The Soviets had encountered both technical and political bottlenecks that had brought their manned space program almost to a standstill.

However, the Apollo spacecraft was entering its eighth year of gestation and was facing some significant problems in moving to its first manned flight. The competition between the two superpowers for the minds of the uncommitted nations was entering its tenth year since the Soviets had surprised the West with their Sputniks. Nevertheless, 1967 proved to be a fateful year for both countries.

Apollo 1: "Fire in the Cockpit!"

The Apollo spacecraft was built in three versions—boilerplate (simply the structural elements used to prove the physical and heat shield integrity) and two manned iterations called "Blocks." Block I was the initial production version that was designed to prove the basic systems, but it did not have the requisite weight or advanced systems to make it lunar capable. The Block I articles were constantly being subjected to revision and redesign of various systems as NASA and North American Aviation (NAA) sought to bring as much improvement (and weight reduction) to the spacecraft as possible.

The change process became quite chaotic, and by late 1965 Maj. Gen. Sam Phillips, who had been recruited from the Air Force to serve as the Apollo Program director, performed a detailed audit of NAA's effort. His findings were critical of company management, and he stated in the

Apollo 1 astronauts Gus Grissom, Ed White, and Roger Chaffee had the challenging task of getting the Block I Apollo CSM ready for its first manned space flight. Courtesy of NASA.

summary memo that NAA was "not giving sufficient attention to the details of the direction and execution of these contracts." He recommended that more attention be paid to the "details and their problems" (Baker 1982, 281). Subsequent changes were instituted that moved NAA's quality control procedures and schedule in a more positive direction but never beyond "average" in overall performance.

Initial flight-testing, which involved von Braun's Saturn I and then IB, was going reasonably well with no significant failures, although the schedule for manned flight continued to slip. The first unmanned suborbital flight of a Block I command and service module (CSM-009) on board the first Saturn IB (AS-201) in February of 1966 was a success. Another suborbital flight, AS-202 (with spacecraft CSM-011) in August of 1966, cleared the way for the next flight to be manned.

Astronaut Gus Grissom was again selected to "shake down" the new spacecraft in the same way he had with the first manned Gemini. Accompanying him was America's first spacewalker, Ed White from Gemini IV, and a rookie, Roger Chaffee. It had been hoped that this first manned flight,

AS-204, would occur before the end of 1966, but there were just too many flaws in the various systems of spacecraft CSM-012. Unlike the conservative three-orbit flight of Grissom's Gemini III, AS-204 was open-ended with a target of fourteen days. The aggressive plan to fly a manned mission after only three flights of the Saturn IB was a part of Manned Space Flight Director George Mueller's policy of "all-up-testing."

CSM-012 arrived at Kennedy Space Center in August 1966 and moved through a series of test programs. Among these was an overall space vehicle "plugs-in" test, which simulated the launch and ascent phases with the space vehicle remaining connected to all the umbilical connections. The second, the "plugs-out" test, simulated the launch to the point where all the electrical connections were ejected and the CSM operated on its own internal power.

A basic design flaw was discovered in the environmental control system's oxygen regulator, and ultimately the entire environmental control system was replaced in late October. The flight slipped into the first part of 1967. Multiple leaks in the water/glycol coolant system caused the fluid to flow over electrical wiring, but by December this problem appeared resolved. On January 18, 1967, the plugs-in test was completed, but discrepancies required that the test be repeated a week later. The plugs-out test was scheduled for January 27, when the spacecraft would be switched to internal power during a simulated launch. Whenever the spacecraft was not directly involved in one of these tests, the technicians from NAA swarmed over it in an attempt to keep up with the more than 623 changes scheduled for implementation since CSM-012 arrived from California. The ability to make changes to many of the systems was hampered by the fact that some of the more than 15 mi. of wiring in the craft had been routed and rerouted so that a specific wire that might have to be spliced could not easily be found. Standards of workmanship and quality control procedures were occasionally inadequate.

Some felt that this situation with the spacecraft was leading to a high-risk condition. Thomas Baron, a quality control inspector (who had a reputation for what some considered "extreme" diligence), spread his alarm to the highest levels. However, his reasoning and evidence were not compelling to upper management, and he was fired from NAA in early January of 1967.

For the plugs-out test, the astronauts were in their space suits and entered CSM-012, where the inner pressure hatch was then installed. This required the astronauts to use a special ratchet tool to remove the six bolts that held it in place. Next the heat shield material hatch followed; it was closed by a simple latch mechanism. Finally, a fiberglass section was put in place as a part of a cover for the entire spacecraft to shield it from the exhaust gases of the escape tower jettison-rocket motor. In all, it took about ninety seconds to egress the spacecraft if a problem should require a rapid exit. A quick-release hatch using explosive bolts had originally been designed but was replaced with the more cumbersome method when Grissom's MR-4 hatch prematurely blew and the spacecraft was lost. Nevertheless, the new design was criticized as not only requiring more time to open but also not readily allowing extravehicular activity. A redesign was in process for the Block II spacecraft. Grissom was concerned about the awkward exit process, and he had scheduled an emergency egress exercise at the end of this plugs-out test.

As the entire prelaunch process was being simulated, the spacecraft was filled with 100 percent oxygen at slightly higher than standard atmospheric pressure to ensure there was no movement of external gases into the space-craft. Despite a cabin filled with oxygen, the test was not considered haz-ardous (no propellants were involved), and therefore the fire and medical personnel were on "standby" rather than "alert."

The test was interrupted by several problems but nothing of a serious nature. Grissom complained about the poor communications between the spacecraft and the blockhouse several times, once commenting sarcasti-cally, "If we can't talk between a few buildings here at the Cape how are we going to talk on the Moon?" The activity wore into the early evening and was nearing completion when, at 6:30 p.m. a small power surge was noticed on AC Bus-2. Ten seconds later Grissom made the fateful discovery: "Fire! We've got a fire in the cockpit!"

It was understood by everyone, including the astronauts, that a fire in these circumstances would be fatal. The flames spread rapidly, feeding on the flammable materials that had crept into the spacecraft over the past few years. White immediately dropped down from his couch and engaged the ratchet into the first bolt. Chaffee stayed at his position and simply relayed what everyone on the outside knew. "We've got a bad fire—let's get out . . . we're burning up in here!"

Within fifteen seconds, the fire had consumed most of what it could with the available oxygen and had created a significant overpressure that caused the spacecraft pressure vessel to rupture. This allowed the flames to move toward the split and to be sucked over the three men.

Chaffee's last words repeated his first plea. "We're burning up!" In less

The charred remains of the spacecraft interior—a powerful testament to an oxygen-rich fire. Courtesy of NASA.

than thirty seconds, it was all over. The fire had burned itself out, the smoke had rendered the astronauts unconscious, and the flame had imparted severe burns. As those on the outside raced to rescue the crew, they were initially driven back by the dense toxic smoke that spewed from the ruptured craft. The large solid-fuel abort rocket motor positioned just above the spacecraft was ignored by those intent on saving the three crewmen. By the time the hatches could be opened, almost five minutes had elapsed, and there was no hope for a miracle (Wendt and Still 2001, 100).

The nation mourned the loss and to some extent was in shock that something like this could happen. The possibility of death accompanied each launch, but a ground test should not have had the prospect of becoming lethal. An outpouring from most of the free world expressed sympathy to the widows and support for the effort of the United States.

As expected, the Soviet Union's response sought to make as much political propaganda as possible. Expressing sympathy for the deceased astronauts and their families, the Soviet press stated that the astronauts were "victims of the space race created by the American space program chiefs" based on "hate" for the people of the USSR. While their use of the word "hate" certainly overstated the basic premise of the United States toward the people of the Soviet Union, the essential argument was correct. In its haste to retake the technological lead, NASA and its contractors had taken many shortcuts that did increase the risk. The Soviet leadership might have been more charitable had they known what was in store for their own program just a few months into the future.

The spacecraft was unmated from its Saturn IB and taken apart piece by piece. What was discovered had already been known—there were many areas in the electrical wiring that presented a hazard, primarily because of the constant access for changes and troubleshooting problems. Although the specific cause of the fire was burned beyond the ability to identify, an electrical arc near the lithium hydroxide panel came closest to being the point of origin. As a part of the investigation into the cause of the tragedy and the prevention of similar catastrophes, the pure oxygen atmosphere was again examined.

A number of investigation boards were established, and a congressional inquiry was undertaken. During James Webb's testimony, he was asked about General Phillips' memo that was critical of NAA's workmanship and other significant problem areas. Webb denied such a report existed. Nevertheless, it did. It had been forwarded to Webb's deputy, Robert Seamans, who had failed to bring it to Webb's attention.

While there was no criminal culpability proved, the lack of management oversight caused several people in NASA and NAA to lose their jobs, including Seamans and Joe Shea. George Low became head of the Apollo Program office in Houston and established a configuration control board to oversee and coordinate all changes—especially computer software that was beginning to become a bottleneck.

Of the triple constraints of project management—cost and performance had generally not been compromised—time was the factor that always appeared to be the culprit. Improvements that had been considered but not incorporated because of the time factor were now evaluated against the safety aspect. There were 125 changes recommended, including a quick-opening hatch.

The Block I spacecraft were discarded and the Block II were revamped with new wiring harnesses built with three-dimensional jigs to avoid the bending of wires—a major cause of stress. The wiring codes and quality practices were upgraded. The new hatch and a series of panels were installed to isolate flame propagation to allow the astronauts more time to escape or combat a fire. Had Apollo been designed and constructed with the awareness that originally surrounded the use of a pure oxygen environment, these improvements would probably have been a part of the original design. However, the success that NASA had enjoyed over the five years of flying manned spacecraft had taken its toll on caution. This situation would arise again twenty years into the future.

More than eighteen months elapsed before the first manned Apollo launch, and the possibility of beating the Russians to the Moon was only a glimmer. But the delay in Apollo for design changes and quality improvements allowed other aspects of the program that were lagging to catch up.

Soyuz 1: "An internal matter"

The development of the three-man Soyuz, like Apollo, played a pivotal role in Soviet space progress. It was their vehicle for perfecting Earth orbit rendezvous, and it provided for circumlunar flights projected for 1967. It was also to be the "mother ship" for the lunar orbit rendezvous lunar lander that was scheduled to fly in 1968.

The first flight-qualified Soyuz spacecraft had gone through a troubled period of testing since its delivery in May of 1966, with more than two thousand discrepancies and changes recorded, which caused its scheduled thirty-day check-out period to extend to four months. Several of the systems had likewise taken an extensive period of development, and some of the less sophisticated segments, such as the parachute recovery system, had proved troublesome.

For the first flight, it was decided to launch two unmanned Soyuz and have them rendezvous and dock. It was an ambitious approach, but the success of the American Gemini program had to be addressed. The first spacecraft, designated Kosmos-133 (to avoid revealing its true nature), was launched on November 28, 1966. The 113 mi. by 145 mi. orbit was less than anticipated as a new R-7 upper stage (being used for the first time) did not provide the expected performance. Within an hour, serious problems with the "mooring and orientation" engine (the equivalent of the Gemini OAMS) had depleted virtually all of its propellant, and the craft was in a slow roll. Recognizing that the situation did not allow for the planned rendezvous and docking, the launch of the second unmanned Soyuz was canceled and attempts were made to return Kosmos-133. After an extended effort to provide the correct orientation, the spacecraft overshot its intended recovery area, and the automatic destruct mechanism activated (Siddiqi 2003a, 571).

The State Commission for Soyuz, headed by Maj. Gen Kerim Kerimov, decided to launch the second unmanned Soyuz as a solo mission and test its systems before committing to another rendezvous. On December 14, 1966, when the count reached zero, the four boosters of the R-7 ignited, but the central core did not. This caused the automatic system to shut down the boosters. The launch crew immediately moved in to safe the giant R-7. They had just begun to raise the various service structures into place when the solid-fuel escape rocket on top of the Soyuz fired, some twenty-seven minutes after the first stage had shut down. The R-7 was destroyed, but, by a stroke of luck, only one of the launch crew perished although many were severely injured.

There have been two reasons published for the mishap more than thirty years after the event. The first is that one of the service towers was raised out of sequence, which, coupled with some gusty winds, caused the R-7 to tilt. The gyroscopic guidance system (which had not yet been shut down) sensed an angle that exceeded the 7-degree tolerance for the first vertical phase of the launch sequence and signaled an abort. The second reason again relates to the gyroscopes but ascribes the out-of-tolerance condition to the rotation of the Earth during the twenty-seven minutes following the aborted ignition sequence. The gyroscopes maintain a rigidity in their position, and as the Earth continued to rotate on its axis, the gyros sensed the tilt. The failure not only destroyed the launch vehicle but also badly damaged the launchpad.

The next Soyuz was launched on February 7, 1967, and was reported as Kosmos-140. It again experienced problems with its attitude control system and sun sensor. The spacecraft was bought down early and landed short on the frozen Aral Sea, eventually breaking through the ice and sinking in shallow water, although it was later recovered. This was yet another setback for the Soviet program.

The first L1 (lunar) version of Soyuz flew an unmanned flight on March 10, 1967, six weeks after the Apollo fire and four weeks after the failure of the third unmanned Soyuz. The L1 did not have the living quarters module. This was to keep its weight to 11,500 lb., the upper limit of the now four-stage UR-500 Proton that boosted the intended circumlunar mission. The new translunar injection stage (called the Blok D) put the L1 in a parking orbit. At the appropriate point, the Blok D fired a second time, sending the spacecraft into a highly elliptical orbit that would have reached the Moon had that been the plan. However, this was an engineering test, and there was no desire to involve the Moon at this point, so its timing ensured that the Moon was not in the vicinity when the spacecraft (named Kosmos-146 to conceal its intent) reached its apogee, some 220,000 mi. out. On its return,

The Soyuz was the Soviet's second-generation spacecraft, whose development was plagued with problems. Courtesy of Energiya.

the craft burned up in the atmosphere since there was reportedly no goal of recovery, although other sources indicate that it was to survive. The Soviet team was ecstatic at the success, as most of the systems performed well and those that did not presented relatively minor problems (which tend to discount the survivability issue).

The second L1 spacecraft (Kosmos-154) was launched a month later on April 6, 1967, but this test required the Blok D to remain in its parking orbit for twenty-four hours before its second firing to send it on to the Moon. The Blok D failed to fire the second time due to a problem that was traced to an incorrect switch setting. The third flight (which was scheduled to be the first manned flight around the Moon) had been tentatively set for July. This schedule was unrealistic, as the first manned Earth-orbiting Soyuz was set to fly within a few weeks, and there were many unknowns yet to be discovered. Since America's Apollo effort was grounded for at least a year because of the tragic fire, the chances were now excellent that the first man around the Moon would be a Soviet citizen.

In spite of the series of problems that plagued the Soyuz program, it was decided to proceed with the manned Earth orbital flight. It was felt that all of the difficulties encountered could be remedied or at least be under the control of the pilot—although Soyuz was designed to be flown as an automated spacecraft with the crew as passengers. This had been the Soviet philosophy from the beginning. As had been previously demonstrated, the presence of the pilot had saved several missions, although the lack of true integration of manned and unmanned systems had revealed the designs of the Vostok and Voskhod to be inadequate; Soyuz sought to remedy this problem.

The inaugural flight of the manned Soyuz promised another Soviet spectacular. Not one but two manned spacecraft would rendezvous and dock. Two of the three cosmonauts from the second ship would spacewalk to the first to join its lone occupant and return in it. It was believed that the prestige lost during the Gemini program could be regained in one flight that demonstrated most of what the Americans had accomplished with ten Gemini flights.

The first manned Soyuz, enclosed in an aerodynamic shroud similar to Vostok, launched on April 23, 1967, with its lone occupant, Vladimir Komarov (who had commanded the first Voskhod flight in October of 1964). Komarov did not wear a spacesuit in the shirtsleeve environment of nitrogen and oxygen at normal atmospheric pressures. With the living module on the front end, isolated from the control module during extravehicular activity, the Soviets believed there was no need for the astronauts to wear spacesuits during normal operations, which included launch and reentry. Following the five-minute powered flight and arrival in orbit, the mission was immediately announced by the TASS news agency as Soyuz 1. Many rumors circulated in the press that a second craft would rendezvous.

As with the American program that had established a manned spacecraft center for Mission Control in Houston, Texas, the new Soyuz flight inaugurated a new chief operations and control group called the Scientific-Measurement Point No. 16, located in the Crimea. It didn't take long before the control group recognized that the left solar panel had not deployed, the backup telemetry antenna was not functioning, and the attitude-control sensor had apparently become fogged over by exhaust gases.

Cosmonaut Vladimir Komarov recognized the high risk that was involved with making the first manned flight of a poorly debugged Soyuz spacecraft. Courtesy of Energiya.

Komarov worked professionally to address each of the problems and effectively communicated his status to the control group. The loss of 50 percent of the electrical power and the problem with the attitude control dimmed hopes for a rendezvous and docking. The second Soyuz launch was canceled as Komarov attempted manually to orient his craft.

Although a series of problems precluded the use of the ionic sensors to help establish the proper attitude, Komarov was able to complete the retrofire sequence after a full day in orbit and was on his way back to Earth after jettisoning the living and instrument compartments. However, his trajectory, a basic ballistic path, caused him to endure 10 gs as opposed to 5 gs, had he been able to use the lift capability of the craft.

No further communication with Soyuz 1 was received after the expected reentry radio blackout period; seemingly, this was not a major concern to the control group. A recovery helicopter got a fleeting visual on the descending craft and then spotted it lying on its side on the ground next to the parachute. As they observed the scene, the soft-landing rockets, which should already have fired to cushion the descent under the parachute, were seen to ignite, and the craft caught fire and was destroyed before a rescue could be attempted. Arriving sooner would not have helped Komarov, as the primary chute had failed to fully deploy, and the backup became fouled in the primary. The ship hit the ground at over 100 mph, and Komarov was killed by the impact (Siddiqi 2003a, 589).

Announcing the failure and the death of Komarov to the world was a bitter pill for the Soviets to swallow, especially after their accusations only two months earlier regarding the Apollo fire. They refused to allow American astronauts Gordon Cooper and Frank Borman (who were in the USSR on a goodwill tour) to attend the state funeral services, citing that it was "an internal matter."

The remains of the reentry module after its high-speed impact and resulting fire caused by the braking rockets. Courtesy of Energiya.

As with the Apollo fire, a lengthy investigation and faultfinding exercise followed. Several areas were uncovered in the parachute-folding and ejection processes that contributed to the malfunction. As for the problems in flight, it was determined that the spacecraft systems had not been sufficiently tested before committing to a manned flight. The desire to catch up with the Americans had caused the Soviets to fly Soyuz before it was ready.

Others Who Succumbed

The fire in the isolation chamber that killed cosmonaut-trainee Valentin V. Bondarenko in 1961 (see chapter 9) and the Apollo fire in January 1967 that took the lives of Grissom, White, and Chaffee demonstrated that the risks of being an astronaut were not reserved for actual space flight. America had lost it first astronaut, rookie Theodore Freeman, in a T-38 accident in October 1964. This was followed by the T-38 crash of Elliot See and Charles Bassett in St. Louis in February 1966. Edward Givens was killed in a car crash in June 1967, and Clifton Williams was killed in another T-38 crash in October of that year. The demise of Komarov in the April 1967 Soyuz flight was the only one of the eleven deaths to that date that had occurred during an actual mission.

In June 1967 Robert H. Lawrence, a graduate of the Flight Test Pilot Training School at Edwards AFB, was selected as an astronaut for planned Air Force Manned Orbiting Laboratory program, becoming the first African American astronaut designee. Unfortunately, Lawrence was killed on December 8, 1967, in the crash of an F-104 Starfighter. He was flying as

The Northrop T-38 was used by the astronauts for proficiency flights as well as business transport. Several astronauts died in T-38 accidents before they had an opportunity to fly into space. Courtesy of NASA.

America's first African American astronaut, Robert H. Lawrence, was killed in a training accident flying the F-104 seen here. Courtesy of Department of Defense.

the instructor-pilot for a flight-test trainee learning the steep-descent glide technique. The pilot flying made such an approach but flared too late. The airplane struck the ground hard, and its main gear failed; it caught fire and rolled. The front-seat pilot ejected upward and survived with major injuries. The back seat's ejection, which delays a moment to avoid hitting the front seat, ejected as the aircraft rolled sideways, killing Lawrence. He likely would have been among the Manned Orbiting Laboratory astronauts who transferred to NASA after the program's cancellation, all of whom later flew on the space shuttle.

Yuri Gagarin

It was no wonder, then, that the Soviets were reluctant to allow a second space flight for the world's first astronaut, Yuri Gagarin. Gagarin was far more useful to the Communist regime as a goodwill ambassador, and some were not displeased to see his physical conditioning and technical competence wane. Gagarin himself recognized that his situation was personally humiliating, and he decided to make a comeback and dropped some unneeded weight. By late 1967 he had completed his dissertation for a graduate degree and had been promised another flight with the completion of his academic work. Cosmonaut chief General Kamanin was concerned with Gagarin's newfound fervor, but until the time came to assign a flight, there was no need to confront the popular cosmonaut.

Gagarin completed a new medical and was cleared for assignment to another space flight but was not permitted to fly high-performance jets solo. He was required to use the two-seat UTI MiG-15 with a senior safety

Cosmonaut Yuri Gagarin was killed flying with a senior instructor pilot while preparing for his second flight into space. Courtesy of Energiya.

pilot. On March 27, 1968, Gagarin and Col. Vladimir Seregin took off in the MiG from an airfield on the outskirts of Moscow for a one-hour proficiency flight. What actually occurred during the fateful last few minutes of the flight remains obscured, but the smoking wreckage of the jet was discovered several hours later after it dove into the ground almost vertically at 500 mph.

The initial official inquiry cited pilot error, but a reinvestigation twenty years later sought to assign some of the blame to procedures of the Soviet airspace system. The most likely cause was a high-speed stall from which the jet was too low to recover. This situation may have been exacerbated by several layers of low clouds that contributed to disorientation. Gagarin, a man who will be forever linked with the conquest of space exploration, was dead. The Soviet people mourned his loss as though he were their own relative. The Americans and other nations were allowed to attend the official state funeral.

The Unfortunate Demise of the Chief Designer

If the early successes of the Soviet space program infused life into Korolev and his team, the long string of failures in the lunar and planetary programs in the years between 1962 and 1965 created an air of despair for this talented group of people. However, it was the inability to fly the second-generation Soyuz in a timeframe to keep the Soviet program ahead of the American program that seemed to have the most negative effect. As for Korolev himself, the emotional impact of these years had taken a significant toll on the man. He had lost his ability to work effectively with many members of his own team, and his physical health, impaired by the severe years of Gulag incarceration, was now aggravated by his incessant long hours of work and the intense conflict among the other chief designers.

Low blood pressure and a serious heart condition continued to affect his health, and he realized that he was in grave trouble. This was evidenced by a comment in a letter to his wife in late 1965 in which he stated, "I am in a constant state of utter exhaustion and stress. . . . I am holding myself together using all the strength at my command." These words are from a man who was only a few weeks shy of his fifty-ninth birthday.

In mid-December he underwent several medical tests, which indicated a bleeding polyp in the intestines. The simple surgery was scheduled for January 14, 1966, and he expected only a few days in the hospital. During the operation it was discovered that he had a large malignant tumor, which was removed with great difficulty and with much bleeding. Following a series of complications, reportedly aggravated by the lack of a competent medical staff, Korolev's heart failed, and the great man died (Oberg 1981, 88).

While some in the Soviet space program, such as his rival, Valentin V. Glushko, were not saddened to see his demise, most of the Soviet scientific community who understood the role he had played in the advancement of Soviet rocketry felt a keen loss. Soviet Party first secretary, Leonid Brezhnev, in approving a state funeral, also allowed Korolev's name to be revealed as "the chief designer," although most in the West still had little or no idea of the true significance of the role he had played.

Korolev's legacy was the R-7. It was created and perfected in a remarkably short time, and its effect on mankind's move into space as well as on the relationship between the two most powerful countries the world had ever seen was profound. The fact that it remained the only manned space booster in the Soviet inventory for over fifty years says much about its engineering. It also reflects on the inability of the Soviet system to provide a viable replacement for either the man or the machine.

17

THE CIRCUMLUNAR GOAL

The Soviet Space Program: After Korolev

A top priority for the Soviet space program was to find Korolev's successor. In reality, there was no one who had the intricate understanding of the workings of his design bureau. No one had his set of relationships (some of which were admittedly in disarray), and certainly no one had the ability to generate a vision with which others could readily identify. His staff was leery of outsiders and promptly requested that Vasily Mishin, Korolev's deputy, be appointed (Siddiqi 2003a, 517). However, his lack of people skills and political connections caused some to oppose him. Nevertheless, his engineering abilities and knowledge of the existing program structure and technology base provided assurance of a degree of continuity for Korolev's work in progress, and Mishin received the appointment—but a critical five months had elapsed. As with the other chief designers, Mishin's status remained a state secret for years to come.

As one of his first priorities, Mishin undertook a review of the manned space program and Voskhod in particular. Three more flights had been planned at that point and, in fact, had been delayed from February 1966 by a series of failures in various subsystems. The twenty-two-day flight in March of Kosmos-110 (a Voskhod-type spacecraft) carrying two dogs was supposed to pave the way for Voskhod III, but the poor condition of the dogs and problems with several life support systems caused the manned flight to be postponed (Siddiqi 2003, 523).

America's success with Gemini loomed large, and many in the Voskhod program were reluctant to proceed with any more flights but wanted to move on to Soyuz—even if it meant a delay of up to eight months. The only real milestone that Voskhod might salvage was a long-duration flight of perhaps three weeks and higher apogees (altitudes) of 600 mi. The world now recognized America's ability to rendezvous, and adding longer duration flights would not impress even the media, which was becoming more knowledgeable of the various technologies. As a result, further Voskhod flights were canceled—additional evidence that the Soviets were now reacting to America's lead.

As 1966 ended the Soviets had finally settled on three primary projects designed to retain their perceived lead in space technology and to continue their efforts to convince the uncommitted nations that communism was the wave of the future. These projects consisted of the three-man Soyuz spacecraft with the ability to rendezvous and dock with another Soyuz (7K-OK), the UR-500 circumlunar flight (UR-500K-L1), and the N1 for manned lunar landing—possibly using both Earth orbit rendezvous (EOR) and lunar orbit rendezvous (LOR) techniques (N1-L3). Rather than develop a new spacecraft for the lunar expeditions, a modified *Soyuz* (less its living compartment) was to be used for both the UR-500 circumlunar flight and the N1 lunar landing. The latter required the development of a lunar lander.

Soyuz: Struggling for Success

The effort to get the basic Soyuz spacecraft cleared for another manned attempt followed completion of the critical recommendations that were made after Komarov's fateful flight. Two more unmanned spacecraft attempted to rendezvous and dock. The first launched on October 27, 1967, with the designation of Kosmos-186, again precluding any association with the Soviet manned space effort. It was the active participant in the rendezvous attempt that was to occur. Following three days in orbit to test and remedy some problems that arose, the second Soyuz, designated Kosmos-188, left the launchpad on October 30.

The Igla radar system on Kosmos-186, the active satellite, and the computer designed to interpret the range and rate information performed flawlessly, and the two vehicles docked together in less than one orbit, as called for by the technique they employed. Televised views of the impressive event were broadcast on the Soviet news, but the angle and clarity of the transmission revealed little of the configuration of the two craft.

It had been more than six years since Yuri Gagarin's flight, and the Soviets had yet to release pictures of either his Vostok or the R-7, and they weren't about to unveil their new spaceship. They were justly proud that they had accomplished another first with the completely automated rendezvous and docking of two satellites. While it was not a manned circumlunar flight,

Kosmos-186/188 was the best the Soviets could provide for their people to celebrate on the fiftieth anniversary of the revolution.

But all was not well with the event (although the world would not know that for another thirty years). The spacecraft were unable to achieve a complete latching into a hard-dock configuration. After several orbits, the two ships separated, and Kosmos-186 returned to Earth on October 31 using a ballistic path when its attitude control was unable to provide a less g-intensive "guided" reentry; however, it touched down safely. Kosmos-188 also had to execute a ballistic return, but its path threatened to fall short of Soviet territory, and it was intentionally destroyed.

Almost one year after the death of Komarov, a revised and better-tested (but still unmanned) Soyuz rose from its launchpad at Baikonur (Tyuratam) on April 14, 1968. Named Kosmos-212, it was followed the next day by the unmanned Kosmos-213. The launch parameters were so accurately timed that the second satellite was less than three miles away from the first at orbital insertion. The automated docking was followed closely by the ground controllers on live television. The two spacecraft undocked after about four hours, and then each proceeded to spend five days in space conducting tests of the various systems. Kosmos-212 de-orbited and performed a guided reentry to a successful landing, followed the next day by Kosmos 213, which performed equally as well.

In spite of the successful flights, the political hierarchy was reluctant to commit to the next mission being manned. The bad press from Soyuz 1 was still fresh in everyone's mind. As a result, Kosmos-238 was a solo flight launched on August 28, 1968, and returned on September 1, finally clearing the way for a manned flight.

Soyuz 2 successfully launched on October 25, 1968, but it was unmanned as there was still reluctance to trust the Soyuz systems. Soyuz 3, with only cosmonaut Georgy Beregovoy on board (the first manned flight since Komarov's death), launched the following day. The automated rendezvous apparatus brought the two spacecraft to within 500 ft., and Beregovoy attempted to perform the docking manually. The lack of coordination between the automated and the manual systems thwarted his attempts.

Although the Soviets publicly claimed that docking had not been one of the objectives, privately most of the blame for the failure to dock was placed on Beregovoy for his inability to properly control the ship. In the process, he almost exhausted the attitude control fuel, and the docking was canceled. (In the postflight debriefing, he complained that the attitude control system was too sensitive and not easily managed in space.)

He spent the remaining time performing experiments and providing three sessions, which were broadcast live on Soviet television, before being brought back safely on October 30. The new "openness" of the Soviet program to television (which still did not reveal the launch vehicle nor the spacecraft configuration) was a direct result of the Apollo 7 flight, which had occurred a week earlier.

Saturn V: The Giant Flies

The American plan to fly the modified Block II Apollo on board the Saturn IB was delayed by the emphasis on improved wiring, the quick-opening hatch, and a reduction of flammables in the cabin. There would not be another manned attempt until the fall of 1968. In the meantime, the effort to fly the first Saturn V took on new meaning. It was now imperative that the big booster perform its first flight virtually flawlessly if the Kennedy commitment was to be satisfied. This was asking a lot of a rocket that was four times as powerful as the Soviet UR-500.

The designations of the Saturn V vehicles began with AS-501 as the first flight article. An inert facilities test assembly (designated Saturn 500-F) was used to verify all of the electrical and mechanical umbilical connections. A renumbering of the Apollo spacecraft also occurred at this time. They were retroactively assigned numbers that recognized the deceased crew of Grissom, White, and Chaffee; it was designated Apollo 1; Apollo 2 and 3 were the boilerplate versions that flew on the Saturn IB AS-201 and AS-203; Apollo 4 sat high atop AS-501. It was a true Block II spacecraft, with virtually all the systems active.

On August 26, 1967, the first flight-ready Saturn V was moved to Kennedy Space Center Pad 39A. It was a media event that caused almost as much excitement as a launch itself. The massive assembly was truly an awesome sight to watch as it moved ponderously along the five miles to the flame pit while thousands of spectators gathered to witness the occasion. It would be another two months before the launch, as more delays crept into the schedule.

On Thursday morning November 9, 1967, the fully fueled rocket weighed almost 6 million lb. As the final seconds of the weeklong countdown ticked by, a computerized progression of events removed the actual firing sequence from human hands at T-3 minutes. The command to start the S-IC engines of the first stage was sent at T-8.9 seconds, when the turbopumps began their high-speed whine to deliver the propellants to the five huge 17

There were eight umbilical causeways between the service tower and the Saturn V for power, data signals, and fluids that had to be disconnected at launch. Another four connecting units extended up from the launchpad to the first stage. Courtesy of NASA.

ft. F-1 thrust chambers. Within a few more seconds, smoke erupted into the flame pit at the base of the rocket. The six umbilical causeways ejected their connections to the rocket and retracted. Then slowly—ever so slowly—the giant lifted.

To this point, there had been no sound, as the nearest official spectators were more than two miles from the rocket. When the acoustical impact of the shock wave traveling at over 1,000 ft./s hit the gathered viewers, it was not just a physical impalement of force but an emotional and, to some, a spiritual experience. The sound literally shook the surrounding ground, structures, and rib cages. Even to the old hands who had witnessed many launches at the Cape, this was different.

Almost immediately, the Saturn V responded to a subtle command to yaw slightly to provide more clearance to the umbilical tower. At T+11.9 seconds a slight pitch and roll maneuver began that aligned the rocket with its intended inclination to the equator and initiated the ballistic trajectory to an orbital path. A full ten seconds passed for the mammoth rocket to clear the 400 ft. umbilical tower while the 7.5 million lb. of thrust devoured almost 3,400 gal. of propellant each second.

As the rocket rose higher into the atmosphere and gained speed, the characteristic "frozen lightning" (condensation trails) appeared, and the exhaust plumes of the five F-1 engines widened as the atmospheric pressure lessened. Mach I (the speed of sound) was passed at T+68 seconds, and this was followed shortly by maximum dynamic pressure as evidenced by a vapor cloud that briefly surrounded the midsection of the rocket. At T+135 seconds into the flight, the center engine shut down as planned to lessen the forces on the rocket when the first-stage burnout would occur at T+150 seconds—40 mi. high and 6,000 mph. The first stage separated, and eight small, solid-fuel rockets fired opposite to the direction of flight to slow it down. This ensured that it would not collide with the remainder of the rocket, which was now experiencing the effect of eight small ullage rockets that were designed to provide a slight acceleration to settle the propellants in the second-stage tanks.

Now, for the first time in flight, the five LH2 J-2 engines of the second stage came to life providing 1 million lb. of thrust. By incorporating two major milestones into a single flight, all-up testing had saved its first $135 million (the cost of a Saturn V launch). Thirty seconds into the burn, the corrugated interstage adapter was released to reduce weight. The superefficient LH2 cluster burned for 360 seconds and accelerated the upper stages

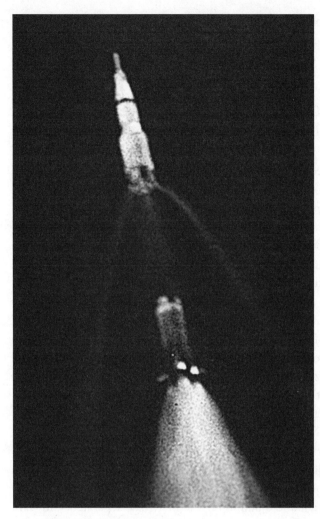

The first stage of the Saturn V separates as the S-II stage ignites and continues to thrust the remainder of the vehicle into space. Notice that there is virtually no visible flame from the exhaust of the second stage, which is essentially water molecules. Courtesy of Department of Defense.

to an altitude of 120 mi. and a velocity of 16,500 mph, parallel to the Earth's surface.

The second stage separated, aided by four forward-firing solid-fuel rockets to again ensure that the spent stage would not run into the remaining rocket assembly. The single J-2 engine in the S-IVB ignited, and its 200,000 lb. of thrust continued to accelerate the Apollo command and service module (CSM) for 145 seconds to an orbital velocity of 17,500 mph, where it "parked" for a little more than three hours.

The objective of the flight now called for re-ignition of the S-IVB to simulate the translunar injection burn. Rather than a full burn, the S-IVB pushed the Apollo up to a speed of "only" 20,000 mph, resulting in an apogee of just over 10,000 mi. The Apollo CSM was then separated from the S-IVB, and its own service propulsion system (SPS) fired to raise the apogee another 600 mi. and to establish a reentry path of 8.75 degrees, simulating a return from the Moon. All-up testing had saved another $135 million.

At 8 hours and 10 minutes into the flight, the SPS fired again for 270 seconds to accelerate the now descending spacecraft to a full lunar return velocity of 25,000 mph. The Apollo 4 command module separated from the service module and proceeded to execute a "double dip" into the atmosphere. It bled off speed (energy) on its first encounter (exposing the spacecraft to about 8 gs) and used the resulting lift to skip back to a higher altitude before performing the second and final phase of the reentry (resulting in 4 gs). External temperatures rose to 5,000 °F while the interior temperatures showed an increase of but 10 °F. The craft was retrieved by the recovery forces in the Pacific.

The first Saturn V test was successful beyond anyone's expectation. In one inaugural flight, it had proved the first and second stages and had fully qualified all of the spacecraft systems on a simulated lunar voyage. While a second planned unmanned flight remained on the schedule, the third Saturn V flight was to be manned—if the first manned Apollo Saturn IB flight, still almost a year away, was successful.

The second test of the Saturn V, AS-502, occurred on April 4, 1968, sending the unmanned Apollo 6 on a similar flight profile to that of Apollo 4 the previous November. This flight revealed several problems beginning at T+133 seconds, when the vehicle experienced longitudinal oscillations (approximately five cycles per second for the last ten seconds of burn), referred to as the "pogo effect." The phenomenon had occurred with the Gemini-Titan vehicle, but instrumentation on AS-501 had not shown any significant problems. The forces exerted on AS-502 caused portions of the spacecraft adapter to break away, but the huge rocket continued its flight.

During the S-II second-stage thrusting, engine 2 shut down prematurely, followed by engine 3. The guidance system was able to control the rocket, and the remaining three engines were commanded to burn longer (using the fuel allocated to the inoperative engines) to compensate for the lack of thrust. However, the energy was insufficient to totally make up for the second stage, only producing 60 percent of its rated thrust, and the S-IVB had to fire 29 seconds longer to put the Apollo CSM in the proper parking orbit. At the appointed time for the second burn of the S-IVB for the simulated translunar injection (TLI), it failed to restart. A contingency was available for the mission, and the CSM SPS provided a long seven-minute twenty-two-second burn to send the CSM 14,000 mi. into space. It then returned to Earth and was recovered from the Pacific.

The problems with AS-502 were immediately addressed. The pogo oscillations were dampened by adding helium to shock-absorbing cavities in the LOX line pre–valve assembly to change its natural resonance. The pogo had not occurred during the first flight, apparently because of its lighter weight. A leaking LH2 igniter fuel line caused the premature shutdown of engine 2 in the S-II stage. The anomaly was detected from thermocouple temperature readings in the tail section of the S-II. These readings indicated that the problem was caused by the flow of cold gas (the leak) resulting in the lowering thrust from the engine. The signal to shut down the engine was sent to the fuel valve of engine 2, and to the LOX valve of engine number 3 as a result of a wiring error—thus the loss of two engines.

Why did the fuel line develop a leak? Exhaustive tests were carried out with special attention to the accordion-like bellows section protected by a metal braid to allow the line to flex. Using higher flow rates, pressures, and temperatures, none of the test items failed. Turning to the possible implication of the conditions in space, eight test items were placed in a vacuum chamber and exposed to vibration while LH2 was pumped through the lines. All eight test items failed within one hundred seconds. When tested in normal atmospheric conditions, the LH2 caused the air around the lines to liquefy, and this liquid was held between the bellows and the braiding, dampening the vibrations—so failures did not occur. Only in the vacuum condition, where there was no air to liquefy, did the vibrations set up resonances that caused the malfunction. The failure of the S-IVB to fire for a second time was due to a failed igniter. That the Saturn experienced these failures and was able to continue its flight was particularly gratifying to the entire MSFC team—especially von Braun.

The next question, then, was when to fly AS-503. If it was simply to verify the corrections being made, it could have flown as early as July of 1968, but with the concept of all-up testing, there had to be more to the mission. NASA management decided not to repeat another unmanned Saturn V flight. As the schedule now appeared, all of the remaining flights would be manned. If AS-503 were to be flown with a crew, the current schedule called for it to provide an Earth-orbit test of the lunar module, LM-3. This flight did not occur until January 1969 because of delays in installing changes. In the interim, Saturn IB AS-205 flew in October of 1968 as the inaugural two-week manned Apollo flight. The question that faced NASA management was what to do with AS-503?

Soviet Manned Circumlunar Efforts

That the Soviets were still committed to a lunar program seemed to be validated just a month after the Soyuz tragedy. Two American astronauts, David Scott and Michael Collins, had an opportunity to speak with cosmonauts Pavel Belyayev and Konstantin Feoktistov at the 1967 Paris Air Show. Collins related that Belyayev spoke openly of Soviet manned circumlunar flights in the near future. On his return to the USSR, Belyayev was soundly reproached for his comments, and the task of toning down his rhetoric was apparently given to Academy Sciences' Leonid Sedov a few months later when, during a press conference, he made a point of saying that "manned flight to the Moon is not in the forefront of Soviet astronautics" (Siddiqi 2003a, 610).

While the Komarov accident dimmed the prospects for a manned circumlunar flight in time for the November celebration of the fiftieth anniversary of the Russian revolution, the Soviets pressed forward with perfecting the UR-500/7K-L1 combination. The first opportunity to test the vehicle occurred on September 28, 1967, when the 2 million lb. thrust rocket left the pad at Tyuratam. Observers noted that the huge rocket appeared to rise slower, and those in the blockhouse knew why: only five of the six engines had fired. The rocket's guidance system was able to sustain a near normal trajectory until sixty-one seconds into the flight, when the abort sensors recognized that the rocket was unable to maintain the desired path and triggered the escape mechanism for the spacecraft. The engine problem turned out to be a rubber plug in a propellant line that had not been removed during manufacture (similar to the first attempt to launch Gemini VI). The failure removed any possibility of a manned circumlunar flight in 1967. There was no space spectacular for the fiftieth anniversary of the Bolshevik revolution.

Another attempt was made on November 22, 1967, and this time it was the second stage that experienced a failure when one of its four engines failed to fire. Again the abort sensors detected the malfunction and the spacecraft was rocketed clear of the booster. During the landing, the solid-fuel rocket motors in the base of the Soyuz fired too early, and the craft experienced a hard landing, so it was fortunate that there were no occupants.

With the fiftieth anniversary behind them, there were no more "political" deadlines, and Korolev's old team, now headed by chief designer Vasily Mishin, had at least a year to complete its goal of a manned circumlunar flight before the Americans could possibly make a similar attempt. Nevertheless, the confidence of the engineers as well as the cosmonauts themselves had been shaken by the series of failures.

The next flight of the UR-500/7K-L1 on March 2, 1968, was made without the Moon as its target in order to expedite the flight without having to wait for the tight time constraints of a lunar launch window. The unmanned spacecraft successfully achieved a high-apogee orbit that sent it out to a distance of 220,000 mi. This time the TASS announcement referred to the ship as Zond 4 (Russian for "probe"), a moniker that had previously been used for smaller interplanetary probes a few years earlier.

Two days into the flight, a midcourse maneuver was attempted. However, due to a problem with the attitude orientation system (which failed to identify correctly the navigation star, Sirius, when employing a low-density optical filter), it was not executed. The burn was not critical, and a second attempt the following day using the high-density optical filter was also unsuccessful. A third attempt on March 7 finally succeeded using a medium-density filter. The trajectory for reentry back into the Earth's atmosphere appeared nominal, and the spacecraft was separated from the equipment module for the fiery plunge.

But the projected path, which was to take the descent module to within 30 mi. of the surface before skipping back out to almost 100 mi., did not occur. Instead, it continued into the atmosphere, exposing the spacecraft to extremely high temperatures and up to 20 gs. Had a crew been on board, they probably would have survived but may have experienced significant injuries. When it became apparent that the "skip" would not take place and the spacecraft would actually land near the west coast of Africa, the decision was made to destroy it for fear of its falling into American hands (Siddiqi 2003a, 618).

It was again believed that the star-tracker opticals had been contaminated by the engine exhaust gases, and provisions were made for subsequent craft to provide a sensor enclosure that was disposed of after the final burn of the Blok D engine on the next attempt on April 22, 1968. The star tracker never had a chance to operate on that flight because the abort-detection system malfunctioned during the first-stage flight, and the emergency escape system was activated.

Frustration was apparent, not only with chief designers Mishin (for the 7K-L1 spacecraft) and Vladimir Chelomey (for the UR-500) but also with their entire teams. The political hierarchy was also showing signs of irritation at the inability of the circumlunar project to show success. Even the tight security that surrounded each launch and subsequent failure was not immune from some leaks through a variety of paths to the American CIA. However, the Western press was still captivated by the Soviet mystique and continually predicted an imminent spectacular.

A Surprising Apollo Schedule Change

The desire of both nations to complete the first manned circumlunar flight was strong. For the Soviets, who were still struggling to achieve the same rendezvous, docking, long-duration, and high-apogee flights that Gemini had demonstrated, circling the Moon before the Americans was critical to maintain the appearance of space technology leadership. As for the Americans, who were finally getting it all together after so many years of playing catch-up, they were not about to lose the race to circle the Moon.

During the summer of 1968 NASA Administrator James Webb was kept informed by the CIA of the pending prospects for a Soviet circumlunar flight. He had been made aware of the Soviet project to build a Saturn V equivalent (the N1) and had alluded to it in congressional testimony. American reconnaissance satellites had regularly photographed Baikonur and had seen the ground-test mock-up of the N1 on the launchpad in the spring of 1968. Because Webb could not reveal his sources (and the available information was rather skimpy), many thought his use of the mythical giant rocket was simply a ruse to pressure Congress into releasing more money for NASA (whose funding was now on the decline). The phantom Soviet rocket was often referred to derisively as "Webb's Giant."

The associate administrator for manned space flight, George Mueller, decided that AS-503 would be a manned flight following the resolution of the problems with AS-502; the next Saturn V would carry a full crew on only its third flight. With the first flight-qualified lunar module (LM-3) not expected to be ready until early 1969 for a test in Earth orbit, what should be the mission of Apollo 8, the first manned Saturn V? To wait for LM-3 would lose a full Saturn V launch cycle of ten weeks, and it was critical that a full five launch cycles be available in the last calendar year of the decade to have a chance of meeting the Kennedy goal.

The summer of 1968 was one filled with foreboding as far as the United States was concerned. Politically, the Vietnam War was going poorly; President Johnson had acknowledged that in March, when he announced he would not seek reelection. Martin Luther King Jr., a principle civil rights advocate, was assassinated in April, and Sen. Robert Kennedy was gunned down in June following his presidential primary campaign win in California.

The American space program had but seventeen months remaining to place a man on the Moon to fulfill the late President John F. Kennedy's pledge of May 1960 to accomplish the task "before this decade is out." Yet the first manned Apollo spacecraft (Apollo 1) had killed its crew of three in a ground test eighteen months earlier. Following an extensive redesign of several systems, NASA was ready to again attempt to orbit three astronauts in October (Apollo 7) using a Saturn IB. The giant Saturn V launch vehicle had experienced several significant problems on its second test the preceding April with the unmanned Apollo 6.

During the first week of August in 1968, George Low, then manager of the Apollo Spacecraft Office, had an incredible conversation with the director of flight operation, Chris Kraft. Low's far-fetched proposal was to use the third launch of the Saturn V (AS-503) to send Apollo 8, a manned CSM, around the Moon before the end of the year (Chaikin 1994, 57).

Low was a naturalized U.S. citizen, son of Austrian immigrants who had fled Hitler's rise to power in 1938. He served in World War II and graduated from Rensselaer Polytechnic Institute with a bachelor of aeronautical engineering degree in 1948. After a brief period with General Dynamics (Convair Division), he returned to Rensselaer to complete a master's degree. He then had joined the National Advisory Committee for Aeronautics as an engineer. The advent of *Sputnik* (1957) and morphing of the National Advisory Committee for Aeronautics (NACA) into the National Aeronautics and Space Administration (1958) brought Low into the Manned Spaceflight Center.

Low's reasoning for his bold proposal was based in part on the "all-up" testing philosophy established by George Mueller in 1963. To beat the Russians to the Moon, some dramatic change had to be made with respect to how flight-testing was performed. Mueller advocated that each launch had to "prove" several major components of the hardware rather than the traditional and conservative method of incremental testing. Thus, the first test of the Saturn V rocket (AS-501) in November 1967 had all three stages live and drove its unmanned Apollo 4 into Earth's atmosphere at lunar return

George M. Low, then Apollo Spacecraft Program Office director, pushed for a circumlunar flight for the third flight of the Saturn V. Courtesy of NASA.

reentry speeds to test its heat shield. The flawed (but still somewhat successful) test of the second Saturn V (AS-502) in April of 1968 involved qualifying the spacecraft's propulsion system in a series of complex maneuvers.

But Low was not only facing the Kennedy deadline of the end of the decade; he was also confronting the challenge of the Russian space program that had been accomplishing space spectaculars since the first *Sputnik* more than ten years earlier. The Soviets had also encountered a tragedy in their manned spaceflight program just months after the Apollo 1 fire when Vladimir Komarov died in the first test of the Soyuz spaceship. However, within the past ten months, the USSR had launched four large unmanned Zond spacecraft that appeared to be the equivalent of Apollo. Although not publicly acknowledged, Zond was a Soyuz adapted for circumlunar flight.

Low's rationale for using the third Saturn V to send the second manned CSM to the Moon was actually simple. Wernher von Braun's team was making changes to the Saturn V to avoid the pogo effect that had almost destroyed the previous flight. In keeping with the all-up testing philosophy, there had to be more significance to the next launch (which now cost about $200 million) than simply verifying that the changes addressed the problems.

The progression of Apollo flights was defined as a series of missions labeled A through G. The objectives of each had to be completed before the

Table 1. Apollo Mission Objectives as of August 1968

Designation	Objective	Status
Mission A	Test all three stages of Saturn V and CSM at lunar reentry velocity	AS-501, Apollo 4, Nov. 1967 AS-502, Apollo 6, April 1968
Mission B	Test LM ascent and descend engines in LEO	SA-204, Apollo 5, Jan. 22, 1968 (Saturn IB)
Mission C	Test first manned CSM	SA-205, Apollo 7, Sched. for Oct. 1968 (Saturn IB)
Mission D	Manned CSM and LM system check in LEO	AS-503, Apollo 8, Sched. for Feb. 1969 (Third Saturn V)
Mission E	Repeat of Mission D in high Earth orbit	AS-504, Apollo 9, Sched. for May 1969
Mission F	Manned CSM and LM system check in high Earth orbit or lunar orbit	AS-505, Apollo 10, Sched. for Aug. 1969
Mission G	Manned lunar landing	AS-506, Apollo 11, Sched. for Nov. 1969

lunar landing could be achieved. Table 1 shows the progress as of August 1968 (Zimmerman 1998, 213).

Despite the problems with completing Mission A with the second Saturn V, von Braun was confident that the changes being implemented such as the helium shock-absorbing cavities in the LOX line prevalve assembly would resolve the pogo problem and the leaking fuel-line igniter. His assessment reflected more risk than his previous ruling in March 1961 regarding the Redstone, which allowed the Soviets to send the first human into space.

The lunar module (LM) was experiencing many development delays and was not expected to be available for flight-testing for another five months. So the question was—what could the next Saturn V accomplish? Mueller understood that the third flight of the Saturn V had to be manned to keep on schedule—but with the LM behind in its delivery, what kind of mission could move the program forward? As it now stood, achieving a manned lunar landing before the end of 1969 was doubtful if they could not fly another Saturn V (AS-503) in the programmed three-lunar-month launch cycle. To Low, the answer was obvious—AS-503 would send the second manned CSM (Apollo 8) around the Moon.

While Kraft did not oppose Low's dramatic proposal, he indicated that key elements of the Apollo program had to be evaluated to determine if such a flight was within the current capability. The entire focus of the program had been the manned lunar landing, and an explicit circumlunar flight had not been a part of the mission objectives. But Low did not just want to loop around the Moon. In keeping with the all-up testing philosophy, he wanted to have the three intrepid astronauts orbit the Moon for several days to simulate some of the tasks required for later flights—including the effect of the lunar mass concentrations (MASCONS) on lunar-orbiting bodies such as the Apollo spacecraft. Low was essentially modifying Mission D, which was to send the CSM and LM into a high Earth orbit. It would not have the LM, and instead of high Earth orbit would go all the way to the Moon. It would accomplish most of the Mission D objectives and would perhaps beat the Russians to the circumlunar goal.

Within days of the proposal, the various NASA centers reported that the tracking, communications, launch vehicle, and CSM were capable of supporting the mission (Kranz 2000, 224). The specialized computer systems and software required were also available.

James Webb, who had been out of the country, was more than annoyed when he returned to find that Low had not only proposed the high-risk mission but had also performed so much preliminary planning. Webb had been a key ingredient in the administration of NASA and the political maneuvering required to get, and keep, the Apollo program funding—especially in light of the costs of the Vietnam War. As he listened to Low's reasoning, he too quickly saw the rationale. Webb was receiving sensitive covert intelligence briefings on Soviet progress and knew of the giant N1 rocket being readied for testing in preparation for a manned landing on the Moon. He also recognized that the Proton booster and its Zond payloads were probably unmanned tests in preparation for a manned circumlunar flight. Although he was quite sure that America was finally outpacing the Soviet program, he did not want America to be on the losing end of yet another "first" in the space race.

However, if the stakes were high, the risk was too. Webb knew that by approving an attempt to send three men to the Moon on only the third test of the Saturn V and only the second flight of the Apollo spacecraft, he was leaving himself open for major political fallout if the attempt failed—especially if the astronauts were lost. He advised President Johnson of the possibility and the risks. The president had no reservations (Zimmerman 1998, 218). He went with Webb's judgment, and the planning continued. However, no commitment could be made until after the Apollo 7 flight and the entire test results were calculated relative to changes generated by

AS-502. Moreover, the plan remained an internal document, not to be disclosed to the press.

Astronaut chief Deke Slayton called in James McDivitt, the scheduled commander of Mission D (Apollo 8), to give him the opportunity to accept the change in the objectives or adjust his flight to become Apollo 9—he opted for the latter. Having invested so much of their training in the CSM/LM flight plan, he did not want to make a change. Thus, the Apollo 9 crew of Frank Borman, James Lovell, and William Anders readily agreed to move their Mission E to Apollo 8 (Borman and Serling 1988, 189). Since that mission had been planned as a high-apogee flight, it more closely suited the initial training of Borman's crew, although he got the CSM-103 (Apollo 9) spacecraft.

In an interview with Jim Lovell years later, he reflected on the event: "We were in Los Angeles at North America when Frank was called back to Houston. When he returned, he told Bill and me that the mission was changed." As conservative as NASA had been, even considering the policy of all-up testing, did Lovell initially feel it was too big a step? "I was delighted! I had no worries about the risk involved. Perhaps I didn't understand the situation." Did he have any reservations about not having the LM available for Apollo 8? "The thoughts about the LM as a lifeboat never entered our minds. It was only after [Apollo] 13 that the additional risk of not having the LM was apparent."

By August 12 a basic flight plan had been formulated, and Low started its circulation through the various NASA organizational structures as a confidential internal document. There were more than a few who felt the proposal was carrying all-up testing beyond reasonable limits, that the risk of a lunar flight on only the second manned Apollo flight was too great. In the meantime, Arthur Rudolph, the man who had headed V-2 production at the Mittelwerk in the Harz Mountains of southern Germany and who was now the manager of the Saturn V Program Office, was directed to assemble AS-503 in the vehicle assembly building. Despite all of the quiet planning that was taking place, presumably to send AS-503 on a lunar mission, it was emphasized to the NASA staff that Apollo 7 had to fly a flawless first manned mission on AS-205.

As the tentative planning for Apollo 8 continued to move forward, another Zond launch on September 15 sent live tortoises and other biological specimens around the Moon and back to Earth. The Soviets had apparently achieved complete success with their launch vehicle and spacecraft. The world now expected the Soviets to launch men to the Moon with the next lunar opportunity. However, the Soviets had not revealed that the reentry was ballistic instead of employing aerodynamic lift—thus placing extremely high g loads on the spacecraft.

Apollo 7: Disgruntled Astronauts

By the time the redesigned Apollo was ready to fly in October 1968, twenty-three months had elapsed since the last successful American manned flight—Gemini XII. Four astronauts had been killed (Grissom, Chaffee, and White in the United States, and Komarov in the USSR), and, for America, the situation on Earth had deteriorated with the escalation of the war in Vietnam and the riots in the streets of several major cities.

The expensive space race seemed to have lost its appeal to many who now considered the resolution of racial strife and achieving peace to be far more pressing national goals than the Moon. President Lyndon Johnson had assumed the presidency following Kennedy's death and had won the presidential election of 1964 over Republican Barry Goldwater. Johnson's presidential banner was "The Great Society," through which he sought to end poverty and provide "negroes" with equal opportunity.

However, his unbridled willingness to continue an unpopular war—one in which the tactics proved difficult for the American military to effectively handle—became his downfall. Faced with opposition, primarily from within his own Democratic Party, Johnson announced in March 1968 that he would not seek reelection the following November. Then, just days later, civil rights leader Martin Luther King Jr. was assassinated. This was followed two months later by another assassination—this time former President Kennedy's brother Robert, who was running in the presidential primaries.

Even the funding for Apollo, which had hit its peak in 1966, was now on the down slope, and thousands of young engineers and technicians who had staked their careers on the space program were being laid off as the Moon program moved closer to its goal. Congress showed little interest in a follow-on manned mission to Mars or a permanent space station or lunar-base program. The dynamic head of NASA, James Webb, had resigned only weeks before the scheduled Apollo 7 flight and was replaced with Thomas O. Paine.

Thus, by the time astronauts Wally Schirra, Walter Cunningham, and Donn Eisele were strapped into the Apollo 7 spacecraft on October 11, 1968, America was a divided and depressed nation. Many of its youth, using the

The first manned launch of the Saturn IB propelled the Apollo 7 spacecraft into orbit. Courtesy of NASA.

Vietnam War as an excuse, had turned to drugs and abandoned the ethical and moral standards that had made the nation great. One of the many counterculture movements professed "God is dead."

Schirra, who had flown the perfect six-orbit mission in MA-6 and demonstrated the first rendezvous with Gemini VI, was tasked to fly the first manned Apollo as mission commander. He was a demanding person who expected only the best from his crew. Cunningham was designated the command module pilot, and Eisele, the LM pilot, although there was no LM for this mission. The primary objectives of the eleven-day flight were the qualification of all the Apollo systems, crew coordination and performance, and rendezvous techniques as well as compatibility with ground support facilities.

The Apollo 7 crew had followed CSM-101 through its redesign, fabrication, and test. Now they sat in it, atop AS-205, the first manned Saturn IB, on Pad 34. The Block II ship was the result of over 1,800 changes since the fatal fire, and it had moved through the ground-testing phases with relative ease as the contractor, NAA, had made noticeable improvements in its quality assurance and fabrication methodologies. Apollo 7 flew on only the fifth flight of the new Saturn IB booster, which had never taken men into space. This was in stark contrast to more than one hundred Atlas firings and thirty-five Titan II launches that qualified those rockets for a manned flight.

At T-3 seconds the eight H-1 engines of the first stage came to life and quickly built to 1.6 million lb. of thrust. The launch was flawless, and the huge rocket blazed a trail through the sky—a sight that had become familiar to space fans for the previous ten years. At T+143 seconds the four inboard engines cut off, followed by the four outboards 4 seconds later. The S-IVB immediately separated, and its single 200,000 lb. thrust LH2 J-2 engine ignited. The escape tower was jettisoned 20 seconds later, and the S-IVB burned for a total of 470 seconds, placing the 36,000 lb. CSM in an orbit that ranged from 142 mi. to 178 mi. The CSM then separated from the S-IVB, and the astronauts proceeded to exercise the reaction control system and SPS propulsion system to perform a series of rendezvous maneuvers with the spent S-IVB over the next few days.

Unlike previous missions, all three astronauts were able to remove their spacesuits after the launch and spent the flight in light blue jumpsuits that were more comfortable. The Apollo was roomier than Gemini and allowed the American astronauts to unbelt and float around the cabin for the first time. In addition, Americans were now able to visit the astronauts through the miracle of television.

Almost from the first, the crew developed severe head colds that were aggravated by the weightless condition, which allowed their sinuses to expand but didn't allow them to drain naturally under the influence of gravity. This condition caused considerable discomfort and made the astronauts less agreeable as the eleven days wore on.

Despite the contentious nature of the crew toward Mission Control, the live televised programs from space were well executed. Courtesy of NASA.

Schirra, in particular, was constantly annoyed by changes to the schedule and new tasks that were communicated up to the crew. He criticized many of the activities that they were asked to perform, as did the other two astronauts. At one point he canceled the first scheduled TV period and remarked to Mission Control, "The show is off! The television is delayed without further discussion. We've not eaten. I've got a cold, and I refuse to foul up my time." Later, when the crew was awakened from a sleep period earlier than planned, Schirra again complained that his request for an extra hour of sleep had been ignored. There were many other heated exchanges, but when subsequent TV periods were scheduled, Schirra conducted them without any hint that he had been angered by various situations, and the American public was excited to be able to see the astronauts at work on live television for the first time. Although Schirra had indicated that this was his last flight, the other two astronauts also never flew again—more than likely because of repercussions from their disagreeable nature.

Apollo 7 was the first and only use of the Saturn IB as a manned launcher during this period. The accelerated NASA schedule immediately used its larger brother, Saturn V, for subsequent Apollo flights. It was never used for any other program despite its efficient payload-carrying capability. It was called upon to support the Skylab space station with three manned launches of the Apollo spacecraft and the single Apollo-Soyuz Test Program of the mid-1970s.

If the head colds and the added tasks impeded the working conditions and the rapport between Mission Control and the crew, the outcome of the flight could not have been better. All of the assigned objectives were met or exceeded. The booster and the spacecraft worked to perfection. It was time to take a bold step forward.

With Richard Nixon set to take over the presidency the following January (1969), Jim Webb formally submitted his resignation. There was speculation that he had suffered too much negative publicity, and the risk involved in Apollo 8 was too great a burden for him.

Soviet Missed Opportunity

As the next UR-500 Proton launch vehicle was being ground tested for the August 1968 lunar launch window, an oxygen tank on the Blok D stage exploded on the launchpad during prelaunch preparations and nearly caused the loss of the entire rocket as well as the 7K-L1 spacecraft. It was only by heroic efforts that more damage was not sustained, although one life was lost of the ground crew.

The Zond 5 spacecraft being prepared for a circumlunar flight. Courtesy of Energiya.

The next UR-500/7K-L1 spacecraft launched on September 15, 1968. This flight contained live tortoises, among the many biological specimens on board. The booster performed flawlessly, and the upper stage Blok D sent the spacecraft toward the Moon with its TLI burn. The mission was designated Zond 5 to the world, and attitude control once again threatened its success.

A constant stream of work-around commands allowed Zond 5 to complete its loop around the Moon, coming to within 1,200 mi. of its surface. It then proceeded back to Earth after taking the first high-quality pictures of Earth, showing it as a complete blue disk suspended in the black void of space. (America's lunar orbiter had taken similar pictures, but the resolution of its TV scanner could not compare with the direct image photos returned by Zond 5.)

Once again, the attitude control was critical for the reentry, and, following another series of difficult maneuvers, the 4,500 lb. Zond 5 splashed down in the Indian Ocean just 65 mi. from a Soviet recovery ship. Although the reentry was ballistic (there was no attempt to "skip" off the upper layers of the atmosphere to lessen the g load), it would have been survivable for a cosmonaut. The flight was a milestone in that it was the first spaceship to be recovered after traveling to the vicinity of the Moon. Soviet space planning was again becoming more solidified.

Zond 6 launched on November 10, 1968, on another unmanned circumlunar mission, and it carried a biological payload similar to its predecessor. It took high-quality pictures of the far side of the Moon and returned to Earth on November 17, using the more sophisticated atmospheric "skip reentry" necessary for a manned flight. Its trajectory allowed it to land in the Soviet Union—only ten miles from the launch site. However, the spacecraft experienced a loss of pressurization that killed most of the biological specimens and caused the parachute recovery system to activate early and to release before the craft had landed. This caused a high-speed impact with the Earth that effectively destroyed Zond 6. Amazingly, the camera film was recovered intact.

Technologically, there were no more hurdles preventing the Soviets from launching cosmonauts on the next flight. They had the equipment necessary, but reliability remained questionable. The next lunar launch window for the Soviets was the second week of December 1968, and there was great speculation in the West that this flight would be manned. Nevertheless, the Soviets realized that they had to skip that window and perhaps look to January 1969. Cosmonaut Chief Kamanin wrote, "I have to admit that we

The unmanned Zond 5 reentry vehicle is retrieved from the Pacific Ocean following a successful flight around the Moon. Courtesy of Energiya.

are haunted by U.S. intentions to send three astronauts around the Moon in December. Three of our un-piloted L1 spacecraft have returned to Earth at the second cosmic velocity, two of them having flown around the Moon. We know everything about the Earth–Moon route, but we still don't think it is possible to send people on that route" (Siddiqi 2003a, 665).

Apollo 8: The Christmas Flight around the Moon

Following the successful flight of Apollo 7, a review board met on November 7, 1968, and declared that all of the elements necessary for the tentatively planned circumlunar flight were complete. In an unusual Sunday meeting with key NASA and contractor representatives on November 10, it was unanimously agreed that Apollo 8 would go to the Moon. One contractor objected to the CSM entering lunar orbit, suggesting that it merely loop around to avoid the possibility of its being stranded should the CSM's SPS engine not restart. However, redundancy of the engine systems along with the fact that the spacecraft carried three fuel cells (and could operate with only one) coupled with two independent oxygen systems gave solid assurance that the risks were minimized. It is not known if the launch of Zond 6, that same day, had any influence on the final decision.

The following day, the new NASA administrator, Thomas Paine, approved the mission scheduled for December 21, 1968. If the Soviets were to attempt a circumlunar flight, their launch window was the second week of December. Even with its bold plan to launch Americans to the Moon on the first manned launch of a Saturn V (and only the third launch of that rocket), America might still lose another opportunity to be first.

Paine met with the press the following day to make the momentous announcement: "After a careful and thorough examination of all the systems and risks involved, we have concluded that we are now ready to fly the most advanced mission for our Apollo 8 launch in December, the orbit around the Moon." While the revelation was met with excitement and anticipation by most, it was also the subject of some critical analysis. Among the detractors was Sir Bernard Lovell, director of Britain's Jodrell Bank Observatory, who noted that, within a few years, missions to the Moon could be done by robotic spacecraft, avoiding the risk to a human crew. He also commented that the mission represented "a dangerous element of deadline beating," an obvious reference to the Kennedy commitment and the race with the Soviets (Baker 1982, 316). Some critics cited the lack of an extensive track record for either the Saturn V or the Apollo spacecraft, and there were a few who predicted outright failure.

If there was a dominant theme in why America should turn its back on the Moon, it was often the cost factor and the reference to the millions of people who lived in poverty in the United States. To this argument, the proponents of a vigorous space program provided statistics to show that the space program budget for 1968 stood at $4.7 billion (down from its peak of $5.9 billion in 1966), which represented about 2.5 percent of federal expenditures. This was less than Americans spent on cigarettes. President Johnson's war on poverty had escalated to more than $20 billion that year— more than four times the NASA budget. The duel in space between the Americans and the USSR was a high-stakes affair that had political, economic, and military implications that demanded both nations offer their best efforts. In the final analysis, most felt that it was mankind's destiny to explore the unknown, even if robots could eventually accomplish the task. Humans needed to satisfy that inner desire "to go where no man has gone before."

Preparation continued for the launch. It appeared to be another situation where one nation would perhaps win by a week or two—not enough to make a significant technological statement. However, when the Soviet

The three Apollo astronauts who became the first humans to orbit the Moon. Courtesy of NASA.

tracking ships returned to port early in December, it seemed obvious that no lunar attempt was imminent.

The Apollo 8 crew entered their spacecraft in the predawn hours of December 21, and the countdown proceeded without any major problems. The excitement of the day was highlighted because of the Saturday launch date and the Christmas school break that allowed millions of Americans to witness the event via television.

The crew included Air Force lieutenant colonel Frank Borman, a veteran of the Gemini VII flight and commander of the Apollo 8 mission; Navy commander James Lovell, who shared Borman's Gemini VII flight and who had flown a second time on Gemini XII, becoming the first man to go into space three times; William Anders, who was making his first flight into space.

The automatic sequencer assumed control of the Saturn V countdown at T-187 seconds. At T-10 seconds, the flame pit was deluged with tons of water that served to cool the launchpad and dampen the acoustical effects of the 7.5 million lb. of thrust. Ignition sequence started at T-8.9 seconds, with each of the engines starting at a slightly staggered interval to avoid the stress of the explosive-like impact of the volatile liquids flaming into life. It took almost six seconds for the engines to stabilize and for the computerized diagnostics to verify their health before the four steel arms that held the rocket to the pad were given approval to move back and allow the rocket to begin its ascent.

The eight umbilical arms that had, until seconds earlier, exchanged electrical signals with the rocket now moved out of the way and were themselves drenched in a cascade of water to minimize the damage from the rocket's exhaust.

The astronauts, as well as the millions of spectators who lined the beaches and roads of central Florida, experienced a soaring degree of excitement as the rocket rose higher into the sky. Still more tens of millions viewing on television had increased heart rates in the vicarious experience of mankind's first attempt to break fully his bonds with Earth.

Lovell observed that they were pleasantly surprised by the ride on the Saturn V compared to the Titan. Maximum G's was only four compared to eight at the burnout of the second stage of Titan. The loading on the upper stages of Saturn V were less than one-G. He humorously referred to the experience as riding the "Old Man's Rocket."

Except for some pogo oscillations in the S-II stage, the first and second stages performed to specification, and the S-IVB finished the job of inserting Apollo 8 into a parking orbit 180 mi. above Earth, 11.5 minutes after liftoff. The crew worked with Mission Control to verify the proper functioning of all of the systems before committing the spacecraft to its journey to the Moon.

Soon the call was communicated to the crew from Mission Control: "Apollo 8, you are 'go' for TLI, over." TLI was the final burn of the S-IVB that provided the needed 24,500 mph escape velocity. Lovell responded, "Roger, understand we are 'go' for TLI." For those who understood the significance of the call, it was a milestone in the human exploration of space (Borman and Serling 1988, 203).

The S-IVB engines lit up, and the CSM began accelerating the final 7,000 mph increment over a period of five minutes and eighteen seconds. Following shutdown, Borman separated the spacecraft from the S-IVB and turned it around to face the spent third stage. Had a lunar module been on board the S-IVB, this was the point at which the Apollo CSM would have docked with the LM and pulled it free. However, the maneuver on this day was just to simulate the activity, and soon another burn of the reaction control

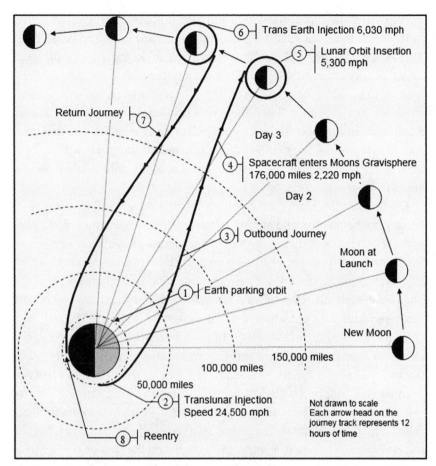

Each key point in the circumlunar flight is defined in numeric order. Courtesy of Ted Spitzmiller.

system provided separation from the S-IVB, which passed by the Moon and entered an orbit around the sun.

The crew removed their spacesuits and performed a series of postlaunch translunar navigation sightings using the guidance and navigation (G&N) system while floating within the cabin. Within hours, all three experienced space sickness, possibly induced by the psychological aspect of floating free. The Mercury and Gemini astronauts had not had a significant problem because of the small confines of those spacecraft. Borman also contracted intestinal flu-like symptoms, but none of these problems appeared to inhibit crew performance or represent a hazardous situation.

The G&N system was critical in determining the path of the spacecraft and the midcourse maneuvers necessary to ensure that the Moon was intercepted at the appropriate point and time in space. The digital computer in the unit contained 38,864 bytes of read-only memory and 2,048 bytes of read/write memory. The computer had a small keyboard with nineteen keys that represented the ten digits and nine function keys. Another element of the G&N system was an optical telescope that served as a sextant.

Eleven hours into the flight, the spacecraft was now traveling at a mere 5,600 mph, having been slowed by the pull of Earth's gravity. The first midcourse firing of the large SPS engine occurred at that point with a short burn of 2.5 seconds, slowing the CSM by 14 ft./s. It would take another 69 hours to reach the Moon.

On Sunday, December 22, Apollo 8 broadcast its first TV session—31 hours into the mission and almost halfway to the Moon. It included a view of the relatively small disk of the Earth. The black-and-white camera lacked the ability to display the brilliant blue that the astronauts described. "It's a beautiful, beautiful view, with predominantly blue background and just huge covers of white clouds," said Borman—who by that time had mostly recovered from his bout with the flu (Borman and Serling 1988, 206).

At 55 hours into the mission, a historic milestone was reached in the annals of humanity when the three crewmembers passed into the gravitational field of the Moon. Two scheduled midcourse maneuvers had not been needed, and only a slight velocity correction by the reaction control system was required. Now, traveling at a mere 2,700 mph, they began to accelerate toward the Moon. They were on a "free-return" trajectory, but the mission called for them to fire their SPS engine on the far side of the Moon to slow their velocity and allow them to be captured by the lunar gravity into an orbit around it.

As the CSM neared the Moon, all the systems and navigation were again verified and the call "Go for LOI" (lunar orbit insertion) was sent up to the

---

Free-Return Path

A free-return path is a flight path from Earth that simply loops around the Moon (because of the lunar gravity) and returns to Earth—if no other velocity changes are undertaken after translunar injection. Other possible trajectories include a lunar impact (if the velocity of the spacecraft is too slow) and lunar fly-by (if the velocity is too great). For lunar orbit insertion to occur, the spacecraft must be slowed to lunar orbital velocity on the far side of the Moon.

Apolune and Perilune

Just as the terms "apogee" and "perigee" apply to the high and low points of an orbit about the Earth, the corresponding points of a space vehicle in orbit around the Moon use the terms "apolune" and "perilune."

astronauts. As the spacecraft disappeared around the far side, it was now out of communication. If the burn were successful, Apollo 8 would emerge after a silence (loss of signal) of thirty-three minutes. If the burn failed to take place, the signal would be acquired early.

When the signal was not acquired early, everyone following the mission sensed the anticipation that indeed the ship was now in orbit around the Moon. Precisely on schedule, the telemetry from Apollo 8 came streaming across the quarter-million-mile distance and a cheer went up from Mission Control. The SPS had slowed the spacecraft by 2,000 mph to 4,500 mph, providing for a perilune (closest approach to the Moon) of 70 miles and an apolune of 195 miles (the farthest from the Moon).

From their viewpoint, the astronauts saw details of the Moon as no human had ever seen—not even through the most powerful telescopes. When

The three astronauts were emotionally shaken by the experience of viewing the Moon from such a close-up perspective. Courtesy of NASA.

asked what the Moon looked like, Lovell responded, "O.K., Houston. The Moon is essentially gray, no color . . . we can see quite a bit of detail. . . . The craters are all rounded off, there's quite a few of them . . . the walls of the craters are terraced, about six or seven different terraces all the way down" (Borman 1988, 211). They were viewing the surface at a passing rate similar to what is experienced on Earth when traveling at high altitude on an airliner. "Seeing the Earth as it really is. Extending my arm, my thumb could completely cover the Earth. It immediately conveyed the thought of how insignificant we are in the universe. A small body orbiting a normal star, tucked away in the outer edges of the Milky Way—only one of the millions of galaxies in the universe" (Baker 1982, 320). Lovell noted in later years that it was an emotional experience.

On Christmas Eve, Apollo 8 sent its last TV transmission from the Moon. One more revolution and they would head back to Earth. Borman began by saying, "The Moon is different to each one of us. . . . I know my own impression is that it's a vast, lonely, and foreboding type existence." Lovell then commented, "The vast loneliness up here . . . is awe-inspiring. . . . The Earth is a grand oasis in the big vastness of space." Anders added about the "terminator" (the area where the sun is rising or setting on the Moon), "[it] brings out the stark nature of the terrain with the long shadows" (Chaikin 1994, 121).

Anders closed out the TV session by saying, "the crew of Apollo 8 has a message that we would like to send to you. 'In the beginning, God created the heavens and the Earth. . . .'" Starting at Genesis 1:1 (the Genesis Factors that are the foundation of rocket science) the three astronauts took turns reading from the Bible the first ten verses of Genesis as the rugged and desolate terrain of the lunar surface moved slowly across the TV screen of the Earth-based audience. It was an emotional occasion for the millions who witnessed it (Zimmerman 1998, 245).

It was time to return home as the spacecraft passed behind the Moon for the last time and Mission Control experienced loss of signal. If the SPS failed to fire, Apollo 8 would remain in its orbit around the Moon, with no hope of return for the astronauts. But right on cue, thirty-seven minutes after disappearing from view, the voice of Jim Lovell confirmed what the telemetry was telling Houston, "Houston, Apollo 8, over." Houston replied, "Hello, Apollo 8, loud and clear." Lovell responded, "Roger, please be informed that there is a Santa Claus." The reference was to a successful SPS burn. They had spent twenty hours orbiting the Moon.

The Genesis Factors

Space flight is based on the four fundamentals of physics: time, energy, mass, and space. Because these factors are revealed in the first verse of the book of Genesis (In the beginning [time] God Created [energy] the heavens [space] and the Earth [mass]), they are referred to as the Genesis Factors.

The sixty-three-hour return trip was uneventful, and thirty-five minutes before reentry the command module switched to its own battery supply and separated from the service module fifteen minutes later. The spacecraft then made its initial flight into the upper fringes of the atmosphere at an angle of 5.5 degrees, dropping down to thirty-five miles from the surface before skipping back up another five miles for its final descent. The 5,000 °F temperature that ionized the atmosphere around the spacecraft caused the expected five-minute communications blackout period. Apollo 8 landed only a few miles from the recovery task force after a trip of more than a half-million miles.

America received worldwide acclaim. Even the Soviet's often-critical response to America's achievements offered its congratulations and added, "The world now stands on the brink of entirely new experiences in interplanetary exploration." The effort to downplay the Soviet's own aspirations was voiced by Leonid Sedov, who was the primary spokesman for their space program: "There does not exist at present a similar project in our program. In the near future we will not send a man around the Moon." He noted the Soviets preferred to use unmanned robots, playing on the comments of Sir Bernard Lovell a few weeks earlier. In fact, the giant N1 Moon rocket was nearing its first critical flight test. Although the Soviets had lost the vital opportunity of circumlunar flight, there was still much that could go wrong in the American program. Kamanin wrote in his diary, "The flight of Apollo 8 is an event of worldwide and historic proportions. This is a time for festivities for everyone in the world. But for us, the holiday is darkened with the realization of lost opportunities and with sadness that today the men flying to the Moon are not named Valeriy Bykovskiy, Pavel Popovich, nor Aleksey Leonov."

The high-risk decision to send Apollo 8 to the Moon dramatically revealed the shift in space exploration leadership. The foresighted George Low and the man on whom the responsibility for failure would have been placed, James Webb, had defied the odds. They had put their confidence in the skills and determination of the Apollo engineers and astronauts that had allowed America to triumph.

The year 1968 had been a difficult one for America with the Vietnam War, racial strife, and assassinations. Many felt that the successful circumlunar mission gave hope that, if humanity could overcome such highly technical problems presented by a flight to the Moon, the troubles here on Earth could be managed as well. However, engineering technical solutions are more easily mastered than human behavior.

18

ONE SMALL STEP

The Soviet Response to Apollo 8

At a meeting of senior Soviet space officials just two weeks after the return of Apollo 8, the agenda focused on how to handle the success of that flight. To this point, the Soviets had used the weightlifting advantage of the R-7 to accomplish an amazing string of firsts. However, *Sputnik I* (which many in the West equated with a technological Pearl Harbor) was now more than ten years in the past, and the remarkable accomplishments of Gemini and then Apollo had come without a comparable response from the Soviets.

The meeting sought to determine both long- and short-term goals that might counter the American success. As for their circumlunar bid that had come so close to success, several months would be required before a comparable Soviet effort could emulate Apollo 8. As it would provide no real increment in capability or technology, it was decided to abandon their manned circumlunar effort. The remaining L-1 spacecraft was used in an unmanned role with "scientific excellence" being touted. Because the "Zond" moniker had never been differentiated between the smaller early unmanned craft and the Soyuz flight-test articles, the world would not know of the Soviet's failure in this area.

As for responding to the possibility of an American manned lunar landing, the Soviets again took their cue from Sir Bernard Lovell—they emphasized unmanned robots and made a point of showing that there was no need for humans to travel to the Moon, and that its exploration could readily be accomplished with less expensive and "no-risk" automated spacecraft. Over the past several years, they had been developing two unmanned lunar landers. The Ye-8 Lunokhod was a wheeled vehicle to explore the Moon and perhaps provide a homing beacon for a manned lander. The Ye-8-5 automated laboratory was designed to return samples from the lunar surface. This could be accomplished with the UR-500 booster, whose reliability was still questionable. The decision was made to accelerate the Ye-8-5 project in an attempt to beat the Americans in returning samples of the Moon. This in itself would be a major triumph.

However, the majority of those present at the meeting still felt that a manned landing was possible in a period competitive with the Americans; they continued with their N1/L-3 manned lunar program. However, its lack of lifting capability (no LH2 upper stages) and the overweight condition of their lunar lander required two launches to assemble the required lunar vehicle.

As the Soviets had not publicly declared the Moon as a national goal, they could still hide the program behind their wall of secrecy if they failed to achieve their objective. Moreover, it was unlikely that the Americans could continue their unbelievable string of successes—it would take only one major failure to set back the Apollo landing program for a considerable time. What was the likelihood of the Americans achieving six consecutively successful Saturn V launches?

There was also some discussion on turning to near-Earth goals of a manned space station. This, too, became a part of their deception—depending on how the year played out. Some thought was given toward an elaborate Mars program culminating in a manned landing in 1977. This as well would never be funded, as the events of the year would take their toll on any advanced interplanetary planning. While money was still a major problem for the Soviet space program, the lack of decisiveness and effective management had likewise impaired engineering progress.

Soyuz 4/5: Another Close Brush with Death

Almost two years after the death of Soyuz 1 cosmonaut Vladimir Komarov, the Soviets were once again ready to try to dock two manned spacecraft. Soyuz 4 was launched on January 14, 1969, with forty-one-year-old lieutenant colonel Vladimir Shatalov as its lone occupant. Following a slight change in his orbit with the onboard maneuvering engine, he held a brief TV session in which he pointed out the two additional seats but made no commitment on how they were to be filled.

The following day, Soyuz 5 launched with a crew of three: lieutenant colonels Boris Volynov and Yevgeny Khrunov and civilian Aleksei Yeliseyev. As with previous Soyuz missions, a single-orbit rendezvous (M=1) had been planned, but for undisclosed reasons the process was extended

Almost three years after Gemini demonstrated docking, the Soviets finally achieved the milestone with Soyuz 4 and 5. Courtesy of Energiya.

over several orbits. The docking was accomplished manually, avoiding the problems that had plagued the previous mission. With a combined weight of over 28,000 lb., the Soviets claimed "the world's first experimental space station." Unlike the Apollo/LM combination that had yet to be flown, the docking apparatus did not allow for an internal transfer of cosmonauts from one ship to the other—they had to exit into space from one ship and then enter the other ship. Khrunov and Yeliseyev moved into the living compartment of Soyuz 5 and donned space suits with the help of Volynov, who then returned to the command module. It was an awkward and difficult task in the limited confines and under weightless conditions.

Following a test of the integrity of the suits, the pressure was lowered in the living compartment, and the hatch was opened. Tethered to the craft, both men made their way over to Soyuz 4 and entered the hatch to its living compartment. The thirty-seven-minute excursion was only the second spacewalk for the Soviets since the historic first walk of Alexey Leonov almost four years earlier. As had been the response from all spacewalkers, the two were amazed and excited by the experience.

After the successful transfer of the two cosmonauts, the spacecraft were undocked and proceeded to complete independent experiments before

Soyuz 4, now containing three crewmembers, returned to Earth on January 17. Volynov, alone in Soyuz 5, was about to experience the most harrowing reentry yet. Following the retroburn, the three segments of the spacecraft were to separate. However, as Volynov could see from his porthole, only the living quarters had jettisoned. The service module was still in place as the antenna on the solar cell "wings" were plainly visible. Volynov reported the problem to Mission Control. Several Vostok missions had experienced a similar problem in that they had only partially separated, and some of the wiring had held the two components together. However, the two segments of Soyuz 5 appeared to be locked together mechanically, and this was essentially a fatal situation.

The reentry began with the two units tumbling end over end as the attitude-control fuel was quickly exhausted in its attempt to stabilize a mass that was twice what it had been designed for—and with a center of gravity that was out of bounds. Only the heat shield had an ablative coating to survive reentry, and this was now embedded between the descent module and the service module. The atmospheric friction began destroying both segments. The hydrogen peroxide tanks in the service module were the first to show the damaging effects of the reentry heat when they blew up. Miraculously, the explosion accomplished the separation, and Volynov realized for the first time that perhaps he might not die.

The center of gravity was now where it belonged, and, despite the lack of attitude control, the spacecraft assumed the proper aerodynamic position to allow the heat shield to do its job. When the parachute opened, the spinning spacecraft started to twist the risers, and he feared that the chute might collapse. Again providence prevailed, and the chute was able to stop the rotation and unwind itself, allowing Volynov to touch down safely—some 400 mi. from his intended recovery point. The TASS news agency reported that the flight ended successfully, without any mention of Volynov's unbelievable brush with death.

On January 20, 1969, another UR-500 Proton rocket with its unmanned L-1 spacecraft was launched, which would have been designated Zond 7, had one of the engines in the fourth stage not shut down twenty-five seconds prematurely. The flight program should have been able to accommodate the problems and continue, but confusion in the abort sensor caused the L-1 escape rocket to fire. This lifted the unmanned spacecraft clear from the launcher, which was then destroyed, and the spacecraft returned to Earth under parachute. On February 19, 1969, another UR-500 booster with the first Ye-8 lunar lander was launched, but as it passed through Max

Q (maximum aerodynamic pressure) at T+54 seconds, the new payload shroud failed and caused the destruction of the rocket. These were difficult failures for the Soviet program—and the people who had worked so hard to prepare these missions.

The N1 Flies: Almost

The first N1 rolled out to the launchpad on February 3, 1969 and was about to take flight. Weighing over 6 million lb. when fully fueled, it was slightly heavier than the Saturn V. A railroad transporter moved it to the launchpad horizontally in a cradle of steel supports. Its success would put the Soviet program back in competition with the United States. Its failure would perhaps doom the Soviet manned lunar program.

The firing sequence began on February 21, 1969, as thirty rocket engines, developing more than 10 million lb. of thrust (one-third more than the Saturn V), came to life in an unbelievable display of sound and fire. Deputy Chief Designer Boris Chertok described the event: "Even if you have attended our Soyuz launches dozens of times, you can't help being excited. But, the image of the N1 launch is quite incomparable. All of the surrounding area shakes, there is a storm of fire, and a person would have to be insensitive and immoral to be able to remain calm" (Siddiqi 2003a, 682).

However, problems had begun even before liftoff. The engine operation control system (KORD) had already sensed trouble with one of the first stage engines (number 12) and had shut it down, along with the corresponding engine on the opposite side of the thrust ring to keep the vehicle in balance. As designed, the mammoth rocket continued to fly. However, more problems awaited the giant as it climbed steadily into the winter sky. At T+70 seconds, all the engines shut down, and the abort system rocketed the unmanned L1S spacecraft clear of the booster (Siddiqi 2003a, 682).

Although the launch had not been a complete success, neither had it been an outright failure. It proved many aspects of the most complex and powerful rocket ever launched by the Soviets—a system that had never even been test-fired with all its engines operating.

As the telemetry results were studied, several problems were revealed that had been working in the giant rocket during its brief flight. The failure of engine 12, detected by KORD just before launch commit, had been an erroneous indication—there had been no need to shut it down or its counterpart number 24. Acoustical vibrations, in particular, had played havoc with connections, and a fire started in the aft end—ultimately proving to be the final source of failure. It was clear that the lack of static testing played a major role in the failure. Although certainly disheartened, the "almost" aspect of the launch still gave hope for the beleaguered members of the space program.

Apollo 9: Lunar Module Earth-Orbital Test

In spite of the eighteen-month delay caused by the Apollo 1 fire, the entire program came together for the first flight of all three modules with the Apollo 9 mission on March 3, 1969. James McDivitt, Russell Schweickart, and David Scott flew the second manned Saturn V and experienced some pogo effect in the second stage but encountered no other problems as they achieved an initial 120 mi. Earth orbit. The plan called for the astronauts to exercise all three modules in a sequence that emulates a lunar flight—while remaining in Earth orbit. This was the most demanding flight yet, but it did not receive as much publicity as Apollo 8 because it remained in Earth orbit.

The CSM was turned around and docked with the LM three hours into the flight. The LM was then released from the S-IVB, and McDivitt moved the CSM/LM combination a safe distance from the empty S-IVB, which was then restarted twice and flown to escape velocity and into an orbit around the sun. The astronauts removed their spacesuits and began a ten-day series of tests, most of them directed toward proving the various LM systems.

With McDivitt and Schweickart on board the LM, the two vehicles undocked. For the first time since Gemini IV (*Molly Brown*), spacecraft were again given distinctive names so that communications between the two could be differentiated. The LM, because of its obvious wide legs, was

---

The Docking Tunnel

The Apollo spacecraft and the lunar module required the internal probe assembly of the docking mechanism be removed once the docking was complete. This allowed the astronauts to move between the two spacecraft, through the docking tunnel, without having to leave the confines of the two pressurized vehicles. This was not the case for the early Soyuz spacecraft.

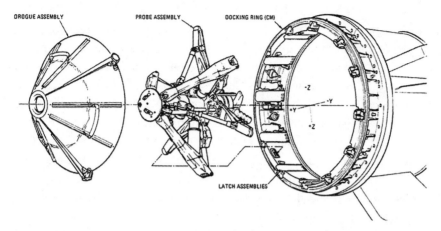

Apollo internal probe assembly of the docking subsystem. After docking, the tunnel between the two spacecraft could be opened up by removing this assembly. Courtesy of NASA.

Spider and the CSM was *Gumdrop* because of its silver cylindrical shape. A series of burns of the descent propulsion system (DPS) and ascent propulsion system (APS) tested the integrity of both systems as well as the ability of either the CSM or LM to function as the "active" participant in the rendezvous process. These maneuvers were critical, as the LM had no way of returning to Earth should the final rendezvous fail.

A brief stand-up extravehicular activity was also conducted to prove the ability of the astronauts to transfer from the LM to the CSM if a lunar return docking failed to be achieved. The lunar spacesuit was also tested on Schweickart during the brief (thirty-seven-minute) extravehicular activity. After completing all of the LM tests, McDivitt and Schweickart returned to the CSM, and the LM was eventually jettisoned on the fourth day of the mission, ultimately to burn up on reentry.

Apollo 9 continued with a series of experiments to complete the ten days in orbit before firing the SPS to return to Earth only 3 mi. from the prime recovery ship. The way was now clear for a second trip to the Moon.

Apollo 10: Rehearsing the Moon Landing

Apollo 10 launched on May 18, 1969, with the intent of flying a lunar-landing mission that was complete in every detail except for the final burn of the DPS to the Moon's surface. Crewed by Tom Stafford, Eugene Cernan, and John Young, it was the most experienced crew to go into space—five

Gemini flights between them . . . no rookies! It was also the roughest ride the Saturn V had yet given its occupants, prompting John Young to comment, "Charlie, you sure we didn't lose Snoopy on the staging?" (Young 2012, 127). The reference was to the CSM being named *Charlie Brown* and the LM named *Snoopy*, for the Charles Shultz comic-strip characters. It was the last of the somewhat flippant names applied to spacecraft by their crews. Patriotism and science dictated future communication monikers.

For the second time in the history of humanity, the call went up to a spaceship: "Go for TLI" (translunar injection). The S-IVB lit up for its second burn to accelerate the CSM/LM vehicle out of its Earthly parking orbit and to escape velocity. Halfway into the six-minute burn, a high-frequency vibration rattled the spacecraft. Stafford weighed the abort option as the vibrations continued to worsen to the point that he could hardly read the instrumentation (Young 2012, 127). However, he had come this far and wasn't about to end a mission if all the numbers still indicated that the rocket was taking them to the Moon. (Later analysis revealed the problem to be with the pressure relief valves, which was remedied for future missions). Following the S-IVB burn, *Charlie Brown* docked with *Snoopy* and pulled him from his "doghouse."

The trip to the Moon was uneventful, with the crew using many rest opportunities to beam color TV sessions back to Earth and the millions of viewers around the world. The camera had broadcast-quality resolution with 525 lines and 30 frames per second. On the fourth day, Apollo 10 slipped behind the Moon and executed the lunar orbit insertion burn of the SPS engine to place the combination into lunar orbit.

Stafford and Cernan entered the LM and completed all the preliminary checks prior to the undocking and initiation of the new set of "Peanuts" call signs. A novel problem was encountered. During the docking three days earlier, there apparently had been a slight rolling motion that caused the docking latches to become slightly skewed. It was determined that the undocking could take place if the alignment was 6 degrees or less; the misalignment was 3.5 degrees. The only caveat was communicated from Houston: "Charlie Brown, we're concerned about this yaw bias in the LM and apparent slippage of the docking ring. We'd like you to disable, and keep disabled; all roll jets until after undocking" (Baker 1982,335). The undocking was uneventful—another indication of the wide-tolerance engineering built into the Apollo systems.

Snoopy then began a series of orbital maneuvers to simulate the initial descent profile, or descent orbit initiation (DOI), to the Moon and at one

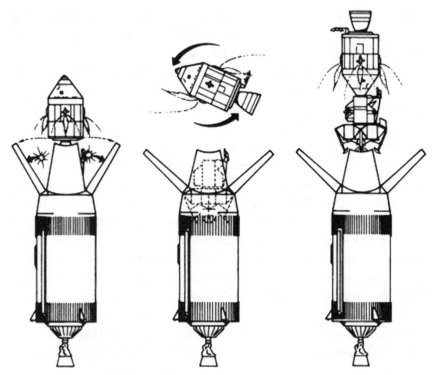

The process of extracting the LM from the S-IVB following TLI. Courtesy of NASA.

point came within 50,000 ft. of the surface. This never-before-seen perspective caused Cernan to again make a light-hearted comment when he made his initial contact with Earth as they came around the Moon. "Houston, this is Snoopy. We is Go, and we's down among 'em Charlie" (Cernan 1999, 215). The spacecraft was now within a few miles of some of the mountain peaks, and, at their velocity, it was a thrilling ride. He then added, "Ah, Charlie, we just saw Earthrise and it's got to be magnificent." Stafford interjected, "There's enough boulders around here to fill up Galveston Bay, too." Cernan emphasized the situation again when he said, "We're low babe, we're low."

The descent engine fired again to raise the orbit, and *Snoopy* prepared to separate the ascent from the descent stage. A switch (automatic verses attitude hold) had been left in the wrong position and caused the ascent stage to automatically seek the CSM communication signal. Thus, when the stages separated, the ascent stage, with Stafford and Cernan on board, executed an unexpected 360-degree pitch causing Stafford to emit the expletive, "Son of a Bitch!" Despite the unprogrammed maneuver, the astronauts were quickly back in control, and the remaining rendezvous and docking was accomplished without problems. Stafford reported, "Snoopy and Charlie Brown are hugging each other" (Cernan 1999, 218).

Speculation in the media suggested that perhaps Stafford and Cernan might just set the LM down—the opportunity to be the first on the Moon overriding the mission profile. What the media neglected to inform its viewers was that the propellant in the ascent stage had been limited to only two-thirds of its capacity to simulate the weight and balance of the final return to *Charlie Brown*. A second consideration was that the LM was the early version, which did not have the weight reductions and could not have achieved the return flight even with full fuel. Had they attempted such a stunt, they would not have had enough propellant to complete the return to lunar orbit—so much for speculation. Perhaps the overriding factor was the personal discipline and integrity that these men brought to their profession. Among its other objectives, Apollo 10 confirmed the ability to rendezvous in lunar orbit—another "space" first.

The history-making first trans-Earth injection burn occurred on schedule, and as the spacecraft came from behind the Moon and into communications, Mission Control was treated to the singer Dean Martin's rendition of *Goin' Back to Houston* from the onboard tape recorder. The last dress rehearsal for a manned lunar mission was over. The LM had passed all of its qualification tests. It was time for the real thing!

One Last Opportunity: The Second N-1 Test

The Soviet program had attempted to move forward during the first half of 1969, following the directives it had established after the Apollo 8 flight. A small core of eight cosmonauts continued to train for a manned lunar mission, but it was obvious to them that, even with total success of the next N1, they would probably not fly until 1970. A CIA assessment predicted that 1972 was a more likely period. There had been many ambiguous statements by both cosmonauts and academicians regarding Soviet intentions. The cosmonauts continued to drop hints of their preparedness while the politically oriented science community followed the official line not only that was there no Moon race but also that Soviet efforts were geared toward automated scientific methods of returning lunar samples.

The head of the cosmonaut office, Kamanin, had written in his memoirs that during the Apollo 10 flight, he was disappointed by the "unrestrained lying" of Soviet officials regarding the intent of the Soviet space program.

His thoughts perhaps reflected not so much the need for national security that lies often protect but that the truth would ultimately prove to be a national humiliation. He then noted, "We have come to the end to drink the bitter chalice of our failure and be witness to the distinguished triumph of the U.S.A. in the conquest of the Moon" (Siddiqi 2003a, 686).

Chief designer Vasily Mishin, who had taken over Korolev's responsibilities following his death, was summoned to a meeting with Communist Party secretary Leonid Brezhnev in April of 1969 to explain the current state of affairs as the American effort moved closer to success. His report simply reiterated what everyone in the space program knew: the organizational bureaucracy, coupled with a poorly coordinated program and a lagging technology infrastructure, had been severely impacted by lack of funding.

Five Ye-8-5 robotic soil-return probes had been prepared, and the first launched on June 14, 1969, in an effort to upstage the Americans. However, the fourth stage of the UR-500 Proton booster malfunctioned, and the vehicle dropped back into the atmosphere to burn up. The UR-500 had now failed five consecutive times, and eight times in fourteen attempts. Intense analysis appeared to point to quality control as the dominant factor. Random failure as opposed to systemic failure indicated that the basic design was sound. Now there was but one lunar window, which coincided with the scheduled Apollo 11 flight. It might just allow the Soviets to pull off the "scoop of the decade" by returning a lunar sample before the Americans.

The second test of the N1 preceded that effort. The giant rocket came to life in the early hours of July 4, 1969. (Because it was just before midnight in Moscow, the official launch date is often given as July 3, 1969.) There was apprehension that accompanied the frantic effort to prepare the rocket. Night launches are always more spectacular than those in the daylight as the intense lighting contrast highlights the unbelievable event. The N1 rose slowly, but just as it was clearing the lightning towers that surrounded the launchpad, the abort sensor fired the escape rocket, and the unmanned L3S lunar module was pulled free. A few seconds later, an explosion enveloped the vehicle, and the equivalent of a small nuclear bomb devastated Launch Complex 110 (Siddiqi 2003a, 691). Safety precautions prevailed, and there were no casualties other than the local wildlife.

Analysis of the telemetry revealed that five engines had failed at liftoff. KORD had detected the fatal anomaly and had attempted to shut down all the engines at T+10 seconds. However, one engine continued to fire, and this caused the rocket to tip almost horizontal until it impacted the ground.

There were several reasons for the failure, but again quality control loomed apparent. The failure was never reported to the world.

Now there was only one opportunity left for the Soviets to recoup some of their sagging prestige. Another UR-500 with a Ye-8-5 had been prepared in one last effort to upstage the Americans. It launched on July 13, 1969. The vehicle performed perfectly, and the payload, Luna 15, was on its way to the Moon at last. Twelve years after the first Sputnik, and eight years after President Kennedy had made his commitment, the Moon race played out over a period of literally hours to the checkered flag—although one participant was a robot.

Apollo 11: "The *Eagle* has landed"

The crew for Apollo 11 was announced on January 9, 1969: Neil Armstrong (mission commander), Michael Collins (command module pilot), and Edwin "Buzz" Aldrin (lunar module pilot). All three were born in the same year—1930. Armstrong had piloted the X-15 before being selected as an astronaut in the second group following the original Mercury Seven. Collins

The three Apollo 11 astronauts: Neil Armstrong, Mike Collins, and Buzz Aldrin. Courtesy of NASA.

The lunar-landing training vehicle was an invaluable tool to practice vertical descents to a simulated lunar surface. Note the jet engine positioned vertically in the center, with the lunar module pilot in the housing to the left. Courtesy of NASA.

and Aldrin were from group three. All had Gemini flight experience—Armstrong's had been the first and only American flight brought down early because of the stuck thruster problem.

To help prepare Armstrong and Aldrin for the final descent to the lunar surface, they were given training in a helicopter and a special vehicle called the lunar-landing training vehicle (LLTV), which was an ungainly looking device with four legs that used a jet engine (positioned vertically) to support five-sixths of its weight to simulate the lunar gravity. Hydrogen peroxide thrusters then allowed the pilot a limited amount of maneuverability equal to the LM. All who flew it agreed that it simulated the final lunar-landing profile effectively. Armstrong ejected from one following an in-flight malfunction on May 6, 1968, narrowly escaping death (Hansen 2005, 329).

At the time of this crew's appointment, there was no way of really predicting that Apollo 11 would actually be the first manned lunar landing. However, by late 1968 that is what the schedule called for. With the unerring success of Apollo 9 and 10, Apollo 11 was to carry out its assigned mission—land on the Moon. From the time that their crew assignment had been made public, and as the date for the launch grew closer, all three recognized that their lives would never again be their own. Now they were public figures, and the situation would only get worse—if they were successful.

LM-5 and CSM-107 were assigned to ride atop AS-506, the first of the "operational" Saturn V rockets that no longer carried a complete set of engineering sensors. Only 1,348 measurements were provided, as opposed to the 2,342 on the Apollo 10 flight. The crawl from the vehicle assembly building to the flame pit of the completed flight vehicle occurred on May 21, 1969, two months before the scheduled flight.

Initially, the first four landings were to be primarily engineering efforts to "prove" the path to the Moon. However, as costs soared and criticism continued to be leveled at the usefulness of the Moon program, it was decided to include more scientific aspects in the early flights. The Apollo Lunar Surface Experiment Package (or ALSEP), deployed by the astronauts, provided some of the science performed from the Moon's surface.

Collins, Aldrin, and Armstrong trained intensely during June, and there was some concern that all might not be ready for the July launch date. A ninety-minute conference call occurred on June 12, which included Deke Slayton, Wernher von Braun, Chris Kraft, and the crew, who discussed the situation. In the final analysis, Armstrong summarized the training situation by saying that the schedule was tight, but they would be ready. With the final input on the booster, the spacecraft, the ground teams, and the crew, Apollo Program Director Sam Phillips made the decision to proceed with the launch on July 16.

Preparations for the media to report the first landing on the Moon, and to accommodate the spectators at Cape Canaveral for the Apollo 11 launch, were greater than any other event in history. Many believed this date might be as famous as Columbus' 1492 discovery of America. It was celebrated in real time as no event in the past had been. The gaudy displays of the Super Bowl halftime had yet to be invented, and there was no basis for comparison. More than eighteen thousand official visitors were invited; many were representatives from around the world. Even within the restricted areas of the launch facility, more than forty thousand people gathered. The real crush of humanity was the hundreds of thousands who jammed the beaches and highways that surrounded the Cape in an attempt to see history in the making. Almost without exception, the assembled masses congregated without incident. During the two months leading up to the launch, even natural deaths declined noticeably; apparently the desire to witness the first man on the Moon was greater than the pull of the Grim Reaper.

Rev. Ralph Abernathy, who had succeeded the assassinated Dr. Martin Luther King Jr. as the head of the Southern Christian Leadership Conference, an organization lobbying for civil rights legislation, arrived at the gate to the Kennedy Space Center with a group of poor people to protest the expensive lunar program. He was met by NASA administrator Thomas Paine, who compassionately but realistically, told them that "if it were possible to not push that button and solve the problems you are talking about, we would not push that button" (Baker 1982, 341). The NASA budget for 1969 was 2.31 percent of the federal budget—far less than the 28 percent that was then being expended on welfare programs.

Hermann Oberth, the man who had envisioned a flight to the Moon some fifty years previous, was present. Willy Ley, the prolific writer and advocate of space flight, had unfortunately passed away just one month before the flight. The wife of Dr. Robert H. Goddard, the man criticized by the *New York Times* for prophesying that man could one day send rockets to the Moon, was also among the distinguished dignitaries.

Many of the Saturn V launch staff had been members of the von Braun team that arrived in the United States twenty-four years earlier as a part of Operation Paperclip. However, the bulk of the staff in Mission Control were young, "homegrown" men in their late twenties and early thirties. Flight Director Gene Kranz, with his crew cut, white shirt, tie, and white vest, epitomized those in the front line of consoles referred to as "the trench" (Kranz 2000, 141). The vest had become a symbol—made special by Kranz's wife, Marta, for each launch. There were thousands of support people who were every bit a critical part of the mission—but only the three men in Apollo 11 had their lives on the line.

As first envisioned by NASA planners in the early 1960s, only one of the two astronauts would actually leave the LM to explore the lunar terrain and gather soil samples. The second crewmember would remain "safely" in the lunar module to avoid any hazards and to be ready to leave if problems arose. By 1968, as the planning took on more detail, it was realized that having both men outside made the setting up of scientific equipment easier and provided the ability to help each other should trouble occur.

The question of who should exit first was also an interesting issue. It was realized that the first man on the Moon would be the name forever in history books. Initial planning called for the lunar module pilot (Aldrin) to exit first. However, the final design of the LM positioned the men in such a way that it would have been difficult for him to squeeze by Armstrong to exit first. Although Armstrong was a quiet and introverted person, he left no doubt that, as mission commander, he would be the first to set foot on the hallowed soil. Aldrin, who would later legally change his first name to Buzz in 1988 because he had become so well known by that moniker, was also somewhat of a loner. He had missed being selected in the second astronaut selection group because he was not a graduate of a test pilot school. He made his mark in the astronaut corps as an academic, immersed in the physics of orbital mechanics. He made some effort to reverse the decision as to who exited the LM first but realized that it was futile (Aldrin and McConnell 1989, 216).

The time had arrived—July 16, 1969. Aldrin recalled that they were awakened early and had breakfast with Dr. Thomas Paine, the NASA administrator. "He told us that concern for our own safety must govern all our actions, and if anything looked wrong we were to abort the mission. He then made a most surprising and unprecedented statement: if we were forced to abort, we would be immediately recycled and assigned to the next landing attempt. What he said and how he said it was very reassuring" (Aldrin and McConnell 1989, 226). Paine wanted the three to know that there was no reason for them to push a bad situation, thinking it was their only opportunity to be first to the Moon. Then, Aldrin recalled, "we . . . began to suit up—a rather laborious and detailed procedure involving many people."

There was the twenty-four-minute ride to the launchpad. The humid tropical heat of that July morning was beginning to build, and the sky was almost cloudless. Aldrin recalled,

> While Mike and Neil were going through the complicated business of being strapped-in and connected to the spacecraft's life-support system, I waited near the elevator on the floor below. I waited alone for fifteen minutes in a sort of serene limbo. As far as I could see, there were people and cars lining the beaches and highways. The surf was just beginning to rise out of an azure-blue ocean. I could see the massiveness of the Saturn V rocket below, and the magnificent precision of Apollo above. I savored the wait and marked the minutes in my mind as something I would always want to remember. (Aldrin and McConnell 1989, 226)

Collins recalled,

> I am everlastingly thankful that I have flown before, and that this period of waiting atop a rocket is nothing new. I am just as tense this time, but the tenseness comes mostly from an appreciation of the enormity of our undertaking, rather than from the unfamiliarity of the situation. I am far from certain that we will be able to fly the mission as planned. I think we will escape with our skins, or at least

I will escape with mine, but I wouldn't give better than even odds on a successful landing and return. There are just too many things that can go wrong. (Collins 1974, 360)

The countdown proceeded smoothly, and, almost before mankind was ready for it, the numbers dwindled—3, 2, 1, Zero! As if in character, the trio on board the roaring beast had few comments as they were thrust into the atmosphere—part of 6,384,400 lb. of gross weight. Unlike AS-505, this Saturn V flight was smooth and presented no pogo effect—the engineering changes were working.

At T+140 seconds the inboard F-1 shut down as planned. Twenty-one seconds later, the S-IC first stage completed its burn, and the S-II began to convert hydrogen and oxygen into superefficient power. The escape tower jettisoned at T+166 seconds. The S-IVB began its job at T+549 seconds, and, at the end of its 2.5-minute burn (10 minutes and 26 seconds after leaving the ground), the 297,000 lb. of mass that remained was in its parking orbit, 120 mi. above Earth.

Tracking data was now analyzed and systems checked to ensure that nothing was left to chance. "GO for TLI" was issued, and two hours and nine minutes into the flight the S-IVB ignited for the second time to propel the space travelers free of the Earth's gravity. The exact velocity of 24,394 mph was achieved at five minutes and forty-seven seconds into the burn. The CSM separated from the S-IVB, turned 180 degrees, and docked successfully with the LM, which was housed, legs folded in the S-IVB. The LM was then released from the S-IVB, and the CSM thrust the two vehicles clear of the spent third stage, which proceeded into an orbit around the sun.

To control the extreme temperatures in space, the CSM/LM was turned broadside to the sun and put into a slow roll of one rotation every two minutes; it was called the "barbeque mode." At twenty-five hours into the flight, Apollo 11 passed the halfway point on its journey as its speed continued to diminish under the pull of the Earth's gravity; it was then traveling at only 6,000 mph. The trajectory was so precise that Apollo 11 would have come within 208 mi. of the lunar surface. This was adjusted by a short 2.9-second burn of the SPS engine to provide a miss of only seventy miles some two days later.

Despite the pending history-making event that they were now a part of, the astronauts ate and slept in relatively normal patterns. Their engineering discipline served them well as they were able to perform all the

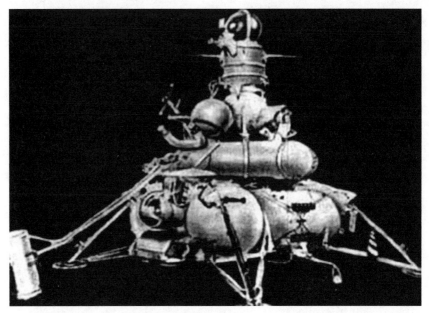

The Soviet Luna 15 automated lunar probe was an impressive "sample and return" vehicle. The sampler arm extends down to the left. Courtesy of Energiya.

required equipment tests and experiments en route. Several TV broadcasts were made using the color camera (although most people viewing still had only black-and-white sets). When the LM was entered for inspection, it was found that, like Apollo 10, there had been some roll component during docking, but the two spacecraft were offset only 2.05 degrees.

Meanwhile, the Soviet Luna 15 successfully fired its descent engine to enter lunar orbit on July 17. The world now wondered, hyped by the media, if the Moon race had actually come down to a virtual sprint to the finish line.

At sixty-one hours into the mission, with more than 40,000 mi. to go and traveling at 2,050 mph, Apollo 11 passed into the influence of the Moon's gravitational field. It then began accelerating toward the waiting target, fourteen hours into the future. Mission Control brought the three travelers up to date on how the world was reporting their adventure. They were told that the Soviet newspaper *Pravda* referred to Armstrong as the "Czar of the ship," to which Collins quipped, "The Czar is brushing his teeth, so I'm filling in for him."

As the spacecraft prepared to loop behind the Moon, the array of flight controllers in Houston were polled for the status of their systems, and a ripple of "Go" responses was heard. With ten minutes to loss of signal,

Mission Control sent up the call, "Eleven, this is Houston. You are Go for LOI [lunar orbit insertion]." Another milestone had been passed.

As the spacecraft looped around the far side, the SPS engine burned for six minutes and two seconds, and Apollo 11 was now in lunar orbit. The telemetry numbers received when the spacecraft reappeared from behind the Moon confirmed that all was well. As had the crews of the Apollo 8 and 10, Armstrong, Aldrin, and Collins were mesmerized by the sight of the lunar landscape passing beneath the ship.

Luna 15 made an orbital correction on July 19 that put it into an orbit with a 70 mi. apolune and 10 mi. perilune. Its radar altimeter, however, showed no smooth areas beneath its orbital track at the perilune. However, preparation for its landing continued.

A review of the LM systems began at T+81 hours and 25 minutes, followed by another night's sleep before the big day. This sleep period was unlike the previous ones in that none of the astronauts appeared to sleep soundly (according to the biomedical data being transmitted from the sensors on their bodies)—could any human have taken this occasion with normal responses?

The day of the landing had now arrived—July 20, 1969, and the astronauts wriggled back into the spacesuits that they had shed after launch. Armstrong and Aldrin floated into the LM for the final checks before separation. Collins remained in the CSM. The LM landing legs were extended and all systems given one final review. "Go for undocking," was the advisory from Mission Control. Now a new set of names, chosen to distinguish the two separate spacecraft, highlighted the political aspect of the mission.

"*Eagle*, Houston. We're standing by, over." *Eagle* was the call sign of the LM. It was the symbolic bird of America—represented in flight on the decorative patch that adorned the space suits of the astronauts. "The *Eagle* has wings," was the response from the LM as it undocked. Now a series of numbers were communicated to update the onboard computer for the DOI burn. "Go for DOI" was issued as the two now independent spacecraft disappeared behind the Moon.

Armstrong and Aldrin made the final preparations for executing the first burn of the DPS. *Eagle*'s engine faced its direction of flight and parallel to the lunar surface so that the thrust reduced its 3,800 mph speed and dropped it out of orbit. The astronauts were initially oriented so they faced the Moon and could identify various landmarks. At ignition, the engine developed only 10 percent of its rated thrust, so there was little detectable g-force. It was then throttled up to 40 percent for thirty seconds before shutting down. The spacecraft was then reoriented so the astronauts faced away from the Moon—allowing them to view only the star-studded blackness of space.

Eagle now fell in a curved path toward the lunar surface. There were no wings and no parachute that could save a fatal impact should the DPS engine fail to restart for the PDI (powered descent initiate), and only an emergency abort using the APS engine was an alternative to an unsuccessful landing. It is also interesting to note that this was the first powered "landing" of a manned rocket—the first time it had ever been attempted!

"*Columbia*, Houston, over," was the call from Mission Control to Collins as the CSM emerged from behind the Moon. The word "*Columbia*," derived from Christopher Columbus' name, had for almost two centuries personified the identity of the United States. It was also a symbolic reference to Jules Verne's fictional *Columbiad* that had carried three travelers to the Moon one hundred years earlier (Collins 1974, 335). These two call signs, *Eagle* and *Columbia*, left no doubt about the intended meaning of the mission—America was winging its way to the lunar surface in response to a deadly serious challenge perceived by its commander in chief some eight years earlier.

As *Eagle* appeared from behind the Moon, it was just seventeen minutes from PDI, and there were some problems establishing communications with it and with *Columbia*. The high-gain antennas on both vehicles had to be pointed at the Earth for the signals to be successfully transmitted. With anticipation of a quick resolution to the situation, the directive "Go for Powered Descent" was passed to *Eagle* by Collins. In *Columbia*, he had a direct communication path with the LM.

The PDI was a twelve-minute burn of the DPS—all the way to the surface. The engine again came to life at 10 percent, and after a short period throttled up to 100 percent. Now feeling the effects of positive gs, the two astronauts were supported by a set of harnesses that positioned them in front of two triangular windows, which still looked out into the blackness of space. It was not until *Eagle* pitched forward (relative to the direction of flight) that the astronauts were finally able to view the surface of the Moon. This motion also brought the radar altimeter into position so that it could begin monitoring the height above the terrain.

A steady banter of communications filled the quarter-million miles between the event and those monitoring the critical systems on Earth. A three-second gap existed between the transmission of a signal and the receipt of the first possible response to that signal.

"*Eagle*, Houston. You are GO . . . you are GO for continued powered descent."

"Roger."

"And *Eagle*, Houston. We've got data dropout. You're still looking good." Houston advised that, although they were not getting a steady stream of telemetry data, what they had showed all factors were nominal.

"PGNCS we got good lock on. Altitude lights out. Delta H is minus 2900." With these cryptic comments, *Eagle* was indicating that the primary guidance and control system (PGNCS) was getting good data from the radar and confirming that the PDI was within specification.

The computers in Houston were crosschecking the performance data with the remaining fuel and "altitude yet to go" to ensure that the event was within the capabilities of the spacecraft.

Eagle then sent the first of several advisories regarding computer alarms. "1202, 1202." This computer code indicated that the onboard computer was unable to keep up with all of the data being presented. However, back in Houston, engineer Jack Garman knew his software codes and was quick to assure flight controller Steve Bales that the computer was able to continue its primary function by ignoring data that was not of immediate value (Mindell 2011, 222).

"We're Go on that alarm," was the response issued by Houston.

As *Eagle* descended through 24,000 ft. traveling at just under 1,000 mph, the DPS engine throttled down. One minute later, *Eagle* passed through 15,000 ft. and the DPS was down to 55 percent of rated thrust.

With three minutes and thirty seconds until estimated touchdown, *Eagle* was 7,000 ft. above the surface and descending at 90 mph. Armstrong could now see the landing site through his window. A new program, called P64, that allowed manual control of the flight was called up on the computer.

"*Eagle* you're looking great, coming up on nine minutes."

Flight Director Gene Kranz, the primary controller, made one more poll of the flight control stations of FIDO (flight dynamics), guidance and control, and the flight surgeon produced the immediate and positive response—"Go!"

Then *Eagle* reported, "1201 alarm . . . 1201."

Astronaut Charlie Duke, assigned as CapCom for the mission, had been relaying the directives from the various Mission Controllers to the crew. He relayed the 1201 override to *Eagle*, "We're Go, PGNCS high, we're Go!" Jack Garman's expertise continued to reassure the controllers (Kranz 2000, 269, 289).

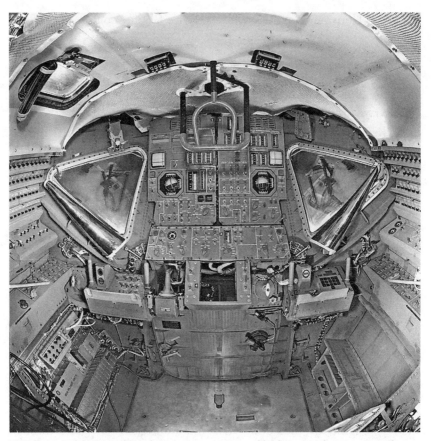

The interior of the LM where the two astronauts stood looking out the two triangular windows during the landing. The hatch that led out to the "porch" is in the bottom center. Courtesy of NASA.

Although Aldrin was designated as the lunar module pilot, it was Armstrong who was guiding the LM toward the lunar surface (Mindell 2011, 191). Armstrong now switched to another computer program, P66, that allowed him to literally hover and move across the lunarscape using his thruster controls while the DPS held altitude. *Eagle* was slowly descending through 350 ft., and its pitch was almost vertical. From this point P66 lowered the spacecraft the final distance to the surface while moving forward about one-quarter-mile. As the descent continued, Armstrong could see that the intended landing area had several large boulders, so he elected to begin manual control. Now the communication was virtually one-way. Both astronauts described what was happening with their responses: "Eleven forward, coming down nicely. Five percent Quantity light. Seventy-five feet, things looking good. Down a half. Six forward." (The numbers represented feet per second.)

CapCom interjected the fuel remaining: "60 seconds." *Eagle* confirmed the warning. "Light's on. Down two and one-half . . . picking up some dust . . . faint shadow . . . four forward."

CapCom sent the next scheduled warning of fuel remaining: "30 seconds." The fuel gage and the telemetry-sending unit only had an accuracy of 2 percent. Thus, at this point the fuel tanks could have run dry. However, the abort sensor was set to detect an interrupt in the supply to the engine— not until that absolute point was reached would the APS initiate an abort.

Armstrong knew that he was flying on borrowed time at that point, but his landing approach continued. "Forward. Drifting right . . . contact light!" The thin 6 ft. rod that hung beneath one landing pad had sensed the surface and communicated the event by illuminating a yellow light in front of Armstrong. His response was immediate: "Okay, engine stop." Now the one-sixth g of lunar gravity settled *Eagle* softly onto the surface as Armstrong continued with the shutdown procedure. "ACA out of detent. Modes control both auto, descent engine command override off. Engine arm off. Four-one-three is in."

CapCom responded, "We copy you down, *Eagle*."

"Houston, Tranquility Base here, the *Eagle* has landed."

"Roger, Tranquility, we copy you on the ground. You've got a bunch of guys about to turn blue. We're breathing again, thanks a lot."

The landing occurred as planned on the Sea of Tranquility, the dark mare referred to as a "sea" by the ancient astronomers; the first manned scientific station was named for that prominent feature—Tranquility Base.

The actual landing was about 1,000 ft. farther west than planned and about 300 ft. to the left of the intended track. For the first time in history, a manned spaceship had set down on another body in the solar system.

A series of events quickly transpired to "safe" the spacecraft, and to prepare it for immediate takeoff—if any systems check discovered a critical flaw that might require a quick return to *Columbia*. Referred to as "stay/ no stay," these contingencies might have to be acted upon if any spacecraft damage was found. The call came from CapCom: "*Eagle* you are stay for T-1" (T-1 was the first two minutes after landing and T-2 was an eight-minute period). T-2 also passed uneventfully a few minutes later, and Tranquility Base was ready for its next phase.

The tense minutes after the landing were agonizing for the Grumman engineers, who observed through the telemetry that high pressure was building in a fuel line. Those who were following the elation of the landing—indeed, even the astronauts themselves—were unaware that only minutes remained before the rising temperature could have caused a tragedy. The event passed as the unanticipated heat was absorbed into the structure and the temperatures remained within tolerance (Kelly 2001, 213).

The Moon Walk

The flight plan called for the astronauts to enter a sleep period before their moonwalk, but it was obvious that, with their heightened state of awareness, sleep was not possible at this point. Besides, it was now early evening in Europe (where countless millions were following the flight), and rather than keeping half of that continent awake through the night, it was decided to allow the surface excursion to begin. Following a brief prayer of thanksgiving from Aldrin and a moment to partake in the sacrament of Holy Communion on Aldrin's part (Aldrin and McConnell 1989, 239), the two ate a quick meal and then each attached their portable life-support systems (PLSS) to their spacesuits. Houston confirmed that they were "GO for depressurization."

The hatch was opened at T+109 hours. Armstrong, down on his knees, began backing out though the narrow 32 in. wide hatch onto the small platform that was referred to as the "front porch." "Okay Houston, I'm on the Porch." As there was no video at this point, Armstrong wanted to keep Mission Control advised of his progress. He then pulled a lanyard that allowed one of the equipment bays on the descent module to swing open, and a camera, directed toward the ladder, began to beam images back to Earth (Chaikin 1994, 208). Because the video had to share the communications channel, the resolution was restricted to 320 lines per frame and 10 frames per second. Combined with the stark lighting conditions, the images received were almost ghost-like.

At first the image was upside down because of the way the camera had been packed, but within seconds the video engineers electronically righted the images, and Armstrong could be seen slowly making his way down the nine-step ladder. As he reached its end, there was still a drop of almost three feet to the foot of the LM—the landing gear had not been compressed to any degree by the soft touchdown. In one-sixth gravity, the fuzzy image could be seen in a slow-motion movement traversing that final distance. He then made a quick hop back up to the last step of the ladder to verify that there was no return problem and commented, "Okay, I just checked

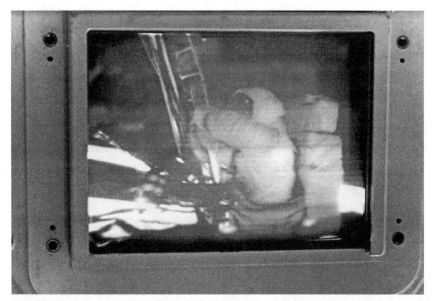

Neil Armstrong takes his "one small step" as millions watched on television. Courtesy of NASA.

getting back up to that first step, Buzz; it's not even collapsed too far, but it's adequate to get back up."

He was aware of the historical importance of his first words as he stepped onto the lunar surface, so he ensured that the world knew when that event happened: "I'm at the foot of the ladder . . . the surface appears to be very, very fine grained, as you get close to it. It's almost like a powder."

Then the momentous words: "I'm going to step off the LM now. That's one small step for a man; one giant leap for mankind" (Hansen 2005, 493). The phrase would live in perpetuity—the impact of live TV.

Armstrong quickly scooped a small quantity of soil into his pocket as a "contingency sample," should an emergency require an immediate return to the LM. Aldrin then followed to the lunar surface. The camera was moved to a tripod so that the viewers could see the complete LM and the activities of the astronauts. The American flag was then planted. A plaque on one of the legs of the descent module that remained on the surface was read: "Here men from the planet Earth first set foot upon the Moon, July 1969 AD. We came in peace for all mankind." President Nixon used the occasion to talk briefly with the astronauts and the world. Scientific instruments were set about. Samples of the lunar soil and rocks were gathered. All too soon, the two-hour and thirty-minute moonwalk was over.

The astronauts returned to the LM with their box containing 46 lb. of lunar samples. After repressurizing and connecting the spacesuits to the ascent-stage environmental system, the LM was depressurized one last time to allow unneeded equipment, including the heavy portable life-support systems, to be left on the lunar surface—reducing the weight to be propelled back into lunar orbit. The act was reminiscent of a scene in the science fiction movie *Destination Moon*—some nineteen years earlier—in which similar actions were taken. Contemporary science continued to stalk science fiction. The inside of the cramped ascent stage was now filled with lunar dust that had clung to their boots and space suits, and the two began a restless and uncomfortable six-hour sleep period.

Luna 15 could no longer beat the Americans back to Earth with the prized cargo of lunar soil, but it could still accomplish essentially the same mission for considerably less dollars and risk. This was what the Soviets were now counting on as its final descent was initiated just two hours prior to the scheduled launch of Apollo 11 from the Moon. The descent was scheduled to take six minutes, but suddenly all communications with Luna 15 ceased at the four-minute mark. Luna 15 had hit the side of a mountain. TASS reported that it had completed its planned flight and had landed on the Moon.

Although the official Soviet response to Apollo 11 was generous, their TV carried no video or film of the launch or the images of Armstrong and Aldrin on the Moon. To the average Soviet citizen, the Apollo 11 lunar landing was just another terse report that was sandwiched between everyday events of the world.

The Return Flight

Now, for the first time ever, a rocket launched from the Moon and into orbit around it. Liftoff occurred 124 hours and 22 minutes into the mission; a little more than 21 hours had been spent on the Moon. The ascent stage of *Eagle*, now weighing only 11,000 lb. (down from the initial weight of 33,000 lb.), lifted from the descent stage and ascended into the black lunar sky. In 7 minutes and 20 seconds, the astronauts had been accelerated to orbital velocity—4,162 mph.

The rendezvous and docking with *Columbia* went smoothly. To avoid getting moondust tracked into the CSM, it was overpressurized slightly with respect to the LM so that the air flowed from the CSM into the LM when the hatch was cracked. In addition, a small vacuum cleaner helped reduce the amount of particulates. This was an important procedure because

The LM, having left its descent stage on the Moon as a launch platform, is now legless, maneuvering to rendezvous with *Columbia*. The magnificent blue ball of the Earth can be seen above the lunar horizon. Courtesy of NASA.

the dirt could clog sensitive filters in the environmental control system. Then the two tired crewmembers of *Eagle* transferred themselves and their precious cargo of moon rocks to the CSM and resealed the hatch. On the twenty-ninth lunar orbit, *Columbia* undocked from *Eagle*. Trans-Earth injection occurred at T+135 hours and 34 minutes. The trio was on their way home.

Because several scientists believed there was the possibility that some form of life might have adapted to the extreme temperatures and radiation on the lunar surface, it was decided to quarantine the astronauts and their cargo of Moon rocks for a period of three weeks to determine if any such microorganisms might exist. This meant that upon their return they were sequestered in a trailer-like mobile quarantine facility. It also required that the recovery force provide the three travelers with coveralls and breathing apparatus during the transfer from the command module to the quarantine facility. It was a less-than-perfect process, but under the conditions, it was a reasonable effort to avoid Earthly contamination.

Splashdown on July 24 in the Pacific Ocean was uneventful. The decontamination suits were quickly passed to the crew, through the momentarily

President Richard Nixon flew to the USS *Hornet* to welcome the astronauts back to Earth. Here he enjoys a few humorous moments with the trio, who spent three weeks in the quarantine facility. Courtesy of NASA.

opened hatch, as they bobbed around on the ocean, and the Apollo 11 spacecraft was lifted to the deck of the aircraft carrier USS *Hornet*. The three astronauts, garbed in decontamination suits and breathing apparatus, walked to the mobile quarantine facility where President Nixon awaited them for a welcoming ceremony.

Mission Control, which had been awash with cheers, now started to empty as the controllers and visitors said their last congratulatory remarks and began departing for splashdown parties or just a good night's sleep. One of the large displays at the front of the room had President Kennedy's 1961 commitment prominently displayed: "I believe this nation should commit itself to achieving the goal before this decade is out of landing a man on the Moon and returning him safely to the Earth." The Apollo 11 emblem was projected next to it with the inscription above it reading: "Task Accomplished . . . July, 1969."

Final Soviet Lunar Efforts

The 13,000 lb. unmanned Zond 7 launched on August 8, 1969, two weeks after the return of Apollo 11, looped around the Moon, and, after taking color photos of Earth and the Moon, successfully returned to the Earth. Zond 8 launched a year later on October 20, 1970, was the last of the Zond series. Following a succession of meetings, the Soviets decided to terminate the L1 manned circumlunar program—without a manned flight. The N1/L3 lunar-landing project was allowed to proceed, but not for long. Plans also called for continued multiple Soyuz Earth-orbital flights, but there was nothing now that could truly recoup the lost prestige, and a final effort was made by the Soviets to deflect the true nature of their failure.

The third secret N1 launched on June 27, 1971, and failed forty-eight seconds into the flight. The fourth and last N1 to be flight-tested rose from the launchpad on November 23, 1972, and all appeared to proceed properly until just seven seconds before the first stage was to have shut down. At that point, an explosion in the tail section destroyed the rocket. Following a considerable period of investigation and discussion, the decision was made to terminate the N1 program in June 1974. There was to be no Soviet manned lunar landing.

Following the successful Apollo 11 mission, America (and the world) congratulated itself on the accomplishment—and on beating the Soviets to the Moon. Those in the West who had fought against the manned lunar program were quick to point out that there had been no race. They believed that the Soviets were intent on exploring the Moon with robots, and the whole idea of Kennedy's challenge was foolishness and a tremendous waste of money. The Soviets themselves fed that premise continually, and even one of America's most ardent enthusiasts of the space program, CBS reporter Walter Cronkite, jumped on the bandwagon and announced to the nation that "there had been no space race." With that kind of endorsement from "the most trusted man in America," many around the world assumed that to be the truth.

The Soviets had so well hidden their intent and their failures behind a veil of secrecy that only by intense scrutiny could anyone have deduced otherwise. For more than twenty years following Neil Armstrong's step onto the Moon, it was assumed by many that the competition had all been one sided.

With the fall of the Iron Curtain in 1989, the truth trickled out. Many who were a part of the cover-up came forward to reveal the details of the

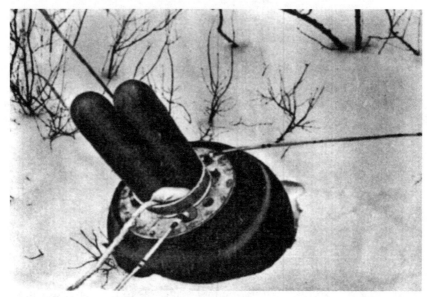

The Soviet Luna 16 rests in a snow bank following its successful mission to return samples of the lunar surface back to Earth in 1970. Courtesy of Energiya.

Soviet Moon project and to provide documentation of the failures, although there was still obfuscation by some to avoid any personal responsibility. It was a bitter pill for a proud nation, which had overcome significant adversity to accomplish spectacular feats. The basic reason given by the primary participants in the Soviet space program for the failure to compete successfully was that they were never able to achieve the required reliability of the UR-500 or the N1.

Insufficient funding (estimates ranged upward of $13 billion [US dollar equivalent]—no more than half the Apollo commitment) led to the decision to flight-test complex systems instead of extensively ground-test them. This was exacerbated by the lack of quality control processes and an undereducated workforce. The results were devastating.

The inability to develop LH2 upper stages resulted in an inefficient N1 that had one-third greater takeoff thrust but could carry only two-thirds of the Saturn V payload. The gross liftoff mass was about the same as the Saturn V, but the empty weight of the stages was more than twice as great.

As for the personalities, the lack of visionary leadership following Korolev's death and the constant conflict among the chief designers ensured no consensus on technical direction and a constant squabble for the meager funds.

When the editorials of the prestigious New York Times that followed the launch of the first two Sputniks in 1957, are reread in light of the events of the next twelve years, it is clear that America's own course of action was negatively affected by what some might categorize as a predictable reaction. The Soviets had not developed a superior scientific base supported by an advanced educational system, as these editorials claimed. That American industry, education, and resourcefulness were able to accomplish an incredible task in such a short time was a testament not only to America's technological and economic superiority but also to a culture that could gather the resources and skills and could organize an assault on a problem so effectively.

The disappointment of the Soviet manned Moon-landing program was a reflection of a problem that plagued many of its large technology projects. From nuclear power plants to the supersonic transport, the Soviet system too often failed to produce sustainable results.

Soviet Lunar Sample and Return Success

Although the Soviet lunar program failed to send humans to the Moon, it was ultimately successful in returning three soil samples with automated probes. The first fruitful return occurred in September 1970 with Luna 16, which provided 101 grams of material, followed by Luna 20 (30 grams) and Luna 24 (170 grams). Although modest in comparison with the Apollo effort, the Soviets touted their unmanned program as far more cost effective than the American program.

19

EXPLORING THE MOON

Well before Apollo 11 set down on the surface of the Moon, there were those who looked on the Kennedy goal as simply the opening round in establishing a Moon base. The scientific community, which had taken a back seat to the political expediency of the space race, now moved forward with determination. However, with the goal of beating the Soviets to the Moon accomplished, many Americans (and many members of Congress) lost interest in the space program.

As early as 1965, the lunar module had been seen as the basis for developing living quarters and as a lunar taxi to allow the establishment of an initial base of operations for extended exploration periods. The modest instrumentation and exploration carried out by Apollo 11 was expanded for subsequent Apollo flights. NASA anticipated the period beyond the first lunar landing and established a modest exploration plan for the nine authorized missions that remained.

Not knowing how many attempts it might take to achieve success, planning and funding were approved for ten lunar flights. Apollo 11 targeted a landing area that represented the least risk in terms of unobstructed terrain and power requirements—within 10 degrees of the lunar equator. Emboldened by the success of Apollo 11 and the continued performance improvements in the Saturn V, two immediate and more demanding objectives were established. The first was to demonstrate the ability of Apollo technology to perform a pinpoint landing. Establishment of a lunar base was not possible if provisions and equipment could not be set down on the Moon in a predetermined location. Even before Apollo 11, tentative planning pushed for the ability to aim for a specific landing area. The second automated Surveyor III spacecraft that had landed in April 1967 provided that target.

The second objective was to explore areas that were more mountainous and those in higher latitudes, which were of more interest to the geologist but required greater power reserves and higher risks. In addition, sufficient weight-carrying margin in the Saturn V and the LM could support up to three days on the lunar surface and return considerably more lunar samples. These capabilities were integrated incrementally into each successive flight.

Apollo 12: Lightning Strike

Since it was not known whether Apollo 11 (AS-506) would successfully complete its mission, AS-507 was assembled in the vehicle assembly building during the summer of 1969 for a targeted launch in September. This provided insurance for the Kennedy goal of a lunar landing "before this decade is out." The success of Armstrong and Collins in July relieved the pressure, and work schedules at the Cape began a more relaxed atmosphere—even a forty-hour workweek. As a result, preparations for Apollo 12 led to a launch in November.

CSM-108 and LM-6 were virtually identical with the previous mission except for the addition of a nuclear-powered electrical generator for the Apollo Lunar Surface Experiment Package (ALSEP). That set of scientific instrumentation increased from the 220 lb. carried by *Eagle* to 365 lb. for the Apollo 12 LM named *Intrepid*.

The actual touchdown point was planned for a quarter-mile from Surveyor III. *Eagle* had set down about four miles from its predetermined landing spot because of inaccurate positioning of the LM at powered descent initiate (PDI). Mission Control in Houston had only seventeen minutes from the time that *Eagle* appeared over the lunar horizon to determine its path and upload the appropriate coordinates to the LM computer for the descent profile. The location of Surveyor, farther to the west on the lunar surface, and the timing of the flight allowed thirty-seven minutes for the task.

To achieve its objective, the mission had to deviate from a "free-return" path early in the flight. However, NASA was confident of the reliability and redundancy of the systems—coupled with the ability to use the LM engine as an emergency backup during the trip outbound.

The weather was questionable on November 14, 1969, the day of the launch. Rain showers and thunderstorms prevailed as Charles Conrad, Alan Bean, and Richard Gordon ascended the elevator to the "white room" forty stories above the sandy terrain. The low overcast and steady rain persisted as the countdown reached zero. Although the ceiling (base of the clouds) was less than 1,000 ft., the mission rules did not preclude the launch.

The giant rocket was quickly enveloped in the low-level overcast, but its fiery tail still licked at the empty launchpad even after it disappeared into the gray mist above. At T+36 seconds into the flight, at an altitude of 10,000 ft., the Apollo 12 crew observed a brilliant white light that enveloped the spacecraft, and caution and warning lights illuminated on their flight display (Chaikin 1994, 236). Those on the ground, including President Nixon, clearly saw a lightning bolt travel down the exhaust plume that still projected from the cloud.

Conrad immediately communicated the status of the ominous red lights. "Okay, we just lost the platform, gang; don't know what happened here; we had everything in the world drop out!" CapCom responded tersely, "Roger," as those on the ground set about trying to understand what had caused the sudden electrical power failure and its immediate implications.

Conrad continued to describe his situation. "Fuel cell lights and AC bus light, fuel cell disconnect, AC bus overload, 1 and 2 out, main bus A and B out."

The possibility of an abort was uppermost in everyone's thoughts. The "platform" that Conrad referred to was the Apollo guidance unit—the electrical interface to the gyroscopic reference had failed. Fortunately, the Saturn V guidance unit in the S-IVB had remained online. Had that been interrupted, the rocket might have met a catastrophic end. Controller John Aaron recalled a glitch from an earlier prelaunch situation and called for an obscure switch labeled "SCE" to be placed in the "AUX" position, ensuring the continued guidance.

As the Saturn V continued along its path toward space, the crew reset circuit breakers to bring the electrical systems back. By the time the first stage had completed its burn, Conrad reported, "We are weeding out our problems here; I don't know what happened; I'm not sure that we didn't get hit by lightning." He added, "I think we need to do a little more all-weather testing." After communicating more status, he noted with some humor, "That's one of the better sims [simulations], believe me!" The astronauts were often presented with "what if" scenarios during their training in the simulators, some that they felt bordered on the ridiculous. Conrad's comment was meant to dispel that notion. CapCom responded, "We've had a couple of cardiac arrests down here too, Pete."

The S-II stage with its five J-2 engines, and 1 million lb. of thrust now continued to power the "stack" toward orbit. The ride was rough, for some reason, prompting Conrad to comment, "She's chugging along here, minding her own business." The reference to the vehicle with a personal female pronoun was often a mannerism of the test pilot.

After a short first burn of the S-IVB, the *Yankee Clipper* was in its parking orbit. Now the crew could settle down and take the time, along with Houston, to evaluate what had happened and determine if the mission in fact should proceed to the Moon or return to Earth. There were voices who expressed concern that there could easily be damage that would not be known until later sequences, and that the astronauts' lives should not be risked. Kennedy's goal had been accomplished—why push the issue? The guidance gyros were caged and the platform successfully realigned. All of the electrical systems appeared normal. The decision was made: "Go for TLI."

The S-IVB was fired a second time after almost two orbits of Earth, and Apollo 12 accelerated to escape velocity. Following the TLI burn, the *Yankee Clipper* separated and extracted the LM from the S-IVB stage. Residual oxygen from the propellant supply then vented through the single J-2 combustion chamber to produce a small thrusting. This changed the path of the S-IVB so that it would not interfere with the assembled spacecraft.

Compared with the lightning strike on launch, the cruise phase to the Moon was uneventful. There was some concern that the crew could not see the exhaust plumes nor hear the low-powered thrusters (25 lb. of thrust) when they fired, but Armstrong, who was in Mission Control, advised the crew that as long as the rate gyros recorded the event, there was no problem—his crew had not heard them either. In contrast, the high-powered thrusters (100 lb.) were definitely audible as Conrad noted, "this thing shakes, rattles, and rolls when you fire the thrusters; it's like being on a jerking train."

They also decided to open up *Intrepid* a day early to make sure there were no surprises waiting within. This was important as the "no-return" burn of the service propulsion system (SPS) motor was scheduled. Several video periods were beamed back to Earth during the eighty-eight-hour trip to the Moon.

On arrival, the SPS fired on the backside of the Moon to place the CSM and its ungainly LM into orbit, and three more men were held spellbound by the opportunity to gaze upon the craggy lunar surface.

Despite Al Bean suffering from a head cold, exacerbated by the weightless conditions and 100 percent oxygen environment, the mission continued. At the appointed time for vehicle separation, care was taken not to

The proximity of the LM (*upper right*) with the Surveyor III provided ample appreciation of the accuracy that could be achieved in a lunar landing. Courtesy of NASA.

disturb the orbit of *Intrepid* so the orbital parameters could be more accurately determined. Real-time video of this event and others provided even more vicarious opportunities for the Earth-bound public.

The powered descent went as planned, and soon the LM was enveloped in billowing dust as the final few feet of the landing were determined by the radar altimeter. Gordon, who remained in the CSM, reported he had a visual on the LM and that it was resting on the rim of the Surveyor crater in the Ocean of Storms.

Conrad and Bean completed preparation for the first of two surface excursions. As Conrad made the traverse down the LM ladder, making the three-foot drop to the foot pad, his remarks were not as profound as Armstrong's but were equally quotable, "Whoopee! Man, that may have been one small step for Neil but that was a long one for me." Looking over into the crater he exclaimed, "Boy, you'll never believe it. Guess what I see setting on the other side of the crater. The old Surveyor." The Apollo program had remarkably demonstrated that humans could not only travel to the Moon but could land at any preselected point!

When Bean was repositioning the color camera, he inadvertently pointed it at the sun and damaged the vidicon tube. The large S-band antenna, set up to allow much-improved video signals, never got the chance to show the millions watching a clearer view of the lunar landscape and the activities of the astronauts. It was a major disappointment for NASA.

The two astronauts spent four hours on their first moonwalk and then returned to the LM for sleep and a replenishment of their oxygen supply. The second moonwalk took them 1.5 mi. around several small craters and finally to the Surveyor. There they took pictures and removed several items from the now inert spacecraft before returning to the LM and preparing it for launch. The return to Earth was uneventful, and there was an air of routine in the sending of men to the Moon. This would change.

Apollo 13: "Houston, we've had a problem"

The crawler-transporter carried AS-508 and its Apollo 13 payload out to pad 39A on December 15, 1969. The unhurried schedule called for a launch in April. Few anomalies marred the testing of the rocket during the interim period. One small but aggravating problem was the inability to empty the No. 2 oxygen tank of the fuel-cell system following a variety of tests. Several attempts and techniques were tried before the technicians were satisfied. The problem was attributed to a poorly fitting fill tube. It was a significant effort to replace the tank, and that would delay the flight until the next lunar launch period. Because it posed no in-flight hazard, it was decided to go with the existing hardware.

The crew of James Lovell, Thomas K. Mattingly, and Fred Haise had been assigned the flight the previous August. A few days preceding the launch, it became known that all three had been exposed to rubella, a form of measles, by Charlie Duke from the backup crew. However, only Mattingly showed no immunity and had to be replaced. A complete crew replacement was not possible at this point, so Jack Swigert from the backup crew flew in place of Mattingly. However, replacing a member of a close-knit Apollo flight team was not a matter to be taken lightly. A series of quick simulator sessions allowed Lovell and Haise to observe Swigert closely as they made the final decision. They approved.

The media no longer considered a flight to the Moon of great importance. News coverage of the flight was lackluster, although all three networks managed to interrupt their regularly scheduled programs to show the Apollo 13 launch live at thirteen minutes past the hour on April 11, 1970. The vibration of the J-2 cluster in the second stage again provided a rough ride—so uneven, in fact, that the center engine shut down more than two minutes early due to low pressure in the liquid oxygen (LOX) line. The guidance system automatically commanded the remaining four engines to burn longer. However, the S-IVB also had to extend its burn nine seconds to help make up the deficit in velocity. This caused a quick recalculation to verify that enough fuel remained for the TLI burn—there was.

Following a checkout of critical systems while in Earth orbit, the S-IVB again fired to provide the TLI, and the astronauts were on their way. It was Lovell's fourth space flight and his second to the Moon, as he had been a part of Apollo 8. Swigert and Haise were rookies. At little more than twenty-seven hours into the flight, the crew passed the halfway point, and a burn of the SPS engine negated the "free-return" path. A fifty-minute video show followed, and the crew turned in for the night.

The next morning, April 13, 1970, the capacity of the No. 2 oxygen tank for the fuel cells—the one that had given problems during the countdown demonstration test—showed 100 percent. The evening before, it had shown 82 percent, which was about what was expected. Following a brief conversation with Mission Control, the consensus was that the gauge was bad. Later that day another video session was beamed down, but it was not carried live by any of the networks—space travel was not as exciting as the *Dick Van Dyke Show*.

Following the closeout of the video, Mission Control sent up a request just short of fifty-six hours into the flight. "13, we've got one more item for you when you get a chance. We'd like you to stir up your cryo tanks."

Swigert replied, "Okay," and drifted over to the console and flipped the four switches that sent electrical current to the small heaters in each tank and the fans used to move the oxygen past the heaters. In the weightless condition of space, there is no heat convection to move the super-cold liquid. A few minutes later, the crew heard a loud report, and the 44 ton vehicle reacted violently.

Swigert immediately keyed the mic, "OK, Houston, we've had a problem here."

CapCom responded, "This is Houston say again please."

Swigert repeated as he scanned the instruments in front of him, "Houston, we've had a problem. . . . We've had a Main B undervolt."

"Roger, Main B undervolt. Okay, standby 13 we're looking at it."

Haise then added, "Okay, right now Houston the voltage is looking good. And, we had a pretty large bang associated with the caution and warning there. And if I recall Main B was the one that had an amp spike on it once before." Within minutes, it was obvious from the instrumentation readings that they had lost fuel cells 2 and 3. As Lovell scanned the oxygen-supply gauges for the fuel cells, he saw that No. 1 was slowly but steadily dropping as well. He looked out the hatch window and reported even more bad news to Mission Control. "And it looks to me . . . that we are venting something. We are venting something out into space." The fuel cells not only provided their electrical energy but the crew's oxygen as well, and the service module was rapidly losing their precious supply. The spewing tank generated enough thrusting to cause the spacecraft to perform a continuous roll.

Flight Director Gene Kranz quickly instructed that the command module's "surge tank" be isolated so that the small oxygen reservoir that it held (about 5 lb.) would not also vent into space. It was their only source of air during the reentry following separation from the LM—several days away. It proved to be a key move.

Kranz, who had just come on duty, asked the question to which he was afraid he already knew the answer: "Okay, lets everybody think of the kinds of things we'd be venting. G&C [Guidance and Control] you got anything that looks abnormal in your system?" G&C replied, "Negative, Flight."

With two fuel cells out, it was obvious that the lunar landing was no longer possible. Kranz immediately realized that they were facing a life-threatening situation. "Okay, now let's everybody keep cool; we got LM still attached, let's make sure we don't blow the whole mission" (Kranz 2000, 315).

Almost immediately, mission control recognized that a short circuit in the heater of No. 2 oxygen tank had caused a fire resulting in the tank, exploding and damaging the other tanks. With the electrical power essentially disabled in the command module, whose call sign was *Odyssey*, the LM, named *Aquarius*, was now the only source of power and oxygen. Actions were quickly taken to bring that resource to bear on the problem. The command module was powered down and only critical systems in the LM powered up. However, the LM ran off batteries designed only for the duration of the lunar excursion, and its oxygen supply and carbon dioxide scrubbers had been configured for only two men for two days.

With significant damage to the service module, it was doubtful that the SPS engine could be used. The two spacecraft were locked together on a path that would take them around the Moon before beginning the journey back toward Earth. The situation, as it developed over the next few hours, presented Mission Control with an almost impossible task—bringing the astronauts back alive.

Forty-five minutes after the explosion, oxygen tank No. 1 was still venting, and its pressure was well below the minimum necessary to generate power. It was obvious that it, too, would soon be depleted, leaving only the LM for power. The decision to move the astronauts to the LM was communicated. "It's still going down to zero, and we're starting to think about the LM lifeboat." The passageway between the two spacecraft was opened.

The guidance unit of *Aquarius* was aligned before the unit in *Odyssey* was powered down. As the procedures continued to transfer the life support and other critical functions to the LM, the various contractors from around the United States were called to confirm capabilities and to inquire about added longevity in certain key components. The spacecraft was just over 200,000 mi. from Earth—and moving away from it. With the inability to use the big SPS engine, Apollo 13 was committed to completing its journey around the Moon.

An important task was to use the LM descent engine to change the trajectory back to a "free return." But with an almost a full load of fuel remaining in the service module, the ability of the descent stage to provide significant velocity changes was limited. There was some thought of jettisoning the service module, but that would expose the heat shield of *Odyssey* to the direct and hostile elements of space. It was not known what degradation might take place in the three-day return to the Earth.

A survey of the consumables was one of the first tasks to be completed, for without a clear understanding of what was available, plans could not be made. The vital supplies consisted of water (for the cooling system), oxygen (for the crew), electrical power to sustain the environmental and communications systems, and the lithium hydroxide units for removing the carbon dioxide from the cabin air.

The LM was designed to provide for two men over a period of 35 hours. It was now being called upon to sustain three for perhaps 100 hours. With the ability to use the oxygen in the two backpacks allocated for the moonwalk, there was approximately 50 lb. available, which could last well over 150 hours. The 345 lb. of cooling water would also be adequate for 110 hours, assuming that the heat generated by the electrical equipment was significantly reduced. The 2,200 amp-hours of electrical battery supply appeared to be a critical entity. The LM normally used 50 amps, but this was immediately cut to between 10 and 15 amps. Some equipment had to remain powered up such as the gyros to ensure that they did not freeze.

From the point the calculation was made of the available consumables, it was going to take 100 hours to return to Earth. That might be too long. It was possible for the LM to provide another shove after they rounded the Moon to shorten the trip by about 10 hours. As it turned out, this represented the critical factor.

Within an hour of powering up *Aquarius*, the astronauts completed the powering down of *Odyssey* and the power from the failing fuel cell was irrevocably terminated. Back in Houston and the Cape, astronauts were powering up the simulators as they prepared new "pad" sheets (revised sequences of activities) that would be communicated to the astronauts.

With a plan to conserve the consumables, only the buildup of carbon dioxide became a pressing issue. The available replacement containers for *Odyssey* were a different size than those of *Aquarius*. A team at Houston, equipped with the same resources available to the astronauts, devised a means of using duct tape and cardboard to adapt the lithium canisters. Normally these canisters are replaced when the CO_2 level reached 7.5 mm/Hg. As this was a conservative figure, it was decided to allow these levels to reach twice that number before replacing them.

At T+61:30, the first burn of the DPS engine restored *Odyssey* and *Aquarius* to a free-return path. The next burn, called the Pericynthion+2, took place two hours after the closest approach to the Moon, accelerating the long journey home, with a reentry at T+142 hours—allowing the most critical consumable, the LM batteries, to perform their lifesaving activity.

As the spacecraft swung around the Moon, the three astronauts pressed themselves to the windows to view a sight that only eleven other humans had ever witnessed—the Moon passing silently beneath their vantage point. A once-in-a-lifetime opportunity for Lovell and Haise to walk the lunar surface had been lost.

Odyssey was now a dark cold shell, and the three astronauts huddled in the tight confines of the *Aquarius*. The temperature quickly dropped to 35 degrees, and the moisture from their breath condensed on the interior surfaces. During sleep periods, *Odyssey* provided the "bedroom" so that any activities by the astronaut on alert in the LM would not disturb those trying to sleep. But sleep did not come easy due to the damp, chilly environment—not to mention the prospects for survival residing within the thoughts of each of them.

The 4-minute 23-second Pericynthion+2 burn at T+79:27 was successful, and now the prospects for a safe return improved considerably. Nevertheless, two-and-a-half days yet remained of the flight, and there was no margin for error or system failure. This may have been the most stressful time for the crew because at this point there was no longer anything that could be done—but wait in the cold, dark, emptiness of space.

The astronauts understood the risks they faced and the extraordinary work being performed by thousands of support people back on Earth. Several times during the mission both the astronauts and Mission Control had offered prayers. Atheist Madalyn Murray O'Hair had been a driving force in removing prayer from the public schools just a few years earlier. But her subsequent effort to ban prayer from being uttered across the federally funded communications that linked the astronauts to their Earthly support people had been rejected quickly by the Supreme Court (Hansen 2005, 487).

As Apollo 13 accelerated back toward the Earth, the tracking data showed the spacecraft reentering at an angle of 6.03 degrees—well within the tolerances of 5.5 to 7.5 degrees. Mission Control was concerned that when the service module jettisoned, it might impart a slight velocity vector to *Odyssey*, so a short burn was scheduled to move the angle more toward the nominal figure. However, because the LM was still attached, the burn could not require a positive velocity trim by the LM since the forward-facing thrusters impinged on *Odyssey*. Therefore, the final twenty-two-second midcourse correction that came at T+137:40 was designed to leave a slightly negative delta-*v*, allowing the aft thrusters of *Aquarius* to make up the difference. The burn was successful.

As the trio started to power up *Odyssey*, the guidance alignment was transferred from *Aquarius*, and they began preparations for setting their "lifeboat" adrift. The Pericynthion+2 burn had been the lifesaving factor, for without the push that reduced the flight by ten hours, the water and battery power would have been depleted.

The next step was to separate Odyssey from the service module and make the slight velocity correction with *Aquarius*. The charges were fired by Swigert in *Odyssey* to sever the connection, while Haise and Lovell used the LM thrusters to make a positive separation.

The three astronauts took to the windows to observe and photograph whatever damage might be visible on the service module. As the long-dead service module drifted into view, Lovell made the first observation, "And there's one whole side of that spacecraft missing!" CapCom replied incredulously, "Is that right?" Lovell continued his description (Kranz 2000, 333).

The damage to Apollo 13's service module is visible after its separation from the command module. Courtesy of NASA.

Inside equipment bay 4, the astronauts could see the damage done, including to the SPS engine nozzle.

With the life-support responsibilities now transferred to the reentry consumables of *Odyssey*, the hatches between the LM and the command module were secured and pressure checked. *Aquarius* was released at T+141:30 with the comment, *"Farewell Aquarius, we thank you."*

The reentry and recovery were without incident—splashing down only a few miles from the prime recovery ship. A dramatic chapter in the annals of space flight had ended. However, as life threatening as it was, several decades passed before much of the world truly understood the magnitude of the problem and the rescue that had been achieved. Flight Director Gene Kranz highlighted the episode in his book, aptly entitled *Failure Is Not an Option.* A quarter-century later, a feature-length movie, *Apollo 13,* gave a more complete portrayal of the dramatic events of the flight. It was perhaps one of the few movies Hollywood ever made that was close to historical and technical accuracy.

As for the cause of the explosion, an in-depth inquiry revealed that the tank heater had been correctly manufactured in 1966 by Beech Aircraft Corporation to the original specifications for 28 volts. However, the voltage was changed a year later to 65 volts—but paperwork and hardware had not been modified. The error might not have caused the explosion, had it not been for the problem encountered in attempting to drain the tank following the ground test at the Cape a few weeks before launch. The tank purge procedure had caused the switch to become "welded" in the closed position, and it subsequently could not perform its function as a protective thermostat. An overpressurization resulted, and the tank exploded. Because of the manner in which all of the various elements of the fuel-cell systems were tightly packaged, extensive collateral damage resulted.

Changes were made to the service module to make it less susceptible to explosive effects by adding a third oxygen tank in another part of the module. Battery reserves were increased, along with several other changes. Fortunately, none of these changes were called upon for the remaining flights.

Apollo 14: Science and Golf

The flight of Apollo 14 was unique for several reasons. It marked the beginning of more intensive science with regard to the expedition; previous missions had concentrated on engineering feedback. It also was the only trip to the Moon by an original Mercury astronaut. Alan B. Shepard had been given the privilege of being America's first astronaut in May 1961, when he rode MR-3 on a simple 115 mi. ballistic arc. Although celebrated at the time, his accomplishment had already been eclipsed by Yuri Gagarin's flight three weeks earlier—which had been a true orbital journey. Following his flight, Shepard was grounded by an ear problem called Ménière disease, which kept him from actively participating in either the Gemini or the early Apollo flights. George Low relates that he and Robert Gilruth met with Shepard to tell him he was permanently grounded.

Unlike some of the other astronauts who left NASA when it was obvious that they not fly, Shepard, along with Deke Slayton, remained. Shepard became chief of the Astronaut Office. However, he never gave up hope of some medical procedure that might return him to flight status. Early in 1969 he became aware of an operation that could repair the ear damage and had it performed in May of that year—without the knowledge of NASA management. The results were more than satisfactory, and Shepard's perseverance paid off (Thompson 2004, 387). After much discussion—and a rigorous flight physical—he was assigned to the Apollo 14 flight team in August of 1969, along with Stuart Roosa and Edgar Mitchell.

The CSM-110 and LM-8 became the spacecraft *Kitty Hawk* and *Antares*, respectively. They were launched on January 31, 1971, following ten months of changes dictated by the almost-fatal Apollo 13 mission. The early shutdown of the center J-2 on the second stage of Apollo 13 had been caused by a return of the "pogo" effect, and that, too, was addressed with modifications to the LOX plumbing.

The powered segment of Apollo 14's Saturn V (AS-509) flight was without problems. Following the TLI burn, Roosa had difficulty docking with the LM. Three times, he tried to nose into the docking collar over a period of more than an hour. As the time required to accomplish the docking wore on, there was concern that the batteries in the S-IVB would be depleted. Should that occur, the S-IVB could not to hold a stable attitude, and the mission would be canceled. With the fifth attempt, which bypassed the soft-dock interim stage, a hard-dock finally was achieved (Kelly 2001, 238). Visual inspection of the mechanism revealed no direct reason for the problem, and the mission continued toward the Moon.

Because of the scientific interest in the Fra Mauro area of the Moon to which Apollo 13 had been targeted, Shepard, Roosa, and Mitchell had their destination changed to land there. To accomplish that task, it was necessary to use the big engine of the CSM to lower the LM closer to the Moon before it began the PDI burn. As the astronauts descended to a pericynthion of only 10 mi. (50,000 ft.), they were in awe. Mitchell commented, "Looks like we're getting mighty low here. It's a very different sight from the higher altitude." CapCom reassured them that they were no lower than 40,000 ft. above the highest point. Mitch responded, "Well, I'm glad to hear you say we're that high. It looks like we're quite a bit lower. As a matter of fact, we're below some of the peaks on the horizon, but that's only an illusion," he said rhetorically.

After separating from Antares, Kitty Hawk fired its big SPS engine to climb back to a higher orbit of 70 mi. and wait for the expedition to return from the lunar surface. In preparing for the landing sequence to initiate, Shepard suddenly announced, "Hey, Houston, our abort program has kicked in." A faulty abort switch was quickly diagnosed by Grumman engineers who were able to program around the problem—another mission saved (Thompson 2004, 417). The astronauts were again cleared for the landing.

Antares then began the PDI that was again almost aborted when the radar could not get a lock on the terrain. Shepard quickly reset a breaker and the descent continued. The daring duo landed an estimated 150 ft. from the designated spot on the lunar surface of Fra Mauro.

This region of the Moon was believed to have material from several miles beneath the surface that had been thrown up by a meteoroid impact.

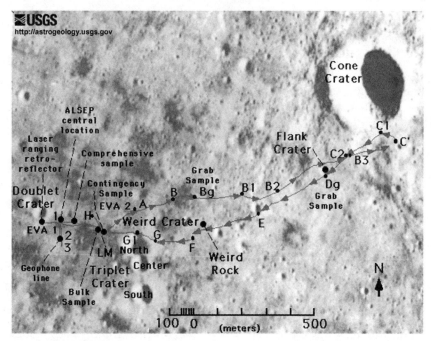

Each of the EVA excursions from the LM were carefully mapped for all Apollo missions. Note the scale along the bottom to provide an indication of the distance traveled. Courtesy of NASA.

The color camera was set up, and Shepard and Roosa began the first of two excursions that lasted four hours. The second extravehicular activity (EVA) involved pulling a small, two-wheeled cart called the modularized equipment transporter on which tools and samples were carried.

One of their objectives was Cone Crater, but the navigation was uncertain and the trek pulling the wheeled modularized equipment transporter proved particularly difficult for the forty-seven-year-old Shepard. In the end, they had to abandon the effort (Thompson 2004, 424).

On the second EVA, Shepard produced the head of a number six golf club that he affixed to the handle of the contingency sampler. Dropping a small white ball to the surface, he proceeded to make the first golf shot on the Moon. As this first shot was not quite as good as he expected, he allowed himself a "Mulligan" with a second ball (Shepard et al. 1994, 318).

The return to the CSM was uneventful as the rendezvous was made using the M=1 technique, first used with Gemini XI. Several scientific experiments were performed during the journey back from the Moon, and, with the splashdown in the Pacific, another successful lunar mission had been accomplished.

Apollo 15: The Moon Buggy

With the completion of Apollo 14, America had demonstrated a commanding technological lead over the Soviets, and many believed that NASA should not send any more expeditions to the Moon. The narrow escape of Apollo 13, coupled with the tremendous cost of each flight, was seen as justification to suspend the remaining three flights (Apollo 18–20 had already been canceled). However, to the scientific community, Apollo was just beginning to return its $20 billion investment.

NASA had recognized that there was more stretch in the hardware and had prepared the J-series missions to wring the most from the capabilities of the Saturn V. Even before Apollo 11 had successfully completed the first landing, preliminary contracts had been let to develop a lunar roving vehicle or LRV—a small "moon buggy" (similar to a "dune buggy") that would allow the astronauts to cover more of the lunar landscape. To accommodate the increased payload weight, the Saturn V had to shed weight and increase its performance. Four of the eight retrorockets used to separate the first and second stages were removed, along with the four ullage rockets used to settle the propellants (residual thrust of the spent first stage was deemed sufficient). A modified trajectory into a lower Earth parking orbit allowed another 660 lb. to be accelerated to the Moon. A slightly higher propellant flow rate in the F-1 engines completed the tweaking of the Saturn V, and the payload for Apollo 15 was increased by more than 5,000 lb.

The lunar roving vehicle was an extremely lightweight, four-wheeled, electrically powered vehicle. Perhaps more impressive than its innovative engineering was the fact that the 10 ft. long buggy weighed less than 500 lb. and could be deployed from one of the four LM equipment bays by simply pulling two lanyards. A longer burn of the LM descent engine, whose propellant tanks were enlarged 6.3 percent, accommodated the 2,600 lb. increased weight of the LM. The longer burn also required the silica ablative coating on the engine to be changed to quartz.

To enable the astronauts to range more than ten times as far from the LM than on previous excursions, the life-support systems of the space suits were enhanced to provide three EVAs and twenty hours of surface time. The suits were improved to allow greater mobility and increased drinking

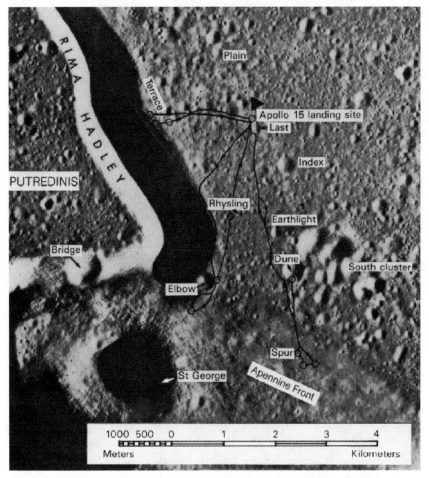

Exploring Hadley Rill was among the objectives of Apollo 15. Courtesy of NASA.

water. The added capability was also needed if the moon buggy should break down at its furthest point from the LM and require that the astronauts walk back.

Apollo 15 was targeted for a most difficult and dangerous landing area. It required a higher descent angle to clear the mountains that surrounded the region called Hadley-Apennine. CSM-112 was teamed with LM-10 atop Saturn AS-510 for the launch on July 26, 1971. The call sign *Endeavour* was chosen for the CSM in recognition of English captain James Cook's 1768 scientific foray to the South Pacific. The LM, in anticipation of an experiment using a falcon feather to take place on the lunar surface, was aptly named *Falcon*.

The launch was uneventful, and David Scott, Al Worden, and Jim Irwin were propelled into an initial lunar trajectory that, for the first time, did not provide for a free return. There was an indication that the SPS propellant valves were in the wrong positions, which would have been potentially serious problem. This was ultimately proven to be a short circuit in an electrical switch and was worked around. A series of other less threatening anomalies were likewise dealt with as the resiliency of the designs of the various systems again showed the strength of the engineering.

As the spacecraft reappeared from the far side of the Moon, Dave Scott reported the successful burn of the SPS engine that now placed them in lunar orbit. "Hello, Houston, Apollo 15. The Falcon is on its perch." Preparations for the landing were completed, and the PDI brought Scott and Irwin into the designated landing area. The touchdown was somewhat more pronounced than the previous landings as the exhaust nozzle of the descent stage contacted the surface and was deformed—as it was designed to do. The LM sat backward at an angle of eight degrees, but this was well within the design tolerance to deploy the rover. An initial indication of an oxygen leak in the LM proved to be an improperly closed urine dump valve.

Scott's egress onto the lunar terrain prompted his comment, "Okay, Houston, as I stand out here among the wonders of the unknown at Hadley, I realize there's a fundamental truth to our nature—man must explore. And this is exploration at its greatest."

The first EVA involved deploying the rover and lunar communications relay unit. This set of antennas allowed the astronauts to transmit voice, data, and video from the rover (and their suits) directly back to Earth when they were out of sight of the LM. With the camera mounted in the rover (and controlled from Earth), viewers could follow along with the explorers. Scientists at Mission Control provided real-time discussions with the astronauts, who served as their remote sensors in picking up samples and describing their characteristics.

The rover drove well in its first test in the lunar environment, although the front-wheel steering failed to function. Since the rover had both front and rear steering capability, this was not a problem. It was recognized that navigation on the lunar surface, with its undulating slopes and deceptive distance estimates, was a problem. A directional gyroscope, an odometer, and a computer that integrated the track and time to show the astronauts where they were and the direction to travel to return to the LM was provided.

Astronaut Irwin, next to the lunar roving vehicle, with some impressive mountains in the background. Courtesy of NASA.

Driving at a maximum speed of about 5 mph, Scott enjoyed the mobility and driving qualities of the rover. "Okay Joe [CapCom], the rover handles quite well. . . . It's got very low damping compared to the 1-G rover [the unit used on Earth for training]. . . . It negotiates the small craters quite well . . . It feels like we need the seat belts . . ." He later added, when queried about the constant movement of the steering, "It's just there are a lot of craters and it's just sporty driving. I've got to keep my eye on the road every second."

One of the first objectives on their traverse was to peek down into the Hadley Rill, a sinuous narrow canyon that meandered across the lunar landscape. It was a spectacular sight for Irwin and Scott as they stood on the rim looking across the gorge estimated to be about 700 ft. deep and about as wide. The telemetry noted an increase in their heart rates and respiration.

At the end of the first EVA, the ALSEP scientific package was deployed along with a laser retroreflector. An attempt to drill two holes 10 ft. into the surface, for a heat-flow experiment, resulted in the drill binding up after only half that distance. Scott noted, "I had the impression we were drilling through solid rock." The deepest core sample went down 8.5 ft. The EVA ended after six hours and thirty-four minutes, during which time they had traveled 17 mi.

Following a sleep period, the two were again out on the surface for the second driving excursion. This time the front steering was restored when a circuit breaker on the rover was reset. Upon returning to the LM after more than seven hours and with another load of lunar samples, they had their second rest period. The physical strain on both astronauts was noted in the biomedical data being reviewed by the medical team in Houston. Both had experienced cardiac arrhythmias on more than one occasion because of the physical exertion.

On the third and final EVA, Scott produced a hammer and a feather and proceeded to demonstrate a legendary scientific principle.

> In my left hand I have a feather. In my right hand, a hammer. I guess one of the reasons we are here today was because of the gentleman named Galileo, a long time ago, who made a rather significant discovery about falling objects in gravity fields. And we thought what better place to confirm his findings than on the Moon. And so we thought we'd try it here for you. The feather happens to be appropriately a Falcon feather . . . and I'll drop the two of them here and hopefully, they'll hit the ground at the same time. (Baker 1982, 422).

As the viewers looked on, Scott released the hammer and the feather at the same time and they dropped to the surface—landing at the same time. It had taken five hundred years for humanity to travel to a location where such a demonstration could be performed in a natural environment.

Following the ceremonial cancellation of some postal stamps (with a special lunar postmark) and the laying of a plaque bearing the names of fourteen Soviet and American astronauts who had lost their lives in the pursuit of space flight, the astronauts returned to the LM. Preparations were completed for liftoff, and, at the appointed time, viewers on Earth were treated to the first televised launching of a rocket from the Moon. The rover was left parked in such a manner that its camera could observe the launch.

The rendezvous was completed, and *Endeavour* was soon on its way toward Earth. Just after passing the equigravisphere, Worden performed a thirty-eight-minute spacewalk, the first to be performed in deep space, to retrieve film canisters from the service module. The descent into the Pacific landing was a bit faster than planned because one chute failed

> ## Equigravisphere
>
> The gravisphere is the spherical region of space dominated by the gravitational force of Earth (or any celestial body). When two celestial bodies are in relatively close relationship, such as Earth and the Moon, there is a point at which the force of the two gravispheres are equal—the equigravisphere. This point is about 210,000 mi. from Earth and 40,000 mi. from the Moon when it is at its apogee.

after deployment; the reaction control system (RCS) propellant that was dumped during the descent had eaten through the parachute risers.

The flight would have been remembered for its scientific success had the trio not succumbed to the temptation to make $150,000 by including more postal covers to be illegally marketed to collectors. An additional indiscretion involved a deal to sell replicas of the small "fallen astronaut" figure that had been left on the lunar surface. None of the three flew again. After voluntarily backing out of all the questionable business arrangements, they were "retired" or moved to other NASA centers to close out their careers.

Apollo 16: "Wow, wild, man, look at that"

John Young, Thomas K. "Ken" Mattingly, and Charlie Duke's flight of Apollo 16 was delayed by several months due to a variety of problems that included some changes dictated by the performance of Apollo 15 and a bout with pneumonia experienced by Duke. Changes to the parachute system of CSM-113 and the reinstatement of the solid-fuel retrograde motors in the first stage of the Saturn V were but some of the modifications to AS-511.

The launch on April 16, 1972, was uneventful, but CSM-113 (*Casper*) and LM-11 (*Orion*) experienced a series of problems that delayed the descent to the lunar surface and could have aborted the mission. The most troublesome of these was an overpressurization in the RCS thrusters of the LM due to an apparent leak in the regulator and oscillations in the yaw gimbals of the SPS engine. If there was a problem with the SPS engine, the crew might need the LM to provide the power to return the CSM back to Earth, as it had with Apollo 13. The problems were experienced after the two vehicles had separated in preparation for the PDI burn, and the descent was put on hold. The LM thrusters were fired several times to reduce the quantity of

Each of the lunar landings included the planting and saluting of the American flag.
Courtesy of NASA.

propellant, allowing more room for the leaking helium as the problem was analyzed. The decision was finally made that neither problem was of a critical nature, and the descent engine came to life more than five hours late.

The lunar landing was again without problems, and the normally subdued John Young was enthusiastic with the comments: "Wow, wild, man, look at that. Old *Orion* has finally hit it, Houston. Fantastic!" If flights to the Moon had become mundane for those left back on Earth, they certainly were anything but that for those who were fortunate enough to make the trip.

The astronauts carried out three EVAs with the assistance of the moon buggy. In moving among the deployed experiments, Young inadvertently walked through the ribbon cable for the heat-flow experiment, tearing it from the ALSEP unit. There was nothing that could be done to repair the damage, and Young was angry with himself (Young 2012, 177). However, the three EVAs were highly successful, with the two spending twenty hours, traveling more than 16 mi. exploring the region, and returning 170 lb. of lunar samples.

Leaking orange juice in the spacesuit combined with the lunar dust made the suiting and unsuiting a more difficult chore in the tight confines of the LM. Despite the loss of the heat-flow experiment, the remainder of the mission was successfully concluded.

Apollo 17: The Last Men on the Moon

There was nothing particularly different about the Apollo 17 launch on December 7, 1972, that set it apart from the preceding six missions to land on the Moon—except that it was the last of the planned manned missions to the Moon. While those who participated in the planning and execution of the mission went about their tasks, as they had for the previous lunar excursions, many did so knowing this mission would mean the end their jobs.

For the Grumman team, who had engineered the lunar module itself, there was no follow-on program to take advantage of the skills and knowledge of the workers. They placed a sign at the LM level of the giant red gantry that read, "This may be our last but it will be our best." Likewise, most of the Boeing employees in Michoud, Mississippi, who had built the giant Saturn V, had sought other employment years earlier as their AS-512 creation, built in 1967, waited in storage five years for its brief eleven-minute flight. Employment in the space program, which reached its peak in 1966 at almost a half-million, dwindled to about 100,000 by early 1973.

The pending Skylab project saw the last Saturn V rocket fly five months later, supported by three Saturn IBs and the Apollo spacecraft originally allocated for lunar missions 18–20. The last Apollo was flown for the Apollo-Soyuz mission in 1975. Then there would be no manned flights in the American space program for six years, until the first shuttle in 1981.

Gene Cernan, a veteran of Gemini IX and Apollo 10, commanded the mission with geologist Harrison Schmitt, a scientist, occupying the LM pilot position. This was an important crew change in that Schmitt was the first nonpilot to fly in the American space program. The desire to get an accredited scientist to the Moon caused Joe Engle to be bumped. Ron Evans rounded out the crew.

The target for Apollo 17 was the Taurus-Littrow area of the Moon, which was thought to have cinder cones indicative of volcanic activity. A wide assortment of scientific equipment was carried, including the unfulfilled heat-transfer experiment of Apollo 16. The CSM, christened *America* by

Cernan, was heavily instrumented to allow Evans to perform a wide variety of scientific observations as he orbited the Moon while waiting for his companions to return from the surface.

A hold at T-30 seconds caused by failure of an automatic sequencer to perform a critical function resulted in a manual workaround and a liftoff that was two hours and forty minutes late but still within the three hour and thirty-eight minute launch window. The area for miles around the launchpad was illuminated brighter than day for this 12:33 a.m. first night launch of a Saturn V. The exhaust plume itself could be seen as far away as 400 mi., as the rocket climbed into the black sky. The parking orbit was achieved, and the condition of the spacecraft was verified prior to receiving the call from CapCom, "Guys, I've got the word you wanted to hear. You are GO for TLI—you are go for the Moon."

A growing confidence in the various systems allowed mission planners to select a less forgiving trajectory for Apollo 17. The previous two missions had been established in a nonreturn path that could be powered into a free-return path by the RCS thrusters for the first five hours of the mission. After that point, the LM engine could provide the needed thrust if the SPS failed. Apollo 17, however, was propelled into a path that would not allow the RCS or the LM engines to return the craft to Earth should the SPS fail. In fact, the initial nonreturn trajectory actually would impact the Moon if the subsequent SPS burn failed. The first 1.7 second burn thirty-three hours into the flight increased their speed to miss the Moon by about 60 mi.

Eighty-six hours into the mission, the SPS fired to place Apollo 17 into orbit around the Moon, followed by a sleep period for the crew. The following day the crew prepared the LM for the lunar landing mission, unfolding its four spindly legs and separating from the CSM, *America*. On the far side of the Moon, the LM, now being referred to by its radio call sign, *Challenger*, executed the twelve-minute descent burn; Cernan and Schmitt were on their way to the lunar surface. As *Challenger* slowly pitched forward, the cratered surface of the Moon appeared to the astronauts, as it had many times in the simulations run back at Houston. However, this time there was an air of reality that the "sim" could not reproduce.

Cernan commented, "OK, there it is Houston, there's Camelot right on target" (a distinctive crater used as an aide to visually identify the landing area). As with all the other Moon landings, the last 50 ft. produced a cloud of dust as the LM backed down until the 4 ft. feeler probe touched the surface and illuminated the "contact" light. "OK, Houston, the *Challenger* has landed."

The size of this boulder is evident by the diminutive astronaut on its left and the moon buggy on the right. Courtesy of NASA.

There was a moment of uncertainty as the rear leg settled into a shallow crater, and the LM tipped back five degrees. Again, it was well within the tolerances of the LM. As they set about verifying the condition of the LM, Cernan was obviously pointing to various gauges and confirming their readings. "That hasn't changed . . . it looks good . . . manifold hasn't changed . . . the RCS hasn't changed . . . ascent water hasn't changed . . . the batteries haven't changed. Oh, by golly, only we have changed!"

Preparation for the first EVA was soon completed, and Cernan was out the hatch and down on the ladder. "I'm on the footpad. And, Houston, as I step off at the surface of Taurus Littrow, I'd like to dedicate the first step of Apollo 17 to all those who made it possible. . . . Oh, my golly! Unbelievable!" (Cernan 1999, 321).

The first of three seven-hour excursions began with the rover deployed, the American flag erected, and an extensive set of scientific devices powered by a nuclear power supply set about the landing area. Experiments to measure the lunar atmosphere (recording gas molecules near the lunar surface), a lunar surface gravimeter, lunar ejecta, and micrometeorite facility, and a lunar seismic profiling experiment were the primary ALSEP apparatus. The holes were drilled for the heat-flow experiment, and all of the connections were completed.

Schmitt was in his element, describing in detail his professional observations of the rock structures as he spoke of "fine grained regolith . . . lighter colored ejectas . . . scarps . . . and penetrating bedrock." He was intrigued by mysterious "orange soil," which he believed might be oxidized material from subsurface venting.

The three EVAs allowed the astronauts to spend a record twenty-two hours exploring the surface, while the rover took them almost twenty-three miles. They took 2,200 pictures and collected 242 lb. of lunar samples. Cernan and Schmitt spent more than seventy-five hours on the Moon. During this time the sun progressed from 15 degrees above the horizon when they had arrived to more than 50 degrees when it was time to leave. They noted how dramatically the lunar landscape changed with the varying sun angle.

Before they climbed back into the LM for the last time, Cernan took a few minutes to note the historic aspect of the event. "To commemorate not just Apollo 17's visit to the valley of Taurus Littrow, but as everlasting commemoration of what the real meaning of Apollo is to the world, we'd like to uncover a plaque that has been on the leg of the spacecraft." It reads: "Here man completed his first exploration of the Moon, December 1972 AD. May the spirit of peace in which we came be reflected in the lives of all mankind." With the signatures of the three astronauts and that of President Nixon, the plaque portrays the outline of Earth's continents.

NASA director James C. Fletcher said a few brief words from Houston, and then Cernan hesitated before ascending the ladder. He was looking at another, smaller plaque placed where the astronauts saw it each time they moved up or down the ladder. He said, "While we've got a quiet moment here . . . I'd just like to say that any part of Apollo that has been a success thus far is probably . . . due to the thousands of people in the aerospace industry who have given a great deal, besides dedication and besides effort and besides professionalism . . . and I would like to thank them. I guess there might be someone else that has had something to do with it too." He then read the engraving on the small plaque, "God speed the crew of Apollo 17 . . . and I'd like to thank Him too."

Cernan then added as he stepped from the surface to the LM footpad, "As I take these last steps from the surface . . . for some time to come, but we believe not too long into the future, I'd like to record that America's challenge of today has forged man's destiny of tomorrow. And as we leave the Moon at Taurus Littrow, we leave as we came, and, God willing, as we shall return, with peace and hope for all mankind" (Cernan 1999, 337).

Following the record stay on the lunar surface and a record number of manned orbits around the Moon, the CSM *America* made the burn on the far side to return the last mission to Earth—almost four years to the day

A frame from the video camera on the moon buggy captures the ascent stage launching from the lunar surface. Courtesy of NASA.

when humans first reached that distant body. Twenty-two men had traveled to the Moon . . . twelve had walked its dusty surface . . . all had returned safely to the Earth. The Apollo program brought into focus the magnitude of the problem of "landing a man on the Moon and returning him safely to the Earth."

The Moon Is Different

The lunar environment presented some interesting surprises with respect to sight, sound, touch, and smell. Shadows on the Moon, for example, are blacker than those on Earth, making it more difficult to see and work in any area not in direct sunlight. On Earth, a shadow is somewhat illuminated by light scattered through the atmosphere. Without omnidirectional scattering, the lunar shadow allows little discernment of objects that lie within its boundary, although there is some reflected sunlight off the terrain.

Neil Armstrong commented on the phenomena as he attempted to do work in the shadow of the LM during Apollo 11: "It's quite dark here in the shadow and a little hard for me to see that I have good footing" (Gray 2013). The astronauts were attempting to access the equipment lockers on the side of the LM. Armstrong continued, "It is very easy to see in the shadows after you adapt for a while." Aldrin noted that "continually moving back and forth from sunlight to shadow should be avoided because it's going to cost you some time in perception ability" (Gray 2013).

As Apollo 14 astronaut Alan Shepard was attempting to unload the ALSEP experiments bolted to a pallet using "Boyd bolts," in which each is released by inserting a special tool into a recess and rotating it to release the bolt, the recessed sleeves filled with moondust, and the tool would not go in far enough. Because of the shadow, Shepard could not see into the recess to recognize that it was moondust that was keeping the tool from being inserted. When the problem was recognized, he was frustrated at not being able simply to blow into the hole to clear it. They had to turn the pallet upside down and shake out the moondust. The astronauts were continually thwarted by the smallest shadows working to obscure any recessed feature or gauge, and this cost them valuable time.

The shadow phenomenon also created a problem estimating in distances and relationships. Apollo 12 astronaut Al Bean related, "I thought it [the Surveyor III] was on a slope of 40 degrees. How are we going to get down there? I remember us talking about it in the cabin, about having to use ropes." As the sun moved slightly higher in the lunar sky, the perspective changed. When they actually went over to it, they discovered the slope was more like 10 degrees.

Apollo astronauts quickly discovered that moondust is finer than beach sand—more like a fine, powdery snow. This resulted in the dust easily invading virtually every possible crevice of the suit. It was also abrasive, cutting through seals and creating air leaks, binding joints and scratching polished surfaces. Because it readily clung to the EVA suits, the astronauts tracked it back into the LM. Eugene Cernan thought its strong odor smelled like gunpowder while John Young thought its taste was "not half bad." Moondust was a significant problem.

Harrison Schmitt experienced hay-fever-like symptoms that developed quickly. The severity subsided with his second and third EVA. It was surmised that the dust from the lunar surface reacted with the warm, humid, oxygen-rich content in the LM to produce the smell and the allergic reaction. Unfortunately, the seal on the containers, which should have provided an airtight closure, were damaged by the sharp edges of the dust. As a result, the lunar samples were prematurely exposed to the Earth's atmosphere on their return and could not be properly analyzed with respect to selected chemical reactions back on Earth.

The weight of lunar samples steadily increased from the 46 lb. returned on Apollo 11 to the 243 lb. on Apollo 17. Likewise, the total duration of

Apollo 11 on the lunar surface was twenty-one hours and thirty-six minutes, while Apollo 17 remained for seventy-five hours. The single EVA of Armstrong and Aldrin of two hours and forty minutes was extended to twenty-two hours over three excursions for Schmitt and Cernan. The Apollo 11 astronauts remained within 800 ft. of the LM while the lunar rover took Schmitt and Cernan more than twenty-two miles across the lunar surface. The CSM/LM structure in lunar orbit grew from 74,087 lb. to 77,433 lb. The Saturn V increased its Earth-orbiting ability from 297,200 lb. of Apollo 11 to 310,000 lb. of Apollo 17.

The End of the Space Race

One of President Kennedy's concerns in establishing the Apollo Moon program as a national objective was that, if the United States and the USSR both achieved that goal within a short period, then being first would not have a significant effect. He was betting the Soviet system would not be able to keep up with the United States over the long run—a compelling argument to demonstrate the fallacy of communism to the uncommitted countries. He was correct. The USSR could not sustain the level of effort necessary to compete. However, by the time the event happened, most of the world had already accepted as fact that there really was no Moon race, and some of its significance was lost.

Post-Apollo planning had called for a manned lunar base, a large orbiting space station, and a manned flight to Mars by 1981. Vice President Spiro Agnew declared, "It is my individual feeling that we should articulate a simple, ambitious, optimistic goal of a manned flight to Mars by the end of this century" (Baker 1982, 361). He was careful not to link his opinion with that of President Nixon, who had reserved feelings about the next step in space. With the Soviets presenting less technological threat following the Apollo program, there was no political motivation to move these projects forward.

However, the Kennedy challenge had already reshaped the emphasis on space exploration for both the United States and the Soviet Union for the two decades that followed. The political climate in the United States during the decade of the 1970s that followed the lunar landing was one of cynicism. The Vietnam War, the Watergate scandal, the perceived oil shortage, and racial unrest continued to take center stage, while the American manned-space-flight effort ground to a halt by 1975. Six years passed before another American astronaut would orbit. For the Soviets, the decade of the '70s was one of frustration as they attempted to move forward with manned space stations.

President Kennedy's own misgivings about manned space flight, expressed in 1962, were echoed by his brother, the Democratic senator from Massachusetts, Edward "Ted" Kennedy, in 1969. Like his brother the late president, Senator Kennedy was not an ardent supporter of a manned space program. He sought to reduce America's commitment to space when he said, "I think . . . the space program ought to fit into our other national priorities." The cost of these extensive adjunct goals beyond Apollo was too much for Congress—which refused to approve an expanded lunar research program.

The Apollo lunar program also proved to be the high point in Wernher von Braun's career. He had moved on to NASA's headquarters in Washington, D.C., to do future mission planning. However, after less than two years of dealing with the ever-encroaching bureaucracy of NASA, von Braun could see that the nation was not supportive of his grand visions for flights to Mars and beyond. The Saturn V was the culmination of his career, although he did participate in some of the later changes that defined the space shuttle.

Von Braun retired from government service in 1972 to become the corporate vice president of engineering and development for Fairchild Industries—a position that moved him far from the creative drafting boards and roaring test stands that had been his domain for the preceding forty years. The preeminent visionary and early pioneer of rocketry died of cancer in 1977 at the age of sixty-five. His questionable involvement with the Nazi Party during World War II, the use of slave labor to build the V-2, and its use as a terror weapon was, essentially, buried with him.

Arthur Rudolph, the former Saturn V project director, was not so fortunate. In 1984, following a reinvestigation of his alleged war crimes, he agreed to leave the United States and renounce his U.S. citizenship in return for not being prosecuted (Bower 1987, 275).

While the successful completion of the Apollo program essentially marked the end of the space race as far as America and most of the world was concerned, the Soviets continued their effort to compete with the United States. Even though they still performed some exciting events in space, after the Moon landing, these efforts were anticlimactic. With a few exceptions, the building apathy of the world toward space exploration continued.

Political Postscript: Detente in Space

Both the Soviets and the Americans were eager to improve their image to the world and to ease the tensions between one another. While neither side was actually changing its fundamental distrust of the other, President Nixon had made significant overtures to the Soviet Union (as well as to the Communist Chinese), and the two countries were engaged in negotiations that ultimately led to the first Strategic Arms Limitation Treaty in 1973. As a part of this initiative toward a more cooperative stance, several suggestions had been put forward to have a joint space project. While the more grand dreams of an international Moon base or a flight to Mars were beyond current consideration, simply having two space craft rendezvous and dock was a first step in a more cooperative effort.

The Nixon administration, after much soul searching, agreed to a joint space flight and the Apollo-Soyuz Test Project was born. On the Soviet side, there was considerable consternation. There was no doubt that Soviet technology was well behind America's at this point, and a joint flight would open up the Iron Curtain enough so that the Americans could make a reasonable determination of how far behind the Soviets actually were. This aspect made for hesitant progress in the negotiations. However, the reverse was also true. By working with the Americans, the Soviets could learn more about the technology and the methods the Americans were using to manage and control the quality of their programs.

The basic plan called for the Americans to fabricate a small docking module compatible with the Soyuz on one end and the Apollo on the other. This ultimately led to a universal adapter that allowed any spacecraft to dock with any other. To this point, there were "male" and "female" receptacles on spacecraft, but it was obvious that there was a need for an "androgynous" docking mechanism. The schedule was set for the event to take place in 1975.

As the Soviet space scientists had feared, the politically motivated Apollo–Soyuz program coerced them into revealing previously classified data about several of their manned failures, including the harrowing experience of Soyuz 18.

With the reality generated by a decade-long struggle to tame the hostile environment of space came a new perspective of how to pursue that struggle. While the Soviets were forced to open the Iron Curtain to some extent by agreeing to the rendezvous and docking with an Apollo in 1975, there was also much gained by the Soviets. The initial visits to the United States by key Soviet scientists, engineers, and cosmonauts as a part of that program were initially accompanied by a high degree of caution and skepticism.

The Soviets' first impression was that the Americans had made elaborate schemes to impress them. Their visits to supermarkets, where the shelves were too well stocked to be real, reinforced this impression, as did the lack of long lines at the checkout counter (they obviously were not taken there during a Saturday afternoon). Nevertheless, the sincerity of their American hosts and the almost unlimited access to American space technology soon convinced the Soviets that America was not what their life-long indoctrination had told them.

At the Johnson Space Center in Houston, the Soviets were given a first-hand look at the management structures and quality assurance techniques that had beat them to the Moon. It was as though the Americans were providing a postgraduate course in technology management. In return, America learned to appreciate the simplicity of the Soviet systems while the camaraderie brought a new understanding of the Soviet culture. Therefore, while most will admit that Apollo–Soyuz was a public relations stunt for both sides, it turned out to be far more profitable than either side had hoped.

The Apollo spacecraft was launched with a module designed to allow the two spacecraft to dock and was used only for this mission. The Saturn IB launch vehicle and CSM were left over from the lunar program. The docking module was retrieved from the S-IVB upper stage by the Apollo module once in orbit in the same manner as the lunar module. The space shuttle used the same APAS-89 docking hardware through the end of the shuttle program to dock to Mir and then the International Space Station, the latter through the pressurized mating adapters.

The docking module was designed as an adapter and as an airlock. The Apollo was pressurized at 5.0 psi using pure oxygen, while the Soyuz used a nitrogen–oxygen atmosphere at sea-level pressure. There were some changes to both spacecraft—the Soyuz had its internal pressure reduced to 10.2 psi to make it easier for the 5.5 psi Apollo to accommodate the difference in the interior pressure between the two spacecraft (Stafford and Cassutt 2002, 169).

One end of the docking module used the same "probe and drogue" docking mechanism used on the lunar module and Skylab, while its other

The configuration of the Apollo-Soyuz Test Project. This display is exhibited in the Smithsonian Museum. Courtesy of National Air & Space Museum.

end had the APAS docking collar, which Soyuz 19 carried in place of the standard Soyuz/Salyut system of the time. Either spacecraft could terminate a docking on own initiative.

The Apollo spacecraft was launched on July 15, 1975, with Thomas P. Stafford in command and Vance Brand as the command module pilot. The third member of the crew was Deke Slayton, an original member of the Mercury Seven. He had been denied a space flight because of an irregular

The crew (*left to right*) of the ASTP, from Apollo (Donald K. "Deke" Slayton, Thomas P. Stafford and Vance D. Brand) and from Soyuz (Alexey Leonov and Valeri Kubasov). Courtesy of NASA.

heartbeat. However, he had endured sixteen years as an astronaut in a ground-based role waiting for just such an opportunity.

It was the last flight of the Saturn 1B launch vehicle and for the Apollo spacecraft, which was not given a designation, although it was inappropriately referred to on occasion as Apollo 18—the designation of a canceled lunar landing mission.

The Soviet Soyuz 19 launched less than eight hours after the Apollo. The commander was Alexey Leonov, the first man to walk in space. It was his second and last flight. The other crewmember was Valeri Kubasov, who was also on his second flight. The two ships rendezvoused and docked for the first of several occasions on July 17. The Soyuz could only handle a two-man crew because of the need to wear spacesuits.

While docked, the three astronauts and two cosmonauts conducted joint scientific experiments, exchanged flags and gifts (including tree seeds later planted in the two countries), signed certificates, visited each other's ships, ate together, and conversed in each other's languages. (Because of Stafford's pronounced drawl when speaking Russian, Leonov commented that there were three languages spoken on the mission: Russian, English, and "Oklahomski") (Shepard et al. 1994, 346). There were also docking and redocking maneuvers, during which the two spacecraft reversed roles, and the Soyuz became the "active" ship. On the fourth day, the two ships separated for the last time and went their separate ways. The Soyuz remained in space for five more days and Apollo for nine.

20

THE FIRST SPACE STATIONS

Early Concepts

The concept of a manned platform or station in outer space is not new. Under the genre of science fiction, the *Atlantic Monthly* magazine in 1870 presented the essential idea of a laboratory in space in the form of a serial story written by an American minister and writer, Edward Everett Hale.

A German mathematics teacher, Kurd Lasswitz, in 1897 produced a novel, *Auf zwei Planeten* (*Two Planets*), which depicted a Martian space station supported by antigravity, which provided a stopover for long-duration flights.

Konstantin Tsiolkovsky's science-fiction work entitled *Reflections on Earth and Heaven and the Effects of Universal Gravitation* (1895) described asteroids and artificial satellites as launching facilities for interplanetary travel. He also provided for artificial gravity on these man-made space stations by rotating them to create centrifugal force—an idea that caught on with subsequent enthusiasts.

Hermann Oberth included a brief description of "observation stations" in his 1923 book and some of their possible uses such as astronomy, Earth observation, military reconnaissance, and use as a refueling station for interplanetary flights. In a subsequent revision of his book published in 1929, *Wege zur Raumschiffahrt* (*Ways to Spaceflight*), Oberth provided more detail. He presented the concept of the space mirror facility as a part of a space station that could redirect solar energy to a single point on Earth as a military weapon—much as a magnifying glass can be focused to create intense heat and burn through a piece of paper. He also thought that, when applied over a wider area, such a mirror could warm northern ports to keep them free of ice in winter or could illuminate large cities at night.

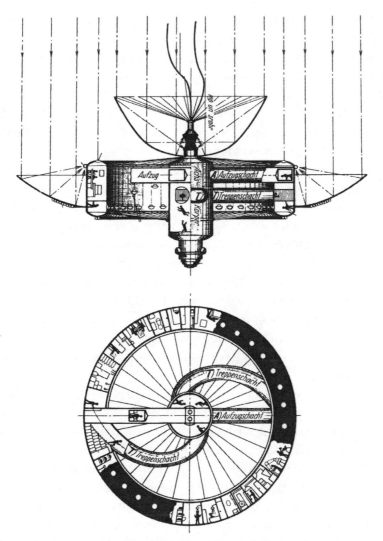

Hermann Oberth's "Observation Station" from his 1923 book. Courtesy of National Air & Space Museum.

A somewhat mysterious writer whose pen name was Hermann Noordung wrote of a giant space station in his science-fiction work published in 1927. He was, in reality, an Austrian Army Officer named Hermann Potočnik. Although Noordung had some obvious flaws in his treatment of physics, his detailed design used a rotating wheel to provide artificial gravity and a central "hub" to provide an airlock to gain entry. Electrical power was provided by solar heating of unspecified fluid compounds.

Potočnik foresaw many uses for a space station, including scientific experiments unimpeded by gravity and astronomy without the interference of the Earth's atmosphere. He recognized that Earth could be examined for meteorological and military applications. In addition, like Oberth, he wrote of the use of a space mirror to focus the rays of the sun on Earth. Because his work was widely read (Wernher von Braun had read Noordung as an enthusiastic teenager), Noordung no doubt galvanized the thinking of many space enthusiasts of the period.

Following World War II, both science fiction and tentative scientific planning produced several concepts of a space station. The postwar writings of von Braun in his book *Mars Project* and his *Collier's* magazine articles of 1952, coupled with the prolific writings of Willy Ley, continued to spread the possible capabilities of the space station. Ley prophesied a 250 ft. diameter wheel with a 22 ft. rim connected to the "hub" by three spokes and manned by a crew of thirty to sixty men (women were conspicuously absent in the male-dominated sciences of the 1950s).

In von Braun's vision, the space station was the key to pursuing most other aspects of space exploration—following quickly after the initial unmanned satellites. His space station at this point was the culmination of many ideas gathered from the visionaries of the previous decades, and he conceived it as being built in much the same manner as a ship of the sea, in parts welded and riveted together—in space. However, there were others, such as Sergei Korolev, who saw this form of fabrication as unrealistic. They saw the space station as built in prefabricated segments, complete with their interior appointments and joined in a simpler manner.

With the advent of Sputnik in 1957, the space station concept became a primary interest for the United States and the Soviet Union. Both Korolev and von Braun felt that the space station and trips to Mars would occur within ten to fifteen years, if not sooner. However, by 1961, with the Moon established as the goal of the "great space race" by President Kennedy, the focus temporarily shifted away from the space station. With the ability to build giant super rockets such as the N1 and the Saturn V, and with the lunar orbit rendezvous technique, there was no need to have an intermediary step before reaching for the Moon.

The military in both countries, having little interest in the Moon race as a national defense priority, continued to pursue the space station. In December 1963, the U.S. Air Force created a project for developing its manned orbiting laboratory (MOL), and the Soviets had Korolev's Soyuz-R, a reconnaissance version of the planned Soyuz spacecraft. Soyuz-R was

An artist rendering of the U.S. Air Force manned orbiting laboratory. It would have used a Gemini spacecraft to return the astronauts to Earth. Courtesy of Department of Defense.

soon followed in October 1964 by a competitive proposal from Vladimir Chelomey for the Almaz (Russian for "diamond"). Like the MOL, Almaz was a relatively small space station manned by two or three cosmonauts. While the 32,000 lb. MOL never flew, the 40,000 lb. Almaz was originally scheduled for launch by 1968 and was to accommodate cosmonauts for up to one year. Both of these goals were highly optimistic. By the end of 1969, the first launch of Almaz was still not ready despite ten stations in varying degrees of fabrication.

Salyut 1: The Rush to Ruin

As the race to the Moon culminated in spectacular American success, and with the secret Soviet lunar landing effort smoldering in ruins, First Party Secretary Leonid Brezhnev was forced to produce a new strategy to address his country's now second-rate technological position. The triple space flight of Soyuz 7, 8 and 9 in 1970 showed the Soviet's ability to put three manned spacecraft into orbit within a few days but demonstrated little else. That the Soyuz 9 crew was unable to stand after their record eighteen-day flight and had to be carried to the recovery helicopters on stretchers indicated that there was yet much to learn about living in space. The effect of weightlessness on the human body was of primary concern. Extensive loss of calcium by as much as 15 percent, the pooling of blood in the head and upper body, and the loss of muscular strength were high on the list of phenomena to be investigated.

The space station, even in the modest forms being designed in both the Soviet Union and America, allowed relatively long-duration flights that could emulate a trip to Mars in thirty-day increments. By rotating crews to the station, the effects of space flight could be gauged on both the human element and the systems required to support them without committing to the hazards of a two- or three-year mission.

For the Soviets, the space station seemed like an ideal choice to recapture the imagination of the world. It could be touted as a scientific and "peaceful" project, and it did not depend on the development of any new large booster. The station, at about 20 tons, could be launched by the UR-500 (Proton) and crewed and provisioned (once in orbit) by the Soyuz atop the continually improved and reliable R-7.

Once again, time was of the essence for the Soviets, as the Americans announced that they would use much of the Apollo hardware to pursue what was referred to as the "Apollo Applications." It included a large space station called Skylab, launched by the Saturn V. As neither the Soyuz-R (a reconnaissance version of the Soyuz) nor the Almaz (a military space station) were making good progress, it was decided to move the Almaz (the larger of the two) to Korolev's old design bureau, now run by Vasily Mishin. This did not please Chelomey, who headed the competitive design bureau from which the project was taken. Almaz had included a two-man spacecraft (almost identical in configuration to the American Gemini) riding into orbit with the space station. But because there was no need to duplicate a capability that Soyuz already possessed, that aspect was canceled to accelerate the project.

This space station did not resemble the classic wheel shape as so many of the visionaries of the past had configured it. With the desire to investigate the effect of weightlessness, these first efforts simply used large cylindrical enclosures that could readily be installed as a payload on existing launch vehicles such as the UR-500.

The project had to be completed well before the launch of Skylab for its impact to be significant. As with many Soviet space projects that preceded it, systems were adapted wherever possible from other projects and with minimal testing. Not only was time a critical factor but the budget had to be kept to a minimum. This meant that some of the more sophisticated scientific equipment was not installed in order to meet the scheduled launch deadline.

The revised Soviet configuration consisted of a large cylindrical work area, adapted from the Almaz project that was 30 ft. long and 14 ft. in diameter, which narrowed to a little over 9 ft. in diameter after the first 9 ft. The space station was named Salyut—meaning "salute" in Russian—in honor of the first Soviet cosmonaut, Yuri Gagarin. To accelerate the completion of the Salyut, portions of the Soyuz design were fabricated into the space station. The Soyuz service module, which contained the on-orbit maneuvering, attitude control, and environmental systems, was a 7 ft. cylindrical section affixed to the large (aft) end, while a docking and transfer compartment was placed on the opposite (forward) end. This docking mechanism was an improvement over the initial Soyuz in that it allowed the crew to transfer directly between spacecraft (through the docking adapter), as had been done in Apollo/LM, without requiring a spacewalk.

Because a Soyuz crew transfer vehicle was docked to the station, a second hatch on the side of the transfer compartment allowed for a spacewalk capability. As with *Voskhod I*, the Soyuz cosmonauts did not wear spacesuits during their launch—these were provisioned within the Salyut for later planned spacewalk activities. Two sets of solar cells (also adapted directly from Soyuz) extended from the fore and aft modules to provide 3.6 kw of electrical power.

Although equipped with an attitude control system (ACS), the space station employed gravity-gradient positioning because the larger, heavier aft end aligned its longitudinal axis toward Earth's center. The ACS was used during some experiments when it was desired to point the station in a particular direction with ion sensors providing the reference.

An "up" and "down" orientation was maintained within the space station to provide a traditional directional reference in the weightless environment for the cosmonauts. Since previous experience with Soyuz flights had shown a significant degradation in the muscle and bone composition of the cosmonauts for even short, two-week excursions, a treadmill allowed the cosmonauts to get daily physical conditioning with rubber bungees providing the artificial gravity for the cosmonaut.

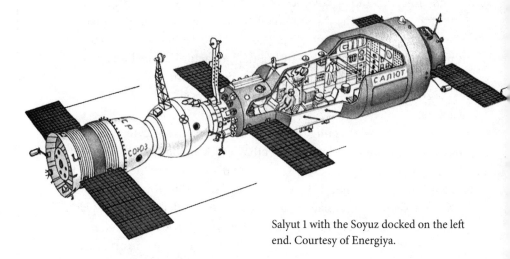

Salyut 1 with the Soyuz docked on the left
end. Courtesy of Energiya.

The UR-500 booster left its pad with Salyut 1 as its payload enclosed in a launch shroud on April 19, 1971—a full two years ahead of the American Skylab. Placed in a 124 mi. by 138 mi. orbit with an inclination of 51 degrees, the 42,000 lb. space station was the largest object the Soviets had ever placed into orbit. The low orbit was a result of its primary role of military reconnaissance and required the periodic firing of the 1,000 lb. thrust orbital maneuvering system engine to raise its orbit to avoid premature reentry.

Only one significant problem plagued the unmanned craft as soon as it had reached orbit. The cover for the main telescope failed to jettison, and this extremely large experiment (it occupied most of the aft portion) could not be used. As soon as the Soviets felt confident that Salyut 1 was in a stable orbit and functioning, they made an unprecedented announcement—not only that they had successfully launched the world's first space station but also the date of the launch of its first crew and their names.

However, the scheduled April 22 launch (three days later) with Lt. Col. Vladimir Aleksandrovich Shatalov, Aleksei Yeliseyev, and Nikolai Rukavishnikov on board Soyuz 10 did not take place. A part of the launch tower failed to retract during the final minutes of the countdown, apparently because of a buildup of ice from heavy rains and the cryogenic temperatures. The launch was canceled. This was a major embarrassment for the Soviets. During the second attempt the following day, the segment again failed to retract. Korolev's former first deputy, Vasily Mishin, as the controlling authority, decided to risk the launch and allowed the countdown to continue.

Whether it was the vibration generated at ignition or some other quirk, the launch tower segment retracted properly and the cosmonauts were on their way.

Following the uneventful launch, the cosmonauts of Soyuz 10 spent the next two days performing the rendezvous with the passive Salyut space station. However, failure of the Soyuz attitude control gyros and the inability of cosmonaut Shatalov to achieve a "hard dock" after several attempts caused the twenty-two-day mission to end after only three days in orbit. The crew returned successfully, and it was determined that the Soyuz docking latches had been damaged during the first docking attempt.

Soyuz 11 was scheduled to provide the second crew for Salyut, but problems began before launch when one member of the crew became ill and the entire crew was exchanged for the backup crew. The replacement crew of Georgi Dobrovolski, Viktor Patsayev, and Vladislav Volkov complicated the situation, as the six spacesuits already on board the Salyut space station were those for the cosmonauts of Soyuz 10 and the prime crew for Soyuz 11. The unannounced launch of Soyuz 11 occurred on June 6, 1971, and its reinforced docking latches appeared to resolve the hard-dock problem.

The three cosmonauts of Soyuz 11 are in good spirits prior to their launch. Note the lack of spacesuits. Courtesy of Energiya.

By now Salyut had been in orbit two months, and when the Soyuz crew entered Salyut, they smelled the distinct odor of what they believed was burned electrical insulation. Not knowing how toxic the atmosphere might be, they quickly replaced six lithium perchlorate cartridges that had failed and turned on the air regeneration system. They elected to return to the Soyuz and spent the first night there, isolated from the Salyut environmental system.

The next day the odor had been eliminated, and the crew returned to begin a series of scientific and military experiments. An extensive physical exercise program using the bungee treadmill and other devices for the cosmonauts was closely monitored. As with their American counterparts, the cosmonauts had a difficult time establishing a good schedule for working and sleeping—trying to stagger the sleep periods was not successful.

With their initial efforts to get good publicity from the Soyuz 10 flight thwarted by the delay in launch and then the inability to rendezvous, the Soviets were more cautious with the information they revealed about Soyuz 11. However, after the first week in orbit, a press conference was held in which extensive videotape of the cosmonaut's activities was shown on Soviet television. A second program two days later proved equally popular, and the Soviet leadership was more than pleased with their newfound "openness."

However, the strong smell of an electrical fire caused an alert on the tenth day in orbit, and the cosmonauts again took refuge in Soyuz while they discussed their course of action. The only member of the crew who had previous flight experience, Volkov, wanted to return to Earth, while the mission commander, Dobrovolski, who, along with Patsayev, returned to Salyut in an attempt to locate the smoldering fire. Because the atmosphere within Salyut was similar to that of the Earth's (75 percent nitrogen and 25 percent oxygen) the fire did not flare, as would have been the case with the early American spacecraft. Eventually, the smoldering cable was located and disconnected, but the obvious conflict among the crew had made the situation more difficult.

The contention between the crew members was not made public at the time, and the Soviet press was bathing the Soviet people (and the world) in the great success of Salyut. Subsequent TV shows continued to demonstrate various aspects of the mission and its experiments, and all seemed serene on board. On June 25, the crew broadcast their last TV show and made it clear that they were eager to return. As with all the previous broadcasts, the sessions were not shown live but were taped and edited.

On June 29, 1971, the crew completed the packing of experiments to be returned and secured Salyut in preparation for the next crew. As they prepared to undock, a warning light illuminated, indicating that the hatch between the descent module and the orbital module had not closed and sealed properly. Volkov, who had been on edge for the entire flight, nervously reported, "The hatch is not pressurized, what should we do, what should we do?" (Zimmerman 2003, 44). His voice reflected the seriousness of the situation since the three cosmonauts were not wearing space suits, and if the hatch failed to hold the internal pressure of the descent module, they would perish.

Mission Control ordered the crew to reopen and reseal the hatch. However, the warning light still did not go out. Next they were told to reopen it and clean the gasket as dirt might have been keeping it from maintaining a good seal. They did this, and finally the warning light went out. Dobrovolski separated the Soyuz from the Salyut space station and, after another orbit, fired the retrorocket to slow the craft for reentry. However, as they jettisoned the orbital and service modules, the vent in the descent module that allowed fresh air into the cabin when they were descending under parachute opened prematurely. Their precious air was being vented into the vacuum of space—slowly but surely.

The three cosmonauts could hear the hissing sound of the escaping life-sustaining pressure, and all three released their seat harnesses in an attempt to locate and close the valve. Their first inclination was that it was the balky hatch. However, the time spent on it wasted precious air. In an effort to silence other sounds so they could hear the hissing more clearly, they turned off the radio communication that was feeding the sound back through their microphones and into their headsets—interfering with their troubleshooting. Mission Control now had no way of knowing what was happening on board the Soyuz descent module.

Within a minute, the pressure had dropped to a value that caused the men to suffer from hypoxia (lack of oxygen), and their efforts were now irrational and uncoordinated. Within another sixty seconds they had passed out—and died a few minutes later. The spacecraft continued on its reentry and achieved a normal touchdown. Mission Control was concerned that they had received no communications but assumed the mission had been completed successfully. When the recovery crews arrived, the descent

The recovery forces apply resuscitation techniques in an effort to revive the three cosmonauts. Courtesy of Energiya.

module was quiet, and on opening the hatch the heartbreaking discovery was made. Films of the recovery forces show the futile attempts to perform mouth-to-mouth resuscitation.

The families and the world grieved at the loss, and the Soviet news organizations found it difficult to manipulate the information of the tragedy to minimize the negative impact on the perception of the Soviet space program. The Soyuz hatch, the separation mechanism, and the pressure-relief valve were reengineered. However, it was more than two years before the next Soyuz flew, and the Salyut itself reentered and burned up just four months after the tragedy. The promise of the space station would have to wait for the next iteration.

Skylab: America's Magnificent Dead End

Following the establishment of NASA in October of 1958, a wide variety of space projects were examined in an effort to ensure that America did not fall further behind the Soviet Union in the hotly contested space race. Among the possible projects was a manned space station. However, President Kennedy's decision to apply America's technology and funding to the first Moon landing delayed any other large-scale projects until that objective was fulfilled. That commitment did not preclude the relatively inexpensive process of preliminary planning.

Throughout the exciting space-race decade of the 1960s, a variety of configurations for a space station were examined, with several efforts being made to validate the concept of the rotating wheel as conceived by many of the early space visionaries. As desirable as it appeared, the strength of the structure and the ability to fabricate such a shape in space dictated a costly project—perhaps more costly than the Apollo lunar landing. A cylindrical structure, similar to the Air Force's MOL that was canceled in 1969, seemed to offer the least expensive first path to a manned space station. However, MOL was a small environment (30 ft. long and 10 ft./ in diameter) with the physical limits being determined by the Titan III launch vehicle. NASA was thinking on a grander scale.

A cylindrical shape, perhaps matching the 22 ft. diameter of the Saturn IB's 60 ft. long S-IVB second stage, would allow a large internal volume and the ability to conduct significant experiments. The medical aspects of long-duration space flights, as necessary for trips to Mars, required a more in-depth understanding of the effects of weightlessness on the human body. It was also clear that the follow-on to the Moon program had to draw heavily on the Apollo hardware to keep development costs and the timetable to reasonable levels.

NASA's Marshall Space Flight Center director, Wernher von Braun, had long pushed for the classical rotating wheel, but as the financial aspect became apparent, he also opted for the cylindrical shape. However, he thought in terms of what was then called the "wet tank" concept. A Saturn IB would put its S-IVB second stage in orbit. Then, teams of "construction astronauts" launched on board subsequent Saturn IBs would purge the remaining fuel in the S-IVB and install various internal and external systems—airlocks, power systems, and attitude control. With the Saturn IB's ability to launch 20 tons, it would require several launches to accomplish the task of building a space station using the wet tank concept, and the hazards of the construction and the need to launch many construction crew ships would escalate the costs.

It was clear that this method of constructing a space station would be difficult to do in the unforgiving environment of space. Thus the "dry tank" concept was born. A complete space station was built from an empty S-IVB stage on Earth and then put into orbit by a Saturn V. Only fifteen Saturn Vs were built, and until it was clear how many were needed to accomplish the first political assault on the Moon, there was no way of guaranteeing their availability for other missions. Von Braun's new vision had to wait for a successful lunar landing.

As the Moon landing moved closer to reality following the first successful flights of the manned Saturn V in late 1968 and early 1969, the possibility of using a Saturn V for the space station, now named Skylab, became more focused. On the return of the Apollo 11 astronauts from the Moon, NASA administrator Thomas Paine announced the plan and schedule for Skylab using a Saturn V in 1972. While this first American space station could be used as a springboard for interplanetary travel, a reusable manned spacecraft for resupply was still too grand for the cash-strapped space program that had been giving up its cherished budget to the Vietnam War and the escalating costs of former President Lyndon Johnson's Great Society programs. (Even though Johnson was by then out of office, the welfare programs Congress had put in place were already showing dramatic escalation.) Any space station had to use existing flight hardware from the Apollo program—there would be no large follow-on project to Mars or to establish a Moon base.

Already there was talk of curtailing the number of previously scheduled flights to the Moon and mothballing some of the Saturn IBs, Saturn V rockets, and Apollo spacecraft. In the end, three flight-ready Saturn Vs became museum pieces for lack of funds to fly them. However, Administrator Paine was able to get President Richard Nixon to approve the dry tank space station. While not the grand 250 ft. diameter rotating wheel featured in the newest science-fiction movie of the day, *2001—A Space Odyssey*, it was an impressive structure.

Skylab was fabricated by using the third stage of the Saturn V (the S-IVB) and adding multiple docking adapters that allowed two Apollo spacecraft to dock simultaneously. It contained an array of solar telescopes and other scientific apparatus that had previously been too large or heavy for either the Gemini or the Apollo spacecraft to carry. Outfitted with the ability to initially support three sets of three-man crews for as long as ninety days in a lavish interior space of 12,700 cu. ft., it weighed an astounding 165,000 lb. (almost four times that of Salyut). The largest single piece of habitable space equipment ever launched, it would hold that record until well into the next millennium.

To stabilize the huge space station, three large gyroscopes (2 ft. wide and 155 lb. each) were oriented to the three axes. With a spin rate of 150 rpm, these held Skylab in a fixed attitude in space by virtue of the gyroscopic

The Skylab space station with the improvised solar heat shield and only one solar panel deployed. Courtesy of NASA.

property of "rigidity in space." A mechanism allowed the gyroscopes to tilt so that the space station could achieve a new selected orientation with the gyroscopic property of precession. An ACS using nitrogen thrusters was also installed.

To provide 10 kw of electrical energy, six large solar cell panels were used. Four panels formed a windmill-like structure in the forward part of the station and served as the telescope mount, which folded to fit inside a large nose shroud that was jettisoned after achieving orbit. The other two panels folded tightly against the side of the spacecraft during launch and would be extended when in orbit.

The launch of the twelfth and last Saturn V to fly carried the Skylab space station into orbit on May 14, 1973—almost a year after its original schedule. The launch appeared flawless, but once in orbit, the telemetry from the yet unmanned space station indicated a problem with the electrical power. The meteoroid and thermal shielding that wrapped around the cylindrical body of the space station had been torn off during the ascent through the atmosphere, and parts of it had apparently jammed the extension of the two solar panels that were folded flat against the sides of Skylab. Although the extent of the problem was not immediately known, the rise in internal temperature of Skylab and the lack of power from the two solar panels were strong indicators of a serious problem aboard the giant craft.

The launch of the first crew, which consisted of veteran Apollo 12 astronaut Pete Conrad and rookies Joe Kerwin and Paul Weitz, scheduled for the following day, was postponed for ten days while NASA tried to determine if the project could be saved.

Following an analysis of the data and the preparation of a set of possible fixes, the crew launched on May 25, 1973. The Saturn IB again performed perfectly, and the rendezvous with Skylab occurred without any problems. On arrival, the astronauts immediately confirmed what the ground controllers had presumed. Conrad reported that solar wing two "is completely gone off the bird. Solar wing one is in fact partially deployed . . . there is a bulge of meteor shield underneath it in the middle and it looks to be holding it down" (Zimmerman 2003, 58). A single metal strap from the shield had become wedged into the solar wing. This had allowed the solar panel to extend only about 15 degrees of its designed 90-degree movement.

A first spacewalk with a large 10 ft. pry bar failed to dislodge the strap after more than an hour of effort. The crew gave up, and it was decided to dock with Skylab. However, that maneuver failed five consecutive times. Finally, the astronauts dismantled the soft-docking mechanism from within the Apollo access tunnel and went directly for a hard dock—which was successful. The crew spent the first night in space in the Apollo without attempting to enter Skylab until the next day. On entering, the crew found the temperatures in the docking adapter and the airlock to be a reasonable 50 °F. But the 130 °F temperature in the large workshop area was stifling.

Using one of the research airlocks on the side of Skylab (that allowed access to the outside), Weitz and Conrad erected a large Mylar "umbrella" sun shade that they had brought with them. The temperature immediately dropped by about 2 degrees per hour. By reducing their electrical needs, the astronauts were able to remain on board and perform useful scientific work—primarily medical experiments—for the next two weeks.

On June 7, during a second spacewalk by Conrad and Kerwin, a wire cutter on the end of a 25 ft. pole was used in an attempt to cut through the metal band that was keeping the solar panel from extending. The problem was getting the right leverage to apply the needed force. After more than an hour of trying, the band was finally cut, and the solar panel abruptly extended another 10 degrees, knocking both Conrad and Kerwin off into space. Fortunately, their 55 ft. tethers held and they were able to pull themselves back. Now they used a cord looped under the panel, and, working together, they pulled until the panel suddenly broke free and extended to its normal deployed position, again propelling the duo into space.

With more electrical power now available, the space station crew continued their scientific work for another two weeks. During this time, they observed phenomena that had not been obvious inside smaller spacecraft. In the weightless environment, it was possible for them to drift out of arms' reach from the side of the spacecraft—thus having no way of propelling themselves. They could actually be "stuck" in the middle of the workspace.

It was also interesting to note that one of the annoyances while trying to sleep was the constant air being blown on them by a number of fans placed around the interior. Without the fans to circulate the air, it was possible (in the weightlessness of space) for the exhaled carbon dioxide to build up in front of their faces—potentially suffocating them.

At the end of a record twenty-eight days, the crew returned to Earth. Because of their dedicated physical conditioning, all three were able to walk from the recovery helicopter. However, they did take about three weeks to return to normal levels of 1-g activity.

The second crew of veteran astronaut Al Bean and rookies Owen Garriott and Jack Lousma launched on July 28, 1973. Almost immediately after docking with Skylab, Lousma experienced space sickness and vomited, and Bean and Garriott became dizzy and disoriented. On top of this, thrusters on the Apollo service module were leaking, and there was some thought that a rescue mission might have to be launched. However, there were four redundant ACSs on Apollo; although two eventually had problems, the other two continued to function properly for the reminder of the mission.

The crew's performance after the first six days of illness was phenomenal. For fifty-nine days they proceeded to perform a wide variety of scientific experiments and observations while doing some outside repair work to replace the solar heat shield with a larger and more durable structure. As with all previous missions starting with the first Apollo, numerous live TV broadcasts were conducted. Inside the cavernous station, a series of tests were performed on several jetpack designs for space walks, and some of these tests were beamed down to the Earthly audience.

In keeping with the tradition of pulling pranks, about six weeks into the flight, Mission Control heard the voice of Garriott's wife coming from Skylab. She told CapCom that "the boys hadn't had a good home cooked meal in quite some time so I thought I'd just bring one up" (Zimmerman 2003, 72). Her statement seemed to fit into the current schedule and was so natural that Bob Crippen, the astronaut on duty at Mission Control, was at a loss for words. Garriott had recorded the session before launch, and the entire crew got a real lift by putting one over on Mission Control. However, during the flight NASA announced that, due to budget cutbacks, a follow-on Skylab II space station had been canceled.

After almost two months in space, the crew returned to Earth on September 25, 1973, and, thanks once again to a good physical conditioning program, they were in good shape—except for the mysterious bone loss that exercise and diet could not control.

The third and last crew—Bill Pogue, Jerry Carr, and Ed Gibson—arrived at Skylab on November 16, 1973. Like the previous crew, they immediately experienced space sickness—Bill Pogue in particular vomited violently. As the station was not in contact with Mission Control at the time, the astronauts held a quick conference among themselves and decided not to reveal the extent of the debilitation to the ground. However, the crew had neglected the fact that all conversations were recorded on tape and downloaded to Houston each day, where they were transcribed to print. When Astronaut Chief Alan Shepard read the hard copy of the discussion the following day, he was livid. "I just want to tell you that on the matter of your status reports, we think you made a fairly serious error in judgment" (Zimmerman 2003, 75). While the crew agreed, the tone for the first month of the mission had been set—one of contention between the crew and the controllers. During this time, the astronauts appeared to be continually behind in the schedule, especially when compared to the exceptional performance of the previous crew.

Following a series of open discussions between the crew and Mission Control, the relationship was finally restored, and the crew performance achieved high levels before returning to Earth on February 8, 1974, after 84 days.

Skylab had been inhabited for a total of 171 days and had produced a wealth of information about the long-term effects of space on the human body. It was obvious that a flight to Mars was feasible except for the possible problem with bone loss, which continued to plague the crews. Tentative plans were made to revisit Skylab (which still had enough oxygen and water for another six months of habitation) when the space shuttle became operational in 1979. However, the space shuttle program was two years late and did not fly until 1981. As a result, Skylab could not be boosted into a higher orbit, and it reentered and burned up in July 1979.

Almaz: Analyzing Military Capabilities

The deaths of the three Soyuz 10 cosmonauts who had crewed the Salyut 1 space station had severely crippled the Soviet space program both techno-logically and politically. As the Americans prepared to launch the Skylab in early 1973, the Soviets had regrouped and improved not only the Soyuz spacecraft but also the Salyut design—the first of which had been hurriedly assembled simply to beat the Americans in 1971. However, a second Salyut launched on July 29, 1972, failed to orbit when its UR-500 rocket second stage malfunctioned at T+162 seconds. By December of 1972 Brezhnev himself decided that the Almaz military station (not the civilian Salyut hy-brid) should be launched—the goal was still to beat the American Skylab.

The first Almaz attempt on April 3, 1973 (just one month before Skylab), achieved orbit and was named Salyut 2 to disguise its true military nature. However, a problem—most likely an electrical fire—caused its pressure hull to rupture, and the planned launch of its first crew was canceled. A third attempt on May 11, 1973, was referred to as Kosmos-557, to avoid another embarrassment if it proved troublesome. The use of the false name was appropriate as a malfunction caused all of its ACS fuel to be spent before ground controllers could respond to the problem.

The Kosmos-557 failure, followed three days later by the successful launch of the American Skylab (four times the weight of Salyut), was yet another blow to the deteriorating Soviet space program. In light of these problems, a major shake-up in the Soviet program occurred—directed from the top echelons of government. Leonid Brezhnev had looked to a new series of space spectaculars to bolster his sagging position not only in the international arena but also at home, where he faced stiff competition for control, and four years of successive failures in the manned space pro-gram was more than he could bear.

Sergei Korolev's original design bureau, headed by Vasily Mishin, was combined with several bureaus others to form a new organization—En-ergiya Scientific Production Association. The new director was Valentin Glushko (Korolev's old nemesis), while Mishin was headed for obscurity. It was at this time that the N1 super booster project was officially terminated. Glushko moved forward with a proposal to build a Soviet space shuttle and a booster for it that would also double as an unmanned heavy launch vehicle. However, work on both Almaz and Salyut continued.

More than a year passed before the next iteration of a Soviet space sta-tion launched on June 25, 1974. Designated Salyut 3, it was in reality another military Almaz configuration. The Soyuz-like service and docking modules were replaced by a single docking port on the large aft end straddled by the orbital maneuvering thrusters. A pair of solar cell panels (larger than those on Salyut 1) was supplemented by two curved panels on the outer surface of the main work area. Total electrical power available was 5 kw. The surveillance cameras were of high resolution and capable of distinguishing objects on Earth as small as 12 in. from orbital altitude. As with Skylab, Salyut 3 used a set of large gyroscopes to stabilize the attitude of the space station.

Besides its primary role as a reconnaissance platform, the military aspect was enhanced by the fact that it was the first satellite that actually carried a weapon—a 23 mm Nudelman rapid-fire aircraft cannon that was fixed to the forward segment and aimed by pointing the entire space station.

Although Almaz was originally intended to be launched with a manned spacecraft attached to its forward end, construction of the three-man Merkur module was still several years from completion. Thus, the station was crewed by linking with Soyuz spacecraft launched only after the Salyut (aka Almaz) had successfully achieved orbit.

The Soviets were not subtle in disguising the intent of Salyut 3. The amount of information about its experiments and daily routine of the crew was severely restricted compared to Salyut 1. Salyut 3 tested several reconnaissance sensors before the first truly successful visit to a Soviet space station was concluded with the completion of the sixteen-day Soyuz 14 mission. Soyuz 15 launched in August for an intended twenty-one-day mission, but problems with the Igla rendezvous radar, coupled with the inability to dock with Salyut 3, caused the cosmonauts to return to Earth without completing their mission.

However, the reconnaissance film capsule that had images from the Soyuz 14 mission was successfully returned in September 1974. The capsule suffered damage during reentry, but all the film was recoverable. On January 24, 1975, trials of the onboard 23 mm cannon were conducted under ground control when no cosmonauts were on board the station. The point of the experiment was to provide defensive armament for the craft should the United States try to board or destroy the space station—a highly unlikely scenario but one that fit the paranoia of the Soviet mentality.

The next day the station was commanded to retrofire to a destructive reentry over the Pacific Ocean. Although only one of three planned crews managed to board the station, that crew did complete the first successful Soviet space station flight.

With Salyut 4, which launched on December 26, 1974, the Soviets returned to the civilian configuration with the Soyuz-like service module on the aft end and the docking port on the bow. The first crew arrived sixteen days later on board Soyuz 17, and this mission, which lasted thirty days, was by far the most successful. However, as the Americans had just completed an eighty-four-day mission—their third and final Skylab mission—there was no propaganda or prestige gained.

The return of Soyuz 17 was not without its moment of concern as the parachute release sequence was late. Georgi Grechko, who had waited almost ten years for his flight into space, was quoted many years after the harrowing experience as recalling, "It was absolutely terrifying . . . sweating, panting like a dog. . . . It was paralyzing, I realized that I had no more than five minutes to live" (Zimmerman 2003, 97). His fears were short-lived when the drogue finally deployed followed by the main chute.

The second crew, two months later on board Soyuz 18, met with a dramatic launch-vehicle failure. As the first stage was nearing the end of its burn, the explosive bolts that held it to the second stage fired prematurely, and the sequence ignited the second stage. The entire rocket rotated in an uncontrollable somersault more than 100 mi. above Earth. Because the escape tower at the top of the rocket had already been jettisoned, the abort sequence called for the Soyuz to be separated from the booster—but the worst was yet to come for the cosmonauts. Following a ballistic arc, the spacecraft experienced more than 15 gs of deceleration during its suborbital reentry. It landed on a steep, snow-covered mountainside in western Siberia. Not knowing how long it might take the recovery forces to find them, the crew exited the now frigid reentry module and built a fire. Fortunately, a search party from a nearby village found the crew after only a few hours.

The failure of Soyuz 18 again put significant pressure on the Soviet space program managers—by Brezhnev in particular—to send a second crew to Salyut 4. After four years of trying, the Soviets had failed to accomplish what the Americans had performed almost flawlessly with Skylab.

Soyuz 18B (the original Soyuz 18 was renamed 18A) launched on May 24, 1975, and was the most ambitious mission yet, with a planned stay of sixty days—almost four times that of Soyuz 12. As with previous experiments, efforts to grow plants in the zero-g environment continued to elude the cosmonauts. Most seedlings died while the survivors were smaller than normal, as was the case for a colony of fruit flies.

Because of the degradation of the environmental controls, water was not effectively removed from the space station's atmosphere, and the inside of Salyut 4 was blanketed with a film of moisture. By the fortieth day of the mission, a green mold had begun to coat the walls, and the cosmonauts requested that they be brought back early. However, Brezhnev did not allow termination of the flight until the completion of the Apollo-Soyuz mission that got under way with the launch of Soyuz 19 on July 15, 1975, followed eight hours later by the Apollo.

The political implications of this edict are not clear. The two spacecraft joined on July 17, and the five men from the two counties performed a series of publicized cooperative experiments. Only after Soyuz 19 undocked from Apollo and returned to Earth safely was Soyuz 18B allowed to close out its mission and return after sixty-three days in space.

Although the intent of the space station was to provide a long-duration facility, all these early Soviet efforts were relatively short-lived. Salyut 4 provided more testing of the station design, its systems and equipment, and the conduct of some scientific experiments for the reminder of its lifetime, but it remained unmanned for its last two years, decaying in February 1977.

Salyut 5 (Almaz 3), launched in June of 1976, was the second successful flight of the Almaz military space station (similar in configuration to Salyut 3). Manned initially by cosmonauts on board Soyuz 21 for a planned sixty days, the crew returned after fifty days when the cosmonauts began having significant disagreements among themselves, and their psychological behavior became a concern to Mission Control. The relationship between the two cosmonauts, Vitaly Zholobov and Boris Volynov, deteriorated so rapidly toward the end that the crew were forced to make a night landing on short notice.

Soyuz 22 was not intended to rendezvous with the space station because, like several others in the Soyuz sequence, there were numerous individual Soyuz missions. Consequently, there are gaps in the numerical sequence of the Soyuz craft sent to the various space stations.

Soyuz 23, launched seven weeks after Soyuz 21 returned, was to have docked with Salyut 5, but its long-distance rendezvous system failed. During its return, the spacecraft landed on Lake Tengiz, which was frozen at the time. The reentry module broke through the ice but floated until recovery forces arrived and attached a flotation collar. However, the two cosmonauts had to wait until daylight to leave the craft.

The third and last flight to visit Salyut 5 arrived four months after the return of the contentious crew of Soyuz 21. Because one of that crew had complained of strange odors, the newly arrived cosmonauts of Soyuz 24 wore breathing apparatus to guard against possible toxic fumes when they entered. However, examination of the air revealed no traces of toxins, and this tended to support the assertion that the previous crew's experience was a delusion.

Soyuz 24 was forced to land on the night of February 25, 1977, in a region where a snowstorm was in progress. Its short, eighteen-day duration was probably a result of its reconnaissance mission and the need to process the film in a timely manner.

Soyuz 25 was planned, but the mission would have been incomplete due to low orientation fuel on board Salyut 5, so it was postponed until Salyut 6. Salyut 5 was the last of the Almaz military configurations and remained aloft for over a year before it was deorbited to a fiery reentry on August 8, 1977.

In a period of six years, only three of six space stations launched by the Soviets had succeeded in receiving cosmonauts. In addition, of the eight crews sent to these stations, three failed to board. It was a telling commentary on the reliability of Soviet technology of the period. For all the failures and near tragedies suffered by the Soviets, the lessons learned had finally given more focus to their space program. It became evident that there was no military advantage to the manned space station, given the level of Soviet technology. This aspect had been determined by the Americans before they had completed the development of the U.S. Air Force MOL, and that project had been canceled before it flew. Almaz continued to fly unmanned missions into the 1980s before the program was concluded.

The Soviet space program had been compelled to include the military in their manned program to ensure continued funding. Now that it had been proven that unmanned reconnaissance satellites were more cost effective and flexible, there was no need to include the military in the manned space flight program if independent funding could be ensured. Brezhnev, like Khrushchev before him, realized the propaganda value of competing with the United States in space—although he had also been bitten by its failures. Nevertheless, unlike Khrushchev, Brezhnev, influenced by President Nixon's emphasis on détente, had moved the Soviets away from confrontational policies and toward cooperation as evidenced by the Apollo-Soyuz flight.

Salyut 6: The Next Iteration

Salyut 6 went aloft in September 1976 and was a civilian configuration similar to Salyut 4 but with an improved water recycling system. Of more significance, however, was a docking port at each end. This arrangement required the relocation of equipment and orbital maneuvering system fuel tanks to the periphery of the aft module. The obvious advantage was the ability to dock two Soyuz spacecraft—or to avoid having to abort a mission if one docking station became unusable. In practice, the second port allowed the replenishing of the Salyut by unmanned Soyuz supply ships called "Progress." These extended the life of the space station by bringing up supplemental food, water, and oxygen as well as the orbital maneuvering fuel needed to reboost periodically to a more stable orbit. The perfecting of the ability to transfer fuel was the result of previous experiments in pumping liquids in the zero-g environment.

While Salyut 6 would prove the most successful of the Salyut series, the first crew to attempt to board it, in Soyuz 25, was unable to achieve a hard dock and had to return after only two days of frustrating attempts. Because the docking mechanism was part of the expendable module jettisoned following the retrograde burn, it disintegrated in the atmosphere, and there was no sure way of determining which part of the latching (Soyuz or Salyut) was faulty. The failure of this all-rookie crew resulted in a policy change dictated by no less an authority than Brezhnev himself—all future crews had to have at least one experienced cosmonaut on board.

Crew assignments were changed for the next attempt with Soyuz 26. Georgi Grechko, a former engineer who knew the construction of the latches, performed the first Soviet spacewalk in six years to inspect and possibly repair them. However, because of another edict that each flight must have a military commander, he was subordinate to rookie Yuri Romanenko. The personalities of the two were considerably different, with the gregarious Grechko in opposition to the militant and younger disciplinarian, Romanenko. That they had only two short months to prepare for the mission did not improve that relationship.

The flight took place in December 1976, and Romanenko successfully docked to the port on the bow of Salyut. They spent the next eight days activating and checking out the various Salyut systems. Then it was time to perform the spacewalk to work on the docking port on the aft end of the space station. They helped each other suit up and then entered the docking transfer compartment and depressurized it before opening the hatch to the outside.

Grechko's inspection revealed no damage to the latches, and both men experienced the exhilaration of floating outside—tethered to the space station by a safety line and the communication link. The suits were of a standard size that allowed any cosmonaut to use them, and the environment was fully self-contained. Previous models for both the Soviets and Americans had been custom made for each cosmonaut.

Both men slept at the same time during this first Soviet long-duration flight. As the Americans had also discovered, the activity of the "on-duty" crewmember tended to disrupt the sleep patterns of the other. After a month in space, and with the assurance that there was no damage to the docking mechanism of Salyut 6, Soyuz 27 arrived in January 1977. The rendezvous and docking were accomplished automatically and without incident. This aspect was critical to the long-range plans for Salyut 6, as the unmanned Progress resupply ships had to use the aft docking port that was occupied by Soyuz 26. Three spacecraft were now joined together for the first time, and some structural dynamics testing was performed to evaluate the rigidity of the structure—the men bounced up and down in unison.

After five days in space, the crew of Soyuz 27—Vladimir Dzhanibekov and Oleg Makarov—returned to Earth in Soyuz 26, clearing the aft docking port for future unmanned Progress supply ships and providing a fresh Soyuz for Grechko and Romanenko since the Soyuz spacecraft, even in its dormant mode attached to the Salyut, had only a two-month life.

Progress spacecraft were similar to the manned Soyuz except the descent module contained the fuel and transfer pumps to replenish the Salyut orbital maneuvering system. The first of these, Progress 1, arrived four days after the visiting crew had left. Grechko and Romanenko then performed the first refueling in space—another milestone envisioned by Konstantin Tsiolkovsky, Robert Goddard, and Hermann Oberth had been achieved. As there was no heat shield, these Progress supply craft burned up when they were deorbited.

After more than two months in space (exceeding the record of Salyut 4), Romanenko and Grechko experienced the most threatening encounter with death—fire in the space station. Without warning, smoke appeared in the main compartment of the Salyut. Romanenko immediately began the evacuation process by grabbing the critical documents and heading to the Soyuz to prepare for undocking. Grechko sought the source of the fire, which was obviously electrical in nature. Because the smoke was so dense, he repeatedly had to return to the Soyuz to get a lung full of fresh air like a swimmer diving under water.

After several forays, he located the burning cable and disconnected it, undoubtedly saving the mission and perhaps their own lives. The air circulation system was turned up to its maximum setting to filter the smoke particles from the space station atmosphere while the two crew members isolated themselves in the Soyuz. As demonstrated in Apollo 1, an incident like that in the all-oxygen atmosphere of the American spacecraft would have been fatal.

While the personalities of the two cosmonauts occasionally clashed, one incident highlighted their professionalism and selflessness. Grechko had noticed that Romanenko was taking large amounts of painkiller. Alarmed, he confronted Romanenko, who admitted that he had an extremely painful toothache. Being the commander, he would not allow it to compromise the mission, which aimed to break the American endurance record of eighty-four days. However, Grechko felt they needed some medical advice, so he convinced Romanenko that they should communicate the problem to Mission Control as though it were Grechko who had the pain. Although the doctors could provide no real relief, the incident solidified the personal relationship between two different personalities.

The arrival of Soyuz 28 at Salyut was noteworthy because of the presence of Vladimir Remek—a Czechoslovakian. While Brezhnev had made positive overtures toward the West during his ten-year tenure, the Soviet hold on its captive Eastern European states was still oppressive. In an effort to mitigate this situation, Brezhnev used one of the seats on each of the replacement Soyuz flights as a propaganda incentive to these caged nations. Remek was the first of many non-Soviets to fly in space, with a total of eight cosmonauts visiting Romanenko and Grechko during their mission.

On March 16, 1977, Grechko and Romanenko returned from space after setting a new duration record of ninety-six days in orbit. To ensure that their physical condition was as good as possible before their return, they spent several weeks performing extensive exercises. However, they were still carried from the spacecraft to the waiting helicopters, and it was months before they returned to a normal physical routine. More importantly for the Russians, the Americans, for whom longer periods in space were no longer of scientific or propaganda value, would not challenge the new record. More than a decade passed before America joined with the Russians and other nations to crew a multination space station. Over the next two years, ten more Soyuz and ten Progress supply ships were launched to Salyut 6.

The next to last crew to visit Salyut 6 was delivered by an improved Soyuz, designated Soyuz-T3, in November 1980. (The use of the new Soyuz "T" spacecraft was intermixed with the standard Soyuz designations, making tracking of the last few flights to Salyut somewhat convoluted.) This version of the spacecraft featured several upgrades, including an improved emergency escape system. The changes now allowed the Soyuz to carry a crew of three with enough room for each to wear a pressure suit. Perhaps more importantly, the Soyuz-T saw a return to the use of solar cells to recharge its batteries and the ability to remain in space for up to four days (independent of the Salyut). The last crew, launched in March of 1981 on board Soyuz-T4, remained in space for seventy-four days before returning in May of that year.

Soviet cosmonauts were little different from their American counterparts when it came to humor and contriving practical jokes. On one occasion, the cosmonauts inflated an empty spacesuit and had it float through the docking hatch to join the two crewmembers during a televised session—it created quite a stir among their Mission Control comrades watching from the ground.

The maturity of Salyut 6 was evident as it hosted six long-duration crews and eighteen manned missions during a five-year period—although it was originally intended to last only two years. The individual endurance record was extended to five months, and a variety of problems, including a serious fuel leak, was successfully addressed by the crew.

Even toward the end, the role of Salyut 6 was not quite finished. The 15-ton Kosmos-1267, a large transport-support module created by Chelomey's design bureau, was lofted by a UR-500 in April 1981. The new craft was the forerunner of an expanded means of supporting the Salyut and was named Kvant (which translates to "Quantum") and included the new Merkur spacecraft, which was unmanned. The ship spent four weeks performing orbital maneuvers, and then the Merkur spacecraft was returned to Earth. The remaining portion of Kosmos-1267 executed more orbital maneuvers before performing an automated docking with the now unmanned Salyut 6 in June of 1981. The combined craft, which weighed over 35 tons, performed several more orbital maneuvers. The space station was deorbited to its destruction on July 26, 1982, using the propulsion system of Kosmos-1267.

Salyut 7: Perseverance Rewards

Salyut 7 provided the Soviet space program with the opportunity to demonstrate what had been learned in over ten years of operating space stations. Launched in April of 1982, Salyut 7 had many improvements to its systems and to the basic operational philosophy. With more electrical capacity from an improved and expanded solar cell array and a more rugged docking collar, the station was designed to support a series of crews for up to one year in duration.

While Salyut 7 represented the pinnacle of Soviet technology, the crew was again mismatched in their personalities and struggled to work together in the isolated and confined quarters. What was rather amazing was that Anatoly Berezovoy and Valentin Lebedev had exhibited significant interpersonal problems during their training before the flight, yet they were allowed to launch in Soyuz-T5.

It was during this first manned period of Salyut 7 (May 14 to August 27, 1982) that the first non-Communist-block cosmonaut arrived—the Frenchman Jean-Loup Chrétien. Brezhnev's desire to influence democratic governments outside the Soviet sphere caused him to make overtures to both India and France—two countries whose strong socialist tendencies were often in sympathy with the Soviets. However, as with the Apollo-Soyuz program, the feedback from the visiting astronauts to their own countries was more negative, as they observed firsthand the lack of freedom and deteriorating economic infrastructure of the Soviet giant.

The arrival of the Soyuz-T6 with French cosmonaut Chrétien was another close brush with disaster. The onboard computer went dead during the final automated docking maneuver—leaving the Soyuz-T6 tumbling as it headed for the Salyut at a closing speed of 15 mph. Soviet cosmonaut Vladimir Dzhanibekov was able to use the small manual thrusters to steer the spacecraft so that it missed the space station by about 30 ft. Eventually he used the manual docking technique to join with Salyut 7.

Following the return of Chrétien and Dzhanibekov to Earth, the next visiting crew in August 1982 on board Soyuz-T7 included the second Soviet female cosmonaut, Svetlana Savitskaya. As the Americans were preparing to launch Sally Ride in the space shuttle, the politburo once again decided it was time to "one-up" the competition. It had been almost twenty years since the first female cosmonaut, Valentina Tereshkova, had been launched as a publicity stunt in 1963. Despite the rhetoric, women in the Soviet Union were woefully behind the West in achieving equal rights with men.

The second female astronaut to fly in space was Svetlana Savitskaya (*left*), who performed a spacewalk (*right*). Courtesy of Energiya.

When the visiting cosmonauts departed, Berezovoy and Valentin Lebedev resumed their bickering. It seemed that long-duration isolation needed the constant invigorating visits of their companions.

However, the Soviets were about to undergo another unplanned change in political leadership with Leonid Brezhnev's death in November 1982. Yuri Andropov took control of a country that was in its own death throes. Twenty-five years of trying to compete with the United States in the high-cost world of military armament and space technology had virtually bankrupted the country. Although Andropov continued Brezhnev's domestic and foreign policies and the space program, the days of the Soviet Union were now numbered.

It was at about this time that cosmonaut Berezovoy began to have severe muscle spasms. However, with only a week to go before setting a new endurance record, neither cosmonaut was eager to return early, after all they had endured. Lebedev was instructed by the mission controllers to give Berezovoy an injection of atropine, which he did, and the pain was eased. The two returned to Earth on December 10, 1982, amid an intense snowstorm that required them to wait some ten hours before they could be airlifted to surroundings that are more comfortable. It has been reported that the two cosmonauts never again spoke to each other.

The next mission to Salyut 7 by Soyuz-T8 began poorly when the dock-ing radar antenna was ripped off by the payload shroud separation. A man-ual docking attempt resulted in a near collision, and the crew of Vladimir Titov, Gennady Strekalov, and Alexander Serebrov were forced to return to Earth when their maneuvering fuel ran low.

Soyuz-T9, with Aleksandr Pavlovich Aleksandrov and Vladimir Lyak-hov, was able to dock successfully in June of 1983. Along with Kosmos-1443, joined the previous March, the space station once again comprised three modules. However, because the station had been boosted into a higher orbit by the Kosmos, only two astronauts could occupy the Soyuz-T9 to reduce its weight for the boost phase into orbit. After it was unloaded, Kosmos-1443 was undocked and deorbited to destruction. Its unmanned Merkur was returned safely to Earth with film and other experimental re-sults, and the two cosmonauts moved their Soyuz from the aft to the bow port so the next Progress could dock.

Then a fuel leak in the Salyut 7 maneuver-thruster supply in September left it drifting helplessly. Although it could be periodically reoriented by the Soyuz or Progress thrusters, moving the great mass of Salyut 7 required excessive fuel from their reserves. With the inability to orient the station so the solar cells could provide the appropriate level of electrical power, the internal temperature soon dropped to 55 degrees, and the humidity rose to virtually 100 percent. This resulted in moisture condensing on the internal walls of the space station. Salyut 7 was in a critical situation. There was no hope of repairing the fuel leak at that time. The space station needed a trained crew to perform a spacewalk to attach the new solar panel previ-ously brought up to restore the higher levels of electrical power.

Because Vladimir Titov and Gennady Strekalov had been trained to at-tach the solar array, they got another chance to fly sooner than either had expected. The two quickly prepared for a launch scheduled for September 27, 1983. However, at T-90 seconds a fire developed in the booster, which quickly burned through the automated abort wiring, rendering it useless. The two cosmonauts tried to activate the manual abort system, but it, too, had been disabled by the fire. They lay in the spacecraft helpless as they could see the orange glow of the flames grow more intense. Now only quick action by the controllers—who had to simultaneously press two buttons located on opposite sides of the control room (to ensure that no one person could abort a mission)—could save the cosmonauts. It was perhaps the longest ten seconds in their lives, but the solid-fuel escape rockets finally

The Salyut 7 space station showed maturing Soviet technology. Courtesy of Energiya.

ignited and pulled the Soyuz spacecraft more than 3,000 ft. into the air just as the booster exploded beneath them.

Now the Soviets faced a difficult challenge—and a tough set of decisions. There was no way that another booster could be readied in time to rotate the Soyuz-T9 that was docked to the space station. If Aleksandrov and Lyakhov remained aloft for the full duration, it meant that a Soyuz spacecraft—with an operational a life of only two months—would be called upon to bring the men back after it had been in space for five months. However, to allow them to stay the full duration, they also had to install the new solar panel extensions without any training. It was decided to make the critical decisions one at a time and to let them perform the spacewalk and see if the panels could be set up.

While they were preparing to suit up, they discovered a tear in the outer layer of one of the suits and had to improvise a repair using duct tape. Finally, they were out in space, and the lure of the fantastic scene was mesmerizing. The two spent more than three hours performing their spacewalk and successfully completed the solar panel installation. Aleksandrov and Lyakhov were able to complete their record-breaking stay in space and returned to Earth three weeks later.

The next crew, launched in February 1984, consisted of three cosmonauts—Leonid Kizim, Vladimir Soloviev, and Oleg Atkov (a cardiologist). With the experience they had thus far gained, the Soviets decided that it might be helpful if the crew was expanded to three—using the new slot to provide more real-time evaluation of the health of the crew with the presence of a medical doctor. It was also determined that perhaps the leaking oxidizer tank could be fixed by a series of space walks.

After they had been in space for more than two months, Kizim and Soloviev began a series of extravehicular activities to the aft end of Salyut, where the leaking tank was located but where there were few handholds. A variety of specially designed tools brought up on the Progress supply ships allowed the two cosmonauts to cut holes in the side of the service module and to disconnect a series of fittings and pipes. It took several space walks over periods of hours to prepare the work area. Included was a ladder-like structure and a TV camera to allow the ground to view the work environment.

Nevertheless, the leak was elusive. Every time they managed to isolate and bypass one leak, another appeared. Finally, they felt they had all but the last leak fixed, and this one required yet another special tool that had to be fabricated on Earth, tested in the neutral-buoyancy water tank, and flown up to Salyut on Soyuz-T12.

For the final fix, the special tool was accompanied by female astronaut Svetlana Savitskaya for her second space flight. The decision to send her was obvious. The Americans were planning for astronaut Kathryn Sullivan to perform a spacewalk as a part of an upcoming shuttle mission. The Soviets saw this as another opportunity to scoop the Americans. Savitskaya launched with Dzhanibekov, who had done the testing of the tool on Earth and instructed Kizim and Soloviev on it use. Savitskaya and Dzhanibekov performed a four-hour spacewalk to do some other maintenance and, after ensuring that Kizim and Soloviev were thoroughly trained on the use of the new tool, returned to Earth.

Kizim and Soloviev made their last spacewalk a week later and sealed the final leak. The space station had been returned to full functionality, and the Soviets had performed an impressive accomplishment. Many years passed before the magnitude of that space repair was fully appreciated, due to the lack of information released by the Soviets at the time. Their secrecy tended to work against them almost as much as it aided them, but significant changes in the openness of the Soviet space program were soon to come.

Kizim, Soloviev, and Atkov returned from space in October 1983 after a record 237 days in orbit. Again, bone loss ranged from 3 percent to 15 percent in density in various bones in the lower body but had remained mostly constant in the upper body. All three were back to normal after a month of recuperation and exercise.

The three cosmonauts had left Salyut 7 in good condition, and it was expected that the next crew would extend the space record to ten months and then a yearlong mission. Before the next crew could be launched, communications with the telemetry on board Salyut 7 suddenly ceased on February 12, 1985. All efforts to regain contact failed, and the Soviets had one large but dead space station in orbit.

A rescue mission was immediately planned, and Soyuz-T13 launched on June 6, 1985, with cosmonauts Vladimir Dzhanibekov and Viktor Savinykh. The rendezvous was flawless, and, because the automated docking system had been removed to save weight and space for repair equipment, Dzhanibekov had to perform as he had in previous missions—with skill and professionalism.

Following the docking, the crew ensured that they had a good seal in the docking adapter and began carefully to pressurize it. The pressure held, and the cosmonauts entered the dark and cold space station (ice coated the inside walls) wearing breathing apparatus to guard against any possible toxic gasses that might be present. However, test showed none, and they soon removed the masks and started to troubleshoot the electrical problem, which was quickly found. A sensor for determining when the batteries needed to be recharged had failed and had simply kept the solar cells from doing their job, and the batteries quickly were depleted. The crew replaced the faulty sensor and the batteries and started to bring the electrical systems back online. The temperature was 14 °F, and it took several days to warm the station back to a livable environment and to remove the water that had accumulated.

After more than three months of work, the space station was again ready to support another long-duration crew, and Soyuz-T14 was launched on September 18, 1985, with Grechko, Alexander Volkov, and Vladimir

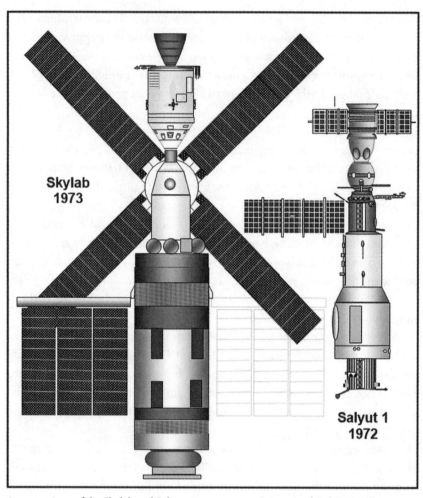

A comparison of the Skylab and Salyut space stations. Courtesy of Ted Spitzmiller.

Vasyutin. After a brief, eight-day visit, Dzhanibekov returned in the Soyuz-T13 spacecraft with Grechko, while Savinykh and Volkov remained on board with Vasyutin.

A large transport-support module, Kosmos-1686, brought 3 tons of fuel for the ACS and more than 5 tons of cargo, and the trio began what was hoped to be a new space endurance record. However, two months later Vasyutin developed a high temperature (104 degrees) and had trouble urinating—the pain was often intense. Despite all attempts to resolve the problem on board Salyut 7, the crew was forced to return on November 21, 1985, due to Vasyutin's inflamed prostate.

Because the cosmonauts had reported that Salyut 7 was still suffering from the effects of its long, cold, deep sleep, the remaining agenda for it was revisited. Plans had called for a large Kvant addition and an all-female crew. However, there was little to be gained at this point since an entirely new space station, Mir, was in the advanced stages of construction. One more visit to Salyut 7 retrieved the remaining experiments and recoverable equipment.

After almost nine years in orbit, Salyut 7, with Kosmos-1686 still attached, had depleted its attitude and maneuvering fuel. Although ground controllers attempted to deorbit to a reentry on February 7, 1991, over the Atlantic Ocean, the timing was slightly off. Following an impressive night-time reentry that lit up the sky, many pieces of the 35-ton space station survived the fiery plunge and fell on Argentina, although there was no reported damage or injuries.

Soviet First Secretary Andropov had passed away after his brief tenure, and his interim replacement, Konstantin Chernenko, had likewise lasted only thirteen months before he died. A younger leader had emerged— Mikhail Gorbachev, who recognized the death throes of the "Evil Empire" (as it had been labeled by the American president, Ronald Reagan). Gorbachev was aware of the critical economic and political situation into which the Soviet Union had fallen. While he wanted to keep the space program as a vital national asset, he understood that the country could not afford the many initiatives undertaken. Unlike his predecessors, he was not enthralled with manned space exploration and personally felt it was a waste of precious resources. There were big changes in store for the Soviet Union and its space program (Zimmerman 2003, 234).

21

DESIGNING A REUSABLE SPACECRAFT

Winged Spacecraft: Early Studies

Direct applications in space for military operations were only conceptual in the early days of the movement into this new frontier, as it was not clearly understood what function the human presence might serve in that hostile environment. The early post-Sputnik thinking pointed toward direct applications of weapons residing in space, and this was one of the leading arguments for allowing the military to control all operations in space.

However, President Eisenhower had rejected that notion and placed NASA in charge of the national space program. Nevertheless, the military was far from being left out. Several military programs of the late 1950s were more perceptive in determining human performance in space, compared to the limited role NASA was deliberating.

The concept of the Sänger-Bredt *Silverbird* of the mid-1930s continued to captivate visionaries in both the United States and the Soviet Union. In the early 1950s Dr. Walter Dornberger and Krafft Ehricke, Peenemünde alumni who had emigrated to the United States after World War II and who were both working for Bell Aircraft, conceived the piloted "bomber-missile" known as BoMi. Launched vertically like a rocket, the upper stage was a winged vehicle to carry nuclear weapons.

A second version configured a manned winged bomber to be carried aloft on a larger delta-wing mother ship that could be reused. These were quite large vehicles with gross weights well over one-half-million pounds. However, the many problems that had yet to be resolved with large liquid-fuel rocket engines, guidance, and the temperatures at reentry caused the Air Force to show little interest. In a final attempt to continue their concept to fruition, Dornberger and Ehricke reworked the bomber-missile into a hypersonic transport in 1957—but found no takers for it.

The Air Force did provide several small contracts during the mid-1950s to study the upper-atmosphere skip-glide method. An outgrowth of this was the rocket bomber. These studies essentially showed that the skip-glide (or boost-glide, as it became known) was not as practical as it first appeared and that it was more effective to use a slightly higher speed and enter orbit—or a fractional orbit trajectory.

Dyna-Soar: A Technology Stretch

Within a week of the launch of *Sputnik I*, the United States Air Force began assembling all of the previous studies on the military aspects of space into a new program. As might be expected, the Air Force perceived operations in space as being simply an adjunct to its winged aircraft. The basic flight scenario was similar to the Sänger antipodal bomber envisioned in the 1930s. The designation given to the project was WS-464L, and its name was Dyna-Soar—a contraction of the words "dynamic soaring," a term that Eugen Sänger himself had coined in the 1930s.

When planning for a weapons system reached the stage where a firm concept could be designed, a study contract was awarded to the Boeing Aircraft Company in March 1958 for a winged vehicle to carry a crew of one. However, as the project progressed, it became obvious that the boost-glide concept had several flaws. Dyna-Soar was modified for low Earth orbit (LEO) operations, and a second crew member was added. Like the ballistic spacecraft being considered for manned spaceflight, the vehicle was launched vertically on a large rocket. However, Dyna-Soar had short delta wings for landing like a conventional aircraft and was to be reused after each flight. A contract for construction was awarded in 1961, and the vehicle itself was assigned the experimental designation of X-20 in June 1962.

The main difference between the X-20 and previous boost-glide concepts was that this craft achieved orbit, and its role was not as a bomber. Instead, it was more like a fighter capable of intercepting possible hostile satellites for inspection and destruction. It could also be used for reconnaissance in a manner similar to the U-2 spy plane. It was to have a payload bay and could place small satellites in orbit.

The 35 ft. long craft had a wingspan of just 22 ft. and weighed about 11,000 lb. The structure used both titanium and molybdenum and had an underside surface protection of an ablative coating reapplied after each flight. As the space plane continued to grow in size and the scope of its mission

The X-20 Dyna-Soar was canceled before the first flight, but it contributed to the technology of winged reentry spacecraft. Courtesy of Department of Defense.

expanded, so did its need for a larger booster. The Titan II, originally slated as the launch vehicle, required the addition of large, solid rocket boosters attached to either side, resulting in the Titan III (Evens 2009, 310).

The project was well along when a critical design review of high-cost military expenditures by President Kennedy's secretary of defense, Robert McNamara, determined that the role it was to perform had not been adequately defined. The Corona spy satellites were producing excellent results, and the ability to intercept and destroy hostile satellites could be handled by unmanned interceptors for less money. On December 10, 1963 (three weeks after President Kennedy's assassination), McNamara ordered the project canceled—with all the schedule slippage that had occurred, it was still three years away from its first flight. The essential concepts of Dyna-Soar were resurrected years later as a larger vehicle—the space shuttle. The Titan III booster continued its funding and became the backbone of heavy lift for both the military and some scientific payloads for the next four decades.

The Search for an Economic Solution

With the Apollo Program's objective of landing a man on the Moon before the end of 1969 a reality, NASA turned its attention to what lay beyond. There had been some preliminary planning, but until the fulfillment of the

Kennedy goal, it had been held in the background. The time had come to look aggressively to the future.

For NASA's "man in space" program, a large space station, a colony on the Moon, and an expedition to Mars were all considered primary objectives. The cost estimates for these adventures were staggering. Whatever the next goal, it would require the ability to lift heavy payloads to LEO. Using the Saturn V, the cost was about $1,000 per pound (about $7,000 per pound in 2015 dollars). To lower this cost, it appeared obvious that some way had to be found to return the various stages of a rocket safely back to Earth so they could be reused rather than being "expended." Wernher Von Braun had foreseen this and had envisioned that his giant "Ferry Rocket" (featured in a 1953 *Collier's* magazine article) had some means of recovering all three stages (Heppenheimer 1984, 246).

As early as 1957, the U.S. Department of Defense had issued study requirements to several contractors to investigate what was required to convert existing rockets, such as the Atlas ICBM and later the Saturn I, with a return-to-launch-site capability. The basic idea was to add wings and turbojets to the vehicle and have it land horizontally like an airplane.

It was also desirable that, at the very least, the manned spacecraft portion should be able to "fly back" from space. Several previous programs had provided considerable research regarding winged reentry, including the X-15, the X-20, and a novel concept called the "lifting body."

Estimates showed that at least two launches per month would be required to make the effort worthwhile. However, the downside of the concept was that the booster might lose perhaps as much as 25 percent (or more) of its payload to the weight of the structures for the recovery effort.

Despite the excitement of the lunar landing, there was still vocal opposition to continuing to send humans into space. The Soviets had demonstrated considerable capability with robotics. With the advent of integrated circuits, the ability to make computers significantly smaller provided extensive onboard programming of events—and even some autonomous decision making. The manned verses unmanned debate continued for decades to come, but for the 1970s, the allure of humans exploring the unknown continued to prevail.

Lifting Bodies

During the 1950s, when pioneering research was being performed on reentry shapes for ballistic missile warheads for ICBMs, it was realized that by contouring the shape of the nose cone (longitudinally forming a half cone), noticeable lift could be generated. This lift was in the order of 1.5:1 (for each foot lost in altitude, the lift moved the object forward 1.5 ft.). These special shapes had to be carefully oriented during the reentry to supply the lifting capability.

This amount of lift is not much compared to a typical airplane with a glide ratio of 8:1. However, it allows the vehicle to achieve a landing footprint of approximately 230 mi. cross range and 700 mi. downrange from an orbital reentry, compared to a pure ballistic trajectory. These half cones had blunt-shaped noses to take advantage of the phenomena of moving the shock wave away from the structure and thereby keeping much of the heat generated during reentry from it. They also had abbreviated aerodynamic fins to provide stability once inside the atmosphere.

By 1958 the "lifting body," as it was called, was considered for the first American manned spacecraft—Mercury. However, more study and development was needed to achieve a stable and controllable vehicle. Because time was of the essence, Mercury became a pure ballistic reentry capsule. Nevertheless, work continued on the concept, and the first test flight of a lifting body was on board Thor rockets as part of an Air Force program called ASSET (Aerothermodynamic/Elastic Structural Systems Environmental Tests). A subsequent series of research activities led to PRIME (Precision Recovery Including Maneuvering Reentry), which concluded with the successful recovery of a small lifting body after it had satisfied all of the objectives of its acronym (Heppenheimer 1984, 38).

However, designing a shape that could withstand reentry while maneuvering to a precise recovery area still did not provide the ability to be flown subsonic to a normal "airplane-like" landing. To address this, NASA's Flight Research Center funded several small radio-controlled models to develop specific shapes and controls to allow the spacecraft to perform a relatively normal landing. Once these had shown clear promise, a manned glider that was 20 ft. long and 14 ft. wide was built of lightweight wood material. Designated M2-F1, the 1,138 lb. craft was towed until airborne behind a special high-performance Pontiac automobile at Edwards AFB in April 1963.

These captive flights were followed by free flight using the venerable Douglas DC-3 to provide an air tow. On these excursions, the M2-F1 was taken to 10,000 ft. and then released to glide to the runway over a period of about four minutes, landing at a speed of 85 mph. To maintain the desired glide speed, the descent angle was on the order of 18 degrees as opposed to the traditional airplane that uses a flatter 3.5 degrees. Thus, the final landing

The M2-FI Lifting Body proof-of-concept glider with its test pilot, Milt Thompson, in 1963. Courtesy of Department of Defense.

flare—the transitioning from the descent to the runway touchdown—was a critical maneuver.

The next step was to explore heavier structures and to install a rocket engine to allow flight speeds up to the transonic region. The all-aluminum M2-F2, built by Northrop Aircraft, used the reliable XLR-11 rocket engine (as first used in the X-1). The craft was 22 ft. long and 10 ft. wide and was carried aloft under the wing of NASA's B-52 that was also used for the X-15.

The first gliding free flights were made in July 1966 from 45,000 ft. and at speeds of 450 mph with test pilot Milt Thompson at the controls. The first powered flight was made almost a year later in May 1967. During this flight, the little craft performed several planned maneuvers but then developed a wild rolling motion. Test pilot Bruce Peterson managed to regain control. However, during the final approach to landing, he became distracted by the position of one of the recovery helicopters, and the timing of the critical flare maneuver was off. The M2-F2 hit hard and bounced back into the air. Peterson lost directional control and the craft proceeded to tumble at over 200 mph down the dry lakebed that was used as a runway. Peterson was severely injured, but, through a series of surgical procedures, he returned to limited flight status. The reconstruction of his body became the central theme of a science-fiction book *Cyborg*, by author Martin Caiden, and a TV series called *The Six Million Dollar Man*. Each week the viewer saw the

The family of lifting-body aircraft (*left to right*) M2-F2, HL-10, X-24A, and X-24B. Courtesy of Department of Defense.

film of the actual crash in the opening credits. The M2-F2 was rebuilt with more effective airflow control as the M2-F3, and by December 1972, it was flying at Mach 1.6 and to altitudes of 71,000 ft.

A similar lifting body designated the HL-10 had been developed in parallel and, following extensive wind-tunnel evaluation, was first flown as a glider in December 1966. Test pilot Bruce Peterson quickly discovered that the aerodynamic qualities were poor, and the craft underwent fifteen months of modifications. The revised HL-10 went supersonic in May 1969 and ultimately flew to Mach 1.86 and 90,000 ft. Further experiments were aimed at determining the feasibility of a small rocket system to flatten the steep 18-degree approach path to 6 degrees. However, it was determined that the added complexity of the system did not justify its possible function.

A third lifting body constructed during this period was the X-24A, which initially looked somewhat similar to the M2-F2. This Air Force–funded program had as its goal a hypersonic cruise aircraft joined to the scramjet (supersonic combustion ramjet). Testing of the scramjet itself was planned for later flights on the X-15. Built by Martin Marietta Corporation, the X-24A was delivered in August 1967 and was again powered by the XLR-11. However, the first glide tests were not made until April 1969, following an extensive series of wind-tunnel and ground tests. The first powered flight was in March 1970, and maximum speeds of up to Mach 1.6 and altitudes of 71,000 ft. were obtained.

The X-24A was then rebuilt as the X-24B, with a more streamlined nose and a 78-degree double-delta-wing plane form. Following glide tests, the first powered flight occurred in November 1973, and tests concluded in 1975 after the X-24B had flown to Mach 1.76 and altitudes of 74,000 ft.

The primary purpose of these lifting-body experiments was to determine the aerodynamic qualities and controllability of the craft as it decelerated through supersonic to subsonic flight. They were not intended to compete with the high-speed hypersonic X-15. Although a follow-on X-24C had been proposed that would have used either the X-15's XLR-99 engine or the yet-to-be-perfected scramjet, neither variant was built. Hypersonic cruise flight reached its apex with the X-15, and, although revisited over the next several decades, no viable power plant or structure was deemed acceptable for the mission.

Birth of a Dream

As work on the Saturn V moved to completion in 1966, the Marshall Space Flight Center was able to spend more of its resources looking to the follow-on. Working with several aerospace contractors, NASA scientists and engineers studied a wide variety of reusable launch-vehicle configurations. However, as in the past, these were all paper proposals since the allocation of money for large projects had to wait for the successful conclusion of the Apollo Project. NASA could not attract congressional funding because of its commitment to the Moon project, and the Air Force continued to lack justification for sending men into space.

By the late 1960s, the objective of a reusable rocket had been seriously studied for almost a decade. The debate began in earnest as to the desirability of a fully or partially reusable rocket. Another consideration, from the view of which cost factor would be more easily justified and absorbed, was whether it should be a single-stage-to-orbit or a two-stage-to-orbit configuration. Initial development costs are high for the fully reusable vehicle, but these are offset if a use factor of two flights per week were contemplated. The partially reusable configuration had the advantage of lower development costs but a higher incremental cost of each flight.

A variety of proposals were considered. A vocal faction believed that any level of reusability did not bring noticeable savings, but might, in fact, result in higher costs. Until the mission of the vehicle was defined, there was no realistic way to determine how many flights might take place during a calendar year or during the projected ten- to fifteen-year life of the vehicle.

However, there was no doubt that a low-cost and reliable launcher was the key to repetitive applications such as a manned lunar base or a large space station, built and maintained by the rocket. Nevertheless, without a specific goal, as with Kennedy's Moon landing, there remained a lack of congressional motivation. In addition, none of the three proposed lunar follow-on projects, including a manned flight to Mars, had the public's attention—let alone that of Congress. The Vietnam War and the racial unrest in the cities had virtually paralyzed America's thoughts on the future. Many preferred simply to wait for both of these problems to abate before venturing farther into space.

However, America's space technology infrastructure, built over the ten years that followed Sputnik, was already in jeopardy. Tens of thousands of skilled engineers and technicians that had filled the pipeline during the heyday of the early 1960s were now being given "pink slips" as there were no large follow-on projects to use their talents. It was not realistic to put the space program on hold for however many years it might take America to resolve its foreign and domestic problems.

In August 1968, then NASA administrator George Mueller made a presentation to the British Interplanetary Society in which he outlined the future of manned space travel and set forth some basic tenets, saying, "there is a real requirement for an efficient Earth-to-orbit transportation system—an economical space shuttle" (Heppenheimer 1984, 93). Although the word "shuttle" had been used in reference to a reusable rocket, this was its first formal introduction, and the phrase caught on. From this point forward, reference to a space transportation system was associated with "space shuttle."

In October 1968 Marshall Space Flight Center went out to industry with a "request for proposal" for another study on the shuttle. However, the basic specifications as to the size of the payload bay and the weight to be carried had yet to be solidified. The Air Force was pressured into agreeing to use the shuttle in order to provide additional justification for its being built. The Air Force wanted a 60 ft. long by 22 ft. wide payload bay that allowed up to 50,000 lb. to be delivered to LEO. NASA was content with a 10 ft. by 30 ft. payload bay and 25,000 lb. However, that was not all—the Air Force needed a landing footprint to permit a launch into polar orbit from Vandenberg AFB in California, with the capability of returning to Vandenberg (or Edwards AFB) after the first orbit. Since the Earth's rotation shifts a LEO track almost 1,300 mi. to the west at that latitude, the military needed a large wing to provide that potential. NASA needed only a modest 300 mi. cross-range capability.

Until specific aspects of its intended support mission were established, even NASA was unsure of the size of the payload bay or the shuttle's cross-range capability. Virtually all of the proposals evaluated were based on the lifting body, although a few used deployable wings after reentry.

November 1968 saw a change in the political administration of the United States with the election of Richard Nixon to the presidency. Nixon's focus was extricating the United States from the quagmire of Vietnam, and he was not a strong supporter of the space program. He appointed Charles H. Townes to chair the Task Force on Space, to help decide the direction taken following the yet-to-be-concluded Apollo Program. Their recommendation came quickly in January 1969, and it was simply to not move forward with the shuttle at that point in time.

February 1969 saw more study contracts awarded and the formation of a new Space Task Group (STG) chaired by Vice President Spiro Agnew. The STG asked both NASA and the Department of Defense to define their requirements and then a joint group to reconcile the needs of each with a specific program.

NASA decided that the shuttle would not only support a projected space station but would also be the prime means of launching all future satellites. This caused their payload requirement to increase to the same 50,000 lb. that the Air Force required, and the 22 ft. diameter needed to match the Saturn S-IVB stage to which future large satellites were being designed (it was expected that the efficient S-IVB would probably be the upper stage of any follow-on rocket). NASA also envisioned a highly automated computerized checkout to eliminate the majority of people who were required to prepare a large rocket for flight.

NASA had initially focused on a fully reusable, two-stage shuttle by August 1969. However, the space shuttle task group, formed to provide input to the STG recognized that there were as many as six possible roles for the shuttle and defined several potential configurations that might be pursued in designing it.

The basic issue of "fully reusable" versus "partially reusable" again arose, as well as whether it would use a piloted fly-back booster or an expendable first stage and if it would use existing or new engines. Further, would sequential staging or parallel burn provide the best arrangement?

The cross-range capability (300 mi. or 1,500 mi.) dictated whether a

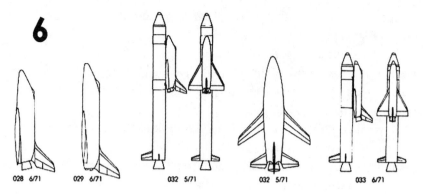

There were many possible configurations that were investigated in the early part of the shuttle transportation system program. Courtesy of NASA.

winged body or lifting body should be employed—the larger footprint was the most costly to develop. Finally, the size of the shuttle was determined by the payload capability of up to 50,000 lb.

The number of anticipated missions fluctuated between a low of 30 per year to as many as 140 per year, depending on the future goals of the American space program. Moreover, since the programs that the shuttle was to support had not been established, there was no definitive means of costing the various options. But it seemed sure that, no matter what configuration or capability was finally determined, the shuttle should drop the cost of going into space by at least a factor of five to fifteen—if the cost estimates were based on realistic assumptions.

By September 1969, with the first Moon landing accomplished, the STG submitted its report. It also presented three new national goals to the president. The first was an $8 billion per year expenditure for a manned Mars expedition, a fifty-man space station, a lunar-orbiting space station, and the reusable shuttle (NASAs budget for 1969 was $4.2 billion, which at that time was 2.31 percent of federal spending). It did not recommend any specific shuttle configuration except to state that this chemically fueled space transportation system should provide for virtually all of America's needs in space for the next twenty years.

The second program was somewhat less costly in that it deleted the lunar-orbiting space station, while the third option (at $5 billion per year) consisted only of the Earth-orbiting space station and the shuttle. President Nixon did not like any of these options. He was, in fact, at odds with his vice president, who strongly advocated for the Mars mission. Planning for the

Mars mission had progressed to the point of determining that the earliest launch window was November 1981—twelve years into the future.

NASA recognized that the Nixon administration was a hard sell. It was also obvious that virtually any future goals were dependent on the shuttle. Therefore, NASA set its sights on that objective. The new administrator (appointed in 1971), James Fletcher, also saw that, with the current state of the federal budget, which was pulled between the Vietnam War and a greatly expanded social program, every effort had to be made to reduce the cost of the Space Transportation System (STS, as it was now officially designated). It was also critical to get the Air Force to become a full partner in the shuttle (Burrows 1988, 293).

The Office of Management and Budget (OMB) oversaw the financial and economic analysis of the shuttle proposal. However, the more the OMB looked into the whole concept of the shuttle, the less economical it appeared to be. Traditionally, American space projects had cost several times their initial projections, especially when moving the state of the art forward. Thus OMB was critical of the numbers that were being presented by NASA and the various proposals submitted by industry.

Even when NASA encouraged other countries to become a part of the STS, there were as many negative aspects as there were positive—especially with the military security requirements. However, an important milestone occurred in 1971, when the Air Force agreed not to develop any new expendable boosters. They would rely on the shuttle for all their future requirements, although they continued to purchase existing expendables in parallel with the shuttle for the first few years of its operation.

A Major Change

By 1971 it was obvious that there were significant technological problems with the shuttle as a two-stage-to-orbit vehicle with a fly-back booster. The Boeing 747 was then the largest aircraft built, and it weighed 300,000 lb. empty and flew at 600 mph. The proposed shuttle first stage weighed 400,000 lb. empty and would "fly" at more than ten times that speed. It was to carry the manned spacecraft (referred to as the orbiter) and perform separation between the two in a region of speed and space for which there was little data but that extrapolated from the X-15 flights. In addition, although considerable progress had been made with heat protection, the specific method to cover either the first or the second stage had yet to be proven. With so many unknowns, it was easy to see why the OMB had little confidence in the costing and schedules for the project. The OMB had essentially given the shuttle project a goal of $1 billion per year. This was less than a third of the initial proposals. Both technologically and financially, the project as defined was beyond what was possible with the state of the art.

New thoughts about how to downsize both the orbiter and its first-stage carrier resulted in the revisiting of expendable tanks. After all, it was the engines and their turbo pumps that were the costly aspect, not the relatively inexpensive tankage. This line of thinking also simplified the problem of possible residual propellants in the tanks. If the tanks were removed from the returning craft, the problems of venting, weight, and thermal protection were significantly eased. At first the liquid hydrogen was a candidate for an external expendable tank because it composed about 75 percent of the propellant volume. However, since it only accounted for 18 percent of the weight, the liquid oxygen was soon included as well. In addition, with the reduced weight, the staging could occur at a lower speed—simplifying the speed problem. It was determined that the optimal staging velocity was about 4,500 mph.

NASA formulated a series of external-tank designs that explored a variety of configurations for the orbiter. This culminated with the most promising (designated MSC-40C) being assigned to several contractors for revising their proposals. The booster itself was the province of the contractor, and it was not decreed that it be reusable. While these efforts reduced the annual expenditures by about one-third, they were still far above what OMB felt the budget could handle.

The thought of using a phased development that delayed some systems to reduce the initial funding impulse was considered. The orbiter could be flown as the second stage of the Saturn V whose development costs had already been absorbed in the Apollo Program. The reusable fly-back booster could then be developed several years down the road, perhaps after the Vietnam War and its expenses ended, easing the funding burden. It was at this time that the possibility of using a simplified liquid-fuel engine, or perhaps even solid-fuel "strap-on" boosters (as used on the Titan III-C), was considered, with recovery made by parachute.

The idea of using solid-fuel boosters with a manned spacecraft had not been actively pursued for several reasons. Solid-fuels did not have the high energy levels (specific impulse—I_{sp}) typically needed for efficient booster performance. Once they begin their burn, they cannot be shut down—limiting the possible escape-system scenarios.

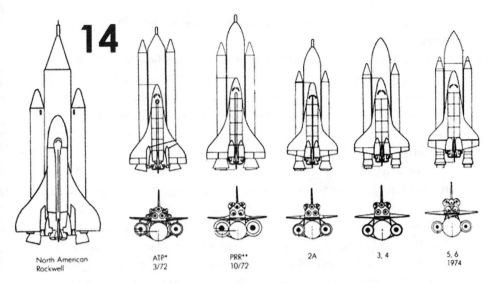

North American
Rockwell

ATP*
3/72

PRR**
10/72

2A

3, 4

5, 6
1974

The final shuttle configuration was arrived at only after many iterations had been evaluated. Courtesy of National Air & Space Museum.

Nevertheless, solids had made significant strides by the early 1970s with the introduction of nitronium perchlorate fuels, and I_{sp} numbers approaching three hundred seconds were attainable. Exceptionally large sizes could be achieved by fabricating solid-fuel rockets in segments that could more easily transported and then "stacked" at the launch site. Motor sizes were measured not only in the thrust they produced but also by the diameter and number of segments that comprised the motor. During the late 1960s significant progress had been made, and motors as large as 260 in. in diameter and five segments in length had been tested. Thrust levels as high as 6 million lb. had been achieved as well (Heppenheimer 1984, 46). The ability to provide thrust vectoring to control direction had also made noticeable improvements. In addition, progress had been made with refurbishing motors so that they could be reused after parachute recovery. The simplicity and relatively low cost made the solid-fuel booster an attractive possibility.

The funding issue was so intense that consideration was given to abandoning the shuttle and using the existing Apollo and Gemini technology. A new study was instituted to review the current progress and make recommendations. Chaired by Alexander Flax, the committee concluded that there was no economic justification for the shuttle—the same conclusion that OMB had come to. Estimated shuttle costs, as defined by NASA, had been almost halved since the first estimates, and yet the capability remained the same. This point was resurrected decades later when shuttle costs were again revisited.

If the Soviets had succeeded with their N1 super booster and a manned Moon landing (even a year or so after Apollo 11), there was no question that the United States would have moved forward then with the shuttle. However, it seemed obvious that the United States had distanced itself technologically from its archrival, and the question was, is the shuttle really needed?

Shuttle supporters held that if the launch of *Sputnik I* in 1957 had taught the United States anything, it should be that the country must continue to be in the forefront of science and technology—cost should not be the overriding issue. In a 1971 meeting between NASA Administrator Fletcher and Deputy Secretary of Defense David Packard, Fletcher was told that NASA was trying to sell the shuttle based on cost, and that was not the correct path. Packard emphasized that the real point of proceeding with the Shuttled had to do with national security and an intangible thing that might be called man's presence in space.

It was the national security aspect that convinced President Nixon to provide more support than he had. Perhaps he was recalling that he had been the vice president under Eisenhower during the Sputnik era and that he had been called on to defend that administration's austere policies during his failed bid for election against Kennedy in 1960. The possible employment factors in key electoral states might also have been an influence on Nixon's opinion of the shuttle as the 1972 elections moved closer.

The Final Configuration

The move toward solid-fuel boosters was strengthened by the recovery aspect. The liquid-fuel systems were more fragile and susceptible to damage, especially from salt water—the prime recovery area. However, Wernher von Braun in particular was not enamored with solids. The use of the Saturn V as a recoverable booster continued to dominate the thinking at Marshall Space Flight Center until it was established that the effort to convert it to a reusable system was not just costly but unrealistic.

The parallel burn of the booster and the recoverable spacecraft had the advantage of being able to ensure the ignition and stable burning of the second stage before launch commit. However, the configuration raised the risk of crew escape from a catastrophic failure.

With respect to the large delta-wing configuration needed to support the Air Force cross-range requirements, NASA finally decided that launch-abort scenarios and some mission profiles could be enhanced with the big wing. The delta wing was by far a more aerodynamic solution that offered additional mission flexibility. NASA decided to use the big wing and parallel staging. Had the planners and decision makers known that the vehicle being configured would be the only path to space for the United States' manned program for almost forty years, they might have had second thoughts.

In December 1971 NASA sent a summary of four possible configurations and their costs (which ranged from $4.7 billion to $5.5 billion) to George Shultz, who was then director of OMB. By that point Shultz had accepted the arguments and probably believed that NASA had squeezed all that was economically possible from the design efforts. On January 5, 1972, President Nixon announced the proposed shuttle funding to Congress; neither he nor Shultz wanted to define the configuration and only stressed that the costs be held as close as possible to the central estimate of $5 billion. Nixon also had the option of naming the project but preferred to use the name that it had become known as—the space shuttle.

To pay for use of the shuttle, NASA established that the using agency fund specific launches. They estimated that over the first twelve-year period (1979–90) 580 flights would take place, almost equally divided between scientific research, the Department of Defense, and commercial satellites. Using these estimates and a per-shuttle launch cost of $11 million, the shuttle would place a pound of payload into LEO for an amazingly low $175 per pound (excluding the research and development costs). This bargain was too good to be true.

22

ENGINEERING THE SPACE SHUTTLE

Once approved by Congress, NASA was quick to move forward with defining those aspects of the shuttle left in limbo. The final configuration consisted of the manned orbiter, a large external tank (ET), and a parallel-boost first stage that using two solid-fuel rocket boosters (SRB). The orbiter contained the high-energy LH2 primary propulsion unit called the space shuttle main engine (SSME)—three engines each producing 375,000 lb. of thrust. The orbiter was protected from the thermal effects of reentry by a combination of heat resistant tiles and high-temperature metals. The ET contained the cryogenic propellants—LH2 and LOX insulated by a spray-on blanket of foam—and was the only expendable item.

Four contractors competed for the orbiter: Lockheed, McDonnell Douglas, North American Aviation, and Grumman. North American's proposal scored the highest points, with Grumman a close second—North American received the initial $2.6 billion contract on July 26, 1972, for the first two orbiters. The Rocketdyne Division of North American had also been selected to build the shuttle's LH2 main engine the preceding year. The awarding of virtually all of the major contracts to a single company resulted in controversy and some legal challenges. The initial specifications called for five orbiters, each capable of five hundred flights over a ten-year life. This was later lowered to one hundred flights per orbiter, and then to fifty-five per year. Following a critical design review in February 1975, the essential engineering of America's Space Transportation System (STS) was complete, and fabrication could move forward.

Solid Rocket Boosters

The parallel first stage used two 144 in. diameter, 149 ft. long SRBs, providing 2.65 million lb. thrust—71 percent of the thrust at liftoff to an altitude of about 150,000 ft. The solid propellant mixture was ammonium perchlorate (oxidizer) and aluminum (fuel) shaped into an eleven-point star. This configuration provided higher thrust at ignition that gradually was reduced by one-third fifty seconds after liftoff to reduce stress during maximum dynamic pressure. The SRBs were performance matched by loading each from the same batch of propellant ingredients.

The SRBs supported the entire weight of the shuttle on the mobile launcher platform. Each booster was attached to the launch platform by four 28 in. long, 3.5 in. diameter frangible nuts that are severed by small explosives at liftoff. The nuts had two holes drilled in them from top to bottom on opposing sides. A NASA standard initiator was installed, which split the nut when fired, causing it to come off the bolt and allow the vehicle to be released from the pad. Traditionally, each crewmember got a set of the ones used for their launch to use as bookends.

A study submitted to NASA by the McDonnell Douglas Company defined their experience with strap-on solid boosters used with the Thor and Delta rockets—foreshadowing future events. They summarized all of the failure scenarios and indicated only one possible fatal malfunction—the failure of the segment joint that results in a "burn through." United Technologies Corporation was the only SRB bidder who proposed a single segment (monolithic) motor, arguing for that safety aspect. However, Thiokol Corporation was awarded the contract for a seven-segment motor that used a standard tang-and-clevis joint and a double O-ring seal.

Only seven tests were conducted between 1977 and 1980 to qualify the SRB for manned flight. None of these tests demonstrated the dynamic conditions of the boost environment or established an ambient operating temperature range. Both of these factors played a role in the pending disaster that crippled the program in 1986.

Mechanical-thrust vectoring using an 8-degree movement of the nozzle was incorporated. Liquid-thrust vectoring had been considered but was not practical due to the large size of the motor. Following burnout of the propellant, the SRBs separated from the ET and descended under three parachutes to a water landing in the Atlantic, some fifty miles downrange, to be recovered and refurbished for reuse.

The SRBs were to include a thrust-termination capability, allowing an abort at any point in the powered phase. This was to be achieved by blowing a hole in the forward-facing end of the rocket to allow the combustion gases to escape and negate the thrust. However, it soon became clear that this technique was fraught with possible catastrophic problems and it was deleted.

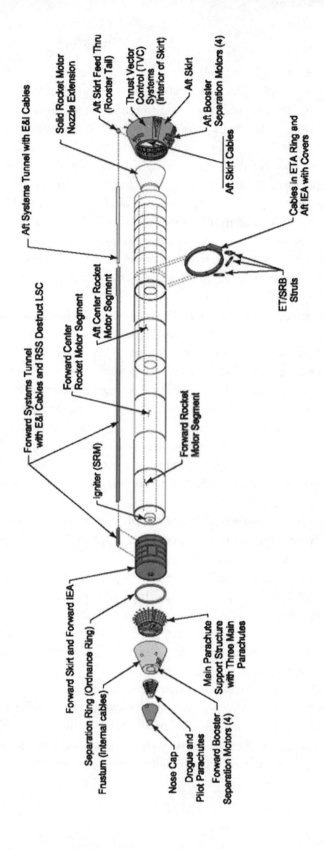

The SRB was a relatively simple booster that was not technologically challenging. Courtesy of NASA.

The Orbiter

The orbiter was the manned component of the shuttle. It housed the main engines and the guidance and control, and contained the payload compartment for the cargo carried into space. The payload bay was 15 ft. by 60 ft. with a capacity of 65,000 lb. to low Earth orbit from Cape Canaveral and 38,000 lb. into polar orbit from Vandenberg AFB in California. The orbiter, in these initial specifications, still retained air-breathing engines to extend its loiter time prior to landing as well as to ferry itself across the country.

A set of solid-fuel rockets mounted to the aft side of the orbiter provided for emergency escape of the orbiter from the rest of the stack during the first thirty seconds of flight. These were deleted from the specification in 1974 as a cost and weight-saving measure. A drag chute to reduce the landing roll was also deleted, as it was felt that the dry lakebeds at Edwards AFB (the prime recovery area) were sufficiently long. This feature was added back when the brakes on the orbiter proved inadequate.

The aerodynamic control surfaces on the orbiter, used following reentry during its glide to a landing, were digital "fly by wire." This meant there were no cables or pushrods between the control stick and the control surfaces. The stick provided input to a computer that determined how much movement to provide to hydraulic actuators based on the speed and altitude of the orbiter. Because the orbiter was an unstable aircraft that required the constant input of the flight controls, the importance of a computer was obvious. Four identical computers worked in parallel, and if one produced an answer that was in conflict with the others, its data was ignored.

The computer initially selected was an IBM AP-101—a 32-bit machine with 32K words of ferrite-core memory and a floating-point instruction set. It weighed 47 lb. and consumed 370 W of power. At $87,000 each, it was a bargain for its time.

The crew was provided with a sea-level atmosphere of 20 percent oxygen and 80 percent nitrogen at 14.7 lb./sq. in. This was a marked departure from the first three generations of American spacecraft (Mercury, Gemini, and Apollo) that used pure oxygen at 5.5 lb. of pressure.

Providing some means for the crew to escape from the vehicle if a catastrophic failure should occur presented a significant challenge. The traditional ejection seat had been developed to a high degree for advanced aircraft such as the SR-71. However, it was recognized that the large crew contingent of seven to ten astronauts needed a more sophisticated mechanism resembling the separable crew compartment of the F-111 fighter or the B-1 bomber prototype. However, these would cost the shuttle not only in dollars but also in weight—and weight is the nemesis of rocketry. It also required that an advanced warning system be developed to ensure that the crew could be propelled away from the possible effects of an explosion of the booster in time to ensure that they suffered no damage. In addition, with parallel staging, there was a time element that might require several seconds to initiate an abort. Catastrophic failures in complex, high-performance rocket systems could rarely be predicted with that much advanced timing.

With these factors in mind, it was determined that, once perfected, the shuttle provided a reliability equal to an airliner and that no escape system was needed. However, this "airliner" reliability became a mindset in the NASA hierarchy that pervaded its thinking about crew escape issues for the next thirty years. It would not, and could not, be substantiated by statistical analysis. Contingency for shutting down a failed SSME engine and flying the vehicle back to the launch site or to an alternate landing site provided for a variety of abort scenarios.

An escape system was needed only during the initial development phases, when a higher degree of uncertainty existed about the various systems. For these first launches, a minimum crew of two used modified SR-71 ejection seats that provided for escape below 100,000 ft. either during the boost phase or following reentry if a landing could not be made on a runway surface.

In the end, it was felt that the added cost and weight of an escape system should be used instead to ensure redundancy and reliability of critical shuttle systems. With a projected reliability of one failure in one thousand, the risks were deemed acceptable. These projections were later lowered to one failure in one hundred, and then to one in fifty-five—statistically, half of the orbiters woud eventually succumb to a catastrophic failure during the life of the program.

Two primary methods of protecting the orbiter during reentry were examined. The first was the use of ablation material, as used on previous manned spacecraft. This presented many problems when the reusability factor was considered.

The second material was a high-temperature reusable-surface insulation in the form of tiles. Advances in materials technology enabled this as a primary solution. High heat areas (650 to 2,000 °F), to include the bottom of the wing, the vertical stabilizer, and forward fuselage, were protected by this material. The thickness of the tile was dependent on the specific

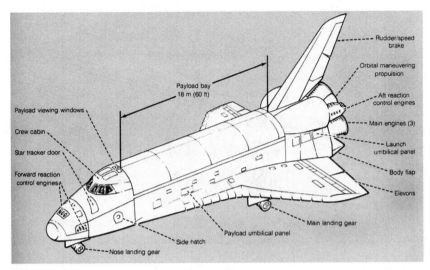

The shuttle orbiter was a large and complex spacecraft that pushed the state of the art. Courtesy of NASA.

location and the temperatures expected to be encountered. The tiles—always considered a considered a high-risk item—were easily damaged and were labor intensive to install and replace.

The tile material was 89 percent porous (i.e., 90 percent air and 10 percent silica) with a basic density of about 15 lb./cu. ft. but manufactured with different densities based on the amount of heat protection required. This represented a weight factor of about 10 percent that of the traditional ablative heat shield. Because of its brittle properties, the tiles could not be fabricated with large surface areas and were applied as a complex set of special shapes—typically no larger than 6 in. square. Because of the thermal expansion qualities of the basic aluminum structure beneath, a felt pad was used between the tile and the aluminum.

The leading edges of the wings and the nose cap experienced temperatures of 2,300 °F. However, because of their shape, they could not effectively be protected with heat tiles. A high-temperature "hot-structure" material—using rare and expensive superalloys—along with a composite of pyrolyzed carbon fibers in a pyrolyzed carbon matrix and a silicon-carbide coating (commonly known as "carbon-carbon") were thus employed. This material was originally developed for the Dyna-Soar project in the early 1960s. Data from other old Department of Defense programs, including ASSET and PRIME, provided invaluable assistance in understanding structural heating. The remainder of the orbiter was covered with a lower temperature

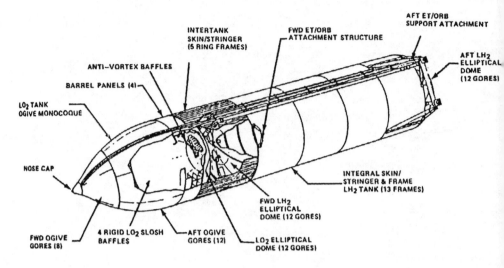

The shuttle external tank assembly. Though a simple component, it had a critical role in joining the three segments that provided the thrust. Courtesy of NASA.

version of the high-temperature reusable-surface insulation tiles and Nomex felt blankets.

As late as 1979 there were still serious questions as to whether the tiles would provide the needed protection when all of the various factors of heat, stress, and vibration were considered. More than 30,709 tiles covered the shuttle. Their application on the first orbiter required three shifts working six days a week for over six months.

The initial program called for five orbiters to be constructed. The first, *Enterprise* (OV-101), did not have all the systems and was used for atmospheric glide tests. The other four were *Columbia* (OV-102), *Challenger* (OV-99), *Discovery* (OV-103), and *Atlantis* (OV-104).

External Tank

The external tank (or ET) represented a significant design challenge, considering its apparently simple role. It was desirable to fabricate it at a low cost (because it was the only expendable item) and at the lowest weight (because it was the largest structure). Accommodating the thrust loads of the two SRBs and the orbiter complicated these factors. Martin Marietta was awarded the contract, in part because of its experience with the Titan III, which involved the attachment of large SRBs.

The ET that resulted was 154 ft. long and 27 ft. in diameter and consisted of the two propellant tanks in tandem, with LOX in the forward tank. Spray-on polyurethane foam covered the entire structure, and an added ablative material was used on the nose and base sections. The nose was exposed to the hottest aerodynamic heating during the ascent, while the base was subjected to the heat given off by the SSMEs.

With previous large cryogenic propellant boosters, like the Saturn V, ice was allowed to form on the outside of the tank as a result of condensation from the humid Florida air and the cold content of the tanks. The ice acted as an insulator during the countdown and was shed by the vibration and aerodynamic forces at launch. This ice was not considered a hazard. However, ice was a potential hazard for the shuttle. Being a parallel staged vehicle, the orbiter was positioned beside the ET, and ice falling from it could impact the wing or sensitive heat tiles. Thus, the insulation performed two important functions—reducing the boil-off during the countdown and ensuring that ice did not form. During the ascent, it also reduced the heat transfer to the propellant from aerodynamic friction.

Space Shuttle Main Engine

The SSME was one of the first components to be defined and was contracted to Rocketdyne in 1971. Along with the heat tiles, it was the pacing item that ultimately delayed the shuttle program. The high-energy cryogenic propellants of LH2 and LOX were the obvious choice because of their high specific impulse. It was also decided that the use of multiple thrust chambers increased reliability while offering more abort options.

To provide for these abort options, the SSMEs had the ability to be throttled up to 109 percent of their rated thrust. In doing so, the shuttle could shut down one of the main engines due to a detected malfunction and continue—either to orbit or to a contingency landing site—by using the added power.

These engines could also be throttled down to 65 percent to minimize Max Q—the greatest dynamic air pressure encountered at an altitude of approximately 35,000 ft. Combined with the propellant perforation design of the solid rocket boosters, which reduced the thrust at Max Q by one-third after fifty seconds of burn, the total stress on the vehicle was kept within the structural design limitations.

As the final shuttle configuration took shape, three 375,000 lb. thrust engines (470,000 lb. in vacuum) were defined for the orbiter. It had been

The SSME (just before launch commit) shows the clear exhaust as it leaves the nozzle. Note the inverted cone of opaque gases a few feet from the nozzles for each of the three engines. Courtesy of NASA.

envisioned that the first-stage booster would also use the same basic engine but with a different nozzle expansion area for its lower altitude burn. However, when the parallel burn configuration was chosen, it meant that these engines had to perform for the entire launch, from sea level to orbital insertion. This 520-second period required that the SSME be optimized for a wide range of atmospheric conditions.

It was originally specified that the engines should have a life of one hundred launches (27,000 firing seconds), but with the longer duration of the parallel burn, this was reduced to fifty-five launches. The engines with the turbopump assembly were 13.9 ft. long with a diameter at the exhaust nozzle of 7.8 ft.

The ability to change thrust was directly related to the propellant supply provided by the turbopump. A mixture of LH2 and LOX in about equal ratios produced a hydrogen-rich steam that powered the turbopump. After driving the turbopump, this steam was routed to the injector where it was mixed with more oxygen (6:1 ratio) for final combustion in the thrust chamber. This feature proved particularly difficult to engineer and, along with the "pre-burn" of propellants, was the key to the high combustion chamber pressures that made the SSME such an efficient power source.

The first full thrust test of the SSME occurred in July of 1975, but continued catastrophic failures of the engine and the turbopump led to a constant series of schedule slippages. A complete flight-qualified engine did not pass the final acceptance test until March of 1980. In order to move forward to the flight test, it was decided to forgo (for the first few missions) the 109 percent capability that had caused significant reliability problems and schedule delays.

Shuttle *Enterprise*: OV-101

As the shuttle began to come together, plans for flight-testing were formulated. Extensive ground, wind tunnel, modeling, and simulator testing were accomplished. Particular emphasis was placed on vibration, resonance such as the pogo affect, and aerodynamic flutter.

With the deletion of the "air-breathing engines" from the shuttle, a new means of transporting it between remote landing locations and the Cape had to be found. Of equal importance were the basic glide tests to determine its aerodynamic flight characteristics. In a manner similar to the X-15, the shuttle was carried aloft and air dropped for these glide tests. After some consideration of building a special aircraft for the purpose, the Boeing 747 was selected as the shuttle carrier aircraft over the Lockheed C-5, primarily because of availability and cost.

There had been some concern regarding the effect of the aerodynamic wake generated by such a large combination. However, wind tunnel tests confirmed the concept of mating, and separation would not produce any problems that could not be overcome with appropriate procedures. The 747 was easily modified using the attachment points that mated the orbiter to the ET. It was determined that the abrupt termination of the orbiter's tail structure produced a pronounced drag component that caused a significant degradation in the speed of the 747, so a drag-reducing tail cone was fabricated for ferry flights.

Construction of the first shuttle orbiter vehicle (OV-101) began in June of 1975. It was to have been named *Constitution* in honor of the two hundredth birthday of the United States as it was "rolled out" in 1976. However, a write-in campaign conducted by aficionados of the television program *Star Trek* caused the White House (President Gerald Ford) to accede to

The shuttle *Enterprise* was carried aloft on a Boeing 747 for transport and for aerodynamic testing. The orbiter has just been released in this first glide test in October 1977, without the tail cone fairing. Courtesy of NASA.

their request, and the first shuttle was named *Enterprise* (Zimmerman 2003, 210).

However, *Enterprise* was not a space-worthy craft. It was used for verifying the ferry capability with the 747, the glide tests, and other activities that required a full-scale model. It did not contain many of the systems, nor did it have the SSME engines or orbital maneuvering system (OMS). *Enterprise* represented the weight and balance and was expected to glide like the real thing. It also contained ejection seats whose presence emphasized that, even in the glide tests, there were hazards for the two crewmembers required to fly the shuttle.

A series of four tests were planned for the mated aircraft. The first was progressively faster taxi tests of the 375,000 lb. combination to determine the effect of the load distribution on the 747 during the initial takeoff and landing phases to ensure that both steering and braking systems were not adversely affected. The gross weight was well below the carrying capacity of the 747.

Next came five airborne flights that began in February of 1977 without the *Enterprise* crew on board; these flights produced no problems. The "captive-active" flights with the crew on board began in June 1977. The *Enterprise* crew noted that, from their cockpit, they could not see any part of the massive 747 that was supporting them.

Finally, the drop tests that allowed the crew to fly the *Enterprise* through a complete landing scenario were performed. Astronauts Fred Haise and Charles Gordon Fullerton were at the controls on August 12, 1977, when the big Boeing 747 with the *Enterprise* perched on its back went into a shallow dive over Edwards AFB in California. With a -7 degree pitch angle, the speed built to 310 mph, and Haise moved the controls of the *Enterprise* to a position calculated to lift the *Enterprise* clear of the 747. At 24,000 ft., he fired the seven explosive bolts that held the mated combination together, and the *Enterprise* flew free.

Haise banked slightly to the right to clear the 747 and then transitioned to a -9 degree pitch that resulted in a 13 degree descent angle. The drag-reducing tail cone had a significant effect on the glide angle. Two 90 degree turns were executed, and then Haise lined up with the runway markings on the lakebed. Coming across the threshold at 213 knots, with the speed brakes out, *Enterprise* touched down smoothly—just five minutes and twenty-one seconds after release. The 150,000 lb. ship coasted to a stop after a roll of 11,000 ft.

Following two additional glide flights with the tail cone, it was time to move to the full drag configuration. The fourth flight was without the tail cone and was equally successful. The drag component represented the flight characteristic of a returning shuttle, but astronauts Joe Engle and Richard Truly easily mastered the 18 degree final approach path.

The final drop test was a bit more exciting when Haise was making the final approach and exercising the gear and speed brakes. Approaching at a higher airspeed of 334 mph, his timing was off for the flare and the *Enterprise* hit hard, bounced back into the air, and experienced some pilot-induced oscillations. However, the shuttle had good, slow speed control, and Haise was able to reposition the craft for a smoother touchdown on the second flare.

The *Enterprise* was subsequently used in a variety of vibration and mating tests at the Marshall Space Flight Center and at the Kennedy Space Center at Cape Canaveral. It was to have been rebuilt as a full-fledged orbiter, but the costs to make the changes were deemed prohibitive, and it was eventually retired to the Smithsonian Air and Space Museum after giving up some of its parts to other orbiters. Over the years, NASA has occasionally returned to the *Enterprise* for a variety of tests, inspections, and evaluations. With the retirement of the shuttle fleet in 2011, the *Enterprise* was moved to the USS *Intrepid* museum in New York City and the *Discovery* was placed in the Smithsonian.

To assist in preparing the shuttle pilots for landing out of a rather unconventional approach profile, four Grumman G1159 Gulfstream II business jets were modified as shuttle training aircraft. These provided the shuttle cockpit display and rotational hand controller for the left-side pilot station while the right side maintained the traditional flight displays and standard control yoke for the safety pilot. By extending the main gear and placing the twin engines in reverse thrust, a steep descent profile of up to 14,000 fpm at 300 kn were achieved. As the shuttle training aircraft itself could not land at these speeds, the touchdown was simulated, and the shuttle training aircraft was cleaned up for a go-around—something the shuttle could not do. The stress placed on these aircraft to perform these "unnatural" maneuvers took a significant toll on their structural life.

Reusability Becomes a Reality

The first flight of the shuttle was manned. This was an unprecedented decision that leaned heavily on the extensive testing and simulations used to validate the various systems. The first flight-ready orbiter, named *Columbia*, arrived at the Kennedy Space Center in March 1979, but mating and testing and rework required another two years before it was ready for the first flight.

The first launch carried an extensive array of recording instruments to transmit data to ground stations during the flight. These measurements were correlated with modeling data to determine how closely the actual spacecraft corresponded to the wind tunnel results, computerized simulations, and assumptions. Of particular interest were the conditions during the various critical phases of flight: maximum dynamic pressure (Max Q), SRB separation, and ET separation.

The shuttle was assembled (stacked) in the same vehicle-assembly building used for the Apollo/Saturn V missions. The SRBs were first raised into position on the launch platform used by the Saturn V—modified to accept the significantly different shuttle configuration. Then the ET was joined to the SRBs in early November 1980, and the orbiter was raised into position and mated on November 26, 1980. The completed Space Transportation System Flight Number 1 (STS-1) was transported to Pad 39A on December 29, 1980.

The flight-readiness firing, a static test of the three SSME LH2 engines—a 20-second firing at 100 percent power—was performed on February 20, 1981. This was the first time that all three engines of a complete flight-ready

Shuttle STS-1 *Columbia* astronauts John Young (*left*) and Robert Crippen (*right*). Their willingness to fly such a massive and complex vehicle on its first test gave new meaning to the word "courage." Courtesy of NASA.

shuttle had been fired. A part of the test was to determine the "twang." Because the position of the SSMEs was somewhat off-center to allow its thrust vector to be aligned with the center of gravity of the total assembly, the entire shuttle rocked forward when the SSMEs were ignited. The movement, which amounted to almost 2 ft. at the nose of the shuttle, was referred to as the "twang." The timing of the SRBs' ignition was set to allow this movement to return the shuttle to the vertical position.

Safety was always a major concern, as the lessons of Apollo were still deeply etched in the culture at this point. NASA came under some criticism for its apparently slow progress, often a result of evaluating possible safety considerations. Despite this cautious approach, there were still some fatal accidents. During a final launch simulation in March, two technicians were asphyxiated when they inadvertently entered the aft section of the shuttle, which was filled with nitrogen to guard against the accumulation of any explosive gases—principally hydrogen. The cautious approach would soon change.

The flight was finally scheduled for April 10, but a software glitch caused a two-day postponement. The crew of astronauts, John W. Young and Robert L. Crippen, brought two of the quietest and most capable men in the astronaut corps together. Young had flown the first Gemini with Gus Grissom on GT-3 in 1965, and again with Michael Collins on GT-10. He was on board Apollo 10, which made the final dress rehearsal around the Moon in 1969, and he walked on the Moon with Charlie Duke on the Apollo 16 mission.

STS-1 was Crippen's first flight. Originally selected for the canceled manned orbiting laboratory, he transferred to NASA in 1969. Now, twelve years after that job change, he finally flew on April 12, 1981—twenty years to the day after the first manned space flight, Vostok I with Yuri Gagarin. Spacecraft had progressed a long way during that time.

The astronauts entered the shuttle two hours before liftoff, and the main side hatch was closed by T-60 minutes. The white room and the pad were cleared at T-30 minutes. The final updates to the various computers were accomplished, and the crew access arm that allowed an emergency escape while the shuttle was still on the pad was retracted at T-7 minutes.

With less than three minutes until launch, the aerodynamic surfaces of the shuttle were exercised to affirm the functioning of the hydraulic system, and the three SSMEs were moved through their entire gimbal travel. The "beanie cap" that removed the vented hydrogen and oxygen gases from the top of the ET was retracted as the count moved to within two minutes, and the propellant tanks were pressurized.

At T-3.8 seconds, a computer sequenced the engine start process for the three SSMEs. The ignition of each was staggered by about one-tenth of a second. The shuttle thrust quickly came up to 1.125 million lb., and the assembly experienced the "twang." As the huge rocket realigned itself with the vertical axis, the SRBs ignited, and the four 3.5 in. diameter explosive hold-down nuts on each were blown. The 2.65 million lb. of thrust from each SRB, added to the SSME, gave a total of almost 6.5 million lb. of thrust, and the vehicle left the pad, balanced on two shafts of white SRB smoke and the virtually transparent plume of the three LH2 engines.

For the first time, the new machine took to the air—a vehicle that had more research and development effort (and cost more) than any flying machine humans had ever created. This being the first flight, public interest was high. (Not quite as many people tuned in to watch the launch as the first Moon shot, however.)

The space shuttle *Columbia* takes flight for the first time. Subsequent flights did not
have the ET painted white but allowed the foam insulation to acquire an orange hue.
Courtesy of NASA.

Because of its greater thrust to weight ratio, the 4,457,111 lb. vehicle accelerated more rapidly than the Saturn V, and the tower was cleared in 6.5
seconds as opposed to the 10 seconds of the Saturn V. At T+11 seconds, a
roll maneuver aligned the orbiter's longitudinal axis toward the northeast
for the orbital inclination of 40.3 degrees.

Forty-four seconds into the flight, the SSMEs were throttled back to 65
percent power to reduce the aerodynamic pressure at Max Q. The throttle-
up began at T+76 seconds as the vehicle arced over to the northeast.

The white exhaust of the SRBs showed an orange color as they reached
burnout two minutes into the flight and were released from the ET at an
altitude of 28 mi. and a speed of just over 3,000 mph. As the SRBs began
their slow tumble back to Earth, the shuttle exhaust was virtually invisible
as it continued to power the orbiter.

At T+265 seconds the point of no return was reached. At this point, if there was a failure of one of the SSME engines, the shuttle could no longer return to the Cape; it would press on to one of the emergency landing sites, the first of which was in Spain. The seven-minute mark was the point where the shuttle could still reach orbit if one SSME engine should fail. In future flights, this was the point where the 109 percent power was needed. However, *Columbia* was lightly loaded on her maiden flight and could reach orbit with the 100 percent power available from only two main engines.

At T+460 seconds the main engines were again throttled back as the weight of the shuttle had been reduced to the point where it would exceed the three-g limitation if it continued at 100 percent.

With the engines at 65 percent power and a speed just shy of orbital velocity for the 172 mi. altitude, the main engines were shut down at eight minutes and thirty-four seconds into the flight. The ET separated and descended back into the atmosphere to burn up over the Indian Ocean.

John Young positioned the orbiter's attitude for the first burn of the OMS—two 6,000 lb. thrust units located on either side of the aft fuselage, accounting for the bulge just beneath the vertical stabilizer. Using monomethyl hydrazine and nitrogen tetroxide as propellants, the OMS unit provided the final push into orbit as well as any subsequent orbital maneuvers for future rendezvous missions. As with all previous manned American spacecraft, the orbiter had a reaction control system for attitude control. This consisted of a set of thirty-eight 830 lb. and six 25 lb. thrusters. These thrusters used the same type propellants as the OMS and had their own supply but could be cross-fed from the OMS tanks if necessary.

The payload bay doors were opened although there was no payload other than the engineering monitors. The inside of the doors contained the cooling radiators for the environmental control system and had to be opened on achieving orbit—and remain opened during the mission.

On-orbit video of those parts of STS-1 that could be seen from crew positions revealed that some tiles were lost from the OMS unit faring. While there was some concern, these tiles were not deemed critical, as the heat generated during reentry to that part of the shuttle was considerably less than to the underbody—for which there was no imaging from the shuttle. To allay those anxieties, confidential channels to the National Reconnaissance Office (NRO) provided both satellite recon resources and Earth-based telescope photos to image the underside. With those assurances, Young and Crippen assumed that the primary heat tiles on the underside were still in place.

The astronauts spent two days and six hours in orbit testing and verifying the various systems, and then it was time to return. A critical operation was the closing of the payload bay doors. If they could not be closed, the shuttle did not have the structural integrity to survive the reentry. Young then positioned the shuttle to traveling backward to the orbital vector, and the OMS thrusters were fired. This reduced the speed of the shuttle to where it descended into the upper reaches of the atmosphere. The shuttle was then reoriented, initially to a 40 degree pitch-up attitude that gradually was reduced as it encountered the more dense layers of the atmosphere. As the shuttle started to feel the deceleration, the g-forces increased, and the heat pulse on the nose and wings of the orbiter generated the ionized sheath of air that enclosed the spacecraft and disrupted radio communications. This occurs from about 400,000 ft. down to about 200,000 ft. For the STS-1 crew, the normal sixteen-minute period was extended to twenty-one minutes by problems with a ground station link.

During the blackout period, the computer displayed the required S-turns that dissipated the energy and kept the shuttle from experiencing the heavier 8 g-forces that previous astronauts had had to endure. *Columbia* and her stable mates eased back into the lower layers of the air, subjecting the occupants to less than 3 gs. The path projected by the computer also ensured that it arrived over Edwards AFB at 60,000 ft., in a position to make a normal landing on the Rogers Dry Lake bed that had, for the preceding thirty-five years, been a safe haven for America's rocket planes.

As *Columbia* arrived over Edwards, she announced her presence with the distinctive double boom, sonic shock wave. As it made its descending turns, *Columbia* was joined by two T-38 chase planes piloted by other astronauts. The descent profile used the computer to manage the shuttle's energy and place it on the final approach. As it neared the ground, Young began a two-phased flare that positioned it over the end of the runway with the gear down. One of the crew of the T-38 chase planes counted the estimated feet to the runway over the radio to aid the shuttle pilot.

Following a smooth touchdown and rollout, there was an extended time when the ground crew had to "safe" several systems and ventilate critical areas that might contain toxic gases. Almost an hour after touchdown, and following some critical comments from the crew about the time it was taking to get them out of the spacecraft, the hatch was finally opened, and two elated astronauts skipped down the stairway and bounded excitedly around the outside of the shuttle for a postflight assessment.

The inaugural flight of the shuttle had been a resounding success, and

The shuttle *Columbia* completes its first flight into space with a landing at Edwards AFB, California. Courtesy of NASA.

great things were expected of the long-delayed and controversial project. However, the next flight did not immediately follow. A total of 16 tiles were lost from *Columbia*, most likely during the acoustical vibration at SRB ignition, and 148 more were damaged. There was more work to be done to ensure greater reliability of the tiles as well as several other systems.

The second flight of *Columbia* (STS-2) occurred eight months later (November 1981) and was crewed by Joe Engle and Richard Truly. The extended turnaround time was needed to analyze the data returned from STS-1 and to make the required changes to various systems. STS-2 marked the first use of the Canadian-built remote manipulator system, the highly dexterous arm that was stowed in the payload bay; the arm was an important part of the shuttle operation. The failure of a fuel cell caused this flight to be shortened from the planned five days to three. A new sound-suppression system that used water to absorb the intense acoustical pulses emitted by the SRBs at ignition was considered successful, as there were only twelve damaged tiles found on *Columbia*'s return.

STS-3 with *Columbia* occurred on March 22, 1982, after a turnaround of four months, and it remained in orbit for eight days with the first Department of Defense payload. Because the Rogers Dry Lake at Edwards AFB was flooded by spring rains, the shuttle was diverted to the contingency landing site at White Sands Missile Range in New Mexico—the only shuttle to land there. However, its landing there was delayed by a day due to high winds that came up just minutes prior to the deorbit burn. During the postflight inspection, it was apparent that the acoustical suppression system did not seem to perform as well; thirty-six tiles were lost and nineteen damaged.

STS-4, in June 1982, also carried a Department of Defense payload in addition to some scientific experiments. Unfortunately, both of the SRBs were lost when the parachute recovery systems failed to function. With the first landing on the hard-surface runway at Edwards, the shuttle was declared operational. Those who truly understood the complexity of the machine realized this was an ominous declaration.

STS-5 was the first spacecraft in which American astronauts flew in a shirt-sleeve environment—no spacesuits. In addition to the two shuttle pilots, the first mission specialists flew into space. Because there was no emergency escape system for these mission specialists, the ejection seats for the shuttle pilot and mission commander (which were eventually removed from *Columbia*) were pinned to disable them. There would be no moral dilemma of saving only two of the crew if catastrophic failure occurred. The mission specialists were astronauts who were not pilots but who were responsible for conducting on-orbit work. STS-5 carried the first commercial satellites: two communications satellites with solid-fuel payload-assist modules that boosted them into geosynchronous orbits after release from *Columbia*'s payload bay.

Geosynchronous Orbits

A geosynchronous orbit has an orbital period of twenty-four hours (one day)—matching Earth's rotation. This synchronization of rotation and orbital period means that a satellite in geosynchronous orbit returns to the same position in the sky at the same time each day. Over the course of a day, the satellite traces out a path (typically in a figure 8) whose characteristics depend on the orbit's inclination and eccentricity.

A special case of geosynchronous orbit is the geostationary orbit—a circular geosynchronous orbit with zero inclination (directly above the equator). The satellite appears stationary, always at the same point. The term "geosynchronous" may loosely be used to mean geostationary. Communications satellites typically use geostationary orbits, so the Earth-based antennas that communicate with them do not have to move.

America's first female astronaut, physicist Sally Ride, flew on board STS-7. She was the youngest American astronaut to fly in space at age thirty-one. Courtesy of NASA.

STS-6 was the inaugural flight of the second shuttle, *Challenger*, in April 1983 and included the first shuttle extravehicular activity (EVA). On board STS-6 was the first of a network of tracking and data-relay satellites that ultimately replaced most of the ground stations built around the world since the days of the Mercury Program. The tracking and data-relay satellites released from the shuttle were subsequently boosted into geosynchronous orbits. They were used by NASA for communications with both manned and unmanned satellites to increase the time spacecraft were in communication with the ground and to enhance the data transmission rates.

STS-7 carried America's first female astronaut, Mission Specialist Sally Ride. STS-8 was the first night launch and included the first African American in space, Mission Specialist Guion Bluford. The flight concluded with the first night landing of the shuttle. STS-9 flew with the first payload specialist and the first six-member crew for the first European-built spacelab housed in the payload bay. It was the last of the original STS numbering system and is now referred to at STS-41-A.

The system of numbering the shuttle missions sequentially was changed in favor of the fiscal year, launch site location (Vandenberg AFB being the other), and letter for sequence. This avoided having an STS-13. Although

America's first African American astronaut, Guion Bluford, flew on board STS-8. He subsequently flew three more shuttle missions. Courtesy of NASA.

NASA is not superstitious, why take chances? The memory of the near disaster of Apollo 13 was a constant reminder.

STS-41-B conducted the first untethered EVA with Mission Specialist Bruce McCandless using the manned maneuvering unit. STS-41-C performed the first shuttle orbital rendezvous with the Solar Maximum Mission (or SolarMax) satellite to retrieve and service it following failure of its attitude control system. STS-41-D was the first flight of *Discovery* and the first to experience an on-pad abort after the main engines had ignited but before SRB ignition. It was rescheduled and launched successfully in August 1984.

STS-41-G in October 1984 saw the first American female EVA with Kathryn Sullivan on board *Challenger*. However, the Soviets upstaged the Americans by having Svetlana Savitskaya perform the first female EVA a few months earlier. on July 1984, from the Salyut 7 space station.

STS-51-C of *Discovery* was the first flight dedicated exclusively to a Department of Defense payload. The satellite it released was boosted into geosynchronous orbit to gather electronic intelligence, although much of the information regarding the payload is still classified. This flight was delayed by a day because of freezing temperatures at the Cape—a foreshadowing of the *Challenger* disaster that took place exactly a year later. It also marked the one hundredth human spaceflight.

With the expected demands placed on the shuttle, the astronaut corps was expanded considerably, to well over one hundred. The roles determined the training and qualifications. Shuttle pilots all had military experience and were graduates of one of the test pilot schools. The payload specialists and mission specialists were not required to have pilot qualifications—although most did. However, all astronauts were required to fly in the sleek, white, twin-engine Northrop T-38 to maintain proficiency, enhance coordination and judgment, and, for the nonpilots, to solidify teamwork. Nonrated pilots were not soloed in the two-seat craft, although several completed their requirements to receive FAA private pilot certification.

23

THE REALITY OF FAILURE

NASA continued to improve the shuttle turnaround time, but problems still plagued the program, and at the end of the fourth year of operation, the shuttle had yet to be launched more than eight times in a calendar year. The time and equipment needed to refurbish each shuttle and return it to flight status caused costs to soar, and the pressure to launch on time was significant.

Challenger: "A Major Malfunction"

The STS-51-L launch had been postponed four times, and the morning of the fifth attempt, January 28, 1986, saw a cold but clear blue sky. The designation of the mission was no longer in synchronization with the number of shuttle flights as some had been canceled or rescheduled (the previous flight has been STS-61-C). It was, in fact, the twenty-fifth flight of the Space Transportation System and *Challenger*'s tenth foray into space. Since its first mission in April 1983, *Challenger* had spent 62 days in space, completed 995 orbits, and had flown almost 26 million miles.

The temperatures had slid into the upper twenties—an unusually cold snap for central Florida but not a record for the date. Water had been sprayed on various parts of the pad to protect some components from the temperatures, and this had resulted in a significant accumulation of ice. NASA was concerned about the ice that had formed on some of the structure of Pad 39B—the first time this launch facility was used for the shuttle. An engineering assessment had been made to consider the effect of the temperature, and an "ice team" had surveyed the launchpad and the *Challenger*. Based on the team's recommendation, the launch had been delayed several hours to allow the temperatures to climb above the freezing point to melt some of the ice.

While the concerns being explored around the pad related to the interference of the ice with the launch equipment, there was also concern expressed by engineers of the Thiokol Corporation (who built the SRBs) as to the effect the cold had on the rubber O-rings that sealed the seven segments of the solid-fuel boosters. Several engineers, Roger Boisjoly of Morton Thiokol in particular, recognized that the ambient temperatures hardened the rubber and would inhibit it from sealing as effectively as it should. They strongly recommended that NASA "launch within the established experience factors," meaning that a launch temperature colder than any previously experienced placed the launch in a high-risk, "unexplored" region (Zimmerman 2003, 224). Half of the previous flights had experienced O-ring erosion, and a few had actually exhibited some "blow-by" of the hot exhaust gases. "Erosion" is the effect of the high temperatures of the solid-fuel combustion, while "blow-by" is actual intrusion of the combustion gases past the O-rings. Even the president of Rockwell, the builder of the orbiter, had expressed his lack of support for a launch.

However, this shuttle carried the first passenger into space—schoolteacher Sharon Christa McAuliffe. Sen. Jake Garn (R-Utah) and Rep. Bill Nelson (D-Fla) had been on board previous shuttle flights, but Christa was to be the first "commoner" to experience space flight. President Ronald Reagan was scheduled to present his State of the Union message that evening, and NASA very much wanted him to be able to refer to her flight. Despite several attempts by lower-level management and engineering personnel to bring the added risk factor to the attention of NASA's upper management, the launch took place at 11:38 EST.

To all observers, there were no indications of any obvious problems as *Challenger* lifted from the pad and began a flight that had become almost routine. At T+19 seconds, the engines were throttled down from their 104 percent thrust to achieve the 65 percent level. The communication "Challenger, you are GO for throttle-up" at T+35 seconds was met with the last response from Commander Dick Scobee: "Roger, Go at throttle-up." Max Q was encountered at T+59 seconds, just three seconds after the SSMEs began to work hard in their gimbals to compensate for the effect of high altitude wind shear. Then a large, violent explosion at T+74 seconds marked the point at which *Challenger* ceased to exist. The "voice" of Mission Control continued reading out the speed and altitude parameters and did not see the event unfolding before the viewer's eyes. When he again glanced at the live picture monitor and saw the huge vaporizing fireball, there was a

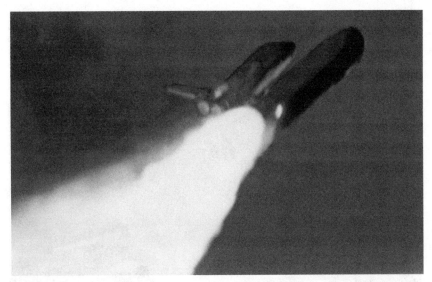

The small bright spot on the lower part of the right SRB is the beginning of the burn-through. Courtesy of NASA.

long pause before he could continue with his job of reporting the status. He simply said what was apparent to all: "obviously a major malfunction."

Although many who were watching knew almost instinctively that the end had come for the seven astronauts, there was still a feeling of incredulity that swept persons in attendance as well as those watching at home on their TV sets. Many assumed that the crew had somehow been propelled to safety by an escape system. However, there was no escape from the shuttle.

The nation mourned the loss of the *Challenger* crew, and one week later the president appointed a commission headed by William P. Rogers (a former secretary of state) to determine the cause. The commission consisted of astronauts Neil Armstrong and Sally Ride, Brig. Gen. Charles "Chuck" Yeager, and several eminent scientists and engineers, including physicist Dr. Richard Feynman. NASA was deep in the process of evaluating the telemetry, film, and videotape when the commission convened. The cause was almost immediately suspected to be the SRB O-rings.

Special cameras that followed each flight to record the intimate details of events readily showed the initial puffs of black smoke at ignition from one SRB joint in particular. This indicated that the O-rings had not seated from the internal pressure but instead had allowed the packing material, a form of putty that protects the rings from the heat of the combustion, to be blown out and expose the rubber rings to the combustion.

The shuttle *Challenger* explodes seventy-four seconds into the flight. Courtesy of NASA.

Following that small puff of smoke at SRB ignition, it is believed the seal actually seated. However, when wind shear at higher altitude caused the joint to flex, the seal was easily broken. At that point, the flame appeared and it grew from there. As with most accident chains, had the wind shear not been present, the SRB joints may have held. In the later frames, the flame could be seen protruding from the segmented joint, enlarging, and impinging on the external tank (ET).

The flame eventually ate through the lower part of the ET and the strut that attached the SRB to the ET. Liquid hydrogen began escaping from the hole burned in the ET, and this was followed almost immediately by the failure of the strut. This caused the right SRB to swing into the top of the ET, destroying it and creating lateral movement that caused the orbiter, traveling at Mach 1.92 at an altitude of 46,000 ft., to experience side loads for which it was not stressed. The orbiter was literally torn apart by the aerodynamic forces.

NASA contended it was doubtful the crew had any impending knowledge of what was to befall them. As the downlink of communications was

Rapid Decompression

When a pressurized vessel (spacecraft or aircraft) experiences an unplanned catastrophic breach that releases the pressure, several events may occur. If the change in pressure is sudden (less than a second), the lungs may not be able to release the air contained within and they may literally explode (barotrauma). If the change in pressure occurs over several seconds, the air can escape from the lungs without physical damage. In either case, unless a pressure oxygen mask is immediately available, the individual will experience hypoxia and become unconscious within about ten seconds.

severed with the explosion and the disintegration of *Challenger*, only one word was captured that indicated an awareness that something was about to go wrong—Shuttle Pilot Michael Smith's last words were "uh oh." When the remains of the crew module were recovered several weeks later, there was evidence that at least two of the crew had the presence of mind to activate the emergency air system provided for them. These were not envisioned for in-flight use but for use on the ground, if toxic fumes should invade the crew compartment.

The pressure vessel of the crew compartment had likely been breached by the destructive forces, so it is probable that rapid decompression immediately followed as the crew compartment continued to soar to over 70,000 ft. before beginning its plunge toward the ocean. However, there was no indication that any of the crew were conscious during the long, two-minute fall into the Atlantic. Death of the crew occurred by "blunt-force trauma" on impact.

The concerns of key Thiokol employees had been proven correct: the failure was the direct result of the inability of the rubber O-rings to properly seat under the temperature conditions experienced. The joint was not well designed, and the bending loads experienced by the SRBs as they rode through strong wind shear aloft accelerated the failure.

However, the problem was deeper because management had been made aware years earlier of the inability of the O-rings to properly seat during cold ambient temperatures. Almost half of the SRBs had experienced some erosion and burn-through. Yet the potentially fatal anomaly was allowed to continue into a pattern that was called "a normalization of deviancy."

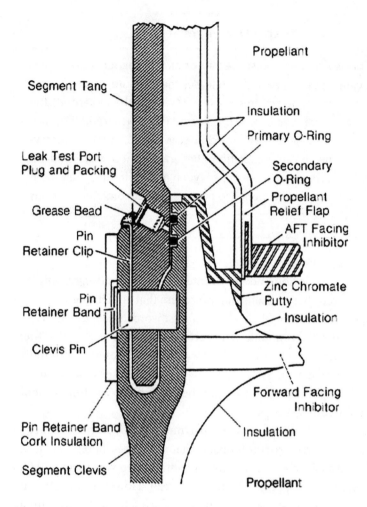

The SRB cross-section shows the position of the tang, clevis, and O-rings. Courtesy of NASA.

Following the recommendations of the commission, the SRB joint was redesigned and recertified, and the management structure and reporting paths were changed. The recommendations were reportedly "indistinct" in order to allow the shuttle to return to flight at the earliest possible date. The reason was obvious—critical "national security payloads" of the Department of Defense now rested solely on the availability of the shuttle. If the shuttle had to await an effective crew-escape system, it probably meant a three- to five-year delay. These reconnaissance satellites had been designed to the size of the shuttle payload bay and its relatively gentle 3-g acceleration into orbit. The available expendable boosters could not accommodate these satellites.

Possible emergency-escape modes were defined in a two-phased approach to the problem. The first, a "bailout" solution for a stable and subsonic gliding shuttle, was implemented; however, it would not directly have helped the *Challenger* crew. The second phase, the use of a survivable escape module, might have. That the crew module emerged from the fireball intact and some of the crew had the physical and mental capacity to activate the emergency air supply caused a more innovative evaluation of escape options. It was possible to rework the cabin and employ a means of separation and a parachute recovery system. However, those who could have moved such a means to reality believed that a shuttle replacement program would be a better investment of the funds.

Nevertheless, the influence of the *Challenger* explosion went far beyond the shuttle. The decision to use the shuttle as a replacement for expendable launch vehicles was revisited and reversed. It was decreed, first by President Reagan's administration and then by federal law, that the shuttle be used only for purposes that "require the presence of man . . . or requires [sic] the unique capabilities of the Shuttle . . . or compelling circumstances" (2465a Space Shuttle Use Policy, 42 Fed. Reg. 2929 [Nov. 16, 1990]).

The entire premise of the shuttle program had been overturned by the *Challenger*'s accident—inexpensive access to space. However, as reviews of the shuttle's performance and costs made clear, the shuttle had not provided inexpensive access. In fact, because of the extensive testing and preparation required for each flight, expenses had constantly exceeded estimates, and new factors continued to add to the cost of each flight. The initial STS development ultimately cost $6.6 billion, $1.6 billion more than the 1973 estimate given at the time the program was approved. Considering inflation and the high-tech complexity of the program, it was not really out of line. However, when the shuttle returned to service in 1988, the per-flight cost of $300 million was a hard figure to reconcile. The cost to orbit one pound of payload had risen to $5,000—not the $175 predicted sixteen years earlier. (As there had been significant inflation over that period, the $175 figure may be adjusted by a factor of five).

Another victim of the *Challenger* disaster was the "deep space" program, which had planned to use the Centaur upper stage launched from the payload bay of the shuttle. This had been considered a high-risk situation, and many in the scientific community had been calling for a rethinking of the wisdom of placing the volatile hydrogen fueled Centaur in the payload bay.

With safety issues dominating the post-*Challenger* period, the use of LH2 was considered too hazardous, and probes such as the unmanned interplanetary missions *Galileo* and *Ulysses* had to be reduced in weight and reprogrammed for different flight profiles in order to use the solid-fuel stage that replaced the LH2 Centaur.

Even the commercial aspects of space flight were changed by the *Challenger* accident. NASA had been competing with the Arianespace Corporation, a company partially funded by the European Space Agency, for contracts to launch commercial satellites. The European Space Agency had been using the French Ariane expendable booster since 1984. In addition, a private U.S. company, Transpace Carriers Inc., had attempted to move into the market using the McDonnell Douglas Delta rocket. There was some legal maneuvering by Transpace to stop Arianespace from receiving contracts from U.S. companies because it was financed in part by the French space agency, Centre national d'études spatiales (a government entity), and represented unfair competition. This action had been dismissed because it was determined that the shuttle presented the same subsidy aspect. However, with the shuttle no longer in the business of launching commercial satellites, the suit was again reopened, and the Office of Commercial Space Transportation began regulating the launching of satellites by issuing licenses.

Nevertheless, the Space Transportation System pressed on. It was, after all, a national asset and represented America at its technological best. Another shuttle, named *Endeavour* (OV-105), was constructed using portions of a test article and spare parts. Although a fifth shuttle had been a part of the original planning, when the final number of four was funded (exclusive of the *Enterprise*), critical spares were constructed that allowed for a fifth shuttle.

For the remainder of the millennium, the shuttle continued to launch research payloads such as the Spacelab, Spacehab, and the Hubble space telescope. Military launchings included reconnaissance, signals intelligence, communications, and early-warning applications. Nine missions were flown to the Soviet space station Mir between 1995 and 1998. Assembly and support of the International Space Station began in December of 1998 with STS-88.

Seventeen years after *Challenger*, the shuttle continued to be the only manned vehicle for America. By January 2003, 112 shuttle flights had been accomplished.

The Hubble Repair Mission

Shuttle *Atlantis* flew as STS-31 orbiting the Hubble space telescope in April 1990. Among the five scientific instruments on board the $2.5 billion satellite was the wide-field and planetary camera, which provided high-resolution imaging for optical observations.

Mission designations were changed again after the *Challenger* disaster in 1986. As the *Challenger* mission was the twenty-fifth and the shuttle facility at Vandenberg AFB was cancelled, NASA could revert from the previous scheme that included the launch facility in the designation to the sequential system without having to designate an STS-13 mission.

Within weeks of the launch, the images being returned from Hubble showed a serious problem with the optical system. Although these images were about as sharp as ground-based viewing, the telescope failed to achieve a final sharp focus, and the best image quality obtained was drastically lower than expected—not the anticipated tenfold increase.

Analysis of the flawed images showed the cause of the problem to be the primary mirror—it had been ground to the wrong shape. It was determined that the conic constant of the mirror (its curvature) was -1.0139, not the intended -1.00229. It was 78 millionths of an inch out of specification!

A critical instrument used by the PerkinElmer company to shape the mirror had been incorrectly calibrated. Two other instruments that were used to crosscheck the precision of the polishing work had detected the problem during the fabrication process, but their findings were ignored since the company believed that these instruments were less accurate than the primary one. Criticism was also leveled at NASA for poor oversight of the quality-control process. In its flawed state, the Hubble could not obtain data any better than ground-based telescopes.

It was not possible to replace the mirror in orbit, or return the telescope to Earth for a repair (the shuttle's maximum landing weight would have been exceeded). However, the Hubble design incorporated provisions for being serviced in orbit, and a plan was devised to install a "corrective lens"—much like fitting a human with a pair of glasses. The design to correct the problem was called the Corrective Optics Space Telescope Axial Replacement (or COSTAR). However, this required that one of the other instruments had to be removed—the high-speed photometer.

This first servicing mission to Hubble required extensive training of the astronauts and the use of many specialized tools. The mission (STS-61) took place in December 1993 and involved installation of several instruments and other equipment with five spacewalks over a ten-day period.

Also replaced were the solar arrays and their drive electronics, four of the gyroscopes used in the telescope pointing system, and the onboard computers. After the repairs and upgrades were completed (and tested), the Hubble was boosted into a higher orbit (368 mi.) by the shuttle. Service Mission 1, with its extensive extravehicular activity (EVAs), was one of the most complex ever attempted. While NASA had taken considerable criticism for originating the problem, it bathed in the accolades that followed when the Hubble began sending the long-awaited images from space—and they were impressive.

Columbia: Foam Insulation

Shuttle *Columbia* had made twenty-seven previous flights into space, and the launch on January 16, 2003, into a 150 mi. orbit was apparently flawless. The crew consisted of seven astronauts commanded by Rick Husband. The primary payload was the Spacehab-DM (double module) research mission, along with several "hitchhiker" experiments.

The sixteen-day (255 orbit) journey was the third longest (STS-80 held the record with seventeen days and nine hours—280 orbits). Because *Columbia* was the first shuttle constructed, it was also the heaviest and was not able to achieve orbital inclinations greater than 40 degrees with a significant payload. It was, therefore, used for missions other than to provide service to the International Space Station, which is at a 51.6-degree inclination.

Within days of the launch, engineers examining the video of the ascent noticed that a piece of the foam insulation from the ET had broken off about eighty seconds after liftoff and impacted either the leading edge of the left wing or the thermal protection system tiles on the shuttle's underside. This began an inquiry as to the possible effects. It was estimated that the piece of debris was about 20 in. long by 16 in. wide by 6 in. thick and weighed about 2.5 lb. It struck the wing at less than a 90-degree angle. An initial analysis was performed by a simulation-modeling program called "Crater." Although the engineers did not have much experience with it, the result indicated a possible breach in the reinforced carbon-carbon leading edge—if that was where the impact occurred.

Because of the conservative nature of the software and the belief that the debris was "just foam," management discounted the effect to some possible localized deformation but decided that it did not represent a "safety of

The shuttle *Columbia* lifts off on its fateful flight. Courtesy of NASA.

flight" issue. However, the 2.5 lb. piece had apparently punched a hole, later estimated to be about 2 in. in diameter, in the leading edge of the left wing. From the point of impact (just eighty seconds into the mission), the crew had almost no hope of returning safely. Although made aware of the debris strike early in the mission, they had no awareness of their impending fate.

It was possible that, because this shuttle had sat on the launchpad for several weeks and endured some significant rainstorms, perhaps water had collected in a break in the insulation, and, with the super-cold temperatures of the cryogenics, the foam became encrusted in rock-hard ice on the day of the launch. Because the video showed a piece of light-colored material striking the wing and exploding into a cloud of white dust or vapor, some experts suspected the material was ice (or ice-coated) as the foam insulation was orange. Moreover, because the piece broke off at the time the shuttle was near Max Q while moving about Mach 2, the impact might have been more destructive than first thought.

Some NASA officials did not agree with the benign impact report and felt it was potentially a more serious problem. Someone in NASA had made a request for the military to inspect the shuttle with its powerful, secret ground-based cameras for damage from the debris (as it had with STS-1). But this imaging request was withdrawn by management.

Management again dismissed the impact of the debris on the shuttle as "inconsequential" in a report issued eleven days into the mission. With respect to the dissenting opinions that predicted greater risk, the shuttle program manager stated (after the fatal reentry) that those concerns weren't brought to his attention.

In the days before *Columbia*'s reentry, concern centered on possible damage to the left wheel well rather than the leading edge. A loss of tiles in this area might allow the temperatures around the left main gear to impede the gear's operation. Several engineers predicted various warning signs that might be exhibited on reentry and took steps to understand the degree of danger—like simulating a landing with blown tires. That simulation took place just hours before *Columbia* began its descent, and some officials were alarmed by this last-minute action when they had been "assured" only a few days earlier that there were no significant risks. Depending on the telemetry indications of the conditions in the wheel well, there was consideration of advising the crew to extend the gear early so that, if it failed to lock into place, they could be prepared to use the "bail-out" capability instituted after the *Challenger* disaster, as a gear-up "belly landing" would probably not be survivable.

The landing was planned for February 1, with the deorbit burn occurring at 8:15 a.m. EST over the Pacific Ocean. The entry interface, an arbitrary point at 400,000 ft. where the first detectable signs of decelerating g-force are encountered, occurred at 8:49 a.m. At 8:52 a.m., *Columbia* crossed the coast of California as it encountered the denser layers of the upper atmosphere and entered Roll Reversal #1 while traveling Mach 20.9 at 224,390 ft. Seconds later, the left wheel well showed an "off nominal" temperature rise followed by a loss of sensor data from the left inboard elevon and several

The shuttle *Columbia* disintegrates and much of the debris is burned up in the reentry.
Courtesy of National Air & Space Museum.

other temperature dropouts. Those engineers who had been analyzing the possible effects of the debris strike now knew that this reentry would not end normally and mentally prepared to advise the crew following the communications blackout period.

At 8:54 a.m. EST, over eastern California, sensors indicated a continued increase in temperature in the left wheel well and mid-fuselage. Two minutes later, while over New Mexico, sensors indicated an increase in drag on the left side, and the flight control systems were automatically compensating. A fuzzy photograph, taken by the Air Force from Kirtland AFB as the shuttle passed over Albuquerque, shows some form of disruption at the leading edge of the wing and a flow trailing off the back, which could have been evidence of plasma vortices from the ever-widening hole.

At 8:59 a.m., the tire pressure sensor caused an onboard alert that was acknowledged by Commander Husband. It is probable at this point that the shuttle pilots understood that something was going wrong with the reentry. The final data from telemetry showed that hydraulic pressure had been lost and that the shuttle's nose yawed left at more than 20 degrees per second (the maximum rate the sensor could determine). The firing of the reaction control system thrusters did not stop the movement, and communication with the crew and loss of telemetry data occurred seconds later.

The vehicle broke up while traveling at 12,500 mph (Mach 18.3) at an altitude of 207,135 ft. over East Central Texas, where it was witnessed by hundreds of people on the ground and videotaped. Debris from the spacecraft rained down on a wide swath across the state. As the visual sighting reports filtered in to Mission Control, and with the inability to regain communications following the anticipated blackout period, it became obvious that the shuttle had met with fatal circumstances.

During the reentry, the hot ionized gases had been allowed to enter the left wing through the hole punched by the foam debris and proceeded to weaken the internal structure until the aerodynamic loads of reentry caused the wing to fail and the shuttle to disintegrate. It was not survivable, and (like the *Challenger*), while the end came quickly, there may have been a period where the life support of the space suits delayed the inevitable. Contrary to some reports, the astronauts, while burned, were not totally incinerated by the residual reentry processes—their remains were remarkably intact. Evidence suggests that a "hardened" crew module, as proposed following the *Challenger* accident, would have allowed the crew to descend to a lower altitude where bailout would have been possible.

While it is speculative to imagine what might have taken place had the imaging request been completed and the results revealed, it is worthy of consideration as there was another shuttle in the preparation stages at the Cape. Under normal circumstances, it was in a twenty-one-day launch cycle. Could *Columbia* have remained in orbit several days longer and the launch cycle accelerated to produce a rescue mission?

Shuttle Pilot and Mission Commander Sid Gutierrez comments:

> Even without a rescue mission, there were other options the crew could have employed had they known the damage was to the leading edge RCC [reinforced carbon-carbon]. To appreciate it, one must understand that although the temperatures on reentry may be very high, the dynamic pressure is usually very low during those same periods—the air is thin. Postflight analysis revealed that had the crew simply stuffed a wet towel into the hole in the wing, the Columbia might have survived. It would have been a complicated EVA, but when your life depends on it, you can do amazing things. The issue was that the lead flight director for the mission assumed nothing could be done and cancelled the imaging request since she assumed the data would not be used. (Gutierrez interview with author, January 20, 2016)

NASA: "A Broken Safety Culture"

The analysis of the *Columbia* disaster revealed that the findings of the *Challenger* accident, seventeen years earlier, had not been institutionalized. The *Columbia* accident report stated that "Space Shuttle Program managers rationalized the danger from repeated strikes on the orbiter's Thermal Protection System ... [and] the intense pressure the program was under to stay on schedule, driven largely by the self-imposed requirement to complete the International Space Station."

The same failures in management awareness of potentially fatal problems—in this case, falling debris—were still prevalent. The debris problem was likely to occur when the ET, filled with its cryogenic fluids, experienced shrinkage. The cocoon of brittle insulating foam could fracture and portions come lose—possibly striking the fragile underside of the shuttle. The loss of even a single tile on the bottom of the shuttle could result in a "zipper effect" that might strip away other tiles and result in a burn-through in the aluminum skin of the orbiter during reentry. There had been many warnings of impending foam damage.

A report in August 1996 by MSFC stated that the leading edge of a wing of *Atlantis* suffered major impact damage in a flight in 1992. In 1997 engineers warned that hardened foam had broken off the ET and had damaged tiles. In October 2002 a piece of foam from the ET of *Atlantis* had broken loose and hit the skirt of a solid-fuel booster rocket. Debris from the ET and SRBs—including ice and hardened foam—had damaged orbiter tiles during the shuttle's first thirty-three flights. The foam that hit the *Columbia*'s wing came from the bipod, the attachment between the external fuel tank and the shuttle's nose.

In the final analysis of the evidence, it was believed that the damage was to the leading edge. Most of the twenty-two panels that made up the leading edge of the left wing had been found, except for the one closest to where the wing was attached to the fuselage. The pattern of the debris that was scattered over more than 400 mi. of central Texas and radar returns from the FAA air traffic control facilities had been used to help determine the sequence of the breakup. The accident report stated that "the five analytical paths—aerodynamic, thermodynamic, sensor data timeline, debris reconstruction [84,000 pieces—34 percent of the shuttle], and imaging evidence" all independently arrived at the same conclusion.

In the beginning of the shuttle program, NASA safety rules said no debris (to include ice and foam) should be allowed to hit the orbiter and possibly damage the fragile heat-resistant tiles. Following the *Columbia* accident, NASA identified more than 170 possible sources of liftoff debris. Engineers recognized that they could not eliminate all risk from debris, but they could do a better job of reducing it. The question must be asked, is foam really that complicated?

The 2001 report of the Aerospace Safety Advisory Panel (ASAP), delivered a year before the *Columbia* accident and presented thorough verbal testimony to both houses of Congress, gave an early warning of what was to come. Excerpts follow:

The purview of the ASAP is safety. Inadequate budget levels can have a deleterious effect on safety. Clearly, if an attempt is made to fly a high-risk system such as the Space Shuttle or ISS [International Space Station] with inadequate resources, risk will inevitably increase. Effective risk management for safety balances capabilities with objectives. If an imbalance exists, either additional resources must be acquired or objectives must be reduced. The Panel has focused on the clear dichotomy between future Space Shuttle risk and the required level of planning and investment to control that risk. The Panel believes that current plans and budgets are not adequate.

Last year's Annual Report highlighted these issues. It noted that efforts of NASA and its contractors were being primarily addressed to immediate safety needs. Little effort was being expended on long-term safety. The Panel recommended that NASA, the Administration, and Congress use a longer, more realistic planning horizon when making decisions with respect to the Space Shuttle. Since last year's report was prepared, the long-term situation has deteriorated. The aforementioned budget constraints have forced the Space Shuttle program to adopt an even shorter planning horizon in order to continue flying safely. As a result, more items that should be addressed now are being deferred. This adds to the backlog of restorations and improvements required for continued safe and efficient operations.

The Panel has significant concern with this growing backlog because identified safety improvements are being delayed or eliminated. . . . Unless appropriate steps to reduce future risk and increase reliability are taken expeditiously, NASA may be forced to choose between two unacceptable options—operating at increased risk or grounding the fleet until time-consuming improvements can be made.

Safety is an intangible whose value is only fully appreciated in its absence. The boundary between safe and unsafe operations can seldom be quantitatively defined. Even the most well-meaning managers may not know when they cross it. Developing as much operating margin as possible can help. But, as equipment and facilities age, and workforce experience is lost, the likelihood that the boundary will be inadvertently breached increases.

Unfortunately, the failure to heed this warning is reflected on pages 104–5 and the conclusion on pages 117–18 of the *Columbia Accident Report* published eighteen months later. Many of the issues raised in the ASAP 2002 report—written before that accident—are also reflected in the *Columbia Accident Report*, including the concern expressed about independence for the safety program. The ASAP argued this issue with NASA throughout the year leading up to the accident, and both finally agreed to disagree.

This was a case where NASA actually did say "No" as opposed to just ignoring the recommendation. Of course this was also a contentious discussion when the report was delivered to the administrator. The Honorable Bob Frances (lead National Transportation Safety Board investigator on TWA 800) delivered this part of the report and cut the administrator off immediately when he started to interrupt him. Sensing it would not go well, the administrator held his comments during Frances' presentation.

Given that the panel had only an advisory function, that clear warnings were ignored by the administration and both houses of Congress, and as there was no indication that the current administrator intended to address the underlying issues, the entire panel resigned en masse during the summer. Sean O'Keefe left and was replaced by Michael Griffin, who invited the original members back to seek their advice. He agreed on the risk inherent in flying the shuttle but thought the better approach was to retire it soon rather than incorporate a crew-escape system. He recommended to President George W. Bush that the shuttle program end. In this case, the roll of the dice worked, and no more astronauts were lost.

Sid Gutierrez notes:

It is correct that the ASAP recommended a crew-escape system every year I was on the Panel. But it would be more accurate to say NASA ignored it. They never really responded. The last year was particularly contentious since the report recommendation was written, printed, and provided to NASA headquarters before the *Columbia* accident but presented in person after the accident. It was particularly embarrassing since we provided a concept for an escape system that would function during all phases of ascent and entry based on 1950s technology and requiring minimal impact to the structure of the shuttle. The administrator was not interested in hearing that presentation and did all he could to stop it. When I eventually delivered it to him in public, it involved what *USA Today* described in a front-page article as a "lively exchange between Sid Gutierrez and the NASA Administrator."

It was obvious to any observer that, had NASA followed our earlier recommendation, the Columbia crew would be alive. Both accidents were eminently survivable—the Soviets were not the only ones interested in putting the best face on dead astronauts/cosmonauts.

The problem with much of the thinking regarding escape systems is that it did not get beyond having all the astronauts sitting in their normal seats. The ASAP envisioned most, if not all, of the astronauts in an escape module in the payload bay replacing the airlock and docking adapter. It had a true full envelope capability with minimum impact on the orbiter. Based on existing escape system data from Martin Baker, I estimate the probability of survival during the *Challenger* or *Columbia* scenario at over 95 percent—standard modern ejection seat numbers.

The crew of the *Challenger* died on impact with the ocean, although they were probably unconscious due to hypoxia long before then—reference the activated emergency compressed air. The crew of the *Columbia* survived the break up at about 200,000 ft. and rode the crew cabin down to about 140,000–160,000 ft. where the aluminum structure failed due to aerodynamic heating and they were literally torn apart. Officially, they died from blunt-force trauma.

To illustrate how survivable the accident was with only minimal escape protection, a PC printer stowed in a Nomex-covered cardboard box about two feet from the head of one of the middeck astronauts was found on the ground. Some of the plastic was broken, but it looked like it was in good shape. Out of curiosity, the recover crew plugged it in and it worked! I saw the printer myself in the recovery hanger. (Gutierrez interview with author, January 20, 2016)

As with *Challenger*, the *Columbia* accident board issued its report on the causes of the *Columbia* disintegration and was especially critical of a "broken safety culture" at NASA that had grown complacent of many risks. It stated: "The organizational causes of this accident are rooted in the Space Shuttle Program's history and culture, including the original compromises that were required to gain approval for the Shuttle, subsequent years of resource constraints, fluctuating priorities, schedule pressures, mischarac-terization of the Shuttle as operational rather than developmental, and lack of an agreed national vision for human space flight." The report went on to say: "Cultural traits and organizational practices detrimental to safety were allowed to develop, including: reliance on past success as a substitute for sound engineering practices (such as testing to understand why systems were not performing in accordance with requirements); organizational barriers that prevented effective communication of critical safety infor-mation and stifled professional differences of opinion; lack of integrated management across program elements; and the evolution of an informal chain of command and decision-making processes that operated outside the organization's rules."

The report concluded that "[the Shuttle] has never met any of its original requirements for reliability, cost, ease of turnaround, maintainability, or, regrettably, safety." Most damningly, it says: "Based on NASA's history of ig-noring external recommendations, or making improvements that atrophy with time, the Board has no confidence that the Space Shuttle can be safely operated for more than a few years based solely on renewed post-accident vigilance."

Astronaut Gutierrez, drawing upon the work of a number of organiza-tions, including the Rogers Commission, the Air Force, independent stat-isticians, and the ASAP, concluded:

NASA actually never projected a failure rate in any scientific or statis-tically accurate manner before the first accident. Although NASA ad-vertised one-in-10,000, that number was pulled out of thin air [They had to be forced] to do a scientific/engineered study of the real risk so they could make informed decisions.

Reluctantly and ever so slowly NASA moved in that direction, but as late as the spring of 2003, when I delivered ASAP's final presenta-tion of a recommendation for a crew escape system to NASA Admin-istrator Sean O'Keefe, I was bombarded by a legion of statisticians at NASA headquarters. They declared I could not call the failure rate one-in-seventy-five based on actual flight data. They said I needed to quantify the failure rate for each component and then integrate it all together. Of course, neither the panel nor I had the resources to do that. In addition, the actual empirical full-up system rate is more credible. The purpose of their effort was to stop the presentation I was going to make. In order to be allowed to give my presentation, I retitled the number used to "Demonstrated Risk of Catastrophic Failure 1/57–1/88." (Gutierrez interview with author, January 20, 2016)

Following a stand-down of almost 2.5 years to address the problem, NASA's contention that it had produced the safest ET in shuttle history was dispelled two minutes into the launch of the shuttle *Discovery*'s return-to-flight mission in August 2005—a 0.9 lb. piece of foam was dislodged and struck the orbiter. The flight, commanded by the first woman, Eileen Collins, could have led to another catastrophe if the piece had ripped away a minute sooner. It forced the immediate suspension of shuttle flights until the problem could again be addressed. Eleven months elapsed before another flight.

With shuttle flights then costing $500 million per launch, and with the shuttle fleet again diminished to three vehicles, the call went out to end the program and replace the aging "superstar" with something more cost-effective and safe. NASA had spent 2.5 years and $14 billion after the *Columbia*'s destruction trying to fix the shuttle's problems.

It had long been asserted that the shuttle was never a cost-effective means of placing cargo or astronauts into space. Much of the work carried out by the shuttle could have been accomplished by unmanned spacecraft for far less money. This is not to say that humans should stop flying into space but that human space flight should be done by means less complex and costly.

Few deny that the shuttle was a modern marvel. With more than 2 million parts, it advanced the state of the art. However, its basic design leaves no room for survival when something goes wrong. The assumption that the shuttle could be made as safe as an airliner was highly optimistic—in fact unrealistic. It is interesting to note that the two catastrophic failures in the shuttle's twenty-five-year career were not the direct result of any of its high-tech features such as the LH2 engines, the exotic thermal protection, or the exacting computer control of its fiery reentry. The failures came from basic components—rubber and foam—that NASA management had been made aware of but had failed to acknowledge.

Following the *Columbia* tragedy, President George W. Bush acknowledged that the time had come to retire the aging relict thirty years after its first flight. It has been asserted that the shuttle continued to be funded long after it had been proven incapable of performing the job for which it had been designed. Its overdue departure was the result of a NASA bureaucracy, which leaned heavily on its effective lobbying power with Congress and successful public relations with the American people. There is perhaps an element of truth in those assertions. However, the shuttle was but another of humanity's attempts to provide an outlet for its intellect, curiosity, and creative engineering.

In a rare admission of error in September 2005, then NASA administrator Michael Griffin, stated that the space shuttle and International Space Station were costly strategic mistakes and raised doubts about America's commitment to getting the shuttle back in flight. When asked if the decision to build the shuttle was a mistake, Griffin said, "It is now commonly accepted that was not the right path. We are now trying to change the path while doing as little damage as we can. My opinion is that it was a mistake." He further added, "It was a design which was extremely aggressive and just barely possible."

Nevertheless, the shuttle was also inextricably tied to the International Space Station, whose construction was dependent on the shuttle's ability to launch large payloads. Was it possible to complete the International Space Station, which NASA's pre-*Columbia* schedule indicated required at least another twenty-eight flights?

The shuttle returned to flight status with *Discovery* (STS-114) in July 2005 and flew another twenty-one missions before it was finally retired in July 2011, with *Atlantis* (STS-135). The flight program had existed for thirty years—about twice the life that was originally envisioned.

24

MIR

A DURABLE SPACE STATION

The Salyut program had been a difficult but rewarding experience for the Soviets. Over the years a wide variety of scientific, biological, military, and fabrication process experiments were tried. However, there had been a fundamental problem in performing many experiments as specialists were needed who could effectively interpret and modify the process in a real-time environment. The American space shuttle was able to take advantage of its ability to fly up to eight astronauts, most of whom were payload or mission specialists, along with the experiment. However, the shuttle was limited to missions of no more than two weeks' duration. To address the Soviet's problem, a more sophisticated follow-on was approved in 1976—Mir. With the new Mir station, crew selection was expanded to include such specialists.

Mir Assembly: Overcoming Problems

The word "Mir" can take on several meanings in the Russian language but is generally interpreted as "peace." The Soviets had struggled since the inception of their country in 1917 to overcome the poor image that their militant and often brutal policies conveyed to the rest of the world. The military focus of much of their space program required that they make a concerted effort to change that image, and the naming of Mir (as well as several other programs) was yet another such attempt.

Mir was designed to be the first continuously manned space station supporting a primary crew of two for nine years (except for two brief periods). The duration of each crew was gradually lengthened from an initial four months to as long as one year. During that period supplies were flown up using improved versions of the R-7 to orbit the unmanned Soyuz called

Progress, and periodic visits were made by visiting crews to exchange the limited life span Soyuz.

The station was assembled incrementally, with a series of five modules (arranged in a "T" shape). It grew to more than 130 tons and measured 107 ft. long with docked Progress-M and Soyuz TM spacecraft, and it stretched 90 ft. wide across its array of solar cells.

The core contained the principal operations area where the crew monitored and controlled the station's systems, science equipment, and facilities. It was also the primary living quarters, with areas allocated to a galley (to prepare and heat food) and personal hygiene—with a toilet, sink, and shower. In a zero-g environment, these normal daily functions were performed in a different manner to prevent water or waste from floating throughout the station.

For the first time, each crew member had a small private living quarters, though not much larger than a closet. This allowed the cosmonauts their own personal space, something recognized as necessary after several crews experienced the interpersonal problems of living in close proximity. These small areas provided a foldout desk, sleeping bag, and porthole. All of the crew areas of Mir had a distinct floor (designated by its carpeting), walls (which were colored) and a ceiling (which was white) as it was believed that humans still required the traditional orientation despite the fact that there is no up or down in space.

The environmental system allowed the water vapor in the air to be removed and recycled into potable (drinking quality) water. Urine was processed and eventually used with an electrolysis process to separate the oxygen to supplement the primary supplies brought up by each Progress ship. The atmosphere was maintained at a normal 14.7 psi (sea-level equivalent) pressure with the normal 80/20 relationship of nitrogen to oxygen. The temperature inside the space station could be adjusted from 64 °F to 82 °F for a shirtsleeve work environment.

As in the United Sates, the tight budgets continued to plague the Soviet space program, but finally the first module of Mir was launched on February 19, 1986 (three weeks after the *Challenger* disaster), while its predecessor, Salyut 7, remained unmanned in orbit. In keeping with Soviet president Mikhail Gorbachev's earlier edict of glasnost, the launch date was released prior to the flight, and the world media was invited to telecast the event. Once again space exploration was providing an opportunity for national policy to be visibly displayed—in this case, a dramatic change.

The timing of the launch was critical in that the Soviets wanted to have both Mir and Salyut 7 in the same orbital plane, as the first crew to visit Mir would also visit Salyut 7. It required a launch window that was only five seconds in length. The first 44,000 lb., 43 ft. long core module, or "base block," of the new space station was placed into an elliptical orbit with the now Soviet standard inclination of 51.6 degrees. The orbit was circularized over the next few weeks to 250 mi. with a period of ninety-two minutes. The core looked almost identical to the Salyut, with two cylindrical sections but with a five-port docking adapter on the forward end. There was also a single aft docking port.

As a part of the service module, two OMS engines and thirty-two smaller attitude-control thrusters were incorporated within the basic structure. These were primarily used to circularize the initial orbit and stabilize the station. They were not intended to be refueled since the Soyuz and Progress ships provided for periodically restoring Mir to its primary orbit. The size of each module of Mir was determined by the lifting ability of the UR-500 and its structural dynamics. Although their heavy-lift Energiya launcher was under development, its first flight was still several years off.

Soyuz-T15 with cosmonauts Leonid Kizim and Vladimir Solovyov launched on March 13, 1986, and the mission was designated EO-1. The term "EO" represented expedition numbers to distinguish between the various crews that typically arrived in one Soyuz and departed in another— making the tracking somewhat ambiguous at times. To add to the possible confusion, some reports used the designation Mir-"x"—thus, EO-26 and Mir-26 represented the same crew. Radio call signs also differentiated the various crews.

After six weeks on Mir, the cosmonauts went back on board the Soyuz-T15, undocked, and rendezvoused with Salyut 7. (At one point in its orbit, Mir passed within five miles of Salyut 7, but there was never any intent for the two space stations to rendezvous.) They made some repairs and re-moved experiments left by the last Salyut crew, which had departed in haste because of the illness of cosmonaut Vladimir Vasyutin. The T15 crew went back to Mir on June 26 and finally returned to Earth on July 26, 1986, after 125 days in space. Salyut 7 was never revisited (Zimmerman 2003, 237).

A new automated docking system, called Kurs, allowed more flexibility during rendezvous so the space station no longer had to be oriented directly toward the approaching spaceship. Computers were playing an expanded role in Mir operations, and seven Strela cargo cranes were installed that could be remotely programmed from the ground. One computer, called EVM, provided the ability to hold a specific attitude without the crew's

The Mir space station was an impressive structure that consisted of large solar arrays and radiant cooling facilities. Courtesy of Energiya.

intervention. As had been done in the American shuttle program, geosynchronous relay satellites called Luch replaced most of the fleet of tracking ships to maintain continuous communications.

Kvant-1, a 19 ft. long astronomy observatory module with a diameter of 14 ft., was launched in March 1987 as the first addition to the core unit. An extravehicular activity (EVA) was performed before docking could be achieved because of a "garbage bag" problem remaining from a previous Progress cargo ship. The bag had become wedged in the collar area and kept the docking apparatus from achieving a hard dock to the aft port.

Over the next nine years, Mir continued to grow as modules were added to the five-port docking adapter. Kvant-2, a 19 ton, 40 ft. long, 14 ft. diameter expansion module (as large as the Mir core itself) was attached in November 1989. It was the first module to be attached permanently to one of the radial ports and essentially doubled the living area. This module had photographic and biotechnology equipment—and its airlock provided EVA capability.

EVA was a common activity, and the Soviets had long surpassed the United States in the number of hours outside of the spacecraft. However, numerous incidents came perilously close to taking the lives of several cosmonauts. In one situation, an outward-opening hatch literally blew open before all of the pressure had been evacuated, and the hatch hinge was damaged. The cosmonauts completed their EVA assignment but then had difficulty sealing the hatch and came within minutes of running out of oxygen. In the end, it was not the bent hinge that kept it from sealing but a hatch-rim gasket cover that had not been removed.

When the airlock was repressurized following an EVA, lithium perchlorate "candles" were ignited to quickly replenish the oxygen. These solid-fuel oxygen generators burned for less than a minute with an intense flame and give off 1.74 lb. of oxygen. (The average cosmonaut consumes 1.5 lb. of oxygen on a typical day.) These, too, proved to be a hazard.

A small crane was attached to the outside of *Mir*. In addition to aiding in EVA activities, it shifted incoming modules from the bow port to one of the radial ports. The aft and bow ports were the only ones to which an initial docking could be made, so incoming modules had to be relocated manually to clear the bow port.

Launched in May 1990—almost two years later than originally planned— the Kristall module was another 19 ton, 40 ft. long structure for biological and materials processing technology. It could produce semiconductor materials that benefit from the microgravity environment of space. The Kristall also had a greenhouse to grow plants and had a spherical universal docking port with two androgynous docking units for the planned *Buran* (Soviet shuttle). Kristall was relocated to the bottom docking port opposite Kvant 2 in the top port.

The lack of funds caused the remaining two modules to wait in storage almost five years before they could be launched. The Spektr module, added in May 1995, contained equipment for atmospheric research and surface studies. The Priroda remote-sensing module, launched in the spring of 1996, contained a variety of infrared radiometers and radar and spectrometers for examining Earth's atmosphere, with specific interest in measuring ozone and aerosol concentrations.

Every two months an unmanned Progress freighter delivered 5,000 lb. of supplies to the station, including food; compressed air and nitrogen, hydrazine fuel, and oxidizer for the attitude control system; supplemental water; and newspapers and mail from home. New scientific experiments, replacement parts, and special tools were also provided. The oxygen-nitrogen atmosphere had to be replenished, as some was lost when airlocks were opened for spacewalks. Refueling could be performed without a crew on board the space station. A radar homing transmitter and TV cameras mounted on the outside of Progress helped guide and monitor each module to the space station.

As a part of many Progress freighters, a small recoverable capsule named *Raduga* (rainbow) allowed the safe return of up to 330 lb. of material—relieving the Soyuz of some of that task. A ton of trash was generated each month and was loaded into the Progress freighter. Following undocking, the Progress was deorbited to burn up over the South Pacific Ocean.

EO-2 with Yuri V. Romanenko (launched in February 1987) recorded the first long-duration mission of 326 days. His arrival back on Earth was amid a snowstorm that had winds of up to 70 mph, which dragged the spaceship several hundred feet (Zimmerman 2003, 251).

In 1988, as the Soviets struggled to regain influence in the Muslim world following their foray into Afghanistan, Abdul Ahad Mohmand, an Afghan, became the first inhabitant of a Soviet spacecraft to recognize the existence of God when his reading from the Koran was broadcast to the public on Soviet television (Zimmerman 2003, 259). He was almost stranded in space when the retrorocket of the Soyuz returning him to Earth failed to provide the required thrust to deorbit on its first two attempts, but a third attempt on the following day was successful.

The Winds of Political Change

An attempted coup to topple Soviet Premier Mikhail Gorbachev in August of 1991 failed while cosmonauts Sergei Krikalev and Anatoli Artsebarski (who performed a Coca-Cola commercial from space) followed the intrigue that was taking place 200 mi. beneath them. The All-Union Treaty that was being hammered out because of the dissolution of the USSR divided the former Soviet Union into fifteen separate nations. This treaty was about to present problems for the Russian space program.

The primary launch site at Baikonur just happened to be located in one of the countries scheduled to get its independence—Kazakhstan. Although it was expected that most of these newly liberated countries would participate peacefully in a new Commonwealth of Independent States (four eventually refused), the leaders in Kazakhstan knew they had a money-making situation that they could use to negotiate with the Russian Republic. Ultimately, an initial twenty-year agreement provided for the Russians to continue to use the Baikonur launch site for an annual fee of $115 million (Zimmerman 2003, 356).

On December 24, 1991, the Soviet flag was lowered for the last time at the United Nations, and the flag of Russia, as the largest surviving member of the USSR, was given its place. Membership in the Security Council also went to Russia. Gorbachev resigned the following day, with Boris Yeltsin eventually assuming the leadership role. It took some time for the details of the structure of the space program to be reconciled, but eventually Glavkosmos (the Soviet space organization) and the Soviet Ministry of General Machine Building were incorporated into the Russian Federal Space Agency (or Roscosmos), which was modeled after NASA. The design bureau that had originally been Vladimir Chelomey's (and had been melded into Energiya) became a semiprivate agency to continue production of the UR-500 (Proton) launcher. Energiya was also semiprivatized and given the responsibility for the Russian manned space program under the direction of Roscosmos. However, this infrastructure for the space program was not in much better shape than the military or civilian sectors—there was no money to pay even salaries. Funding eventually came from a source that had been the prime antagonist of the USSR—the United States.

In an effort to ensure that its former enemy did not degenerate into a massive civil war that might return a totalitarian government to power, the United States embarked on a program to provide significant financial support (Zimmerman 2003, 367). The basic goal was to ensure that the thousands of nuclear weapons that were scattered across the now-independent fifteen states could be controlled and that engineers and scientists in critical positions did not flee to renegade countries such as Iran or Iraq in search of employment. There were several problems with America's financial investment in Russia—not the least of which was the gross misappropriation of the incoming dollars into the pockets of high officials.

In another attempt to secure hard currency, Roscosmos brought more than two hundred items to Sotheby's Auction House in New York in December 1992. These items, which ranged from spacesuits to watches, brought $6 million. One of Grechko's spacesuit gloves went for $10,925.

The lack of money caused a delay in Mir replacement crews into the spring of 1992. When cosmonaut Sergei Krikalev finally returned to Earth in March 1992 after 313 days in orbit, he landed in the prescribed recovery area—but it was now in a different country—Russia, not the USSR.

In 1992 presidents Bush and Yeltsin agreed to a further cooperative endeavor in space with the American space shuttle rendezvousing with Mir and performing a series of crew exchanges. This eventually paved the way for the higher level of interoperation required for the proposed International Space Station. This agreement was further expanded by the Clinton administration to provide $400 million to the Russians for using Mir to gain knowledge and techniques in using the space shuttle to service a space station. Mir was ultimately visited by nine shuttle flights beginning with STS-63. All but the first docked with the space station.

Russian cosmonaut Valeri Polyakov recorded the longest stay aboard Mir—438 days in 1994–95. Initially launched on January 8, 1994, on board Soyuz TM-18, he was also among the first to celebrate publicly the Russian Orthodox Christmas from Mir on January 8, 1995. The godless Communist regime, despite seventy years of trying, had not destroyed the faith of the Russian people.

Elena Kondakova became only the third female to fly in the Russian program (ten years after Svetlana Savitskaya's flight in 1984), when she went aloft in Soyuz TM-20 in October 1994 and returned with Soyuz TM-20 in March 1995. She made a second flight on STS-84 in May 1997 as a part of the crew of the space shuttle *Atlantis* for a two-week stay.

Shuttle *Discovery* (STS-63) was the first to rendezvous with Mir in February 1995. No attempt was made to dock although a close "station keeping" was performed to see how the shuttle's thrusters impinged on Mir. On board the shuttle was cosmonaut Vladimir Titov, Michael Foale (who occupied Mir on a subsequent journey), and Eileen Collins (who eventually became the first female to command a shuttle). The crew on board Mir commented on how precise Shuttle Commander James Weatherbee was in piloting the 100 ton craft alongside the 100 ton Mir. A subliminal Coca-Cola commercial was performed with the dispensing of Diet Coke and regular Coke for a taste test on board *Discovery*.

The first docking of the shuttle with Mir occurred with the flight of *Atlantis* in June 1995. The combined weight of the station, the shuttle, and a docked Soyuz was 250 tons with ten crew members.

Americans on Mir

Expedition EO-18 saw the first American, Norman Thagard, fly on board a Soyuz (March 1995) en route to Mir with cosmonauts Vladimir N. Dezhurov and Gennady M. Strekalov. Both Strekalov and Thagard were experiencing their fifth flight into space. Thagard's flight as a part of a Russian crew highlighted the considerable differences between the two cultures. Although he had trained with the Russians, it was not until he came on board that he discovered that, when "on the job," they became bellicose and condescending—especially the mission commander. These observations were borne out by subsequent Americans who encountered the problems that multicultural crews experienced in the future (Zimmerman 2003, 376).

Thagard's flight was also another foreshadowing of the encroaching bureaucracy that was befalling NASA. Perhaps because these early flights to Mir by American astronauts were a "political deal," Thagard received poor support from NASA's own Mission Control. He returned after 181 days by way of STS-71, which delivered the first Russians to Mir by shuttle. He reported Mission Control's lack of responsiveness to his requests to none other than the NASA administrator, Dan Goldin, who acknowledged the problem but failed to make any changes (Zimmerman 2003, 381). The *Challenger* disaster of almost ten years earlier had not improved the management ability of NASA to address critical issues in the space program.

Shannon Lucid, who launched in March 1996 on board STS-76 and returned in September 1996 by way of STS-79, found the same belligerence with the Russian crew and neglect from NASA, but she handled the problem better than did her male counterparts. Her personality and her ability to receive a one-hour CNN weekly news update, and periodically to email her family, eased the isolation. It was during her flight that the cosmonauts performed their Pepsi commercial. She had to avoid any involvement because of the strict limitations NASA had on possible personal/business gain from space flight activities.

American astronaut John Blaha attended the Defense Language Institute in Monterey, California, to learn the Russian language and completed the Russian training program at the Cosmonaut Training Center in January 1995. In September 1996, by way of STS-79, he arrived on Mir, where he spent four months with the EO-22 cosmonauts conducting material science, fluid science, and life science research. Blaha returned to Earth on STS-81 in January 1997. When he left Mir, he remarked to his replacement,

The space shuttle *Atlantis* docked with Mir. Courtesy of Energiya.

Jerry Linenger, "Don't expect any help from the ground. You're up here on your own" (Zimmerman 2003, 412).

Linenger was the fourth NASA astronaut to crew Mir, launching on board STS-81 in January 1997 and returning with STS-84 in May 1997. His statements were even stronger than his predecessors on his return: "The [Shuttle-Mir] program was not primarily concerned with doing good science or advancing our expertise in space operations, but rather was conceived and thrust down NASA's throat by the Clinton administration as a form of foreign aid to Russia" (Zimmerman 2003, 382). As a result, NASA's indifference to their obligation to support the program was a way of NASA's "getting even" with the politically appointed Clinton administrators (ibid.).

However, Linenger had more than just a crusty Russian crew and ignorant bureaucrats at NASA to worry about during his stay. On February 23, a few weeks after Linenger arrived, cosmonaut Aleksandr Lazutkin was in the Kvant module, where he lit a lithium perchlorate candle to increase the oxygen content of the atmosphere. With six people on board, the oxygen regeneration system had to be periodically augmented. Instead of producing the intended gas, it burst into flames—as one had in 1994. Unlike the previous fire, which was immediately smothered with a cloth, this one burned furiously. The fire alarm was triggered, and a call was made for a fire extinguisher as thick smoke billowed out.

Linenger was in the *Spektr* module and was initially unaware of the emergency. He heard the audio alarm, but because it had previously triggered when there was a problem with the electrical power, he didn't expect that it was a fire. He proceeded to save the work he was doing on a laptop computer. When he passed into the central docking adapter to see what was happening, he was met by one of the crew who was yelling "fire!" As he moved to view the interior of the Kvant module, he saw the flames and smoke, which now blocked any access to the second Soyuz that was moored to the aft docking port. Only the one Soyuz attached to the central docking adapter was available for abandoning Mir, and it could accommodate only three of the six crew.

The crew donned respirators, but the first one Linenger tried failed to work—as did the first fire extinguisher brought out. There were several other extinguishers immediately within reach, but they had been bolted to the walls to secure them during the launch phase several years earlier, and no one had thought to remove the bolts. Linenger positioned himself in the docking tunnel and handed another extinguisher to cosmonaut Valery Korzun. However, as the cosmonaut pulled the trigger, the discharge of its content exercised Newton's third law of motion and he was propelled backward away from the flames. Linenger had to grab Korzun's legs and steady himself against the bulkhead so Korzun could direct the stream of precious extinguishing agent against the base of the fire (Zimmerman 2003, 319).

After what was estimated to be fifteen minutes, the fire went out. Other than Korzun, who had several fingers burned, there were no injuries. It took a full day before the environmental system could clear the smoke and the respirators could come off. However, it took a while before they had all the soot cleaned up. Another tragedy had narrowly been averted. The Russian crew and Roscosmos suppressed the intensity and seriousness of the fire, and NASA was content to accept the Russian assessment despite Linenger's report to the contrary (Zimmerman 2003, 420).

Less than two weeks later, failure in the automated docking system caused yet another near collision with a Progress supply ship. Again the

Russians refused to acknowledge the potentially fatal accident, and NASA was unconcerned. Linenger became so upset with Mission Control that he finally decided not to communicate with them over the voice channels and simply used email.

Astronaut Michael Foale replaced Linenger in May 1997 and found the same conditions that his American predecessors had been complaining. Foale's personality was, like Shannon Lucid's, more forgiving and easygoing, and he found it easier to brush off the cultural affronts of the current Russian contingent, Vasili Tsibliev and Aleksandr Lazutkin, and get on with his experiments.

A docking test was scheduled toward the end of June to perform a manual procedure. The Progress supply ship was undocked from the aft port, and the maneuvers began. However, the commander lost sight of Progress at a critical moment, and before he could reestablish its location, cosmonaut Lazutkin observed it closing at a high rate of speed. He yelled, "My God, here it is already!" Within seconds, the Progress hit Mir between the Kvant module and the docking port. It then somersaulted into the Spektr module, and the crew could feel the pressure inside the space station change as their ears began popping—the pressure hull had been breached (Zimmerman 2003, 431).

Inside the Spektr module, the crew could hear the sound of air escaping into the void of space. It should have been a simple task to isolate that module. However, the hatchway was blocked open with eighteen air ducts and electrical cables laid between the module and the rest of the space station to provide a variety of connections to other equipment and power. Lazutkin grabbed a knife and tried to cut through the cables but this proved futile. He then entered the Spektr module and disconnected them. With the path finally cleared, the hatch was closed, and the module was finally isolated from the rest of the space station.

Some of these disconnected cables, that now lay floating inside Kvant, had provided power from Spektr's solar cells to the rest of the space station. Availability of electrical power dropped significantly. In addition, the impact had caused the station to begin a slow rotation that was too forceful for the gyros to stabilize, and the attitude control system had shut down. With the station no longer positioned for its solar cells to receive maximum sunlight from the remaining solar cells, even more electrical power was lost. For more than a full day, Mir drifted with no lights or fans. During the forty-five minutes of each orbit when they were in total blackness, they could do nothing, and Foale noted that the three of them gathered "in front of the big window, looking at incredibly complex, swirling auroras with the galaxy showering down on them, with nothing else to do" (Zimmerman 2003, 433).

While the Russians pondered their next course of action, Foale convinced them that they could regain control using his observations of the proximity of Earth from the window relayed to Tsibliev, who would work the attitude controls in the attached Soyuz. It took a while, but eventually the crew got the solar panels pointing toward the sun, and enough power was restored to begin recharging the batteries (Zimmerman 2003, 433).

While this was obviously the key to saving Mir, the availability of electrical power once again allowed the toilets to work after almost two days, and the crew was "relieved." However, the stress was too much for Tsibliev and Lazutkin, who returned to Earth on August 15 and were replaced by a new crew before their scheduled time had been completed. Back on Earth, a new hatch was quickly fabricated that had electrical plugs on both sides so that the power cables could be plugged in to restore the connection to Spektr's solar cells. This was flown up on the next Progress, and an "internal" spacewalk allowed its installation.

It was somewhat paradoxical that the Russians became more efficient and improved their quality control in their space endeavors using many techniques observed from NASA. Their American mentor (NASA), who had accomplished virtual miracles in space with Apollo and the creation of the shuttle, had degenerated into a squabbling bureaucracy that was indifferent to the needs of its astronauts. This arrogance and complacency eventually crippled the American space program and brought it to a virtual standstill in the first few years of the new millennium.

Russian Capitalism

When the head of Energiya, Valentin Glushko, passed away in January 1989, Yuri Semenov took the helm; his first objective was to improve funding to enable Mir and the rest of the Soviet space program to continue. At this point only one module had been added to the core in almost three years (Kvant 1—a 19 ft. long astronomy observatory). Even the cost of the bimonthly Progress ships was a drain, and the station remained unmanned for a significant period. This occurred with the departure of A. Volkov, Krikalev, and Polyakov in April of 1989. A Progress freighter then boosted Mir into a higher orbit to extend its life.

Maintaining Low Earth Orbit

Although the FAI definition of space begins at 62 mi., molecules of air exist at extended distances past 1,000 mi.—albeit at greater lessening densities. Each time the spacecraft impacts a molecule, it loses energy and drops into a lower orbital altitude. With the decrease in altitude, the density of the air molecules increases and the orbital decay is exacerbated. To avoid premature reentry, and to enable extended orbital periods in low Earth orbit (less than 1,000 mi.), the spacecraft must be periodically reboosted back up to its original orbit.

As the Soviet Union disintegrated in 1989, the space program, crippled by the lack of funding, was given permission to engage in a series of financial deals with other countries and Western corporations. The decision was an outgrowth of one of Gorbachev's initiatives (called the Enterprise Law) to transform state-run industries into private companies. This included commercials for Coca-Cola and Pepsi to be broadcast from Mir. Mission Control in Moscow suddenly sprouted advertisements hanging from the walls.

Cosmonauts returned to Mir in February 1990 on board Soyuz TM-9, partially funded by the producers of the television documentary program NOVA. The launch vehicle now sported advertising logos, and the launch complex had a 150 ft. high sign for an Italian insurance company.

Another endeavor used the faculties on board Mir to manufacture protein crystals for pharmaceutical companies. An estimated 25 million ruble profit resulted from the sale of 220 lb. of these products. The cosmonauts also produced other organic and inorganic substances whose structures were improved when made in the weightlessness condition on board Mir. This included glass, metal alloys, and semiconductor materials. Unfortunately, most of these proceeds were from other Russian "businesses," so it was simply moving money from one pocket to another and not generating hard cash from outside the country.

The Russians also planned to earn money by charging businesses for testing materials in the Kristall laboratory and by sending "tourists" to Mir for about $10 million per person for a two-week stay. The tourist plan continued into the twenty-first century, after the end of the Mir era, with several space-tourist flights on board Soyuz transports to the International Space Station.

While Energiya and Khrunichev State Research and Production Space Center were marketing spacecraft capabilities (the French paid $36 million for three crew positions, and the European Space Agency paid $60 million for two Germans), the launch vehicles were contracted to foreign agencies to launch satellites by Glavkosmos. These activities provided a significant improvement in needed capital to rebuild and sustain the Russian space program, and the foreign income reached more than $500 million by 1996.

The End of Mir

Over its thirteen-year life span of habitation, almost sixteen thousand experiments were conducted, with the emphasis being the adaptation of humans to long-duration space flight. Mir also experienced more than sixteen hundred system failures—some were serious, such as the onboard fire in 1997. By 1989 almost half the scientific equipment on Mir was no longer functioning, and by 1992 the Kvant 1 astrophysics laboratory module was simply used to store equipment and garbage (Zimmerman 2003, 231). There were several near-fatal collisions with Progress cargo ships, including the one that left the Spektr module unusable. The solar panels degraded at about 5 percent per year and by the end of the century were producing only about 15 kW of electrical power. The windows had become almost opaque from the constant impacts of micrometeoroids. The geosynchronous communications satellites used to maintain an almost constant communication with Mir were also in disrepair, as were the few tracking ships that remained.

However, even the near disasters and numerous problems due to the age of Mir were not enough to deter the Russians. The Soviet/Russian space program had proven to be resilient and resourceful. Ultimately, the cost of upkeep was more than the Russian economy could handle in light of their commitments to the upcoming International Space Station.

NASA finally determined that not only was Mir a hazard, but it was also draining its own budget of funds needed for the International Space Station (ISS). When cosmonaut Valeri Ryumin returned to space in 1998, more than twenty years after his last flight to Salyut 6, his impressions of Mir were very down to Earth. "I don't know how they live up here," he was quoted as saying. "This is worse than I imagined. This is unbelievable. This is unsafe" (Zimmerman 2003, 442).

EO-27, with cosmonauts Viktor M. Afanasyev, Jean-Pierre Haigneré of France, and Ivan Bella of Slovakia, launched in February 1999. They returned to Earth in August 1999, completing the final planned Mir mission of 188 days in orbit. When they left Mir on board Soyuz TM-29 on August 28, 1999, the space station was unmanned for the first time in ten years.

The station was to be deorbited in early 2000, but MirCorp, an American–Russian international group based in the Netherlands, wanted to use Mir for commercial and tourist purposes. An automated Progress mission in February 2000 boosted Mir's orbit 40 mi. higher, while the Russians sought to determine its fate.

EO-28, with cosmonauts Sergei Zalyotin and Aleksandr Y. Kaleri, launched April 4, 2000, in Soyuz TM-30. They spent seventy-two days in space in a flight funded by MirCorp—the only privately funded mission.

On March 23, 2001, the 120 ton Mir was deorbited to a fiery end over the Pacific, where an estimated 1,500 pieces impacted without incident. One hundred and four men and women had inhabited it, representing twelve countries; four of the men had spent more than one year in space in various time increments. Mir had traveled more than 2.2 billion mi. in 86,000 orbits of the Earth. Sixty-six Progress supply ships had delivered more than 80 tons of cargo, as opposed to the 13 tons that had supplied Salyut 6 and the 15 tons sent to Salyut 7.

Cosmonaut Anatoli Solovyov completed sixteen spacewalks, spending 78 hours outside during his five expeditions to Mir. More than 780 hours of EVA were accomplished. Mir had proved to be a significant step in the conquest of space.

25

A PERMANENT PRESENCE
IN SPACE

The oil embargo and resulting inflation of the late 1970s caused a recession that plagued the Jimmy Carter presidency. President Ronald Reagan faced significant budget deficits during his first administration. With the political reality of tight funding and lukewarm public support in the early 1980s, NASA had accepted that the shuttle had to precede the space station. For America, whose Skylab effort in 1975 had been cut short by a lack of enthusiasm and dollars, a large, continuously manned space station as envisioned by Tsiolkovsky, Oberth, Goddard, and von Braun could finally proceed when the shuttle was declared "operational" in 1982.

In an effort to galvanize the American people behind such a project, President Reagan announced his vision during the annual State of the Union message to Congress in January 1984. Leaning on President Kennedy's "before this decade is out" theme, Reagan proposed that NASA "develop a permanently manned space station and do it within a decade." Unlike Kennedy, who wanted to compete with the Russians during the Cold War, Reagan saw this as a step in the direction of stimulating commerce in space and requested $8 billion to accomplish the task (Zimmerman 2003, 208). However, some in Reagan's own administration were opposed to the project—there was yet no proven need for the "permanent presence" of humans in space.

The president was not satisfied with how America was proceeding into space; the bloated and bureaucratic NASA was the antitheses of the private enterprises that he preferred to develop space initiatives. Nevertheless, in his view, now was time for the space station.

International Space Station: Alpha

Reagan's announcement was actually the culmination of many years of planning by NASA. However, the announcement did not provide a specific goal or functionality for such a space station—which was not an end unto itself. Thus began more years of design and redesign of the station. A bureaucratic power struggle occurred between the various NASA centers that fought each other with the vigor that they had previously applied to the technical challenges of the Apollo program twenty years earlier. Each had specific design goals that, coupled with congressional oversight and micromanagement, made progress slow. No less than three distinct configurations were formulated over the next six years as the country attempted to define just what functions the space station should have and how it should be constructed and manned.

The first design that emerged in mid-1984 was a 450 ft. long truss to which five habitable modules were attached. Large solar panels supplied 75 kW of electrical power, and a crew of six operated the gravity-gradient-oriented station. However, this layout was criticized by scientists as not allowing for effective microgravity experiments and for having too little electrical power.

A second arrangement situated the modules in the center of a 508 ft. truss and used mirrors to provide a steam-powered electrical generator—recalling the concepts of the visionaries of fifty years earlier. By this time the Japanese and European space agencies were included as partners, and two of the five modules were built by them—reducing the cost to the American taxpayer. It was at this point that the term "International Space Station" was applied—the ISS. However, this design almost doubled the initial estimates, and another redesign occurred. A reshuffling of the structure and the deletion of the expensive steam turbines left the station with only 45 kW of power and an estimated $12 billion price tag.

The name Freedom was applied to the space station, and the suggestion was made that, to save money, it need not be continuously manned. Many of the experiments could be conducted without constant tending by a crew. The main truss was shortened to 353 ft., and some of the solar-power panels were eliminated, which reduced the power capacity to only 30 kW. The attitude-control fuel was changed from hydrogen peroxide to hydrazine, and the closed-loop environmental system was eliminated—requiring the shuttle to supply water periodically. The design effort alone had cost the nation almost $4 billion by 1991, and yet nothing had actually been fabricated, let alone launched into space.

Although a spectacular machine, the shuttle could carry only about 50,000 lb. of payload in its cargo bay—about one-fourth that of the Saturn V, which had been abandoned a decade earlier. It would take four or five shuttle flights just to put a space station the size of Skylab into orbit.

With little progress made by the time Reagan's successor, President George H. W. Bush, left office in January of 1993, the Bill Clinton administration had to pick up the ISS political hot potato. The Soviet Union was in disarray following its economic collapse, and it was obvious that, despite their accomplishments in long-duration human space flight, the Soviet Union was no longer a serious military threat in space. Their Mir space station, although relatively successful, would not be followed by Mir-2 for lack of funding unless the Russians became a partner in the ISS.

NASA was on its seventh design iteration of the ISS, which continued to grow smaller as the U.S. budget deficit grew larger. Although President Clinton endorsed the ISS, it was nearly canceled by Congress on a vote of 216 to 215 in June of 1993. With its capabilities being severely reduced over the years, it was under fire from the scientific community as being inadequate, and its political meaning had been virtually eliminated by the end of the Cold War. At that point the ISS provided for a crew of only four while the Russian Mir-2 design provided for a crew of six. Clinton urged NASA to finalize its negotiations with the Russians for them to become a full partner. By September Clinton agreed to go with the larger crew accommodation of the Mir-2, and by October 1993 the Russians (there was no longer a Soviet Union) agreed to an amalgamation of their Mir-2 with the ISS. The name Freedom was dropped, as Clinton did not want the ISS to be associated with the Reagan administration, and the somewhat less inspiring name Alpha was substituted (Zimmerman 2003, 332). In December 1993, NASA and Roscosmos formally announced the Russian inclusion into the joint manned space station project.

However, with the inclusion of the Russians, the orbital inclination of the ISS had to be increased to 51.6 degrees to be easily accessible by rockets launched from the Baikonur spaceport. (With the dissolution of the USSR, Baikonur was now being leased by the Russian Federation from new Republic of Kazakhstan). This higher orbital inclination effectively reduced the shuttle's net payload. To improve the shuttle's performance, NASA developed a new super-lightweight aluminum-lithium propellant external tank (ET).

As finally defined, the ISS consisted of thirty-one modules built by five countries. Each of the modules was constructed on Earth, launched into orbit, and assembled together by the astronauts with extravehicular activities. A 354 ft. long truss provided stability to the structure and acted as a wiring conduit for communications and power.

As originally designed, the ISS weighed almost 1 million lb. and had 110 kW of electrical power (the equivalent of about forty homes) furnished by an extensive array of eight solar cell "wings." These arrays track the sun with two gimbaling systems to ensure that maximum power is generated. During the nighttime, the arrays are positioned to reduce the atmospheric drag (called the "Night Glider mode"). During the nominal forty-five minutes of night, the rechargeable nickel-hydrogen batteries ensure a continued supply of electricity.

To accomplish the launch task, more than fifty flights were originally envisioned, with the space shuttle performing thirty-nine and the Russian UR-500 Proton and European Space Agency's (ESA) Ariane providing the remainder. As with Mir, the Russians provided regular unmanned Progress supply vessels.

Assembling the ISS

The first section of the ISS placed in space was Russia's 42,500 lb. Zarya ("sunrise") control module, also known as the functional cargo block. It was launched into a 250 mi. high orbit atop a UR-500 Proton rocket in November 1998. The $200 million module was funded by NASA and built by Khrunichev State Research and Production Space Center in Moscow. A competitive bid by America's Lockheed Corporation came in at over $400 million and was rejected.

The Russian Zvezda ("star") service module, the third major component, launched from Baikonur on July 12, 2000. It is the primary Russian contribution to the ISS and provides the life-support systems and living quarters (sleeping facilities, galley, and toilet). It also has the primary docking port for Progress supply ships and provides attitude control and orbital maneuvering capability for the early station elements.

The first Progress M-1 supply ship (NASA used ISS-1P as its nomenclature) was launched in August 2000. It remained docked to the service module for seventy-three days. During this time, *Atlantis* (STS-106) was launched in September 2000 and ferried supplies to the station in preparation for the first resident crew. The mission included two spacewalks to connect power and communications cables between the Zvezda and Zarya modules.

The crew of Expedition 1 was launched to Earth orbit on board a Soyuz 1 in October 2000 and docked to the space station two days later. The three-man crew consisted of the expedition commander, American astronaut Capt. William M. "Bill" Shepherd (USN), Soyuz commander Russian cosmonaut Col. Yuri Gidzenko (Russian Air Force), and Flight Engineer Sergei Krikalev.

Discovery (STS-102) in March 2001, delivered the 9,000 lb. Leonardo Multipurpose Logistics Module (MPLM) built by the Italian Space Agency (Agenzia Spaziale Italiana). The MPLM is an unpiloted, pressurized module that allows equipment, experiments, and supplies to move between the ISS and the shuttle. *Discovery* also delivered Expedition 2 and returned Expedition 1 to Earth in March 2001. With the second expedition, the command switched to Russian cosmonaut Yuri V. Usachev with Americans James S. Voss and Susan J. Helms rounding out the crew.

Endeavour (STS-100), launched in April 2001, ferried the Canadian space station remote manipulator system, a second MPLM, and the ultrahigh frequency antenna. Soyuz 2 (April 2001) carried Dennis Tito, an American businessman. Tito paid $20 million for the flight as the first ISS space tourist and created a political flap between America and Russia. NASA was determined that Tito not be allowed to fly, but the Russians boycotted their American training until Tito was allowed to participate. The flight delivered a new Soyuz spacecraft to the station for use as an emergency crew-return vehicle. The taxi crew returned to Earth in the old Soyuz vehicle. Fare-paying visitors were finally given official status with the title of "space flight participant" as construction moved into high gear.

Soyuz 3 (October 2001) delivered a new Soyuz spacecraft to the station for use as an emergency crew-return vehicle. Because they have a limited life of about 120 days, the Soyuz are swapped out with each visit so that the returning astronauts leave the newly arrived Soyuz and take the one that has been attached.

Soyuz 4 (April 2002) transported Space Flight Participant Mark Shuttleworth and a new Soyuz spacecraft to the station. Shuttleworth, of South Africa, paid $20 million for the flight as the second space tourist. After eight days, the taxi crew (with Shuttleworth) returned to Earth in the old Soyuz vehicle.

With a nominal (near-circular) orbital altitude of 250 mi., the ISS loses about 300 ft. each day. Thus, the ISS must be reboosted periodically (about

Table 2. International Space Station Structural Assembly of Major Elements in Orbit 1998–2003

Date	Delivery Vehicle	Unit	Payload	Comments
November 1998	Russian Proton-K	Zarya	42,500 lb.	control module, adapted from Mir
December 1998	*Endeavour* (STS-88)	Unity	25,500 lb.	Two pressurized mating adapters provide a six-sided docking hub and passageway to major ISS sections
May 1999	*Discovery* (STS-96)	Strella	12,000 lb.	Space crane; first shuttle docks
May 2000	*Atlantis* (STS-101)	Supply & reboost		First glass-cockpit shuttle flight
July 2000	Russian Proton-K	Zvezda	19,051 lb.	Service module
October 2000	*Discovery* (STS-92)	ITS-Z1 & PMA-3	20,973 lb.	Z-1 Truss & pressurized mating adapter
October 2000	Soyuz 1	EO-1		First expedition
November 2000	*Endeavour* (STS-97)	P6 Truss	17,430 lb.	8-solar arrays, life support, and RCS
February 2001	*Atlantis* (STS-98)	Destiny	14,515 lb.	U.S. laboratory
March 2001	*Discovery* (STS-102)	ESP-1	12,700 lb.	External stowage platform
April 2001	*Endeavour* (STS-100)	SSRMS	4,899 lb.	Canadarm2
July 2001	*Atlantis* (STS-104)	Quest	13,000 lb.	Joint airlock (EVA access) and high-pressure gas assembly
August 2001	Discovery (STS-105)			Deliver EO-3, return EO-2
September 2001	Russia Progress-M	Pirs, DC-1	3,580 lb.	Docking compartment 1
December 2001	*Endeavour* (STS-108)			Deliver EO-4, return EO-3
April 2002	*Atlantis* (STS-110)	S0, MT	27,000 lb. 1,950 lb.	Integrated truss structure, mobile transporter allows arm to travel truss
June 2002	*Endeavour* (STS-111)	Mobile Base Sys	1,450 lb.	Deliver EO-5, return EO-4
October 2002	*Atlantis* (STS-112)	ITS	14,120 lb.	S1 integrated truss segment
November 2002	*Endeavour* (STS-113)	ITS	14,000 lb.	P1 truss, deliver EO-6, return EO-5

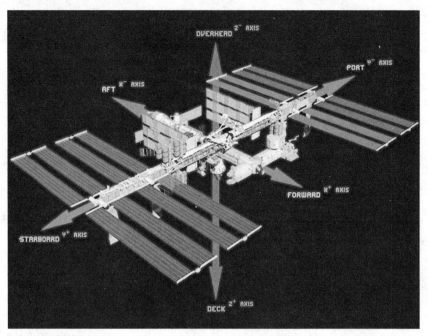

An illustration of the International Space Station showing the orientation axes. Courtesy of NASA.

once each month) to avoid atmospheric reentry. The shuttle orbital maneuvering system was used when it was servicing the ISS; subsequently, the Russian Progress spacecraft are employed. Its four smaller thrusters are used to avoid placing any significant g-force on the ISS structure. These thrusters have about 30 lb. of thrust each, and fire for about 900 seconds—resulting in a delta-V of about 6.5 ft./s—not a big change when the basic orbital velocity is 25,500 ft./s., but a necessary one to keep the orbit stable.

To reboost the ISS typically requires adding velocity at two points in its orbit. The first adds velocity at one point in the orbit that results in an increase in altitude 180 degrees from the point of added delta-V. When the ISS reaches that new apogee (on the opposite side of the world), a second burn increases the altitude to essentially circularize the orbit.

The attitude of the ISS is maintained by several single-axis control-momentum gyroscopes consisting of a spinning rotor and one or more motorized gimbals to change the rotor's angular momentum. When the rotor is tilted, the angular momentum causes the gyroscopic torque (rigidity-in-space characteristic) to rotate the ISS. Thus, the attitude can be adjusted for specific experiments—or simply to achieve the lowest drag profile.

The fiery plunge of space shuttle *Columbia* on February 1, 2003, ended the routine visits of the shuttle to the ISS. NASA recognized that it would be some time (ultimately two and one-half years) before another shuttle crew visited the ISS. While the ISS remained well behind its initial schedule, significant progress had been made with its assembly. This all changed with the loss of *Columbia*.

Soyuz TMA-2 launched on a taxi flight in April 2003. Crew Commander Yuri Malenchenko and Science Officer Ed Lu delivered themselves to the station in a new Soyuz, TMA-2, to become the Expedition 7 crew. They were the first two-person ISS crew and the first primary crew to travel to the space station on a Russian Soyuz spacecraft. By reducing the crew from three to two, ISS supplies lasted longer. The reduced crew maintained systems on board the station, made repairs as needed, and carried out the science experiments. Further construction of the station was postponed. Soyuz TMA-2 remained docked with the station as their lifeboat during their six-month ISS tour that lasted until October 2003. The Expedition 6 crew returned to Earth in Soyuz TMA-1, which had been docked with the ISS as their lifeboat.

Medical and industrial research is the central theme for ISS activities. The Microgravity Science Glovebox allows astronauts to perform a wide variety of materials, combustion, fluids, and biotechnology experiments in weightlessness. The Pulmonary Function System, delivered in July 2005, provides information about the cardiovascular system in the weightless environment. The Kubik incubator performs biological experiments, while human physiology is explored with the Cardiocog, the Neurocog, and the eye-tracking device. The PK-3 device is used for complex plasma experimentation.

Repercussions of Shuttle *Columbia*

When *Columbia* was lost, NASA again grounded all shuttle operations (as it had following the *Challenger* loss seventeen years earlier) while it investigated the problem. *Columbia*, being the first and heaviest shuttle, could not achieve the high orbital inclination required to reach the ISS with the same payload capability as the others and had been used for non-ISS work. However, the problem that resulted in the loss of *Columbia* was just the start of a new series of problems for NASA.

The ISS was far more expensive than originally projected by NASA, and the General Accounting Office was highly critical of NASA's accounting practices, which hid many costs. At the time of the *Columbia* accident, ISS was expected to be finished in 2010—sixteen years behind schedule—and to cost about $100 billion over its lifetime for construction and support. However, the scientific community, for whom the station was built, continued to express its discontent with the ISS, estimating that, when completed, it would have considerably less capability than first planned. That NASA had initially planned to abandon the ISS after only seven years of full operation (by 2017) caused many to again question its viability.

With completion of the ISS closely tied to the availability of the shuttle, a growing sentiment in Congress and the space industry believed that the ISS might never be completed—at least by the United States. The declaration by President George W. Bush that the shuttle program was to end by 2010 and the admission by then NASA administrator Michael Griffin that the decision to build the shuttle and the ISS, in their present configurations, was fundamentally flawed led to speculation that the United States might simply back out of the ISS program.

The Russians indicated that, if America abandoned the ISS, they would continue to operate their part, which has a certain amount of autonomy. The ESA commitment to the ISS is represented by only five of its sixteen member states because of concerns about the expense or simply a lack of interest. ESA contributions include the 40,000 lb. Columbus science laboratory module, which was eventually orbited in 2008 after NASA's space shuttle returned into service.

A General Accounting Office report stated that NASA's logistical support of the ISS with the shuttle was not practical and that it should consider making use of alternative commercial services. An Alternate Access to Station study, a part of NASA's Space Launch Initiative program, was begun in 2000 to determine if other methods of resupply to the ISS could be more efficient than either the shuttle or Progress. The ESA and the Japanese space agency moved forward to develop such a craft.

The ESA constructed a space freighter for the ISS, the automated transfer vehicle, with a cargo capacity of 8 metric tons. It began flights in 2008. After the shuttle's retirement in 2011, and until its replacement is available, the automated transfer vehicle along with Progress and Soyuz are the only link between Earth and the ISS.

However, there were other options. The growing disenchantment with the NASA bureaucracy left the door ajar for private corporations to step in to not only provide supply flights but also create a new manned space vehicle more efficient than the shuttle.

When the shuttle was grounded in 2003, crew rotation and supply of the ISS was carried out by the Russians using the venerable derivatives of Sergei Korolev's R-7 and the Soyuz and Progress spacecraft. However, the science that was being conducted was severely curtailed as the reduced supply flights initially allowed only three persons to operate the station instead of the planned six.

Cosmonaut Pavel Vinogradov, the Soviet commander of Expedition 13 to the ISS lamented that servicing the onboard systems required 62 percent of the crew's time, and 15-percent to personal needs, leaving only 23-percent to science.

There are few in Congress or industry who doubt that the future in space is important and that the ISS can and should play a role in expanding the horizon of the human presence. Nevertheless, for the past thirty years, NASA, like virtually all large bureaucratic institutions, has fallen far short of providing the services for which the American taxpayer is being billed. Periodically there is a strong call for removing NASA from the passenger and payload business and returning its focus to exploration and science.

Because NASA continued to falter in its efforts to resume human space flight following the *Columbia* disaster, it was forced to purchase services from Roscosmos for an initial $44 million for Soyuz flights to the ISS. One of the problems in the arrangement was that Congress had restricted commercial trade with the Russians because of the Iran Nonproliferation Act, which it passed in 2000 when it was determined that the Russians were assisting Iran in developing a nuclear weapons program.

It was humiliating for America, which prides itself on its technological superiority, to admit in 2012 that it no longer had a viable manned space flight capability—and that it would be many years before that capability returned.

Discovery STS-114: Return to Flight

Following a twenty-nine-month hiatus resulting from the *Columbia* disaster, the shuttle program resumed its flights with a much-anticipated launch of *Discovery* in July 2005. During that stand-down period, most of the shuttle systems had undergone careful scrutiny, and several significant problems were addressed. However, because of the national security aspects, principally the need for the shuttle to orbit intelligence and reconnaissance payloads, not all issues were resolved to everyone's satisfaction before flights resumed.

Despite all of the care taken to ensure that a foam insulation separation did not occur, at 127 seconds into the flight (and 5.3 seconds after solid rocket booster separation), a large (36 in. × 11 in. × 7 in.) piece of "debris" was seen by a remote video camera, tearing off the ET. While it weighed only an estimated 1 lb., about half the mass of the piece that had caused the *Columbia* damage, it did not strike the orbiter. However, a smaller piece separated twenty seconds later and did strike *Discovery*'s right wing. NASA later estimated that the piece exerted only one-tenth the energy needed to cause damage.

Cameras on the ET recorded the incident, and photos taken of the ET after separation by the *Discovery* crew (before the orbital maneuvering system burn to achieve orbit) showed several areas of the ET that had lost insulation. With STS-114, the orbiter's Canadarm (a series of robotic arms) was fitted with orbiter boom sensor system that extended its reach an added 50 ft. With it, sensors and cameras scanned the leading edge of the wings and the nose cap (critical parts of the thermal protection system) on all subsequent flights to confirm there was no significant damage.

The revelation that the apparent "fixes" had not resolved the insulation-separation problem caused subsequent shuttle flights to be delayed by another year before STS-121, in July 2006. This "out of sequence" numbered flight was an additional test in the "return to flight" program that focused on the debris problem and an engine-cutoff sensor issue. There were no major issues during launch, and the orbiter boom sensor system found no damage to the thermal protection system.

ISS: Assembly Resumes

With the shuttle returned to flight status, the next manifests were all in support of the ISS except for STS-125 in May 2009. This flight of *Atlantis* was a twelve-day service mission to the Hubble space telescope.

With the termination of shuttle flights, the remainder of the ISS assembly continued with the Russian Proton and Soyuz and the U.S. Space-X Falcon. ISS is scheduled to be completed in 2019. While the assembly has been without any major incidents, there is still some concern about the safety of the ISS itself—particularly the prospects of either a meteor or space debris puncturing one of the modules. Russian engineers estimated that there was a 23 percent chance over its lifetime that the latter might occur. If it should, there is the possibility that the fracture of the pressure hull could cause the

Table 3. International Space Station Structural Assembly of Major Elements in Orbit 2003–15

Date	Delivery Vehicle	Unit	Weight (lb.)	Comments
July 2005	*Discovery* STS-114	Raffaello, ESP-2	2,676	MPLM
July 2006	*Discovery* STS-121	Leonardo MPLM		MPLM
September 2006	*Atlantis* STS-115	P3/P4, SA	15,900	Solar arrays
December 2006	*Discovery* STS-116	P5 Truss	1,118	
June 2007	*Atlantis* STS-117	S3/S4 Truss, SA	15,900	Solar arrays
August 2007	*Endeavour* STS-118	S5 Truss, ESP-3	12,598	
October 2007	*Discovery* STS-120	Harmony, P6 Truss	14,288	Relocation of P6 truss
February 2008	*Atlantis* STS-122	Columbus	12,800	European lab
March 2008	*Endeavour* STS-123	Dextre SPDM		Japanese logistics module (ELM-PS)
May 2008	*Discovery* STS-124	JEM-PM, RMS	15,900	Japanese pressurized module, robotic arm
March 2009	*Discovery* STS-119	S6 Truss, SA	15,900	Solar arrays
July 2009	*Endeavour* STS-127	JEM-EF	4,100	Japanese exposed facility
November 2009	Russian Soyuz-U	Poisk MRM-2	3,670	
November 2009	*Atlantis* STS-129	ELC-1, ELC-2	12,500	ExPRESS logistics carrier 1 & 2
February 2010	*Endeavour* STS-130	Cupola, Tranquility	14,000	Node 3 (Tranquility) wt (lb.) 12,247
May 2010	*Atlantis* STS-132	Rassvet, MRM-1	5,075	
February 2011	*Discovery* STS-133	Leonardo PMM	9,896	ExPRESS logistics carrier 4
May 2011	*Endeavour* STS-134	AMS, OBSS	6,731	Alpha Magnetic Spectrometer, ExPRESS 3
July 2011	*Atlantis* STS-135	Raffaello MPLM	9,500	Final flight of the space shuttle program

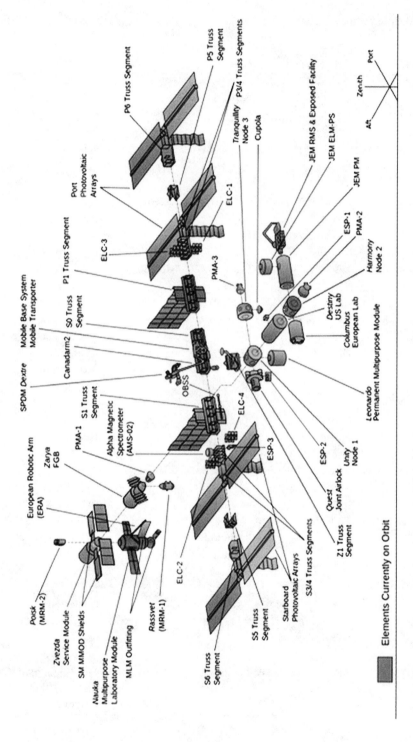

The configuration of the ISS as it had progressed with the last flight of the shuttle in 2011. Courtesy of NASA.

Posk
(MRM-2)

Zvezda
Service Module

SM MMOD Shields

Nauka
Multipurpose
Laboratory Module

MLM Outfitting

European Robotic Arm
(ERA)

Zarya
FGB

Rassvet
(MRM-1)

SPDM Dextre

Mobile Base System
Mobile Transporter

P1 Truss Segment

S0 Truss
Segment

Canadarm2

PMA-1

S1 Truss
Segment

Alpha Magnetic
Spectrometer
(AMS-02)

OBSS

PMA-3

ELC-1

ELC-3

Port
Photovoltaic
Arrays

P6 Truss Segment

P5 Truss
Segment

P3/4 Truss Segments

Tranquility
Node 3

Cupola

JEM RMS & Exposed Facility

JEM ELM-PS

JEM PM

ESP-1

PMA-2

Harmony
Node 2

Destiny
US Lab

Columbus
European Lab

Leonardo
Permanent Multipurpose Module

ELC-4

ESP-2

Unity
Node 1

Quest
Joint Airlock

Z1 Truss
Segment

Starboard
Photovoltaic Arrays

S3/4 Truss Segments

ESP-3

ELC-2

S5 Truss
Segment

S6 Truss
Segment

Zenith

Port

Aft

Port

Aft

Elements Currently on Orbit

An interior view that shows astronaut Leroy Chiao inside the Destiny space lab module gives an impression of the lab's complexity. Courtesy of NASA.

module to explode like a punctured balloon, resulting in the breakup of the station. NASA contends this is not a probable event.

In its first sixteen years of operation (2000–15), forty-five expeditions have inhabited the ISS, with 221 individuals, including 33 women (376 visits), for an average stay of 160 days. Eighteen different countries have been represented. One cosmonaut has made 5 trips; 5 U.S. astronauts have made 4 trips. More than 144 spacewalks have been conducted by the U.S. astronauts and 48 by Russian cosmonauts. Five Americans have spent more than a year in space.

The Russians have continued to send space tourists (Dennis Tito having been the first) to the ISS on board its Soyuz—whenever the third seat has not been allocated. As of 2016 seven of seats have been sold through a private corporation and have brought as much as $52 million for each flight. These individuals remain on the ISS only until the transfer of crew and cargo are complete (typically seven to ten days) and then return on the Soyuz that is being replaced.

The ISS interior provides almost 32,000 cu. ft. of pressurized space for equipment and living area for six crew members—about the volume of a Boeing 747. The diversity of experiments has been impressive, ranging from biology and biotechnology to Earth and space science. The unique environment of microgravity has been a key element in the selection of experiments.

While the lifetime of the ISS has been periodically extended and is now estimated to be 2024, some components cannot be easily replaced, such as the seals between the modules; others, like the solar panels, are more readily swapped. With exposure to the impact of micrometeors, the viewing windows become opaque and pitted, as do some of the sensor surfaces. Also of concern is the final disposition of such a large and potentially dangerous piece of space debris when it is time to deorbit the ISS.

What follows the ISS has been a subject of considerable discussion, and there are many who doubt there will be a NASA-directed follow-on. However, the availability of the pending heavy-lift space-launch system could provide for a less complex space station, if the science being done on the ISS substantiates continued effort.

26

UNFULFILLED SOVIET EFFORTS

Soviet Space Shuttle: *Energiya-Buran*

Following the cancellation of the N1 superbooster in the early 1970s, the Soviets concentrated their efforts on Earth-orbiting projects—the Salyut space station in particular. The UR-500 launch vehicle and improved versions of the venerable R-7 allowed the Soviets to pursue a variety of scientific and military programs. However, the need for a heavy-lift booster in the Saturn V class continued to be studied.

As America moved forward with the development of its space shuttle in the mid-1970s, Soviet paranoia continued to see this as a threat that had to be addressed in the same way that America had perceived the Sputniks of the late 1950s as a threat. Perhaps the need was driven more by those in the rocket-development business and by politicians than by military planners. Several proposals to develop a "reusable space plane" and its attendant launcher were rejected, apparently because Soviet leadership did not perceive this as being as capable as the American shuttle, which the Soviets had been closely watching since it had been announced.

They carefully studied not only the shuttle design and capabilities but the economics as well. They arrived at the same conclusion that shuttle opponents had: the cost numbers used by NASA to justify the program did not add up. The Soviets believed that expendable boosters were still more economical (by far). However, Soviet first secretary Leonid Brezhnev felt strongly that if the United States thought it economical, then it must be. The cross-range capabilities of the American shuttle were of particular interest because of their military implications. Despite the Soviets' belief that the shuttle was not economically justifiable, and without a clear mission for such a heavy-lift capability, they embarked on the development of a large reusable rocket and a space plane similar in concept to the American space shuttle. It was to be the most expensive Soviet space project.

By February 1976 a reusable space system was approved to provide for a payload of up to 65,000 lb.—about equal to that planned for the American shuttle. While the orbiter, called *Buran* ("snowstorm" or "blizzard"), looked strikingly like its American counterpart and the launch vehicle employed similar parallel staging, there were important fundamental differences.

At first glance, *Buran* looked virtually identical to the shuttle, with a bulbous nose and double-delta wing configuration. It used a similar underbody composed of over thirty-eight thousand heat-resistant tiles. The primary difference between the two craft was that *Buran* did not contain the main engines—they were a part of a "universal launch system" called Energiya (Zimmerman 2003, 261). *Buran* housed only the orbital maneuvering engines, which provided the final push into orbit—similar to the orbital maneuvering system of the American shuttle.

The Energiya launch vehicle for *Buran* was composed of two parts. The central core was powered by four LH2 and LOX thrust chambers that provided almost 1.5 million lb. of thrust. Four LOX-kerosene-powered, "strap-on" parallel booster units of 1.6 million lb. of thrust each completed the assembly. Both the core and the booster units ignited at launch in the same manner as the solid rocket boosters of the American shuttle, providing a total of 7.8 million lb. of thrust—more than the Saturn V.

The four booster units were actually a derivative of the first stage of another Soviet launch vehicle called Zenit. These units each contained four RD-170 thrust chambers, which burned for about two minutes before being separated for their descent back to Earth under parachutes for recovery. The central core then continued to provide power until just short of orbital speed. It then separated and was destroyed on its plunge back into the atmosphere. The payload, the *Buran* or a large unmanned satellite, then provided the small added velocity to achieve orbit.

Unlike the American test program, a series of suborbital flights were first made between 1983 and 1986 with small-scale models of *Buran*. A full-scale, jet-powered version was used for twenty-four glide tests starting in 1984 and continuing into 1988.

As envisioned, the Energiya launcher was to have been used with several unmanned payload vehicles besides *Buran*. This allowed for most of the payload weight to be allocated to the satellite, which then needed only a small thrusting unit to achieve orbit. The first launch of the Energiya was

The Soviet *Buran* was an impressive reusable "shuttle-like" spacecraft that flew only one un-manned flight before it was abandoned. Courtesy of Energiya.

employed in this manner, with an attempt to orbit a satellite called *Polyus* in May 1987.

As a part of the Soviet "Star Wars" program, the huge, 120 ft. long, 200,000 lb. *Polyus* (Skif-DM) was constructed in the early 1980s. Most authorities believe that it contained an antisatellite laser system. Launched by the first Energiya vehicle, it may have been the heaviest object ever placed in orbit—almost. Although the Energiya performed successfully, the attitude control

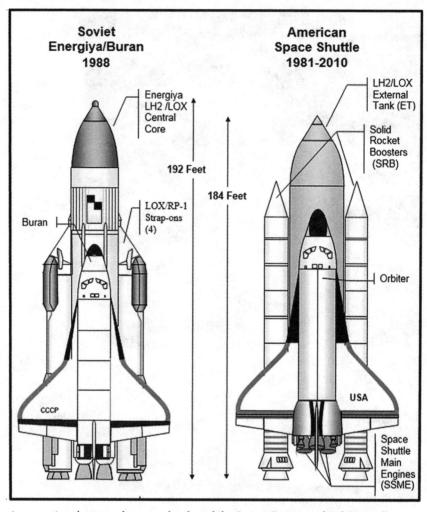

A comparison between the space shuttle and the *Buran*. Courtesy of Ted Spitzmiller.

of *Polyus* failed, and the space tug that accompanied the satellite was unable to provide the final thrusting to achieve orbit. The huge satellite reentered and burned up.

With the *Buran* as a payload, the second Energiya was launched in November 1988. *Buran* was unmanned because many of the systems needed to support a crew had yet to be completed. The vehicle was flown in a totally automated manner and flew two orbits (at 156 mi. and an inclination of 51 degrees) before reentering and landing perfectly. It was an impressive first flight—but it was *Buran's* only flight (Newkirk 1990, 354).

Plans called for the number of booster modules attached to the sides of the core to be variable between two (Energiya M) and eight (Vulcan), depending on the weight and mission of the payload. Energiya II was to be completely reusable with the core employing wings to glide it to a landing. However, this was not to be. The development of the promising vehicle was cut short by the disintegration of the Soviet Union and was officially terminated in 1993. Two additional *Buran* shuttles under construction were never finished. In an unfortunate ending to the remains of the program, a hanger in which *Buran* was stored, along with a full-size model of Energiya, collapsed in 2002, destroying the priceless artifact and killing eight workers.

However, components of the Energiya live on. The strap-on, liquid-fuel RD-170–powered boosters are the basis for the Zenit rocket. A less powerful derivative, the RD-180, powers America's Atlas V—a Russian engine powering an American rocket. The cost of developing the Energiya and the *Buran* along with attempts by the Soviets to maintain military parity with the United States are often cited as contributing factors for the Soviet Union's economic collapse.

The Russian Kliper: Cancelled

In early 2004, the Russian Energiya Rocket and Space Corporation revealed the Kliper manned spacecraft project, which it said had been under development since 2000. The proposed Kliper (the English translation of Клипер is the word "Clipper"), was designed to replace the then fifty-year-old Soyuz design and to reduce the cost of human space flight. It was to be a partly reusable craft with a disposable service module to house some of the support systems and to be capable of twenty-five flights before retirement.

The Kliper employed aerodynamic lifting-body characteristics that allowed it to perform reentries that exposed the occupants to less g-forces than did the Soyuz. The short-winged delta planform would land like the shuttle. The Kliper provided for a launch-escape system that allowed abort during the powered ascent. In many respects, it had a similar size and performance to the long-cancelled U.S. Air Force Dyna-Soar program of the 1960s.

The Kliper was primarily a manned spaceship that could carry up to six cosmonauts, with payloads limited to 1,600 lb. of equipment, to low Earth orbit. This capability highlights the change in space transportation

The Russian Kliper was a lifting-body design that was not carried forward to a test flight. Courtesy of Energiya.

philosophy accepted in the American planning for a shuttle follow-on—the cargo is separate from the human crew and uses a different launch vehicle. The Kliper design provided for up to 15 days of independent operation with an on-orbit life of up to one year when docked to the ISS. This was a clear advantage over the Soyuz, which has an on-orbit life of only 180 days.

During reentry, Kliper's lifting-body design allowed some maneuvering, and with the advances in computer guided de-orbit techniques, it permitted targeting preselected major airports (i.e., those with a 10,000 ft. runway). Its relatively small size (compared to the shuttle) allowed it to be easily transported back to its launch base in any of the large cargo-carrying aircraft such as the Antonov An-124.

The European Space Agency showed interest in collaborating with the Russians to produce the craft, which was similar in size and configuration to the canceled European minishuttle called *Hermes*. The biggest impediment to European Space Agency participation was that the Russians wanted them to be simply an industrial contributor and not involved with the design.

The Kliper would have weighed about 30,000 lb. (beyond the lifting ability of the R-7, which has a current limitation of about 16,000 lb.). This required the use of the new Angara-A3 or the Zenit. However, in an effort to become less reliant on the new designs, Roscosmos reworked the Kliper so that it could use the R-7 derivatives.

In the redesign, the Kliper was to be a part of a two-launch assembly. Dividing the weight between the two components allowed the Russians to retain the ability to use the R-7—now referred to as the "Soyuz vehicle." The second part, a true "space tug," called the Parom, was to service several successive Klipers. The Kliper would rendezvous, dock, and use the Parom to provide orbital maneuvering and to boost the Kliper to higher orbits. The Parom would remain in orbit until it exhausted its propellant supply and then deorbited to destruction, and then would be replaced by another Parom tug.

The Kliper program was cancelled in 2006.

Soyuz Evolution

While emphasis of the Soviet program shifted to space-station operations since the end of the space race, the enabling technology has been the Soyuz spacecraft. The basic failure to produce a follow-on with the *Buran* and Kliper have dictated a long and useful life for the Soyuz design. Conceived originally by the Korolev team in 1963, the basic structure has undergone many changes, and it has been used in both a manned and unmanned role in a variety of programs. The first manned version of Soyuz flew eight missions as an independent spacecraft (Soyuz 1, 3–9). There were twenty-eight flights in support of both the Salyut (10–13, 17, 18, 20, 25–40) and Almaz (14, 15, 21, 23 and 24) space stations. Finally, three flights (Soyuz 16, 19 and 22) were accomplished as a part of the Apollo-Soyuz Test Program; only the last of these was the mission that made the rendezvous with Apollo.

The Soyuz-T was modified for use as a manned transport for the Salyut 6 and Salyut 7 space stations. The final Soyuz-T15 was a Mir mission. The Soyuz-TM craft allowed 500 lb. more payload because of weight reductions and could carry either two or three cosmonauts. Twenty-nine were launched to Mir (TM-2 to TM-30). The Soyuz-TMA included changes that were required by NASA for use as a "lifeboat" for the International Space Station and that provided new cockpit displays, an improved parachute system, and the ability to accommodate a wider height spectrum of crew members.

An older variation of Soyuz, the 7K-L1 (1967–69), was used in a direct launch for circumlunar flight and was called Zond but was never manned.

Soyuz craft modified as unmanned cargo vessels for delivering supplies to the space stations are called Progress (with later variations designated M and M1). Most have no thermal protection system and are destroyed upon reentry.

The Soyuz design has flown more times into space than any other spacecraft. It is currently the primary support vehicle for the International Space Station.

27

RETURN OF
THE EXPENDABLE ROCKET

As the space age moved into the early 1980s, America's expendable boost-ers, most of which had been based on weapons of the 1950s, were nearing the end of their life. NASA had decreed that all future commercial satel-lites and space probes be launched on the shuttle (and the military was just completing its own shuttle launch facility at Vandenberg AFB on the coast of California). The existing stockpile of mothballed military hardware (Atlas, Titan, and Thor), carefully stored away as it was retired from active service in the 1960s, was about depleted. Had the shuttle operated with its predicted economies, it would have claimed the market for launching most of the foreign and domestic satellites, and doomed most of the existing and proposed foreign and domestic launch vehicles.

With the *Challenger* disaster and the realization of its true cost, came a new edict that the shuttle be used only for purposes that "require the presence of man . . . or requires the unique capabilities of the Shuttle . . . or compelling circumstances" (2465a Space Shuttle Use Policy, 42 Fed. Reg. 2929 [Nov. 16, 1990]). Thus, virtually all commercial and military satellites and space probes represented opportunities for a revived aerospace indus-try. As a result, the evolved expendable launch vehicle (EELV) program was born—a rocket whose core unit could be augmented by attaching solid-fuel rockets to match larger payload requirements.

This dramatic change in strategy had a profound effect on the survivors of the Cold War aerospace industry in America. The complexion of that industry had changed considerably from the hectic years of the dawn of the space age in the late 1950s. Many of the giant aerospace corporations had merged or simply closed their doors.

The first Atlas V flew in 2002 and is a candidate as a launch vehicle for the NASA Commercial Crew Development Program CST-100. Courtesy of United Launch Alliance.

United Launch Alliance: Atlas V

The newest version of the Atlas is an Atlas in name only, as it is a completely new structure and first-stage propulsion unit. Most remarkably, Atlas V uses the Russian RD-180 engine. The RD-180 is marketed by RD AMROSS in a partnership between Pratt & Whitney of the United States and NPO Energomash of Russia. It is a two-chamber, gimbaled derivative of the four-chamber RD-170 of the improved R-7. The RD-180 produces 860,200 lb. of thrust at sea level. It has a specific impulse of 337 seconds and can be throttled between 47 percent and 100 percent. The first stage is called the common core booster, as there are several variants that can be built around it.

This new series can use solid rocket boosters (SRB) built by Aerojet Rocketdyne that provides 250,000 lb. of thrust each. Various vehicle configurations employ one, two, three, or five SRB boosters. The boosters ignite at liftoff and burn for about ninety seconds. Total liftoff weight varies with the configuration to a maximum of 735,000 lb., with a total length of 195.9 ft. depending on the payload fairing.

The second-stage Centaur can be configured with the standard two RL-10 LH2 engines, each producing 15,000 lb. of thrust, or a single RL-10, depending on the mission. They have endured for half a century.

Because the Atlas V can be configured with a different number of solid-fuel boosters and Centaur LH2 engines, a three-digit vehicle naming designator distinguishes the various versions. The first digit defines the payload fairing diameter in meters (four or five) while the second digit specifies the number of solid-fuel boosters. The third digit defines the number of Centaur engines. Thus, a 401 has a 4 m payload fairing, no solid-fuel boosters, and a single RL-10 engine. It can lift 27,500 lb. into low Earth orbit (LEO). The 552 is the most capable configuration, with the 5 m payload fairing, five solid-fuel boosters, and two RL-10 LH2 engines. It can lift 45,200 lb. to LEO—almost equal to the payload of the space shuttle. Atlas V is operated by the Lockheed Martin–Boeing joint venture United Launch Alliance.

A major controversy occurred when Russia exhibited aggression in annexing the Crimea in 2014 and subsequently participated in a civil war in the Ukraine. The U.S. Congress, in recognizing that a major U.S. launch vehicle was dependent on engines (RD-180) from Russia, passed legislation that restricted its use. As if this writing, Atlas V is slated to be the launch vehicle for the manned Boeing CST-100.

Boeing: Delta

The origin of the Delta family of expendable launch vehicles extends back to the Thor intermediate-range ballistic missile developed in the mid-1950s for the U.S. Air Force. However, as with the Atlas, the new family of Delta launch vehicles is not related to its namesake.

The Delta IV was developed by Boeing as a part of the EELV program. All configurations use the Boeing common booster core first stage powered by a Rocketdyne RS-68 LH2 engine that develops 650,000 lb. of thrust. The RS-68 is the first large liquid-fuel booster engine developed in the United States since the space shuttle main engine over thirty years ago, with a significantly reduced parts count.

The Delta IV launched an unmanned Orion in December 2014. Courtesy of United Launch Alliance.

The Delta IV second stage uses the Pratt & Whitney RL10B-2 with two propellant capacities and a carbon-carbon extendible nozzle that enables the RL10B-2 to achieve 465 seconds of specific impulse with a thrust of 24,750 lb.

Delta IV vehicles can launch payloads weighing 9,300–28,100 lb. to geosynchronous transfer orbit and can lift 50,000 lb. to LEO, depending on the configuration. The five versions are medium, medium+(4,2), medium+(5,2), medium+(5,4), and heavy. The first digit in parenthesis relates to the payload shroud diameter (in meters) while the second digit indicates the number of 60 in. strap-on GEM solid-fuel booster units.

The Delta IV Heavy uses two additional common core boosters, instead of GEMs, to provide a total of 1,950,000 lb. of thrust. It is capable of launching 50,000 lb. to LEO and 28,950 lb. to geosynchronous transfer orbit.

SpaceX: Falcon

The Falcon is a family of multiuse rocket-launch vehicles developed and operated by Space Exploration Technologies Corporation (SpaceX). The vehicles include the Falcon 1 and Falcon 9. The Falcon 1 made its first successful flight in September 2008 after several failures on the initial attempts. The larger EELV-class Falcon 9 flew successfully into orbit on its first launch in June 2010.

The SpaceX Falcon 9, seen here in a launch of the unmanned Dragon to the ISS, is scheduled to be the launch vehicle for the manned Dragon V2. Courtesy of SpaceX.

The Falcon 9 v1.1 rocket uses nine Merlin 1D engines with 155,000 lb. thrust each as its first stage (1.4 million lb. total). The design is intended to compete with the Delta IV and the Atlas V, along with launchers of other nations as well. Falcon 9 v1.1 made its maiden flight in September 2013 and has a payload capability to 28,990 lb. to LEO from Vandenberg AFB.

The Falcon 9 v1.1 features stretched first and second stages and a new octagonal arrangement of the nine Merlin-1D engines on the first stage (replacing the square pattern of engines in v1.0). It also provides for several aspects of reusable launch-vehicle technology. This includes engines that are throttleable and restartable on the first stage and that structurally accommodate the addition of landing legs. This allows the first stage to return to a floating landing pad. The reusable launch system technology is being developed for both the Falcon 9 and the Falcon Heavy. The reusable first stage is now being flight tested by SpaceX.

Space Launch System

Under the Obama administration, NASA canceled the Constellation program and redirected its efforts, dropping the Ares I launch vehicle. The

An artist's rendering of the Space Launch System as configured for the Orion spacecraft. Courtesy of Boeing.

Ares V, however, continued development as the Space Launch System. It relies heavily on shuttle technology for its two SRB boosters (nonrecoverable) affixed to the sides of an LH2 central core. The core stage is 28 ft. in diameter, and initial flights will use modified RS-25D engines left over from the space shuttle program; later flights will switch to a less expensive engine not intended for reuse.

The core stage's structure consists of a modified space shuttle external tank with the aft section adapted to accept the rocket's main propulsion system and the top converted to host an interstage structure. NASA can use this as a single launch vehicle for both the manned component and cargo, as was done with the shuttle.

The Space Launch System has a takeoff thrust about 20 percent greater than the Saturn V. The initial Block 1 version will lift about 154,000 lb. to LEO using the interim cryogenic propulsion stage—a single RL-10 engine. A follow-on Block 1B will have the exploration upper stage consisting of four RL-10C engines to provide for a payload of 232,000 lb. to LEO. The next iteration, the Block 2, will replace the SRB boosters with a more advanced design and will have a LEO capability of 300,000 lb. The Block 2 will provide for escape velocities to enable lunar, asteroid, and even Mars missions. The estimated initial launch date of the Space Launch System is November 2018.

28

THE NEXT GENERATION
OF MANNED SPACECRAFT

Shuttle Replacement Technology: An Elusive Goal

In the early 1980s, President Ronald Reagan proposed that the United States embark on building not just a successor to the French Concorde supersonic airliner but a revolutionary craft. He envisioned a scenario in which a space plane could, by the end of the 1990s, take off from Dulles Airport and accelerate up to twenty-five times the speed of sound, attaining low-Earth orbit or flying to Tokyo within two hour. Coined "The New Orient Express," the craft was intended to fill two roles—a super-high-speed transport for long-distance flight and a single-stage-to-orbit spacecraft. The implications of such a craft are significant.

From a military point of view, the roles of reconnaissance and weapons delivery stand out—as a successor to the SR-71 Blackbird and a follow-on to the B-2 stealth bomber. This aspect appeared quite prominent when the initial funding showed the Defense Research Projects Agency (DARPA) picking up 80 percent as a part of a project called Copper Canyon that ran through 1985. NASA, presumed to acquire a successor to the space shuttle from this project, did not want to devote any significant part of its budget to the program.

A single-stage-to-orbit vehicle that could operate in the same way as an airliner was also a key part of the Strategic Defense Initiative for a space-based ballistic-missile defense system. However, just as the space shuttle was not able to achieve the "airliner turnaround" promise, the National Aerospace Plane (NASP) seemed destined not to achieve its goals either.

The primary enabling technologies involved (the scramjet and the thermal protection system), when coupled with the logistics of maintaining and

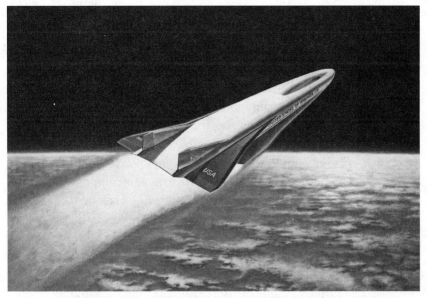

One of many concepts of what the National Aerospace Plane might have been. Courtesy of NASA.

launching a hydrogen-fueled vehicle, appeared to doom the project. While the concept of a single-stage-to-orbit craft is still appealing, its complexity, cost, and volatility—like that of the Shuttle—seem beyond reason.

Further analysis reveals a gap between the velocity provided by the air-breathing power plant and orbital velocity. A second-stage rocket is required. In addition, the promised technology of thermal protection and the efficiencies of the scramjet never materialized—at least publicly. The myth persists that the military have continued the push and that the breakthrough has been made. The presence of the supersecret Area 51 in the Nevada desert continues to reinforce credibility to the presence of a manned hypersonic aircraft.

The NASP never advanced to the building of hardware, despite the designation X-30 given to the craft. The Hypersonic Systems Technology Program (HySTP), was initiated in 1994 to consolidate the research in hypersonic technologies generated by the NASP program and others into a long-term development effort.

Another of NASA's efforts to develop a follow-on to the shuttle grew out of its Space Launch Initiative program of the mid-1990s that continued the quest for a reusable launch vehicle. The X-33 was a cooperative project between NASA and the Lockheed Martin Aeronautics Company to build a technology demonstrator for a single-stage-to-orbit vehicle. Announced in 1996, the unmanned X-33 was scheduled for a series of suborbital test flights by 1999. However, it ran into several technical difficulties that escalated the costs and extended the schedule.

The X-33 design was based on a lifting-body shape with two innovative "linear aerospike" rocket engines and a metallic thermal protection system. The lifting-body configuration was to be supplemented by abbreviated horizontal control surfaces coupled with short vertical fins. The vertical takeoff, single-stage-to-orbit, fully reusable craft struggled against several technologies as it fought to achieve a significant lowering of the cost to low Earth orbit. Its proposed lightweight structure and components emphasized the importance of weight control to achieving performance that had not been previously attainable. If successful, Lockheed Martin was poised to grow the X-33 into a commercial product called the VentureStar—a true single-stage-to-orbit airliner—and take it to space by 2003.

Like its predecessor, the X-33 ran into numerous technical difficulties, not the least of which was the aerospike engine. The ability to achieve a stable, gliding platform of the wedge-shaped craft at the wide range of speeds was also a major challenge as its weight continued to grow.

The X-34 unmanned vehicle was another attempt to demonstrate reusable technology. More conventional in appearance than the X-33, having

The Lockheed Martin X-33 was yet another effort to achieve single-stage-to-orbit. Courtesy of Lockheed Martin.

The Orbital Sciences X-34 was an intermediate program as a technology demonstrator. Courtesy of Orbital Sciences.

short stubby wings that spanned 27 ft., the 58 ft. long ship was to reach speeds of Mach 8 and altitudes of 50 mi. A joint project between NASA and Orbital Sciences Corporation, the X-34 was a test bed for a full-sized vehicle that could be supported by a smaller ground crew for servicing and that could provide a two-week turnaround time between flights.

A review of the X-33 and X-34 programs in 2000 determined that there was no end in sight for several of the technology-intensive problems, and the costs were beyond the ability of NASA to absorb. In 2001 NASA discontinued both programs after a combined investment of $1.5 billion. These and other projects seemed to point out that a single-stage-to-orbit vehicle was not yet possible as a follow-on to the space shuttle. There was no "space liner" to replace the shuttle—the VentureStar project would not fly—at least not in the near future.

Orion: Crew Module

The situation with a shuttle replacement took on critical proportions with the failure of the X-30, X-33, and X-34 projects. In the first few years of the new millennium, the aging shuttle continued to be pressed to shoulder the

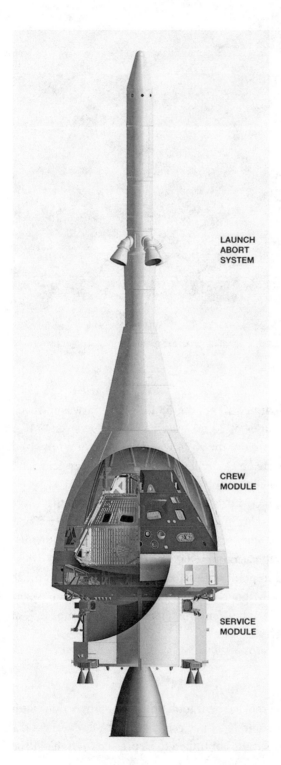

LAUNCH
ABORT
SYSTEM

CREW
MODULE

SERVICE
MODULE

The Lockheed Martin Orion spacecraft does not use lifting-body aerodynamics but rather an off-set CG ballistic return. Note that the spacecraft is enclosed by the launch abort system protective shroud during the powered phase of flight. Courtesy of Lockheed Martin.

burden of constructing and supporting the ISS. However, the *Columbia* disaster in February 2003 brought the problem to a head.

Following the loss of *Columbia* in 2003, it was decided to end the shuttle program—and to develop a long-delayed successor. The first major decision was that an integrated and reusable launch vehicle and spacecraft—such as the shuttle—would not be pursued. The cost and uncertain technology elements presented too high a risk factor. Thus, separate development programs for each were established. The effort was part of the canceled Constellation program.

With respect to capability, the shuttle is a hard act to follow, with its ability to carry the crew, payload, and the primary propulsion system—and to return to Earth like an airplane. However, the cost of that capability and the lack of a crew abort system caused NASA to revert to a ballistic recovery design for America's next-generation (the fifth) spacecraft.

The Orion crew module (CM) is a truncated cone–shape design, similar to but larger than the Apollo and is reusable for up to ten flights. To keep the development costs to a minimum, Orion uses some of the onboard shuttle systems and an expendable launch vehicle to achieve low Earth orbit. Orion is capable of returning to either land or water. The contractor, Lockheed Martin, was selected by NASA in 2006 after intense competition with Northrop Grumman and Boeing.

The Orion CM has various configurations that can carry up to six persons to Earth orbit—or beyond. To accommodate the various configurations, a cylindrical service module (like Apollo) provides for the various life support, communications, orbital maneuvering, and attitude-control functions. The service module is built by Airbus Defence and Space for the European Space Agency.

In place of the Apollo-like fuel cells, the CM uses deployable solar panels for electrical power generation. An abort system (an escape tower, as in Mercury and Apollo) allows the crew to survive a launch-vehicle malfunction during ascent.

The CM's first unmanned test flight, Exploration Flight Test 1 (EFT-1), was launched atop a Delta IV Heavy rocket in December 2014. The flight lasted four hours and twenty-four minutes, landing at its target in the Pacific Ocean. However, the first mission to carry astronauts is not expected to take place until 2021—a long development period. The Orion is not intended to support the ISS.

By returning to a modular approach for the overall configuration, Orion can grow with future missions planned for lunar, asteroid, or Martian

The Orion spacecraft will use solar arrays for electrical power to enable extended-duration flights that would not be possible with cryogenic fuel cells. Courtesy of Lockheed Martin.

exploration. The spacecraft itself provides for the basic functions, including abort during launch, a command center during long-duration missions, and Earth reentry return from deep-space velocities.

Using a combination of Earth orbit rendezvous and lunar-orbit rendezvous, Orion could rendezvous and dock with a mission module such as a surface access module for lunar or Martian exploration and an Earth departure stage to achieve escape velocities.

The surface access module docks to the forward end of the CM for extended missions and provides the extra consumables and living space. The Earth departure stage attaches to the aft end. The configuration is similar to the Apollo CSM/LM. A lunar mission, for example, might be composed of modules launched independently by at least three expendable vehicles.

By stacking all the stages in tandem rather than in parallel (as was the case with the shuttle), the CM concept avoids debris (such as ice) from one stage impinging on another. It also allows for more abort scenarios.

Private Enterprise: A Profit Incentive

A second American manned spacecraft program (apart from Orion) called the Commercial Crew Development (CCDev) emerged from NASA's desire to seek privately developed spacecraft—occasionally referred to as "crew vehicles"—to support the ISS. This category of spacecraft is to be an incremental improvement over the capabilities provided by the Russian Soyuz but does not compete directly with the Orion—which is envisioned as a deep-space venture. The CCDev capabilities include transportation of a minimum of four crew members and their equipment to the ISS, provision of an emergency return (or twenty-four-hour sanctuary) for the ISS crew in case of a catastrophic ISS event, and the capability to remain docked for up to 210 days.

Five companies originally began the design competition, which was winnowed down to two. In September 2014 NASA chose Boeing (CST-100) and SpaceX (Dragon V2) as the two companies to be funded to develop transport of crews to and from the ISS.

SpaceX: Dragon V2

The crewed variant of the cargo-carrying SpaceX Dragon is called Dragon V2 (for version 2). It carries up to seven astronauts, or some combination of crew and cargo, to and from low Earth orbit (the ISS) and varies considerably from an all-cargo-carrying version. Dragon V2, which is launched with the Falcon 9 booster, could make its first unmanned flight in late 2016 and fly with astronauts by 2017.

The "integrated launch-abort system design" for the Dragon V2 differs significantly from the "tractor tower" escape method used with Mercury, Apollo, and Soyuz. In May 2015 a pad abort test for the Dragon V2, using its abort engines, was successfully demonstrated. It reached an altitude of 3,894 ft. and returned safely to an ocean recovery under parachute.

The launch-abort system is a part of the Draco maneuvering system currently used on the unmanned Dragon for on-orbit maneuvering and de-orbit burns. It includes a set of four side-mounted thruster pods and eighteen Draco hypergolic liquid rocket engine thrusters that are dual-redundant for attitude control on all axes.

The heat shield is a phenolic-impregnated carbon ablator called PICA-X and is capable of providing reentry for velocities from lunar missions.

The SpaceX Dragon V2 spacecraft. Courtesy of SpaceX.

Initial flights of the Dragon will land in the ocean, and but plans are to perform a terrestrial propulsive landing.

Boeing: CST-100

A Boeing design in response to the CCDev program is the seven-person CST-100 Starliner spacecraft. The number 100 represents the 100 km (62 mi.) height of the Kármán line, which the Fédération aéronautique internationale defines as the boundary of space. It has both human and cargo configurations and is reusable up to ten times.

With a diameter of 15 ft., it is slightly larger than the Apollo CM but smaller than Orion. The CST-100 can remain on-orbit for up to seven months. It uses the NASA docking system and the Boeing lightweight ablator for its heatshield.

Aerojet Rocketdyne provides the service module and launch-abort propulsion—the RS-88, a liquid-fueled rocket engine (ethanol and liquid oxygen) that has 50,000 lb. (220 kN) of thrust.

The development contract includes at least one crewed flight test with one NASA astronaut on board. Once the Starliner achieves NASA certification, the contract requires Boeing to conduct at least two and as many as six crewed missions to the space station. A pad abort test is scheduled for February 2017. A thirty-day unmanned test mission to the ISS is planned for April 2017. A fourteen-day crewed test mission to the ISS tentatively scheduled for July 2017.

The Boeing CST-100. Courtesy of Boeing.

The CST-100 will be launched atop an Atlas V 401 rocket from Space Launch Complex 41 at Cape Canaveral, Florida.

Space Tourism: Is It Real?

On a smaller scale of human exploration are the efforts of several companies to develop access to the space environment for tourism (distinct from the Russian endeavor to sell available seats on Soyuz to the ISS). To this end, the Ansari X Prize provided a financial motivating factor. A $10 million award was offered in 1996 by a consortium of benefactors to encourage the space tourism industry. In the same way that the Ortiz Prize offered in 1919 sparked competition for the flight by Charles Lindbergh from New York to Paris in 1927, the Ansari X Prize was established to reward the first privately financed team who could build and launch a spacecraft to an altitude of 62 mi. Initially called the "X Prize," it was renamed in 2004 in recognition of the contributions from Iranian-born entrepreneurs Anousheh Ansari and her brother-in-law Amir Ansari, who now reside in the United States. The prize required that a manned spacecraft capable of carrying three people had to be launched twice within a two-week period—obviously returning safely each time.

Scaled Composites, a company headed by the renowned aircraft designer Burt Rutan, announced publicly in 1996 that it would compete for the prize. Among the more than twenty companies who registered, Scaled Composites' effort was financed by Microsoft cofounder Paul Allen and built in secret by a bevy of volunteers. The completed vehicle was rolled out in April 2003. The spacecraft was one of two parts—the first being a twinjet-powered mother ship (the *White Knight*) that carried the space-craft for an air launch. The 8,000 lb. winged *SpaceShipOne* spacecraft was powered by a 17,000 lb. thrust hybrid motor that used a solid-fuel core and a liquid oxidizer that produced specific impulse 250 of seconds.

Launched from the *White Knight* carrier at about 40,000 ft., *Space-ShipOne* climbed into the vertical, attaining a velocity of just over Mach 3 during the eight-seven seconds of powered flight. It then coasted to its maximum altitude of 65 mi., where it executed a physical metamorphosis into what is called the feathering system or "shuttlecock" mode. The twin tails of the ship rotated to present a high drag configuration to ensure that the reentry was aerodynamically controlled and the speed buildup was minimized.

When the ship reached the denser layers of the atmosphere, the tail tran-sitioned back to a normal configuration, and it performed as a glider with the pilot executing an unpowered landing in the conventional manner. Sixteen unpowered and powered test flights were made to test the various systems before first the X Prize attempt on September 29, 2004, with pilot Mike Melvill at the controls. The second flight, which secured the prize, was made one week later, on October 4, 2004, piloted by Brian Binnie.

The Virgin Galactic *White Knight Two* carrying *SpaceShipTwo*. Courtesy of Virgin Galactic.

This combination was designed to prove the concept and collect the prize, but it was not the final configuration for space tourists. For that, the larger *SpaceShipTwo*, launched by *White Knight Two*, began tests in October 2010. Following thirty-one glide tests and three powered flights, the spaceship (called *VSS Enterprise*) broke up eleven seconds into the powered portion of a high-performance test in October 2014. Copilot Michael Alsbury was killed and pilot Peter Siebold was seriously injured. Apparently the shuttlecock mode had been deployed prematurely because of pilot error. A National Transportation Safety Board review concluded that inadequate design safeguards, poor pilot training, lack of rigorous federal oversight, and a potentially anxious copilot without recent flight experience were the causal factors. Virgin Galactic, who sponsored the effort, was criticized by many for its cavalier attitude and overzealous desire to promote space tourism. For the enthusiast however, the future of space tourism is not in doubt.

Several other private organizations have made inroads into space tourism, to include Blue Origin's New Shepard vertically launched suborbital rocket and the Lynx spacecraft of XCOR Aerospace. As of 2016, there are no firm dates when space tourists will make the first trip.

China: A Sleeping Giant Awakens

With the orbiting of Shenzhou 5, in October 2003, with taikonaut Yang Liwei, China became the third country to send humans into space. The Chinese astronauts are known as "yǔ háng yuán" (space navigating personnel). Their English nickname, taikonaut, is derived from the Chinese word for space—*taikong*. The launch followed four unmanned test flights of the Shenzhou spaceship between 1999 and 2003. Yang was followed two years later, in October 2005, by two taikonauts, Fei Junlong and Nie Haisheng, on board the Shenzhou 6 spacecraft for a five-day mission.

Recognizing the importance of public relations, the Shenzhou 6 mission was broadcast live on national Chinese television. It was reported that the loudspeakers set up for visiting dignitaries viewing the launch serenaded them to the theme music from *Battlestar Galactica*. Yang's flight, however, had been launched without any publicity.

Although much of its starting technology was imported from Russia and illegally appropriated from the United States (technology transfer is severely limited by Congress), China is no stranger to innovative engineering. The configuration of the Shenzhou follows the basic design of the Soyuz. It consists of an orbital module that contains scientific equipment, the crew-carrying ascent/decent module, and a service module with attached solar panels to provide electrical power, attitude control, and orbital maneuvering. Because Shenzhou is a close replica of the Soyuz, it was designed with the similar ability to carry a crew of three. It is launched by the Long March 2F booster from China's Jiuquan Space Launch Center in northwestern Gansu Province.

One difference in the Shenzhou its use of the orbital module. As it is equipped with its own set of solar panels, it becomes an independent unmanned satellite after separation from the descent module and continues in orbit, providing scientific data for several months.

China achieved its first spacewalk with Shenzhou 7 in 2008, its first rendezvous and docking with Shenzhou 8 (which was unmanned), and China's first female taikonaut on board Shenzhou 9 in 2012. As of this writing (2016) the most recent human spacecraft launched by China was Shenzhou 10 in 2013.

As with the former Soviet Union, China's initial efforts were hindered by its poor economic conditions. However, this is changing since America has become an important trade partner—boosting the economy significantly over the past two decades. While China has been actively pursuing space exploration with unmanned satellites since 1970, the manned project did not receive state approval until 1992.

China has moved toward a more permanent presence in space with the orbiting of the first component of a manned laboratory, Tiangong-1 in 2011. This 18,700 lb. solar-powered unit was launched by a Long March 2F/G rocket into a 230 mi. orbit. It was visited by the unmanned Shenzhou 8 two months later and then by the three crew members of Shenzhou 9 in June 2012, for a ten-day period. Shenzhou 10 launched a year later, and its crew had a short stay of two weeks.

Chinese officials have indicated that their space program will include orbiting larger modules into a more permanent space station by 2023 as well as subsequent lunar exploration. It is possible that China's entry into human space flight will have far-reaching effects. It will be interesting to see if China's move into space represents another iteration of the space race.

The essence of the Chinese space program was created in the late 1950s by a U.S.-educated visionary—Tsien Hsue-Shen. Tsien was a cofounder of the Jet Propulsion Laboratory in Pasadena, California, in the 1940s. At the end of World War II he was considered one of the few "rocket scientists" capable of understanding the significance of the technology of the V-2 created in Nazi Germany.

The Shenzhou series of manned spacecraft have moved China forward in
what appear to be carefully measured steps. Courtesy of China National
Space Administration.

In 1950 Tsien's security clearance was revoked because of his close ties
with the Communist Party. When it became clear that he would no longer
be allowed to work on U.S. defense programs, he was permitted to return to
China in 1955. Tsien led China's space program until his retirement in 1991
(Seedhouse 2010, 14).

Launching an unmanned satellite is no longer considered a significant
technological achievement and can now, along with the creation of an
ICBM, be accomplished by relatively impoverished countries (North Ko-
rea, for example). Human spaceflight, however, is an order of magnitude

greater in effort, both technologically and financially. China's commitment to a manned space program says much about its political leadership as well as its financial strength.

As China has fifty years of observable experience from the Soviet and American programs, they have been able to avoid some of the pitfalls that cost lives and money. However, whether they have the will to endure a long and costly program remains the question. If China moves toward democracy, the needs of its people could undermine an extensive and costly space exploration program.

Judging from the wall-to-wall "Made in China" products marketed in the United States by megachains, China may become a formidable economic giant to rival the United States. If that potential is achieved, it is conceivable that the huge sums necessary to create a vast space program might propel China into technological leadership. Like the former Soviet Union and the United States, China recognizes that to be acknowledged as a world power, it must participate in space exploration and its commercial exploitation (Seedhouse 2010, 49).

EPILOGUE

WHAT WILL THE FUTURE BRING?

The advances in technology of the past century in some respects have often eclipsed even the dreams of the visionaries. The simple conveniences that we now take for granted, such as the smartphone, were beyond the imagination of the science fiction writer of a few decades past. Who could have conceived a wireless pocket phone allowing the receipt of email and real-time video while transmitting digital pictures and providing GPS navigation? In just the past sixty years, the electronic storage of a single binary digit (bit) of information has progressed from the several cubic inch vacuum tube to a chain of molecules a trillion times smaller. Would any reputable scientist have foretold of such advances in 1955?

However, there have also been plateaus, as in the case of commercial aviation, which allowed travel at 600 mph a half-century ago yet has not increased over that period. Likewise, the speed and altitude capability of military aircraft have not advanced during that time—the F-106 fighter currently (2016) holds the speed record for a single-engine turbojet that was set in 1959. We walked the Moon in 1969 but have not returned or traveled any further in the years since. Thus, while some enabling technologies such as integrated circuits have outdistanced the visionary, aerospace has not realized such dramatic advances.

The period since Sputnik has seen phenomenal exploration and exploitation of the space environment. No doubt the primary unmanned applications (communications, weather, Earth resources, and navigation) will continue. They are now inseparable from our economy and culture. These near-Earth satellites have proven their usefulness. Likewise, the unmanned probes to the far reaches of our solar system have been highly successful—but what about humans in space?

NASA has plans to return to the Moon sometime before 2030 and has overtures toward asteroids and Mars. However, many do not believe that the Moon and Mars have any scientific or commercial value to warrant the expense of human voyagers. While it is difficult to predict the future, the prospects are made more apparent with an understanding of the three primary enabling forces composed of the economic, technological, and motivational factors. Until these three aspects are aligned for a given program, it will not happen.

Recent human space initiatives put forward all have been abandoned—often before any significant funding was expended. Why have we failed to build on the legacy of the Apollo program?

The Apollo program was unique for several reasons. It was born of the space race that saw the apparent technological superiority of the Russians posing a threat to the existence of the free world. It was projected over short time frame (eight years—"before this decade is out") that allowed intervening milestones to be played out on a periodic basis—virtually every few months. Thus, a new "first" in space was achieved at relatively close intervals. The first spacewalk, rendezvous, and docking as well as the first space emergency all took place within a year. This kept the American people (who were paying the tab) interested and involved. Apollo was also able to weather the politics of three administrations, the financial squeeze of the Vietnam War, LBJ's Great Society, and the turmoil of the inner cities.

The public appears to be somewhat indifferent (some might say fickle) with respect to space exploration. While the Smithsonian Air & Space Museum is the most visited museum in the world, interest in the Moon landings that followed Apollo 11 (the first landing) gradually drew less and less enthusiasm. Few Americans have any idea that the ISS has been inhabited continuously since November 2000, and almost no one has any awareness of its goals or its operational life span.

The exciting possibilities that were presented by the proposed follow-on to the shuttle—an aerospace plane—have not found funding, in part because of the lessons of the shuttle. NASA promised more than it could deliver and revealed that pushing technology is an expensive endeavor that uncovers unknown areas that often require uncomfortable compromises. Current human spacecraft development separated the propulsion from the spacecraft returning us to the perspective of the 1960s. Orion is viewed as simply an upgraded Apollo.

At the advent of the new millennium, NASA reevaluated the prospects for developing advanced propulsion systems in the near term and at one point defined four phases to reduce the costs. It had placed the shuttle as being the "first generation" and positioned each of the successive generations as reducing the cost by a factor of ten while increasing the reliability by a like order of magnitude (the shuttle ultimately was defined as one loss in each one hundred launches). The fourth generation was seen as an operational spaceship in the year 2040, capable of taking off and landing horizontally and making ten thousand flights per year. Clearly, NASA still projects space flight to become as commonplace as airliner flights of today. However, is "inexpensive access to space" an oxymoron? NASA's track record is not encouraging. If a nation achieves this milestone it will most likely become the dominant space and terrestrial power.

A return to the Moon finds little support from the general populace. Its closeness makes it more desirable than Mars, but the "been there, done that" syndrome works against popular support. A Mars mission planned for twenty years into the future will not only be yawned at but will have a tough battle keeping any funding through three or four administrations. In addition, a price tag that stretches into the hundreds of billions will not find favor with those who feel we have more pressing Earthly agendas than more deficit spending. However, a mission to Mars has far more popular support than the return to the Moon. Mars represents a truly challenging opportunity to explore a planet that may have harbored life.

The justification for the pursuit of an aggressive human space program has many themes. These range from spreading life throughout the universe to ensuring the survival of our species in the event of a global calamity. Some estimates are that, over the next one hundred years, an asteroid-like object has a one-in-a-hundred chance of impacting in the ocean and creating a tsunami that could destroy one or more major cities along the coast. There is a one-in-a-thousand chance of an impact ruining our civilization and killing a large percentage of the population—and a once-in-a-million likelihood of a mass extinction impact.

A remote space colony could "reverse-colonize" the Earth and restore human civilization following such a catastrophic event. Motives that are more "down to Earth" include the desire to preserve the terrestrial environment as well as to create new business opportunities for economic gain. However, it may be that only a survival-based motivation coupled with a set reasonably priced short-term goals has a chance for succeeding in the next millenia.

The controversy is not simply within the United States; Russia, China, Japan, and the European community may figure prominently in any new space initiative. The costs involved in these projects may be more than one nation can afford—even the United States.

Where should a space colony be built? Mars has the greatest following because of the possibilities that it might have harbored life and has resources that might allow a colony to create a closed biosphere, which is a critical technology to the habitation of space—yet a completely closed system has yet to be achieved. No long-term plans can be made unless a functioning and robust system can be adequately demonstrated.

However, Dr. Gerald K. O'Neill, a space enthusiast of some standing, asked the pivotal and thought-provoking question of whether the surface of a planet is really the right place for an expanding technological civilization. He proposed the Lagrangian L-5 point. If a large space city were constructed at L-5, it could serve as a relatively isolated outpost to prove and improve the closed life-support system—yet be within the range of help from the Earth until it could achieve total independence. A manned Mars mission is essentially a multiyear task that requires a closed biosphere—until transformation of Mars' abundant carbon dioxide can provide not only methane but also oxygen and water. If the decision were made to embark on a manned Mars mission, it could accelerate development of those technologies needed for permanent space habitation.

While these initial proposals can be achieved with existing technology, any further steps will require advanced propulsion and power systems. Chemical rockets will continue to be the primary means to low Earth orbit for the next fifty years. However, virtually all plans to explore, colonize, or mine extraterrestrial assets will require a more economical access to low Earth orbit.

Ion, nuclear, solar thermal, mass drivers, and beam power all hold some potential for human space flight but have technological hurdles. Moreover, none has the ability to produce velocities approaching 1 percent of the speed of light. As a result, when interstellar flight is contemplated, only a paradigm of travel in dimensions and by means yet to be discovered will make this possible. Even if the known laws of physics have a loophole that can allow for velocities at the speed of light—the prospects for future human space flight face significant technological, economic, and political hurdles.

Despite the boundaries of the Genesis Factors (time, energy, mass, and space), we must not allow the imagination to be constrained by our current understanding of the laws of physics if we are to move forward.

ACRONYMS AND ABBREVIATIONS

ABMA	Army Ballistic Missile Agency
ACS	attitude control system
AFB	Air Force Base
AIS	American Interplanetary Society
ALSEP	Apollo Lunar Surface Experiment Package
AMAL	Aviation Medical Acceleration Laboratory
AMU	Astronaut Maneuvering Unit
APS	ascent propulsion system
ARPA	Advanced Research Projects Agency
ARS	American Rocket Society
ASAP	Aerospace Safety Advisory Panel
ASTP	Apollo-Soyuz Test Program
ATDA	Augmented Target Docking Adapter
CapCom	Capsule Communicator
CCDev	Commercial Crew Development
CM	crew module
CSM	command and service module
DFS	German Institute for Sailplane Flight
DOI	descent orbit initiation
DPS	descent propulsion system
EELV	evolved expendable launch vehicle
EOR	Earth orbit rendezvous
ET	external tank
EVA	extravehicular activity
FAI	Fédération aéronautique internationale

GALCIT	Guggenheim Aeronautical Laboratory California Institute of Technology
GDL	Gas Dynamics Laboratory
GIRD	Gruppa Isutcheniya Reaktivnovo Dvisheniya
G&N	guidance and navigation
GT-x	Gemini-Titan launches (number 1–12)
HALO	high-altitude low-opening
HRE	hypersonic ramjet engine
HSFRS	High-Speed Flight Research Station
ICBM	intercontinental ballistic missile
IGY	International Geophysical Year
I_{sp}	specific impulse
ISS	International Space Station
JPL	Jet Propulsion Laboratory
KORD	engine operation control system (N1)
LEO	low Earth orbit
LES	launch escape system
LH2	liquid hydrogen
LM	lunar module
LOR	lunar orbit rendezvous
LOX	liquefied oxygen
MA-x	Mercury-Atlas flights (number 1–7)
MASTIF	Multiple Axis Space Test Inertia Facility
Max Q	maximum dynamic pressure
MLP	mobile launchpad
MOL	manned orbiting laboratory
MOUSE	minimum orbital unmanned satellite of Earth
MPLM	multipurpose logistics module
MR-x	Mercury-Redstone flight numbers (1–4)
MSFC	George C. Marshall Space Flight Center
NAA	North American Aviation
NACA	National Advisory Committee for Aeronautics
NASA	National Aeronautics & Space Administration
NASP	National Aerospace Plane

NBC	National Broadcasting Company
NGS	National Geographic Society
NRL	Naval Research Laboratory
OAMS	orbital attitude and maneuvering system
OMB	Office of Management and Budget
OMS	orbital maneuvering system
PDI	powered descent initiate
PPS	primary propulsion system
RCS	reaction control system
REP	radar evaluation pod
scramjet	supersonic combustion ramjet
SPS	service propulsion system
SRB	solid-fuel rocket booster
SSME	space shuttle main engine
STG	space task group
STS	Space Transportation System
STS-x	Space Transportation System flight numbers
TDA	target docking adapter
TLI	translunar injection
VfR	Verein fur Raumschiffahrt

BIBLIOGRAPHY

Aldrin, B. M., and Malcolm McConnell. 1989. *Men from Earth*. New York: Bantam Books.

Baker, D. 1982. *The History of Manned Space Flight*. New York: Crown Publishers.

Bergaust, Erik. 1976. *Wernher von Braun: The Authoritative and Definitive Biographical Profile of the Father of Modern Space Flight*. Washington, D.C.: National Space Institute.

Bilstein, R. E. 2003. *Stages to Saturn: A Technological History of the Apollo/Saturn Launch Vehicles*. Gainesville: University Press of Florida. Previously published by NASA, History Office, 1996.

Bizony, Piers. 2006. *The Man Who Ran the Moon: James E. Webb and the Secret History of Project Apollo*. New York: Thunder's Mouth Press.

Borman, Frank, and Robert J. Serling. 1988. *Countdown: An Autobiography*. New York: Morrow.

Bower, Tom. 1987. *The Paperclip Conspiracy: The Hunt for the Nazi Scientists*. Boston: Little, Brown.

Brzezinski, Matthew. 2007. *Red Moon Rising: Sputnik and the Hidden Rivalries that Ignited the Space Age*. New York: Times Books.

Burrows, William E. 1986. *Deep Black: Space Espionage and National Security*. New York: Random House.

Carpenter, Scott, and Kris Stoever. 2002. *For Spacious Skies: The Uncommon Journey of a Mercury Astronaut*. Orlando: Harcourt.

Cernan, Eugene. 1999. *The Last Man on the Moon: Astronaut Eugene Cernan and America's Race in Space*. With Don Davis. New York: St. Martin's.

Chaikin, Andrew. 1994. *A Man on the Moon: The Voyages of the Apollo Astronauts*. New York: Viking.

Clary, David A. 2003. *Rocket Man: Robert H. Goddard and the Birth of the Space Race*. New York: Hyperion.

Collins, Michael. 1974. *Carrying the Fire: An Astronaut's Journey*. New York: Farrar, Straus and Giroux.

Dornberger, W. 1954. *V-2*. New York: Viking.

Evens, Ben. 2009. *Escaping the Bonds of Earth: The Fifties and the Sixties*. Chichester: Praxis Publishing.

FAI. (2016). *FAI: The World Sport Federation*. Retrieved January 24, 2016, from http://www.fai.org/about-fai/history.

Glenn, John. 1999. *John Glenn: A Memoir*. With Nick Taylor. New York: Bantam Books.

Gray, Mark. 2013. "Apollo 11 Moonwalk Part 2 of 4." YouTube, April 8. https://www.youtube.com/watch?v=cVAGjO2dtUA.

Gray, Mike. 1992. *Angle of Attack: Harrison Storms and the Race to the Moon*. New York: Norton.

Griggs, D. Introduction to *Preliminary Design of an Experimental World-Circling Spaceship*. Santa Monica, Calif.: Rand Corporation, 1946. http://www.rand.org/pubs/special_memoranda/SM11827.html.

Hansen, James R. 2005. *First Man: The Life of Neil A. Armstrong*. New York: Simon & Shuster.

Harland, David M. 2004. *How NASA Learned to Fly in Space: An Exciting Account of the Gemini Missions*. Ontario: Apogee Books.

Hartford, James. 1997. *Korolev: How One Man Masterminded the Soviet Drive to Beat the Americans to the Moon*. New York: Wiley.

Heppenheimer, T. 1984. *The Space Shuttle Decision: 1965–1972*, vol. 1 of *History of the Space Shuttle*. Washington DC: Smithsonian Institution Press.

Janson, D. 1957. "Balloonist Lands Safely; Aloft a Record 32 Hours." *New York Times*, August 21, p. 117.

Jordan, W. 1957. "Soviets Fire Earth Satellite into Space." *New York Times*, October 5, p. 3.

Kelly, Thomas J. 2001. *Moon Lander: How We Developed the Apollo Lunar Module*. Washington DC: Smithsonian Institution Press.

Killian, James Rhyne. 1977. *Sputnik, Scientists, and Eisenhower: A Memoir of the First Special Assistant to the President for Science and Technology*. Cambridge, Mass.: MIT Press.

Kleinknecht, K. S. 1963. *Project Mercury Summary including Results of the Fourth Manned Orbital Flight: May 15 and 16, 1963*. Washington, D.C.: NASA.

Kraft, Christopher C. 2001. *Flight: My Life in Mission Control*. New York: Dutton.

Kranz, Gene. 2000. *Failure Is Not an Option: Mission Control from Mercury to Apollo 13 and Beyond*. New York: Simon & Shuster.

Lardier, Christian, and Stefan Barensky. 2010. *The Soyuz Launch Vehicle: The Two Lives of an Engineering Triumph*. New York: Springer.

Lasser, David. 2002. *The Conquest of Space*. Burlington: Apogee Books.

Launius, Roger D., and Dennis R. Jenkins. 2002. *To Reach the High Frontier: A History of U.S. Launch Vehicles*. Lexington: University Press of Kentucky.

Lawrence, Richard Russell. 2005. *The Mammoth Book of Space Exploration and Disasters*. London: Robinson.

Lawrie, Alan. 2005. *Saturn*. Ontario: Apogee Books.

Ley, Willy. 1959. *Rockets, Missiles and Space Travel*, rev. ed. New York: Viking Press.

Lovelace, W. R. 1943. *Memorandum Report Engineering Division*. Dayton: Department of the Army.

Mindell, David A. 2011. *Digital Apollo: Human and Machine in Space Flight*. Cambridge, Mass.: MIT Press.

Neufeld, Michael J. 2007. *Von Braun: Dreamer of Space, Engineer of War*. New York: A. A. Knopf.

Newkirk, Dennis. 1990. *Almanac of Soviet Manned Space Flight*. Houston: Gulf Publishing.

Oberg, James E. 1981. *Red Star in Orbit*. New York: Random House.

Raibchikov, E. 1971. *Russians in Space*. Garden City, N.Y.: Doubleday.

Ryan, Craig. 1995. *The Pre-Astronauts: Manned Ballooning on the Threshold of Space*. Annapolis, Md.: Naval Institute Press.

Seedhouse, Erik. 2010. *The New Space Race: China vs. the United States*. Chichester: Praxis.

Shepard, Alan B., Donald K. Slayton, Jay Barbree, Howard Benedict, and Neil Armstrong. 1994. *Moon Shot: The Inside Story of America's Race to the Moon*. Atlanta, Ga.: Turner Publishing.

Siddiqi, Asif A. 2003a. *The Soviet Space Race with Apollo*. Gainesville: University Press of Florida.

———. 2003b. *Sputnik and the Soviet Space Challenge*. Gainesville: University Press of Florida.

Sietzen, Frank, and Keith L. Cowling. 2004. *New Moon Rising: The Making of America's New Space Vision and the Remaking of NASA*. Ontario: Apogee Books.

Stafford, Thomas P., and Michael Cassutt. 2002. *We Have Capture: Tom Stafford and the Space Race*. Washington, D.C.: Smithsonian Institution Press.

Swanson, Glen E. 2012. *"Before This Decade Is Out": Personal Reflections on the Apollo Program*. New York: Dover.

Thompson, Neal. 2004. *Light This Candle: The Life and Times of Alan Shepard, America's First Spaceman*. New York: Crown.

Wendt, Guenter, and Russell Still. 2001. *The Unbroken Chain*. Ontario: Apogee Books.

Wolfe, Tom. 1979. *The Right Stuff*. New York: Farrar, Straus & Giroux.

Yeager, Chuck, and Leo Janos. 1985. *Yeager, an Autobiography*. New York: Bantam.

Young, John W. 2012. *Forever Young: A Life of Adventure in Air and Space*. With James R. Hansen. Gainesville: University Press of Florida.

Zaehringer, Alfred J., and Steve Whitfield. 2004. *Rocket Science: Rocket Science in the Second Millennium*. Ontario: Apogee Books.

Zimmerman, Robert. 1998. *Genesis: The Story of Apollo 8*. New York: Four Walls Eight Windows.

———. 2003. *Leaving Earth: Space Stations, Rival Superpowers, and the Quest for Interplanetary Travel*. Washington, D.C.: Joseph Henry.

INDEX

TED SPITZMILLER began his professional career in the military at the U.S. Army's Ordnance Guided Missile School in Huntsville, Alabama. He went on to nuclear weapons training at Sandia Base in New Mexico, where he taught for two years in the Atomic Weapons Training Group. He has worked for IBM, INTEL, and the Los Alamos National Laboratory from which he retired in 2001 to teach at the university level. Paralleling his profession in computing (MS/Computing Information Systems), Ted has always maintained an intense interest in aeronautics and astronautics. He holds an FAA commercial pilot certificate for Airplanes, Single and Multi-engine Land and Sea, with Instrument privileges. Ted is a flight instructor (CFII) who has logged over 4,500 hours in more than 80 different types of aircraft. Combining his skills in writing with his knowledge of aerospace, he has previously published six books and a number of articles in all major aviation magazines over the past 35 years. He has lectured at wide variety of venues around the country. Ted and his wife Donna have two children, TJ and Susie, and five grandchildren, Stephanie, JJ, Crystal, Baylee, and Megan.